OPTIMIZATION

A Theory of Necessary Conditions

Lucien W. Neustadt

OPTIMIZATION

A Theory of Necessary Conditions

Princeton University Press 1976

Published by Princeton University Press, Princeton, New Jersey
In the United Kingdom: Princeton University Press, Guildford, Surrey

Library of Congress Cataloging in Publication Data will
be found on the last printed page of this book
Printed in the United States of America
by Princeton University Press, Princeton, New Jersey

to Hilda and Adoph Neustadt

CONTENTS

PREFACE ix

SUMMARY OF NOTATION xiii

CHAPTER I: Mathematical Preliminaries 3

CHAPTER II: A Basic Optimization Problem in Simplified Form 63

CHAPTER III: A General Multiplier Rule 96

CHAPTER IV: Optimization with Operator Equation Restrictions 144

CHAPTER V: Optimal Control Problems with Ordinary Differential Equation Constraints 211

CHAPTER VI: Optimal Control Problems with Parameters and Related Problems 263

CHAPTER VII: Miscellaneous Optimal Control Problems 315

APPENDIX: Volterra-Type Operators 361

NOTES AND HISTORICAL COMMENTS 384

REFERENCES 413

SUBJECT INDEX 422

PREFACE

THE PURPOSE OF this book is to give a comprehensive development of necessary conditions for optimization problems. This is done in the context of a general theory for extremal problems in a topological vector space setting. A brief summary of the mathematical background necessary to understand the material in the text is presented in Chapter I. On the assumption that the reader is familiar with the fundamentals of real analysis and basic elements of measure theory and integration, pertinent definitions and results needed for the subsequent analysis in the linear topological space setting are given. Included also in Chapter I is a rather complete discussion of various types of differentials for functions from a linear (vector) space into a topological vector space. In Chapter II is found a generalized Lagrange multiplier rule for abstract optimization problems with a finite number of equality and inequality constraints. It is shown that application of this multiplier rule to a particular class of optimization problems defined in terms of operator equations in a Banach space yields a maximum principle which solutions of the problems must satisfy. Sufficiency of these conditions is discussed under certain convexity hypotheses on the problem data. Chapter III is devoted to a development of an extremal theory that leads to a generalization of the multiplier rule given in Chapter II. These generalizations involve a weakening of the hypotheses on the underlying set on which the optimization is carried out and a relaxation on the allowable constraints to permit a considerably more general type of "inequality" constraint.

In Chapter IV the fundamental multiplier rules developed earlier are used to treat the general optimization problem: Given a family $\mathscr{W}$ of continuously differentiable operators $T: A \to \mathscr{X}$, where A is an open subset of the Banach space $\mathscr{X}$, choose $x \in A$ satisfying (i) $Tx = x$ for some $T \in \mathscr{W}$, (ii) certain equality and generalized "inequality" constraints, and which is in some sense optimal. The formulation here is such that not only are the usual necessary conditions for restricted phase coordinate optimal control problems with ordinary differential equation restrictions obtained as special cases (the subject matter of Chapter V), but many other general optimal control problems can also be easily treated as special cases. This is discussed in Chapter VI, where results are given for control problems with parameters and control problems with mixed control-phase inequality constraints. In Chapter VII necessary conditions using the framework of Chapter IV are obtained for control problems governed by such diverse systems as functional differential equations (differential-difference equations being a special case), Volterra integral equations, and difference equations.

An appendix contains fundamental results (existence, continuation, uniqueness, continuous dependence) for equations defined in terms of the Volterra-type operators used in the formulation of certain of the problems discussed in Chapter IV. A concluding chapter (Notes and Historical Comments) comprises an extensive literature survey in which the development of necessary conditions and sufficient conditions in modern optimization theory is outlined and comments are made on the relationship between the differing approaches of various contributors to the literature.

We in the Division of Applied Mathematics at Brown University were fortunate to have Lucien Neustadt spend the 1971–1972 academic year on sabbatical leave with us. During that period much of the material in the present book was presented by Lucien in a year-long advanced seminar. The main text of the book (Chapters I–VII and the Appendix) was finished by the summer of 1972 but at the time of Lucien's death in October 1972 the question of a publisher was still unsettled. In addition, work on the Notes and Historical Comments chapter was still in a rough draft and incomplete form. In the winter of 1972–1973 Joe LaSalle and I undertook the responsibility for having the manuscript reviewed by several publishers and then made recommendations to the executor of Lucien's estate concerning choice of a publisher. Due to a number of unfortunate delays, it was not until the spring of 1974 that legal matters were finally cleared up and agreement for publication by Princeton University Press was completed. I agreed to undertake the task of completing the work on the Notes and Historical Comments chapter and putting the entire chapter into a polished form. In addition I accepted the usual author's responsibility concerning copy editing, reading of proofs, etc. The present text of Chapters I–VII and the Appendix are unchanged from Lucien's final manuscript except for minor corrections. The referencing system chosen by Lucien has been maintained throughout the text, the appendix, and the historical comments. It is easily understood. Chapters are denoted by Roman numerals and are each divided into a number of sections. These sections in turn consist of collections of numbered paragraphs and formulae. In references the number (II.4.13) refers to paragraph 13 of Section 4 of Chapter II while (I.7.13, 16, and 19) refers to paragraphs 13, 16, and 19 of Section 7 of Chapter I. Chapter numbers are omitted when referring to entities within the same chapter. For example the reference to (2.14) within Chapter II refers to paragraph 14 of Section 2 of that Chapter. The same procedure is followed in referencing paragraphs within a given section (i.e., the reference to (14) in a section of a chapter refers to paragraph (14) of that section of that chapter).

Much of the material of the chapter Notes and Historical Comments came either directly from Lucien's rough draft material or from my own summaries of the references on which Lucien had intended to comment (as indicated by his incomplete rough draft bibliography). But both Lucien's Notes and Historical Comments rough draft and his bibliography were unfinished at the time of his death and I have tried to complete the bibliography (along with relevant paragraphs in appropriate places in the Notes and Historical Comments chapter) in the way in which I judge, to the best of my ability, Lucien would have done so had he lived to complete the task. However, I accept all responsibility for any errors of commission or omission with respect to the bibliography and the final chapter Notes and Historical Comments.

Lucien undoubtedly would have wished to express his appreciation to a number of people for their helpful comments during the lengthy period he worked on this book. It is impossible for me to guess the names of all such individuals and I will therefore mention by name only those who are due thanks (on the behalf of both Lucien and myself) for help since I became involved with completing and publishing of the manuscript. These include Professors N. Nahi (who helped in locating the rough draft notes for the bibliographical section among Lucien's effects), J. A. Burns (who during the summer of 1974 helped with some of the literature search necessary to complete the Notes and Historical Comments chapter) and J. P. LaSalle. Special appreciation is due S. Spinacci and K. Avery for their excellent typing both before and after Lucien's death. Finally I wish to express my personal appreciation to John W. Hannon of Princeton University Press for his patience and help with publication details that were without doubt more difficult than they would have been had Lucien been alive.

I am certain that all of us feel that the present book more than justifies our own meagre efforts and feel privileged to have helped in some small way in the completion of a book that is indeed a lasting memorial to a man who contributed so much to the intellectual and spiritual lives of all who were associated with him.

H. T. Banks
Providence, Rhode Island
May, 1975

SUMMARY OF NOTATION

$\mathbf{B}(\mathscr{Y},\mathscr{Z})$ (I.5.1)
Ch (IV.5.2)
Ch (IV.5.15)
co A (I.1.27)
$\overline{\text{co}}$ A (I.4.24)
co co A (I.1.29)
$\overline{\text{co co}}$ A (I.4.24)
cone A (I.1.28)
$\overline{\text{cone}}$ A (I.4.24)
$\mathscr{C}(I)$ (I.6.6)
$\mathscr{C}^n(I)$ (I.6.5)
$\hat{\mathscr{C}}^n$ (IV.4.5)
$D\varphi$, $D\phi$, etc. (I.7.1)
$D_1\varphi$, $D_1\phi$, etc. (I.7.51)
$\mathscr{D}\chi$ (I.7.58)
$\mathscr{D}_1\chi$ (I.7.60)
F_u (V.1.22)
$\mathscr{F}$ (IV.3.22)
$\mathscr{F}_1$ (IV.3.33)
$\hat{\mathscr{F}}_1$ (IV.4.23)
$\mathscr{G}$ (IV.3.9)
$\hat{\mathscr{G}}$ (IV.4.7)
H (II.2.44)
int (I.3.5)
$\mathscr{K}\begin{cases}\text{(III.2.21)}\\ \text{(II.1.7)}\end{cases}$
lim (I.3.7)
$L_p(I)$ (I.6.23)
$L_\infty(I)$ (I.6.24)
$L^n_\infty(I)$ (I.6.30)
$NBV(I)$ (I.6.10)
R (I.1.3)
R_- (I.1.18)
R_+ (I.1.18)
R^n (I.1.3)
R^n_+ (I.1.21)
R^n_- (I.1.19)
$\bar{R}_+$ (I.1.18)
$\bar{R}_-$ (I.1.18)
$\bar{R}^n_+$ (I.1.21)
$\bar{R}^n_-$ (I.1.20)
$\mathscr{S}_0$ (IV.2.3)
$\mathscr{S}_1$ (IV.2.3)
tancone $(A; y_0)$ (I.4.25)
TV (I.6.11)
$T_{x^*,F}$ (IV.3.28)
$T_{x^*,G}$ (IV.3.15)
$\hat{T}_{\hat{x}^*,\hat{F}}$ (IV.4.29)
$\hat{T}_{\hat{x}^*,\hat{G}}$ (IV.4.14)
$\hat{T}^\pi$ (IV.2.3)
$\mathscr{T}$ (IV.1.1)
$\mathscr{T}_0$ (IV.1.2)
$\mathscr{T}_1$ (IV.1.2)
$\mathscr{T}^u$ (IV.3.7)
$\hat{\mathscr{T}}$ (IV.2.2)
V_2^κ (III.1.18)
$\mathscr{V}$ (IV.3.3)
$\mathscr{V}_1$ (IV.3.38)
$\mathscr{V}_2$ (IV.3.39)
$\mathscr{V}'_2$ (IV.3.40)
$\hat{\mathscr{V}}$ (IV.4.5)
$\hat{\mathscr{V}}_1$ (IV.4.35)
$\hat{\mathscr{V}}_2$ (IV.4.36)
Z_2^κ (III.1.17)
(φ,ϕ,Z)—extremal (III.1.14)
(ϕ,Z)—extremal (III.1.29)
$\leq_W$ (I.2.30)

OPTIMIZATION

A Theory of Necessary Conditions

CHAPTER I

Mathematical Preliminaries

In this chapter we shall present some mathematical background material which will be necessary in order to understand the remainder of this book. For the most part, we shall confine ourselves to stating definitions and to theorem statements. The proofs of many of the theorems are immediate. References will be given for those proofs which are less obvious but which are widely available in the literature. (In this regard, the reference numbers refer to books listed at the end of this chapter.) Proofs will be given for those lemmas and theorems which either are novel or are not widely known.

The material in this chapter is, of course, sketchy at best, and in many cases definitions and theorems are not presented in their full generality. Indeed, in those cases where full generality does not provide any additional insight and is not required in the remainder of this book, we shall confine ourselves to restrictive definitions and theorems.

Since linear topological spaces form the framework of much of what we have to say, most of this chapter deals with such spaces and with background material necessary to understand these spaces. The last section of this chapter, which contains the major part of the novel material in this chapter, is devoted to a study of a wide variety of differentials.

For the sake of completeness, we have included some quite elementary material, such as the properties of finite-dimensional Euclidean spaces. However, the reader is expected to be familiar with the fundamentals of analysis on the real line. In addition, in subsequent chapters—as well as in Section 6 of this chapter—we shall make use of a number of concepts and theorems from the theory of measure and integration on the real line (and, on a few occasions, on real finite-dimensional spaces). Because this material is so widely and easily accessible in many books, including, e.g., [4] and [7], we shall dispense with presenting any background material on it ourselves.

1. Linear Vector Spaces

A *group* is a nonempty collection $\mathscr{Y}$ of elements together with an operation called addition which assigns to each ordered pair (y_1, y_2) of elements from $\mathscr{Y}$ a new element y_3 of $\mathscr{Y}$—which will be written as $y_1 + y_2 = y_3$—subject to the following restrictions:

(i) $y_1 + (y_2 + y_3) = (y_1 + y_2) + y_3$ for all y_1, y_2, y_3 in $\mathscr{Y}$ (the *associative law*);

(ii) there is in $\mathscr{Y}$ a unique 0 (which we shall call the *identity*) such that $0 + y = y + 0 = y$ for all $y \in \mathscr{Y}$;

(iii) for every $y \in \mathscr{Y}$ there exists a unique element, denoted by $(-y)$, such that $y + (-y) = (-y) + y = 0$.

If addition in $\mathscr{Y}$ obeys the *commutative law*, i.e., if $y_1 + y_2 = y_2 + y_1$ for all $y_1, y_2 \in \mathscr{Y}$, then the group $\mathscr{Y}$ will be said to be *Abelian.*

1 A *linear vector space* is an Abelian group $\mathscr{Y}$ together with an operation called scalar multiplication which assigns to each pair α, y, where α is a real number and $y \in \mathscr{Y}$, a new vector $y_1 \in \mathscr{Y}$—which will be written as $y_1 = \alpha y$—subject to the following restrictions:

(i) $\alpha(y_1 + y_2) = \alpha y_1 + \alpha y_2$ for all real α and $y_1, y_2 \in \mathscr{Y}$ (a *distributive law*);

(ii) $(\alpha + \beta)y = \alpha y + \beta y$ for all real α, β and $y \in \mathscr{Y}$ (a *distributive law*);

(iii) $\alpha(\beta y) = (\alpha\beta)y$ for all real α, β and $y \in \mathscr{Y}$ (an *associative law*);

(iv) $1y = y$ for all $y \in \mathscr{Y}$.

What we have referred to as a linear vector space is often referred to in the literature as a *real* linear vector space, because linear vector spaces other than the kind that we have introduced may be defined. Since we shall not deal with these other spaces, we shall dispense with the adjective "real."

2 Although the symbol 0 denotes both the identity in a linear vector space (the word *origin* is often used in place of identity) and the number zero, there should be no confusion about this in the sequel, since it will always be clear from the context what 0 stands for, even when we discuss two different vector spaces.

We shall adopt the usual convention of writing $(y_1 - y_2)$ for $[y_1 + (-y_2)]$.

It is easy to see that, in any linear vector space $\mathscr{Y}$, for all $y \in \mathscr{Y}$, $0y = 0$ and $-1y = -y$, that $\alpha 0 = 0$ for all real α, and that the following "cancellation" laws hold:

(i) $y_1 + y_2 = y_1 + y_3$ implies that $y_2 = y_3$;

(ii) $\alpha y_1 = \alpha y_2$ and $\alpha \neq 0$ imply that $y_1 = y_2$;

(iii) $\alpha y = \beta y$ and $y \neq 0$ imply that $\alpha = \beta$.

3 The real numbers, where addition and scalar multiplication are defined in the obvious way, clearly form a linear vector space, which will be denoted by R or by R^1. The set of all n-tuples of real numbers, where n is an integer greater than 1, with addition defined by the rule $(\xi^1, \ldots, \xi^n) + (\eta^1, \ldots, \eta^n) = (\xi^1 + \eta^1, \ldots, \xi^n + \eta^n)$ and scalar multiplication by $\alpha(\xi^1, \ldots, \xi^n) = (\alpha\xi^1, \ldots, \alpha\xi^n)$, also make up a linear vector space which will be denoted by R^n. If $\xi = (\xi^1, \ldots, \xi^n) \in R^n$, then $\xi^1, \ldots, \xi^n$ will sometimes be called the *coordinates* or *components* of ξ.

4 Two linear vector spaces $\mathcal{Y}'$ and $\mathcal{Y}''$ will be said to be *isomorphic* if there is a one-to-one correspondence between the elements of $\mathcal{Y}'$ and $\mathcal{Y}''$, denoted by $\leftrightarrow$, such that if $y'_1 \leftrightarrow y''_1$ and $y'_2 \in y''_2$, then $\alpha y'_1 + \beta y'_2 \leftrightarrow \alpha y''_1 + \beta y''_2$ for all real numbers α, β.

5 If $\mathcal{Y}_1, \ldots, \mathcal{Y}_m$ are linear vector spaces, the *direct product* (or *Cartesian product*) of $\mathcal{Y}_1, \ldots, \mathcal{Y}_m$, written as $\mathcal{Y}_1 \times \mathcal{Y}_2 \times \cdots \times \mathcal{Y}_m$, is the linear vector space of all m-tuples $(y_1, \ldots, y_m)$ where $y_i \in \mathcal{Y}_i$ for $i = 1, \ldots, m$, with addition defined by the rule $(y_1, \ldots, y_m) + (y'_1, \ldots, y'_m) = (y_1 + y'_1, \ldots, y_m + y'_m)$ and scalar multiplication by $\alpha(y_1, \ldots, y_m) = (\alpha y_1, \ldots, \alpha y_m)$. The space $\mathcal{Y} \times \cdots \times \mathcal{Y}$ (m times) will simply be denoted by $\mathcal{Y}^m$. (It is clear that our notation R^n introduced above is consistent with this convention.) If A_i is a nonempty subset of $\mathcal{Y}_i$ for each i, then the direct product of $A_1, \ldots, A_m$, written as $A_1 \times A_2 \times \cdots \times A_m$, is the set $\{(y_1, \ldots, y_m): y_i \in A_i \text{ for each } i\}$ in $\mathcal{Y}_1 \times \cdots \times \mathcal{Y}_m$. The set $A \times \cdots \times A$ (m times) in $\mathcal{Y}^m$ will be denoted by A^m.

In the remainder of this section, $\mathcal{Y}, \mathcal{Y}_1, \ldots, \mathcal{Y}_m$ will denote linear vector spaces.

If A and B are nonempty subsets of $\mathcal{Y}$, $y_0 \in \mathcal{Y}$, and α and β are real numbers, then we define the subsets αA, $A + B$, and $y_0 + A$ of $\mathcal{Y}$ as follows:

6 $$\alpha A = \{\alpha y: y \in A\},$$

7 $$A + B = \{y_1 + y_2: y_1 \in A, y_2 \in B\},$$

8 $$y_0 + A = \{y_0 + y: y \in A\}.$$

The sets $A - B$, $y - A$, and $A - y$ are defined in an obviously analogous manner. Note that $1A = A$ and that $0 + A = A$, but that, in general, $A + A \neq 2A$. Further, $\alpha(A + B) = \alpha A + \alpha B$, $\alpha(\beta A) = (\alpha\beta)A$, $(\alpha + \beta)A = \alpha A + \beta A$, and $y_0 + (y_1 + A) = (y_0 + y_1) + A$ for all real α, β, nonempty subsets A and B of $\mathscr{Y}$, and elements $y_0, y_1 \in \mathscr{Y}$.

9 A nonempty subset A of $\mathscr{Y}$ will be called a *linear manifold* if $\alpha y_1 + \beta y_2 \in A$ whenever $y_1, y_2 \in A$ and α and β are any real numbers. Note that $A \subset \mathscr{Y}$ is a linear manifold in $\mathscr{Y}$ if and only if A is itself a linear vector space (or a *linear subspace* of $\mathscr{Y}$).

10 Any set of the form $y_0 + A$, where $y_0 \in \mathscr{Y}$ and A is a linear manifold in $\mathscr{Y}$, will be called a *flat* (or a *linear variety*) in $\mathscr{Y}$.

11 A nonempty set $A \subset \mathscr{Y}$ will be said to be *convex* if $\lambda y_1 + \mu y_2 \in A$ whenever $y_1, y_2 \in A$, $\lambda \geq 0$, $\mu \geq 0$ and $\lambda + \mu = 1$. It is easy to see that A is convex if and only if $\sum_{i=1}^{m} \lambda^i y_i \in A$ whenever $y_1, \ldots, y_m \in A$, $\sum \lambda^i = 1$, and $\lambda^i \geq 0$ for each i, where m is an arbitrary positive integer.

12 A nonempty set $A \subset \mathscr{Y}$ will be said to be a *cone* if $\alpha A \subset A$ for all $\alpha > 0$.

13 A set $A \subset \mathscr{Y}$ which is both convex and a cone will be said to be a *convex cone*. It is easy to see that A is a convex cone if and only if $\alpha A + \beta A \subset A$ for all $\alpha > 0$ and $\beta > 0$.

14 A cone (or convex cone) A in $\mathscr{Y}$ will be said to be *pointed* if $0 \in A$.

15 Note that every linear manifold is a pointed convex cone as well as a flat, and that every flat is convex. In particular, the set $\{0\}$ (consisting only of the origin in $\mathscr{Y}$) is a linear manifold and pointed convex cone. It is also easily verified that the intersection of any number of linear manifolds (respectively, flats, convex sets, cones, convex cones, or pointed cones) is either empty or a linear manifold (respectively, a flat, a convex set, a cone, a convex cone, or a pointed cone).

16 Similarly, if A_i is a linear manifold (respectively, flat, convex set, cone, convex cone, pointed cone) in $\mathscr{Y}_i$ for $i = 1, \ldots, m$, then $A_1 \times \cdots \times A_m$ is a linear manifold (respectively, flat, convex set, cone, convex cone, pointed cone) in $\mathscr{Y}_1 \times \cdots \times \mathscr{Y}_m$.

17 Further, if A and B are linear manifolds (respectively, flats, convex sets, cones, convex cones, pointed cones) in $\mathscr{Y}$, then $\alpha A + \beta B$ is also a linear manifold (respectively, flat, convex set, cone, convex cone, pointed cone) in $\mathscr{Y}$ for all real α, β. Finally, if A is a convex set (respectively, a flat) in $\mathscr{Y}$ and $y_0 \in \mathscr{Y}$, then $y_0 + A$ is also a convex set (respectively, a flat) in $\mathscr{Y}$.

18 The set of all negative numbers, which we shall denote by R_-, is clearly a convex cone in R, as is the set of all positive numbers, which we shall denote by R_+. The set of all nonpositive numbers, which we shall denote by $\bar{R}_-$, and the set of all nonnegative numbers, which we shall denote by $\bar{R}_+$, are evidently pointed convex cones in R.

Let us denote by R^n_- and $\bar{R}^n_-$ the following sets in R^n:

19 $$R^n_- = \{(\xi^1, \ldots, \xi^n): \xi^i < 0 \text{ for each } i = 1, \ldots, n\},$$

20 $$\bar{R}^n_- = \{(\xi^1, \ldots, \xi^n): \xi^i \leq 0 \text{ for each } i = 1, \ldots, n\}.$$

21 Then it is easily seen that R^n_-, which we shall call the *negative orthant* in R^n, and $\bar{R}^n_-$, which we shall call the *nonpositive orthant* in R^n, are convex cones, with the latter pointed. The *positive orthant* R^n_+ and the *nonnegative orthant* $\bar{R}^n_+$ are similarly defined (and are also convex cones).

22 A finite collection $\{y_1, \ldots, y_m\}$ of vectors in $\mathscr{Y}$ will be said to be *linearly dependent* if there are real numbers $\alpha^1, \ldots, \alpha^m$, not all zero, such that $\alpha_1 y_1 + \cdots + \alpha_m y_m = 0$. A finite subset of $\mathscr{Y}$ will be said to be *linearly independent* if it is not linearly dependent. Vectors $y_1, \ldots, y_m$ in $\mathscr{Y}$ will be said to be in *general position* if the relations $\sum_{i=1}^m \alpha^i y_i = 0$ and $\sum_{i=1}^m \alpha^i = 0$ imply that $\alpha^i = 0$ for all $i = 1, \ldots, m$. It is easily seen that $y_1, \ldots, y_m$ are in general position if and only if the vectors $(y_2 - y_1), \ldots, (y_m - y_1)$ are linearly independent.

23 If $y_1, \ldots, y_m$ are vectors in $\mathscr{Y}$, then every vector of the form $\alpha^1 y_1 + \cdots + \alpha^m y_m$ (with $\alpha^j \in R$ for each j) will be called a *linear combination* of $y_1, \ldots, y_m$. The set of all linear combinations of $y_1, \ldots, y_m$, which is easily seen to be a linear manifold, will be called the *span* of $y_1, \ldots, y_m$. Conversely, if M is a linear manifold in $\mathscr{Y}$ and M coincides with the span of m vectors $y_1, \ldots, y_m$ of $\mathscr{Y}$, then $y_1, \ldots, y_m$ will be said to *span* M. More generally, if A is any set in $\mathscr{Y}$, then the *span* of A, written as span A, will denote the intersection of all linear manifolds that contain A, i.e., the span of A is the "smallest" subspace of $\mathscr{Y}$ that contains A [see (15)]. It is not hard to see that the span of A consists of all linear combinations of all finite subsets of A.

24 A linear combination $\alpha^1 y_1 + \cdots + \alpha^m y_m$ of vectors $y_1, \ldots, y_m$ in $\mathscr{Y}$ will be called a *convex combination* of $y_1, \ldots, y_m$ if $\alpha^i \geq 0$ for each i and $\sum_1^m \alpha^i = 1$.

25 If $y_1, \ldots, y_m$ are linearly independent vectors in $\mathscr{Y}$ which also span $\mathscr{Y}$, then $y_1, \ldots, y_m$ will be said to form a *basis* for $\mathscr{Y}$. In this case, it is not hard to show that every basis for $\mathscr{Y}$ contains exactly m elements, and $\mathscr{Y}$ will be said to have *dimension m*. On the other hand, if no finite subset of $\mathscr{Y}$ forms a basis for $\mathscr{Y}$, then $\mathscr{Y}$ will be said to be *infinite-dimensional*. The dimension of any linear manifold in $\mathscr{Y}$ (whether finite or infinite) may obviously be defined in the same way, since every linear manifold is also a linear vector space. In this way, one can also define the dimension of a flat in an evident manner.

26 Inasmuch as $(1,0,\ldots,0), (0,1,0,\ldots,0), \ldots, (0,0,\ldots,0,1)$ obviously form a basis in R^n, R^n has dimension n. It is also interesting to note that every n-dimensional linear vector space $\mathscr{Y}$ is isomorphic to R^n. Indeed, if $y_1, \ldots, y_n$ form a basis in $\mathscr{Y}$, then every vector $y \in \mathscr{Y}$ may be uniquely represented in the form $y = \sum_1^n \alpha^i y_i$, and we may thus set up the one-to-one correspondence

$$y = \sum_{i=1}^{n} \alpha^i y_i \leftrightarrow (\alpha^1, \ldots, \alpha^n)$$

between $\mathscr{Y}$ and R^n. It is easily seen that this correspondence "preserves" addition and scalar multiplication, as is required in (4).

27 If $A \subset \mathscr{Y}$, then the *convex hull* of A, written as co A, is the intersection of all convex sets that contain A, and is therefore the "smallest" convex set in $\mathscr{Y}$ that contains A [see (15)]. It is not hard to see that co A consists of all vectors y of the form $y = \sum_{i=1}^m \lambda^i y_i$ where $y_1, \ldots, y_m \in A$, $\sum_1^m \lambda^i = 1$ and $\lambda^i \geq 0$ for each i, with m an arbitrary positive integer; i.e., co A consists of all convex combinations of all finite subsets of A.

Note that the convex hull of a cone is also a cone.

28 The *conical hull* of A, written as cone A, is the cone which is the intersection of all cones that contain A, or, equivalently, the set $\{\alpha y \colon y \in A, \alpha > 0\}$.

29 The *convex conical hull* of A, written as co co A, is the convex cone which is the intersection of all convex cones that contain A. It is easy to show that co co A consists of all vectors $\sum_1^m \lambda^i y_i$ where $y_1, \ldots, y_m \in A$ and $\lambda^i > 0$ for each i, with m an arbitrary positive integer. Evidently cone $A \subset$ co co A.

The following lemma, whose proof is left to the reader, will be needed in Chapter III.

30 LEMMA. *If* $A \subset \mathscr{Y}_1$ *and* $B \subset \mathscr{Y}_2$, *where* $\mathscr{Y}_1$ *and* $\mathscr{Y}_2$ *are linear vector spaces, and if* $0 \in A$ *and* $0 \in B$, *then* $\text{co co } (A \times B) = (\text{co co } A) \times (\text{co co } B)$.

31 A set $S \subset \mathscr{Y}$ will be called a *simplex* if S is the convex hull of a finite number of points $y_1, \ldots, y_m$ in $\mathscr{Y}$ which are in general position. In this case $y_1, \ldots, y_m$ will be called the *vertices* of S (the vertices of S are uniquely determined by S), and S will be said to have *dimension* $m - 1$. It is easily seen that S consists of all y of the form

32 $$y = \sum_{i=1}^{m} \lambda^i y_i \quad \text{with} \quad \sum_{i=1}^{m} \lambda^i = 1 \quad \text{and} \quad \lambda^i \geq 0 \quad \text{for each} \quad i.$$

33 The numbers λ^i in (32), which (because of the general position of the y_i) are uniquely determined by y, are called the *barycentric coordinates* of y. It is trivial to verify that if $\mathscr{Y}$ has dimension n, then $m \leq n$.

34 We shall say that a convex set A in $\mathscr{Y}$ has *dimension* m if there is a simplex S contained in A which has dimension m, but there is no simplex of dimension $(m + 1)$ contained in A. A convex set A will be said to be *infinite-dimensional* if there are simplices of arbitrarily large dimension contained in A.

35 The convex hull of any finite subset of $\mathscr{Y}$ will be called a *convex polyhedron*. Clearly, every simplex is a convex polyhedron, but the converse is false, unless $\mathscr{Y}$ has dimension one.

36 It is useful to note (and easy to verify) that if P is a convex polyhedron in $\mathscr{Y}$, then the span of P is a finite-dimensional linear manifold.

37 A subset A of $\mathscr{Y}$ will be said to be *finitely open* if, for every $y_0 \in A$ and every finite subset $\{y_1, \ldots, y_m\}$ of $\mathscr{Y}$, there is a number $\varepsilon > 0$ (possibly depending on $y_0, y_1, \ldots, y_m$) such that

38 $$y_0 + \sum_{i=1}^{m} \lambda^i (y_i - y_0) \in A \quad \text{whenever} \quad 0 \leq \lambda^i < \varepsilon \quad \text{for each} \quad i.$$

39 A nonempty subset A of $\mathscr{Y}$ will be said to be *finitely open in itself* if, for every $y_0 \in A$ and every finite subset $\{y_1, \ldots, y_m\}$ of A, there is a number $\varepsilon > 0$ (possibly depending on $y_0, y_1, \ldots, y_m$) such that (38) holds.

40 The reader should carefully note the difference between the preceding two definitions. Note that every nonempty finitely open set, as well as every convex set, is finitely open in itself.

Further, the intersection, as well as the direct product, of two finitely open sets (respectively, of two sets finitely open in themselves) is finitely open (respectively, finitely open in itself).

The following lemma is left as an exercise to the reader.

41 LEMMA. *If $A \subset \mathscr{Y}$ is finitely open in itself, and $y_0 \in A$, then cone $(A - y_0) = \text{co co } (A - y_0)$, so that cone $(A - y_0)$ is convex.* [*See* (8), (28), *and* (29).]

42 If $A \subset \mathscr{Y}$ and $y_0 \in A$, then y_0 will be called an *internal point* (or a *core point*) of A if, for every $y_1 \in \mathscr{Y}$, there is a number $\lambda_0 > 0$ (possibly depending on y_0 and y_1) such that $y_0 + \lambda y_1 \in A$ whenever $0 \leq \lambda \leq \lambda_0$. Note that if y_0 is an internal point of A, then cone $(A - y_0) = \mathscr{Y}$.

2. Functions from One Linear Vector Space into Another

1 Let A and B be arbitrary sets and let f be a function which assigns to every element of A a (unique) element of B. Then we shall say that f is a function from A into B (written as $f: A \to B$), and that f has *domain* A. We shall occasionally use the words *mapping* and *map* as synonyms for function. If A_1 is a subset of A, and B_1 is a subset of B, then we shall denote by $f(A_1)$ and $f^{-1}(B_1)$ the following sets:

2 $$f(A_1) = \{y: y = f(x) \text{ for some } x \in A_1\},$$

3 $$f^{-1}(B_1) = \{x: f(x) \in B_1\}.$$

The set $f(A)$ will be called the *range* of f. If the range of f coincides with B, then we shall say that f is *onto* B. On the other hand, if, for every $y \in f(A)$, there is a *unique* $x \in A$ such that $y = f(x)$, then we shall say that f is *one-to-one*. If f is one-to-one and onto, then we shall denote by f^{-1} the map which assigns to every $y \in B$ the (unique) $x \in A$ such that $f(x) = y$; i.e., $f^{-1}(y) = x$ if and only if $f(x) = y$. The map f^{-1} will then be called the *inverse* of f.

4 If $A_1 \subset A$, $f: A \to B$, $f_1: A_1 \to B$, and $f(x) = f_1(x)$ for all $x \in A_1$, then we shall say that f_1 is the *restriction* of f to A_1, and that f is an *extension* of f_1 onto A. If $f: A \to B$ and $g: B \to C$ (or if $g: f(A) \to C$), then the function from A into C which assigns to each $x \in A$ the element $g(f(x))$ in C [or, as we shall write in the future for such situations, the function $x \to g(f(x))$] will be called

the *composite* of g and f, and will be denoted by $g \circ f$; i.e., $g \circ f(x) = g(f(x))$ for all $x \in A$.

5 If $A_1, \ldots, A_m$ are arbitrary sets, then the set of all m-tuples $(x_1, \ldots, x_m)$ such that $x_i \in A_i$ for each $i = 1, \ldots, m$ will be called the *direct product* of $A_1, \ldots, A_m$, and will be denoted by $A_1 \times \cdots \times A_m$. If $A_i = A$ (for some set A) for each $i = 1, \ldots, m$, then we shall also write A^m for this direct product. If f_i, for each $i = 1, \ldots, m$, is a function from some set A_0 into A_i, then the function $x \to (f_1(x), \ldots, f_m(x)) \colon A_0 \to A_1 \times \cdots \times A_m$ will be denoted by $(f_1, \ldots, f_m)$.

Functions from a subset of a linear vector space into a linear vector space will occasionally be referred to as *operators*.

Throughout the remainder of this section, $\mathscr{Y}, \mathscr{Y}_1, \ldots, \mathscr{Y}_m, \mathscr{Z}, \mathscr{Z}_1, \ldots, \mathscr{Z}_m$ will denote linear vector spaces.

6 A function $f \colon \mathscr{Y} \to \mathscr{Z}$ will be said to be *linear* if

$$f(\alpha y_1 + \beta y_2) = \alpha f(y_1) + \beta f(y_2) \quad \text{for all} \quad y_1, y_2 \in \mathscr{Y} \quad \text{and} \quad \alpha, \beta \in R.$$

Linear functions from one linear vector space into another will often be referred to as *linear operators*. Note that $f(0) = 0$ whenever f is a linear operator.

7 If $A \subset \mathscr{Y}$, a function $f \colon A \to \mathscr{Z}$ will be said to be *linear* if f is the restriction to A of a linear function from $\mathscr{Y}$ into $\mathscr{Z}$. In this case, we shall sometimes permit ourselves the abuse of notation of using the same symbol to denote both the function on A and its (linear) extension to $\mathscr{Y}$ (even though there may be infinitely many such extensions).

8 A function f from $\mathscr{Y}$ (or from a set $A \subset \mathscr{Y}$) into R will be called a *functional*. If f is, in addition, linear, then f will be called a *linear functional*.

9 The linear operator E from $\mathscr{Y}$ into $\mathscr{Y}$ which satisfies the equation $Ey = y$ for all $y \in \mathscr{Y}$ will be called the *identity operator* on $\mathscr{Y}$.

10 The set of all linear operators from $\mathscr{Y}$ into $\mathscr{Z}$ with addition and scalar multiplication defined by the relations

$$(f_1 + f_2)(y) = f_1(y) + f_2(y), \quad (\alpha f)(y) = \alpha f(y),$$
$$\text{for all} \quad y \in \mathscr{Y} \quad \text{and} \quad \alpha \in R,$$

is easily seen to form a linear vector space. For the case of $\mathscr{Z} = R$, this space will be denoted by $\mathscr{Y}^+$; i.e., $\mathscr{Y}^+$ consists of all linear functionals defined on $\mathscr{Y}$.

11 It is easily verified that $\ell \in (\mathscr{Y}_1 \times \cdots \times \mathscr{Y}_m)^+$ if and only if ℓ can be represented as the sum of functionals in $\mathscr{Y}_i^+$ in the sense that there exist linear functions $\ell_i \in \mathscr{Y}_i^+$ for $i = 1, \ldots, m$ such that $\ell(y_1, \ldots, y_m) = \ell_1(y_1) + \cdots + \ell_m(y_m)$ for every $(y_1, \ldots, y_m) \in \mathscr{Y}_1 \times \cdots \times \mathscr{Y}_m$. From this it follows at once that $(\mathscr{Y}_1 \times \cdots \times \mathscr{Y}_m)^+$ is isomorphic to $\mathscr{Y}_1^+ \times \cdots \times \mathscr{Y}_m^+$.

12 If $\mathscr{Y} = R^n$, then every $\ell \in (R^n)^+$ is easily seen to be representable in the form $\ell(\xi^1, \ldots, \xi^n) = \sum_{i=1}^n \beta^i \xi^i$ for some fixed $\beta = (\beta^1, \ldots, \beta^n) \in R^n$, which immediately implies that $(R^n)^+$ is isomorphic to R^n.

13 An important particular function defined on $R^n \times R^n$ is the so-called *dot product* (or *inner product*) which assigns to each pair $(\xi, \eta) \in R^n \times R^n$, where $\xi = (\xi^1, \ldots, \xi^n)$ and $\eta = (\eta^1, \ldots, \eta^n)$, the real number $\sum_{i=1}^n \xi^i \eta^i$. This function will simply be written as $\xi \cdot \eta$. It follows from (12) that every $\ell \in (R^n)^+$ can be represented in the form $\ell(\xi) = \beta \cdot \xi$ for some fixed $\beta \in R^n$. An important functional defined on R^n is the function $\xi \to (\xi \cdot \xi)^{1/2}$, which will, for brevity, be written as $|\xi|$.

14 LEMMA. *Let $\ell_1, \ldots, \ell_m$ be elements of $\mathscr{Y}^+$ such that the relations $\ell_i(y) = 0$ for $i = 1, \ldots, m-1$, $y \in \mathscr{Y}$, imply that $\ell_m(y) = 0$. Then ℓ_m is a linear combination of $\ell_1, \ldots, \ell_{m-1}$.*

Proof. See [4, Lemma 10 on p. 421].

15 COROLLARY. *Let $\ell_1, \ldots, \ell_m$ be linearly independent elements of $\mathscr{Y}^+$, and let $\gamma \in R$. Then there is an element $y \in \mathscr{Y}$ such that $\ell_j(y) = \gamma$ for $j = 1, \ldots, m$.*

Proof. It follows from Lemma (14) that there are elements $y_1, \ldots, y_m$ in $\mathscr{Y}$ such that $\ell_j(y_i) = 0$ if $i \neq j$ and $\ell_j(y_j) \neq 0$ for each j. If we set $y = \gamma \sum_{i=1}^n (1/\ell_i(y_i))y_i$, then $\ell_j(y) = \gamma$ for $j = 1, \ldots, m$. |||

16 A function $f: \mathscr{Y} \to \mathscr{Z}$ will be said to be *affine* if there are a linear operator $f_0: \mathscr{Y} \to \mathscr{Z}$ and an element $z_0 \in \mathscr{Z}$ such that $f(y) = f_0(y) + z_0$ for all $y \in \mathscr{Y}$. In this case, f_0 (which is clearly completely specified by f) will be called the *linear part* of f.

17 It is obvious that every linear function $f: A \to \mathscr{Z}$ is affine, and that if $f: \mathscr{Y} \to \mathscr{Z}$ is affine, then $f(\lambda y_1 + \mu y_2) = \lambda f(y_1) + \mu f(y_2)$ whenever $\lambda + \mu = 1$ for all $y_1, y_2 \in \mathscr{Y}$. Further, if $f_i: \mathscr{Y} \to \mathscr{Z}$ are affine for $i = 1, 2$, then $\alpha f_1 + \beta f_2$ is affine for all $\alpha, \beta \in R$. Also, if $f_1, \ldots, f_m$ are affine (respectively, linear) functionals from $\mathscr{Y}$ into

linear vector spaces $\mathscr{Z}_1, \ldots, \mathscr{Z}_m$, respectively, then $(f_1, \ldots, f_m)$ is affine (respectively, linear). Finally, if $\mathscr{X}$ is a linear vector space, and the operators $g: \mathscr{X} \to \mathscr{Y}$ and $f: \mathscr{Y} \to \mathscr{Z}$ are affine (respectively, linear), then $f \circ g$ is affine (respectively, linear).

18 Suppose that $f: \mathscr{Y} \to \mathscr{Z}$ is affine. Then it is easy to see that, if A and B are convex sets (respectively, flats) in $\mathscr{Y}$ and $\mathscr{Z}$, respectively, then $f(A)$ and $f^{-1}(B)$, if it is not empty, are convex sets (respectively, flats) in $\mathscr{Z}$ and $\mathscr{Y}$, respectively. If P is a convex polyhedron in $\mathscr{Y}$—say $P = \text{co}\{y_1, \ldots, y_m\}$—then $f(P) = \text{co}\{f(y_1), \ldots, f(y_m)\}$ and is therefore also a convex polyhedron. If $f: \mathscr{Y} \to \mathscr{Z}$ is linear and A and B are cones (respectively, convex cones, linear manifolds) in $\mathscr{Y}$ and $\mathscr{Z}$, respectively, then $f(A)$ and $f^{-1}(B)$, if it is not empty, are cones (respectively, convex cones, linear manifolds) in $\mathscr{Z}$ and $\mathscr{Y}$, respectively.

19 If A and B are subsets of $\mathscr{Y}$, then a linear functional $\ell \in \mathscr{Y}^+$ will be said to *separate* A and B if $\ell \neq 0$ and if there is a $\gamma \in R$ such that $\ell(y_1) \leq \gamma \leq \ell(y_2)$ for all $y_1 \in A$ and $y_2 \in B$. In this case we shall say that A and B can be *separated*. Our most important results in the body of this book are based on the fact that certain convex sets can be separated. These results are known as *separation theorems*, of which we shall now state two.

20 THEOREM. *Let A be a convex set in R^n such that $0 \notin A$. Then there is a nonzero vector $b \in R^n$ such that $b \cdot \xi \leq 0$ for all $\xi \in A$.*

Proof. See [1, Lemma 2, p. 162].

21 THEOREM. *Let A and B be disjoint, convex sets in a linear vector space $\mathscr{Y}$, and suppose that A has an internal point [see (1.42)]. Then there is a nonzero linear functional $\ell \in \mathscr{Y}^+$ that separates them.*

Proof. See [4, Theorem 12, p. 412].

22 A functional f defined on a convex set $A \subset \mathscr{Y}$ will be said to be *convex* if $f(\lambda y_1 + \mu y_2) \leq \lambda f(y_1) + \mu f(y_2)$ for all $y_1, y_2 \in A$ and all $\lambda, \mu \in \bar{R}_+$ such that $\lambda + \mu = 1$. It is easy to see that $f: A \to R$ is convex if and only if, for every positive integer m, $f(\sum_1^m \lambda^i y_i) \leq \sum_1^m \lambda^i f(y_i)$ whenever $y_1, \ldots, y_m \in A$, $(\lambda^1, \ldots, \lambda^m) \in \bar{R}_+^m$ and $\sum_1^m \lambda^i = 1$.

23 It follows at once from (17) that every affine functional, and therefore every linear functional, defined on $\mathscr{Y}$ is convex. Also, if f_1 and f_2 are convex functionals defined on a convex set A in $\mathscr{Y}$, then $\alpha f_1 + \beta f_2$ is a convex functional for all $\alpha, \beta \in \bar{R}_+$. Finally,

if $\mathscr{X}$ is a linear vector space, $g:\mathscr{X} \to \mathscr{Y}$ is affine, and $f:\mathscr{Y} \to R$ is convex, then $f \circ g$ is convex on $\mathscr{X}$.

24 If f is a convex functional defined on a convex set $A \subset \mathscr{Y}$, then it is trivial to verify that the set $\{y: y \in A, f(y) \leq \gamma\}$ is either empty or convex for every $\gamma \in R$. Further, the functional f_1, defined on the set $y_0 + A$ through the relation $f_1(y) = f(y - y_0) + \gamma$, is then also convex for every $y_0 \in \mathscr{Y}$ and $\gamma \in R$.

We have the following "separation theorem" for convex functionals.

25 THEOREM. *Let A be a convex set in a linear vector space $\mathscr{Y}$, and let f_1 and f_2 be convex functionals defined on $\mathscr{Y}$ and A, respectively. Further, suppose that* (a) $0 \in A$, (b) $f_1(0) = f_2(0) = 0$, *and* (c) $-f_1(y) \leq f_2(y)$ *for all* $y \in A$. *Then there is a linear functional $\Lambda' \in \mathscr{Y}^+$ such that*

26 $$-f_1(y) \leq \Lambda'(y) \quad \textit{for all} \quad y \in \mathscr{Y},$$

27 $$f_2(y) \geq \Lambda'(y) \quad \textit{for all} \quad y \in A.$$

Proof. Let the sets V and W in $\mathscr{Y} \times R$ be defined as follows:

$$V = \{(y,\lambda): y \in \mathscr{Y}, \lambda < -f_1(y)\},$$
$$W = \{(y,\lambda): y \in A, \lambda > f_2(y)\}.$$

It follows at once from our hypotheses that the sets V and W are convex and disjoint, and that $(0,-1) \in V$.

With the aim of making use of Theorem (21), let us show that $(0,-1)$ is an internal point [see (1.42)] of V. Thus let $(y_1,\lambda_1) \in \mathscr{Y} \times R$ be arbitrary. Since f_1 is convex and $f_1(0) = 0$,

$$f_1(\lambda y_1) = f_1(\lambda y_1 + (1 - \lambda)0) \leq \lambda f_1(y_1) \quad \text{whenever} \quad 0 \leq \lambda \leq 1.$$

If we choose λ_0, $0 < \lambda_0 < 1$, such that $f_1(y_1) + \lambda_1 < 1/\lambda_0$, we see at once that $(0,-1) + \lambda(y_1,\lambda_1) \in V$ whenever $0 \leq \lambda \leq \lambda_0$, which means that $(0,-1)$ is indeed an internal point of V.

By Theorem (21), (11), and (12) with $n = 1$, we conclude that there are a linear functional $\ell \in \mathscr{Y}^+$ and real numbers γ_1 and γ_2, with ℓ and γ_1 not both zero, such that

$$\ell(y_1) + \gamma_1\lambda_1 \leq \gamma_2 \leq \ell(y_2) + \gamma_1\lambda_2$$

whenever

$$y_1 \in \mathscr{Y}, \quad y_2 \in A, \quad \lambda_1 < -f_1(y_1), \quad \text{and} \quad \lambda_2 > f_2(y_2).$$

28 But this is possible only if $\gamma_1 \geq 0$ and if

$$\ell(y_1) - \gamma_1 f_1(y_1) \leq \gamma_2 \leq \ell(y_2) + \gamma_1 f_2(y_2)$$
$$\text{for all} \quad y_1 \in \mathscr{Y} \quad \text{and} \quad y_2 \in A.$$

Since $0 \in A$ and $0 = \ell(0) = f_1(0) = f_2(0)$, $\gamma_2 = 0$. If $\gamma_1 = 0$, then (28) implies that $\ell(y_1) \leq 0$ for all $y_1 \in \mathscr{Y}$, which is possible only if $\ell = 0$, contradicting our previous conclusion that ℓ and γ_1 do not simultaneously vanish. Hence, $\gamma_1 > 0$, and, if we set $\Lambda' = (-1/\gamma_1)\ell$, so that $\Lambda' \in \mathscr{Y}^+$, then (26) and (27) follow at once from (28). |||

29 If f is a convex functional defined on $\mathscr{Y}$ and $y_0 \in \mathscr{Y}$, then a functional $\Lambda \in \mathscr{Y}^+$ will be called a *subgradient* of f at y_0, or a *support functional* to f at y_0, if $\Lambda(y) \leq f(y_0 + y) - f(y_0)$ for all $y \in \mathscr{Y}$. If f is linear, then f is a subgradient of itself at every $y_0 \in \mathscr{Y}$ (and there are no other subgradients). If $\mathscr{Y} = R^n$, so that a subgradient, as an element of $(R^n)^+$, can be represented by some element β in R^n [see (12)], then we shall also refer to the vector β as a subgradient.

We shall now generalize the concept of a convex functional in such a way that we can talk about convex functions, even when such functions take on their values in an arbitrary linear vector space. In order to do this, we first must generalize what is meant by an inequality.

30 Let W be a pointed convex cone in $\mathscr{Z}$ [see (1.14)]. Then if z_1, $z_2 \in \mathscr{Z}$, we shall write $z_1 \leq_W z_2$ to mean that $z_1 - z_2 \in W$. Note that $z \leq_W z$ for all $z \in \mathscr{Z}$, and that the relations $z_1 \leq_W z_2$ and $z_2 \leq_W z_3$ imply that $z_1 \leq_W z_3$ (i.e., the relation $\leq_W$ is reflexive and transitive). Further, if $z_1 \leq_W z_2$ and $\alpha \geq 0$, then $\alpha z_1 \leq_W \alpha z_2$; if also $z_3 \leq_W z_4$, then $z_1 + z_3 \leq_W z_2 + z_4$.

31 Note that if $\mathscr{Z} = R$ and $W = \bar{R}_-$ [see (1.18)], then $\leq_W$ is equivalent to $\leq$. If $\mathscr{Z} = R^\mu$ and $W = \bar{R}^\mu_-$ [see (1.20)], then $(\xi^1, \ldots, \xi^\mu) \leq_W (\eta^1, \ldots, \eta^\mu)$ if and only if $\xi^i \leq \eta^i$ for each $i = 1, \ldots, \mu$.

32 If W is a pointed convex cone in $\mathscr{Z}$, and A is a convex set in $\mathscr{Y}$, then a function $f: A \to \mathscr{Z}$ will be said to the *W-convex* if $f(\lambda y_1 + \mu y_2) \leq_W \lambda f(y_1) + \mu f(y_2)$ for all $y_1, y_2 \in A$ and all $\lambda, \mu \in \bar{R}_+$ such that $\lambda + \mu = 1$. It is easy to see that $f: A \to \mathscr{Z}$ is W-convex if and only if $f(\sum_1^m \lambda^i y_i) \leq_W \sum_1^m \lambda^i f(y_i)$ whenever $y_1, \ldots, y_m \in A$, $(\lambda^1, \ldots, \lambda^m) \in \bar{R}^m_+$ and $\sum_1^m \lambda^i = 1$.

33 Note that if $\mathscr{Z} = R$ and $W = \bar{R}_-$, then $f: A \to \mathscr{Z}$ is W-convex if and only if f is convex. If $\mathscr{Z} = R^\mu$ and $W = \bar{R}^\mu_-$, then $f = (f_1, \ldots, f_\mu): A \to \mathscr{Z}$ is W-convex if and only if f_i is convex for each $i = 1, \ldots, \mu$.

34 We point out that [see (17)] every affine function (and therefore every linear function) $f: \mathscr{Y} \to \mathscr{Z}$ is W-convex, no matter what the pointed convex cone $W \subset \mathscr{Z}$. Also, if f_1 and f_2 are W-convex functions from a convex set $A \subset \mathscr{Y}$ into $\mathscr{Z}$, then $(\alpha f_1 + \beta f_2)$ is a W-convex function from A into $\mathscr{Z}$ for all $\alpha, \beta \in \bar{R}_+$. Finally, if $\mathscr{X}$ is a linear vector space, $g: \mathscr{X} \to \mathscr{Y}$ is affine, and $f: \mathscr{Y} \to \mathscr{Z}$ is W-convex, then $f \circ g$ is W-convex on $\mathscr{X}$ [compare with (23)].

35 If $f: A \to \mathscr{Z}$ is W-convex, then the function f_1 defined on the set $y_0 + A$ through the relation $f_1(y) = f(y - y_0) + \zeta$ is also W-convex for every $y_0 \in \mathscr{Y}$ and $\zeta \in \mathscr{Z}$ [compare with (24)]. Further, if $W_1, \ldots, W_m$ are pointed, convex cones in $\mathscr{Z}_1, \ldots, \mathscr{Z}_m$, respectively, and A is a convex set in $\mathscr{Y}$, then a function $f = (f_1, \ldots, f_m): A \to (\mathscr{Z}_1 \times \cdots \times \mathscr{Z}_m)$ (where $f_i: A \to \mathscr{Z}_i$ for each i) is $(W_1 \times \cdots \times W_m)$-convex if and only if, for each i, f_i is W_i-convex.

36 Notice that the bigger the pointed, convex cone W, the larger the class of W-convex functions, i.e., if $W_1 \supset W_2$, then every W_2-convex function is also W_1-convex. In particular, if $W = \mathscr{Z}$, then evidently every function $f: A \to \mathscr{Z}$ is W-convex. An affine function $f: \mathscr{Y} \to \mathscr{Z}$ is clearly W-convex for any pointed, convex cone $W \subset \mathscr{Z}$. The following lemma is, roughly speaking, a converse of this assertion.

37 LEMMA. *If A is a pointed, convex cone in $\mathscr{Y}$ and $f: A \to \mathscr{Z}$ is W-convex, where $W = \{0\}$, then there is an affine function $f': \mathscr{Y} \to \mathscr{Z}$ such that f is the restriction of f' to A. If also $f(0) = 0$, then f' is linear.*

Proof. Let $f_1: A \to \mathscr{Z}$ be defined by $f_1(y) = f(y) - f(0)$ for all $y \in A$, so that f_1 is also W-convex [see (35)] and $f_1(0) = 0$. We shall construct, as is evidently sufficient for our proof, a linear function $f'_1: \mathscr{Y} \to \mathscr{Z}$ such that f_1 is the restriction of f'_1 to A.

If $y \in A$ and $0 \leq \alpha \leq 1$, then $f_1(\alpha y) = f_1(\alpha y + (1 - \alpha)0) = \alpha f_1(y) + (1 - \alpha) f_1(0) = \alpha f_1(y)$. If $y \in A$ and $\alpha > 1$, then $f_1(y) = f_1(\alpha^{-1}(\alpha y)) = \alpha^{-1} f_1(\alpha y)$ (by what we have just shown), so that $f_1(\alpha y) = \alpha f_1(y)$ for all $\alpha \geq 0$ and all $y \in A$. If $y_1, y_2 \in A$ and

$\alpha \geq 0, \beta \geq 0$, and $\alpha + \beta > 0$, then

$$\begin{aligned} f_1(\alpha y_1 + \beta y_2) &= f_1\left((\alpha + \beta)\left(\frac{\alpha}{\alpha + \beta} y_1 + \frac{\beta}{\alpha + \beta} y_2\right)\right) \\ &= (\alpha + \beta) f_1\left(\frac{\alpha}{\alpha + \beta} y_1 + \frac{\beta}{\alpha + \beta} y_2\right) \\ &= \alpha f_1(y_1) + \beta f_1(y_2). \end{aligned}$$

Since A is a convex cone, the span of A, as is easily seen, is $A - A$. If $y \in \operatorname{span} A$, so that $y = y_1 - y_2$ with $y_1, y_2 \in A$, we set $f'_1(y) = f_1(y_1) - f_1(y_2)$. It is then straightforward to verify that f'_1 is well-defined and linear on span A and that $f_1(y) = f'_1(y)$ for all $y \in A$. It only remains to extend f'_1 to $\mathscr{Y}$. But this extension is possible by virtue of the following theorem.

38 THEOREM. *If g is a linear operator from a subspace $\mathscr{Y}_0$ of $\mathscr{Y}$ into $\mathscr{Z}$, then there is a linear operator $g_1 : \mathscr{Y} \to \mathscr{Z}$ such that g_1 is an extension of g.*

Proof. See [8, Theorem 1.71–A, p. 40]. |||

We close this section with the following easily verified and useful lemma.

39 LEMMA. *Let $A \subset \mathscr{Y}$, let y_0 be an internal point of A, and let $\ell \in \mathscr{Y}^+$ be such that $\ell(y) \leq \ell(y_0)$ for all $y \in A$. Then $\ell = 0$.*

3. Topological Spaces

1 Let X be an arbitrary set. A family $\mathscr{A}$ of subsets of X is called a *topology* on X if the empty set as well as X belong to $\mathscr{A}$, if the union of every subfamily of $\mathscr{A}$ belongs to $\mathscr{A}$, and if the intersection of every finite subfamily of $\mathscr{A}$ belongs to $\mathscr{A}$. The pair $(X, \mathscr{A})$ will be called a *topological space*, although in many cases, when $\mathscr{A}$ is understood, X will itself be called a topological space.

2 If $\mathscr{A}_1$ and $\mathscr{A}_2$ are two topologies on X, then $\mathscr{A}_1$ will be said to be *stronger* than $\mathscr{A}_2$, or, equivalently, $\mathscr{A}_2$ will be said to be *weaker* than $\mathscr{A}_1$, if $\mathscr{A}_1 \supset \mathscr{A}_2$.

3 The sets which belong to $\mathscr{A}$ will be called *open* sets in the topology $\mathscr{A}$. If $U \in \mathscr{A}$ and if $x \in U$, then U will be called a *neighborhood* of x; i.e., a neighborhood of a point $x \in X$ is any open set containing x. If $A \subset X$ and $U \in \mathscr{A}$ are such that $A \subset U$, then U will be called a *neighborhood* of A, i.e., a neighborhood of a subset A of X is any open set containing A.

4 Any subset of X whose complement is open will be called *closed*. Evidently, every intersection of closed sets as well as every finite union of closed sets is closed.

5 If $A \subset X$ and $x \in A$, then x is said to be an *interior point* of A if there is an open set O such that $x \in O$ and $O \subset A$. The set of all interior points of A will be called the *interior* of A and will be written as int A. It is easily seen that A is open if and only if A coincides with int A.

6 If $A \subset X$, then x will be said to be a *limit point* of A if every neighborhood of x meets A. (Note that a limit point of A need not itself belong to A.) It is not difficult to see that A is closed if and only if every limit point of A belongs to A. Further, x is said to be a *boundary point* (or *frontier point*) of A if every neighborhood of x meets both A and the complement of A.

7 A sequence of points $x_1, x_2, \ldots$ (which will sometimes be written as $\{x_j\}$) in X is said to have limit x (in X), written as

$$\lim_{j\to\infty} x_j = x \quad \text{or} \quad x_j \xrightarrow[j\to\infty]{} x,$$

if, for every neighborhood U of x, there is a positive integer j_0 (possibly depending on U) such that $x_j \in U$ whenever $j > j_0$. In this case, we say that the sequence $\{x_j\}$ is *convergent*, and *converges to* x.

8 The intersection of all closed sets that contain a given set $A \subset X$ will be called the *closure* of A, written as $\bar{A}$. By (4), $\bar{A}$ is closed, and, in fact, $\bar{A}$ is the "smallest" closed set that contains A. Also, A is closed if and only if $A = \bar{A}$. It is also easy to show that $\bar{A}$ is the union of A with the set of all limit points of A. Finally, we note that $\bar{\bar{A}} = \bar{A}$, $\overline{A \cup B} = \bar{A} \cup \bar{B}$, and $\overline{A \cap B} \subset \bar{A} \cap \bar{B}$.

If A and B are subsets of X and $\bar{A} \supset B \supset A$, then we shall say that A is *dense* in B.

9 If $A \subset X$ and $\mathscr{A}$ is a topology on X, then the topology

$$\mathscr{A}_1 = \{B \colon B = A \cup O, O \in \mathscr{A}\}$$

will be called the topology *induced* on A by $\mathscr{A}$ (or the *relative topology* of A induced by $\mathscr{A}$). When the meaning is clear, we shall often in this case speak about $\mathscr{A}_1$ as being simply the *induced topology* on A (or the *relative topology* of A).

10 If $\mathscr{A}$ is a topology on a set X, then a subfamily $\mathscr{B}$ of $\mathscr{A}$ will be called a *base* for the topology $\mathscr{A}$ if, for every $x \in X$ and every

neighborhood U of x, there is a set $B \in \mathscr{B}$ such that $x \in B$ and $B \subset U$. A subfamily $\mathscr{B}$ of $\mathscr{A}$ will be called a *base at a point* $x \in X$ if x belongs to each set in $\mathscr{B}$, and if, for every neighborhood U of x, there is a set $B \in \mathscr{B}$ such that $B \subset U$.

If $\mathscr{B}$ is a base for a topology $\mathscr{A}$ on X, then the open sets in $\mathscr{A}$ are precisely those sets which are unions of subfamilies of $\mathscr{B}$. Thus a topology on X can be specified by giving a base for the topology.

11 THEOREM. *Let $\mathscr{B}$ be a family of subsets of X and let $\mathscr{A}$ be the family of all unions of subfamilies of $\mathscr{B}$. Then $\mathscr{A}$ is a topology on X and $\mathscr{B}$ is a base for this topology if and only if*

(1) *for every pair B_1, B_2 in $\mathscr{B}$ and every $x \in B_1 \cap B_2$, there is a B_3 in $\mathscr{B}$ such that $x \in B_3$ and $B_3 \subset (B_1 \cap B_2)$;*

(2) *for every $x \in X$, there is a $B \in \mathscr{B}$ such that $x \in B$.*

12 If $\mathscr{A}_1$ and $\mathscr{A}_2$ are two topologies on X with respective bases $\mathscr{B}_1$ and $\mathscr{B}_2$, then $\mathscr{A}_1$ is stronger than $\mathscr{A}_2$ if and only if, for every set $B_2 \in \mathscr{B}_2$ and each $x \in B_2$, there is a set $B_1 \in \mathscr{B}_1$ such that $x \in B_1 \subset B_2$. Thus, $\mathscr{A}_1 = \mathscr{A}_2$ if and only if (a) whenever $B_1 \in \mathscr{B}_1$ and $x \in B_1$, there is a set $B_2 \in \mathscr{B}_2$ such that $x \in B_2 \subset B_1$, and (b) whenever $B_2 \in \mathscr{B}_2$ and $x \in B_2$, there is a $B_1 \in \mathscr{B}_1$ such that $x \in B_1 \subset B_2$.

13 If $\mathscr{A}$ is a topology on X, and if, for every $x \in X$, $\mathscr{B}_x$ is a base at x, then the union of all the families $\mathscr{B}_x$ is clearly a base for $\mathscr{A}$. Thus, a topology on X can be specified by giving a base at each point $x \in X$.

14 If $A \subset X$, where X is a topological space, then any collection of open subsets of X whose union contains A will be called a *covering* of A. A set A will be said to be *compact* if every covering of A contains a finite subcollection which is also a covering of A. A set will be said to be *conditionally compact* (or *relatively compact*) if its closure is compact. A set A will be said to be *sequentially compact* if every sequence contained in A has a convergent subsequence (the limit of this subsequence need not, however, belong to A, although it must, of course, belong to X).

15 It is easily seen that the intersection of a compact set and a closed set is compact (so that every closed subset of a compact set is compact), and that any finite union of compact sets (respectively, conditionally compact sets) is compact (respectively, conditionally compact).

16 Suppose that X and X' are topological spaces, that $A_1 \subset A \subset X$, $x_0 \in X$ and $x' \in X'$, that f is a function from A into X', and that f_1 is the restriction of f to A_1. Then we shall say that $f(x)$ *tends* (or *converges*) *to* x' *as* x *tends to* x_0 [or that x' is the *limit* of $f(x)$ as x tends to x_0], written as

17 $$f(x) \to x' \quad \text{as} \quad x \to x_0,$$

if, for every neighborhood U' of x', there is a neighborhood U of x_0 such that $f(x) \in U'$ whenever $x \in U \cap A$. Note that if $x_0 \notin \bar{A}$, then $f(x) \to x'$ as $x \to x_0$ for every $x' \in X'$. On the other hand, if $x_0 \in \bar{A}$ and X' is a Hausdorff space [see (29)], then if $f(x) \to x'$ as $x \to x_0$ and $f(x) \to x''$ as $x \to x_0$, we must have that $x' = x''$. [Of course, there may be no $x' \in X'$ such that $f(x) \to x'$ as $x \to x_0$.] If $f_1(x)$ tends to x' as x tends to x_0 [whether or not $f(x)$ does], then we shall write

18 $$f(x) \to x' \quad \text{as} \quad x \to x_0 \quad \text{and} \quad x \in A_1.$$

For the special case when $A \subset R$, we shall write

19 $$f(x) \to x' \quad \text{as} \quad x \to x_0^+ \quad (\text{respectively, as } x \to x_0^-)$$

to mean $f(x) \to x'$ as $x \to x_0$ and $x \in (x_0 + R_+)$ [respectively, $x \in (x_0 + R_-)$].

20 If X, X', x_0, and x' are as in the first sentence of (16), Y is an arbitrary set, $A_1 \subset A \subset X \times Y$, $Y_1 \subset Y$, and f is a function from A into X', then we shall say that $f(x,y)$ *tends* (or *converges*) *to* x' *as* x *tends to* x_0 *uniformly with respect to* $y \in Y_1$, written as

21 $$f(x,y) \to x' \quad \text{as} \quad x \to x_0 \quad \text{uniformly w.r. to} \quad y \in Y_1,$$

if, for every neighborhood U' of x', there is a neighborhood U of x_0 such that $f(x,y) \in U'$ whenever $(x,y) \in (U \times Y_1) \cap A$. Clearly, if $f(x,y) \to x'$ as $x \to x_0$ uniformly w.r. to $y \in Y_1$, and if we denote (for any fixed $y \in Y_1$) the function $x \to f(x,y)$ by $g_y(x)$, then $g_y(x) \to x'$ as $x \to x_0$ for every $y \in Y_1$ (although the converse is generally false). We shall write

22 $$f(x,y) \to x' \quad \text{as} \quad x \to x_0 \quad \text{and} \quad x \in A_1$$
$$\text{uniformly w.r. to} \quad y \in Y_1$$

with a meaning that should now be clear, and, in the special case where $A \subset R$, we shall write

23 $$f(x,y) \to x' \quad \text{as} \quad x \to x_0^+ \text{ (or as } x \to x_0^-) \\ \text{uniformly w.r. to} \quad y \in Y_1$$

with an again evident meaning.

In accordance with the usual conventions, we shall often write either

$$f(x) \xrightarrow[x \to x_0]{} x' \quad \text{or} \quad \lim_{x \to x_0} f(x) = x'$$

in place of (17), shall often replace (18) by either of the following two notations:

$$f(x) \xrightarrow[\substack{x \to x_0 \\ x \in A_1}]{} x', \quad \lim_{\substack{x \to x_0 \\ x \in A_1}} f(x) = x',$$

and shall often write

$$f(x,y) \xrightarrow[x \to x_0]{} x' \quad \text{uniformly w.r. to} \quad y \in Y$$

in place of (21). Further, we shall often also replace (19), (22), and (23) by similarly modified expressions.

24 A function f from some set A in a topological space X into a topological space X' will be said to be *continuous at a point* $x_0 \in A$ if $f(x) \to f(x_0)$ as $x \to x_0$. Evidently, f is continuous at x_0 if and only if, for every neighborhood U' of $f(x_0)$, there is a neighborhood U of x_0 such that $f(U \cap A) \subset U'$. If $A_1 \subset A$ and the restriction of f to A_1 is continuous at each point of A_1, then we shall say that f *is continuous on* A_1. If f is continuous at each point of A, then we shall simply say that f is *continuous*.

The following three important theorems may be easily verified.

25 THEOREM. *Let f be a function from A into X', where $A \subset X$ and X and X' are topological spaces. Then the following three assertions are equivalent:*

A. *f is continuous,*

B. *$f^{-1}(O')$ is open in the induced topology on A whenever O' is open in X',*

C. *$f^{-1}(C')$ is closed in the induced topology on A whenever C' is closed in X'.*

26 THEOREM. *Let f be a continuous function from A into X', where $A \subset X$ and X and X' are topological spaces, and suppose that C is a compact subset of A. Then $f(C)$ is compact in X'.*

27 THEOREM. *Let X, X', and X'' be topological spaces, let $A \subset X$ and $A' \subset X'$, and let $f: A \to X'$ and $g: A' \to X''$ be functions such that $f(A) \subset A'$. Then, if f is continuous at a point $x_0 \in A$ and the restriction of g to $f(A)$ is continuous at $f(x_0)$, $g \circ f$ is continuous at x_0. In particular, if f is continuous and g is continuous on $f(A)$, then $g \circ f$ is continuous.*

28 If X and X' are topological spaces, $A \subset X$, and $A' \subset X'$, then A and A' will be said to be *homeomorphic* if there is a continuous, one-to-one function f from A onto A' such that f^{-1} is also continuous. In this case, f will be said to be a *homeomorphism* of A onto A'.

29 A topological space X will be said to be *Hausdorff* (or *separated*) if for every pair of distinct points x_1 and x_2 in X, there are neighborhoods U_1 of x_1 and U_2 of x_2 such that U_1 and U_2 are disjoint, i.e., have no points in common.

The following lemma may easily be verified.

30 LEMMA. *In a Hausdorff space, every finite set is closed as well as compact, and every compact set is closed.*

31 COROLLARY. *In a Hausdorff space, the intersection of compact sets is compact, and each subset of a compact set is conditionally compact.*

Proof. Corollary (31) follows at once from Lemma (30), (4), and (15).

32 LEMMA. *Let X and X' be topological spaces, with X' Hausdorff, let A be a compact set in X, and let $f: A \to X'$ be a continuous, one-to-one function. Then f is a homeomorphism from A onto $f(A)$.*

Proof. By Theorem (25), it is sufficient to prove that, for every closed set $C \subset X$, $f(C \cap A)$ is closed. But if C is closed, then, by (15), $C \cap A$ is compact, so that, by Theorem (26), $f(C \cap A)$ is also compact. It then follows from Lemma (30) that $f(C \cap A)$ is closed. |||

33 If $X_1, \ldots, X_m$ are topological spaces, then we shall define the *product topology* on the direct product $X_1 \times \cdots \times X_m$ by taking as base all sets of the form $O_1 \times \cdots \times O_m$ where O_i is open in X_i for each i. [It follows from Theorem (11) that this defines a topology.] The topological space consisting of the direct product $X_1 \times \cdots \times X_m$ paired with the product topology, will be called the *direct product* (or *Cartesian product*) of the topological spaces $X_1, \ldots, X_m$. In this case, it is easily seen that the following

assertions hold for the direct product:

34 1. If $A_i \subset X_i$ for $i = 1, \ldots, m$, then $\overline{A_1 \times \cdots \times A_m} = \bar{A}_1 \times \cdots \times \bar{A}_m$, and $\text{int}\,(A_1 \times \cdots \times A_m) = (\text{int}\,A_1) \times \cdots \times (\text{int}\,A_m)$.

35 2. If A_i is a closed (respectively, open, compact, or conditionally compact) subset of X_i for each $i = 1, \ldots, m$, then $A_1 \times \cdots \times A_m$ is a closed (respectively, open, compact, or conditionally compact) subset of $X_1 \times \cdots \times X_m$. (See [4, Theorem 5, p. 32].)

36 3. If $X_1, \ldots, X_m$ are all Hausdorff spaces, then so is $X_1 \times \cdots \times X_m$.

37 4. If X_0 is a topological space, $x_0 \in X_0$, f_i, for each $i = 1, \ldots, m$, is a function from a subset A of X_0 into X_i, $A_1 \subset A$, and $x_i' \in X_i$ for each $i = 1, \ldots, m$, then $f_i(x) \to x_i'$ as $x \to x_0$ (respectively, as $x \to x_0$ and $x \in A_1$) for each $i = 1, \ldots, m$ if and only if $(f_1, \ldots, f_m)(x) \to (x_1', \ldots, x_m')$ as $x \to x_0$ (respectively, as $x \to x_0$ and $x \in A_1$). A similar assertion may be made for uniform limits.

38 5. If X_0, x_0, A, A_1, and $f_1, \ldots, f_m$ are as in (37), then $(f_1, \ldots, f_m)$ is continuous (respectively, continuous on A_1, continuous at x_0) if and only if $f_1, \ldots, f_m$ are all continuous (respectively, continuous on A_1, continuous at x_0).

As usual, if $X_1 = \cdots = X_m = X$, then we shall write X^m for $X_1 \times \cdots \times X_m$.

39 As an example, let us consider the usual topology on R. Namely, as a base for the topology on R let us take all sets of the form $\{\lambda : |\lambda - \lambda_0| < \varepsilon\}$ where $\lambda_0 \in R$ and $\varepsilon \in R_+$. It is easy to see that in this way we obtain a Hausdorff topology. In this topology, all sets of the form $\{\lambda : \lambda_0 < \lambda < \lambda_1\}$, where $\lambda_0, \lambda_1 \in R$ are clearly open and will be denoted by (λ_0, λ_1); all sets of the form $\{\lambda : \lambda_0 \leq \lambda \leq \lambda_1\}$, where $\lambda_0, \lambda_1 \in R$ (we do not exclude $\lambda_0 = \lambda_1$) are clearly closed and will be denoted by $[\lambda_0, \lambda_1]$. The sets $[\lambda_0, \lambda_1)$ and $(\lambda_0, \lambda_1]$ are defined in an analogous manner (and are neither open nor closed). The sets of the form $\{\lambda : \lambda > \lambda_0\}$, where $\lambda_0 \in R$, are open and will be denoted by (λ_0, ∞). The sets $[\lambda_0, \infty)$, $(-\infty, \lambda_0)$, and $(-\infty, \lambda_0]$ are defined in an analogous manner. By an *interval* we shall mean any subset of R of the just-described types.

40 Since R^n is the same as $R \times R \times \cdots \times R$ (n times), we may define the product topology on R^n, with the just-indicated topology for R. It is not hard to see that, for any $\xi_0 \in R^n$, we may

then take as base at ξ_0 the family of all sets of the form $\{\xi : |\xi - \xi_0| < \alpha\}$ where $\alpha > 0$. This topology is known as the *Euclidean topology* for R^n and is, by (36), Hausdorff. Henceforth, R^n will always be considered to be a topological space (with the Euclidean topology).

41 It is easily seen that R^n_+ and R^n_- are open in R^n, and that $\bar{R}^n_+$ and $\bar{R}^n_-$ are the closures of R^n_+ and R^n_-, respectively.

The following results are of importance for the spaces R^n ($n = 1, 2, \ldots$).

42 1. The function $(\xi,\eta) \to \xi + \eta$ from $R^n \times R^n$ into R^n is continuous, i.e., vector addition is continuous in R^n.

43 2. The function $(\alpha,\xi) \to \alpha\xi$ from $R \times R^n$ into R^n is continuous, i.e., scalar multiplication is continuous in R^n.

44 3. A set in R^n is open if and only if it is finitely open [see (1.37)].

45 4. A set in R^n is compact if and only if it is closed and bounded. (A set A in R^n is *bounded* if there is an $\alpha_0 \in R_+$ such that $|\xi| \leq \alpha_0$ for all $\xi \in A$.) This result is known as the Heine-Borel Theorem [7, Theorem 10.10, p. 53]. A set in R^n is conditionally compact if and only if it is bounded. (This follows because the closure of a bounded set in R^n is bounded.)

46 It follows from (45) and Theorem (26) that if C is a compact subset of a topological space X, and f is a continuous map from A into R, where $X \supset A \supset C$, then there are points x_1 and x_2 in C such that

$$f(x_1) = \inf\{f(x) : x \in C\} \text{ and } f(x_2) = \sup\{f(x) : x \in C\},$$

so that f is bounded from above as well as from below on C, and attains both its infimum and its supremum on C.

4. Linear Topological Spaces

1 A *linear topological space* is a linear vector space $\mathscr{Y}$ together with a topology on $\mathscr{Y}$ such that (a) with this topology, $\mathscr{Y}$ is Hausdorff, (b) vector addition in $\mathscr{Y}$ is a continuous function from $\mathscr{Y} \times \mathscr{Y}$ (with the product topology) into $\mathscr{Y}$, and (c) scalar multiplication is a continuous function from $R \times \mathscr{Y}$ (in the product topology) into $\mathscr{Y}$.

As an example, it follows from (3.40, 42, and 43) that R^n, with the Euclidean topology, is a linear topological space.

Throughout the remainder of this section, unless the contrary is stated, $\mathscr{Y}$, as well as $\mathscr{Y}_1, \ldots, \mathscr{Y}_m$, will denote linear topological spaces.

2 LEMMA. *For every positive integer m, the map* $(y^1, \ldots, y^m, \alpha^1, \ldots, \alpha^m) \to \sum_{i=1}^m \alpha^i y_i : \mathscr{Y}^m \times R^m \to \mathscr{Y}$ *is continuous (with the product topology in* $\mathscr{Y}^m \times R^m$*).*

3 COROLLARY. *For every finite subset* $\{y_1, \ldots, y_m\}$ *of* $\mathscr{Y}$ *and every neighborhood N of* 0 *in* $\mathscr{Y}$*, there is a number* $\delta_0 > 0$ *such that* $\sum_1^m \alpha^i y_i \in N$ *so long as* $|(\alpha^1, \ldots, \alpha^m)| < \delta_0$*. For every finite set* $\{\alpha^1, \ldots, \alpha^m\}$ *of real numbers and every neighborhood N of* 0 *in* $\mathscr{Y}$*, there is a neighborhood* N_1 *of* 0 *(in* $\mathscr{Y}$*) such that* $\sum_1^m \alpha^i N_1 \subset N$.

4 COROLLARY. *For every positive integer m, and every neighborhood N of* 0 *in* $\mathscr{Y}$*, there are a neighborhood* N_1 *of* 0 *(in* $\mathscr{Y}$*) and a number* $\delta_0 > 0$ *such that the sets* $(N_1 + \cdots + N_1)$ *(m times) and* $\sum_1^m \alpha^i N_1$ *are both contained in N, so long as* $|(\alpha^1, \ldots, \alpha^m)| < \delta_0$.

5 LEMMA. *For any (fixed)* $\lambda_0 \in R$, $\lambda_0 \neq 0$, *and* $y_0 \in \mathscr{Y}$*, the maps* $y \to \lambda_0 y$, $y \to y_0 + y$, *and* $y \to y_0 - y$ *are homeomorphisms of* $\mathscr{Y}$ *into itself.*

6 COROLLARY. *If N is a neighborhood of* 0 *in* $\mathscr{Y}$*, then* $\lambda_0 N$ *also is, for every* $\lambda_0 \in R$, $\lambda_0 \neq 0$.

7 COROLLARY. *N is a neighborhood of* 0 *in* $\mathscr{Y}$ *if and only if* $y_0 + N$ *is a neighborhood of* y_0 *(for any* $y_0 \in \mathscr{Y}$*).*

8 It follows from Corollary (6) that $\mathscr{B}_y$ is a base (for the topology on $\mathscr{Y}$) at a point $y \in \mathscr{Y}$ [see (3.10)] if and only if $\mathscr{B}_y = \{y + B : B \in \mathscr{B}_0\}$, where $\mathscr{B}_0$ is a base at $0 \in \mathscr{Y}$. Thus, a topology on a linear topological space can be specified by giving a base at 0 [see (3.13)].

9 A base at 0 for a linear topological space $\mathscr{Y}$ will be called a *local base* for $\mathscr{Y}$, or a *local base for the topology* on $\mathscr{Y}$.

10 It follows from (3.12 and 13) that if $\mathscr{Y}$ is a linear topological space under two topologies $\mathscr{A}_1$ and $\mathscr{A}_2$, with respective local bases $\mathscr{B}_1$ and $\mathscr{B}_2$, then $\mathscr{A}_1$ is stronger than $\mathscr{A}_2$ if and only if, for every set $B_2 \in \mathscr{B}_2$, there is a set $B_1 \in \mathscr{B}_1$ such that $B_1 \subset B_2$. Thus $\mathscr{A}_1 = \mathscr{A}_2$ if and only if, for every $B_1 \in \mathscr{B}_1$, there is a $B_2 \in \mathscr{B}_2$ such that $B_2 \subset B_1$, and, for every $B_2 \in \mathscr{B}_2$, there is a $B_1 \in \mathscr{B}_1$ such that $B_1 \subset B_2$.

11 THEOREM. *Let* $\mathscr{Y}$ *be a linear vector space, and let* $\mathscr{B}$ *be a nonempty family of nonempty subsets of* $\mathscr{Y}$ *with the following properties:*

(1) $\lambda B \subset B$ *for all* $B \in \mathscr{B}$ *and all* $\lambda \in R$ *such that* $|\lambda| \leq 1$,

(2) *for each* $B \in \mathscr{B}$ *and every* $y \in \mathscr{Y}$*, there is a number* $\lambda_0 > 0$ *such that* $\lambda y \in B$ *whenever* $|\lambda| \leq \lambda_0$,

(3) *for every* $B \in \mathscr{B}$, *there is a set* $B_1 \in \mathscr{B}$ *such that* $B_1 + B_1 \subset B$,

(4) *for every* B_1 *and* $B_2 \in \mathscr{B}$, *there is a* $B_3 \in \mathscr{B}$ *such that* $B_3 \subset (B_1 \cap B_2)$,

(5) *if* $B \in \mathscr{B}$ *and* $y \in B$, *there is a set* $B_1 \in \mathscr{B}$ *such that* $y + B_1 \subset B$, *and*

(6) *only* 0 *belongs to every set* $B \in \mathscr{B}$.

Then there is a unique topology for $\mathscr{Y}$ *such that* $\mathscr{Y}$ *is a linear topological space with* $\mathscr{B}$ *as a local base.*

Proof. See [8, Theorems 3.3–F and 3.3–G, pp. 125–126].

12 Two linear topological spaces $\mathscr{Y}_1$ and $\mathscr{Y}_2$ will be said to be *topologically isomorphic* if there is a linear operator [see (2.6)] from $\mathscr{Y}_1$ onto $\mathscr{Y}_2$ which is a homeomorphism [see (3.28)]. Clearly, if $\mathscr{Y}_1$ and $\mathscr{Y}_2$ are topologically isomorphic, then they are isomorphic as linear vector spaces [see (1.4)] and homeomorphic as topological spaces.

13 THEOREM. *Every n-dimensional linear topological space is topologically isomorphic to* R^n.

Proof. See [8, Theorem 3.3–H, p. 127].

14 COROLLARY. *The Euclidean topology* [*see* (3.40)] *is the only topology on* R^n *under which* R^n *becomes a linear topological space.*

15 LEMMA. *If* $\mathscr{Y}_0$ *is a linear manifold in* (i.e., *subspace of*) *a linear topological space* $\mathscr{Y}$, *then* $\mathscr{Y}_0$ *is a linear topological space under the induced topology* [*see* (3.9)] *on* $\mathscr{Y}_0$.

16 It follows from Corollary (3) that every open set in $\mathscr{Y}$ is finitely open and therefore finitely open in itself [see (1.37–40)]. Thus, in particular, the intersection of an open set in $\mathscr{Y}$ with a convex set in $\mathscr{Y}$ is finitely open in itself.

17 If $\mathscr{Y}_1, \ldots, \mathscr{Y}_m$ are linear topological spaces, we may consider the linear vector space $\mathscr{Y} = \mathscr{Y}_1 \times \cdots \times \mathscr{Y}_m$ [see (1.5)], and we may impose on $\mathscr{Y}$ the product topology [see (3.33)]. It is easily verified that, with the product topology, $\mathscr{Y}$ becomes a linear topological space, which will be referred to, as usual, as the *direct* (or *Cartesian*) *product* of the linear topological spaces $\mathscr{Y}_1, \ldots, \mathscr{Y}_m$. In this case, when we write $\mathscr{Y}_1 \times \cdots \times \mathscr{Y}_m$, it is to be understood that this denotes the just-indicated linear topological space. It is easily seen that if $\mathscr{B}_1, \ldots, \mathscr{B}_m$ are local bases for $\mathscr{Y}_1, \ldots, \mathscr{Y}_m$, respectively, then the family

$$\mathscr{B} = \{B: B = B_1 \times \cdots \times B_m, B_i \in \mathscr{B}_i \text{ for each } i\}$$

is a local base for $\mathscr{Y}$.

18 A convex set in $\mathscr{Y}$ with a nonempty interior will be called a *convex body*.

19 LEMMA. *If K is a convex body in* $\mathscr{Y}$, *then* int K *is convex,* $\overline{\text{int } K} = \bar{K}$, *and* $\lambda(\text{int } K) + \mu\bar{K} = \text{int } K$ *whenever* $0 < \lambda \leq 1$ *and* $\lambda + \mu = 1$. *If K is in addition a cone, then* int K *is also a cone.*

Proof. See [4, Theorem 1 and its proof, pp. 413–414].

20 LEMMA. *If y is an interior point of a set* $A \subset \mathscr{Y}$, *then y is also an internal point of A.*

Proof. This follows at once from Corollaries (3) and (7).

21 THEOREM. *If P is a convex polyhedron (and, in particular, a simplex) in* $\mathscr{Y}$, *then P is compact. If P is a simplex, then the map* $y \to (\lambda^1, \ldots, \lambda^m)$ *from P into* R^m, *where* $\lambda^1, \ldots, \lambda^m$ *are the barycentric coordinates of y, is continuous.*

Proof. Let $P = \text{co}\,\{y_1, \ldots, y_m\}$. If P is a simplex, we shall suppose that $y_1, \ldots, y_m$ are the vertices of P. We first observe that (as is easily verified) the set $\Lambda = \{\lambda : \lambda = (\lambda^1, \ldots, \lambda^m) \in \bar{R}^m_+, \sum_1^m \lambda^i = 1\}$ is closed and bounded, and hence compact, in R^m [see (3.45)]. The map $(\lambda^1, \ldots, \lambda^m) \to \sum_1^m \lambda^i y_i$ from Λ onto P is continuous by Lemma (2), and, in case P is a simplex, is one-to-one [see (1.31–33)]. Our desired conclusions now follow at once from Theorem (3.26) and Lemma (3.32). |||

The following two lemmas may trivially be verified.

22 LEMMA. *If* $A_1, \ldots, A_m$ *are convex bodies [respectively, finitely open sets, sets which are finitely open in themselves—see* (1.37 *and* 39)] *in* $\mathscr{Y}_1, \ldots, \mathscr{Y}_m$, *respectively, then* $A_1 \times \cdots \times A_m$ *is a convex body (respectively, finitely open set, set which is finitely open in itself) in* $\mathscr{Y}_1 \times \cdots \times \mathscr{Y}_m$.

23 LEMMA. *If K is a convex set (respectively, cone, convex cone) in* $\mathscr{Y}$, *then* $\bar{K}$ *also is. If K is a cone, then* $0 \in \bar{K}$, *so that every closed cone is pointed.*

24 If A is an arbitrary set in $\mathscr{Y}$, then $\overline{\text{co}}\, A$, $\overline{\text{cone}}\, A$, and $\overline{\text{co co}}\, A$ will denote the closures of co A, cone A, and co co A, respectively [see (1.27–29)]. It follows from Lemma (23) that, for any $A \subset \mathscr{Y}$, $\overline{\text{co}}\, A$, $\overline{\text{cone}}\, A$, and $\overline{\text{co co}}\, A$ are, respectively, a convex set, a cone, and a convex cone in $\mathscr{Y}$.

25 If A is an arbitrary set in $\mathscr{Y}$ and $y_0 \in A$, then we shall call the *tangent cone of A at* y_0, denoted by tancone $(A; y_0)$, the set of all $\tilde{y} \in \mathscr{Y}$ with the following property: For every $\varepsilon > 0$ and every neighborhood N of 0 in $\mathscr{Y}$, there exist a positive number $\varepsilon_1 < \varepsilon$ and an element $y \in N$ such that $y_0 + \varepsilon_1(\tilde{y} + y) \in A$.

The following two lemmas may easily be proved. [To prove the second, we make use of Lemma (1.41)].

26 LEMMA. *For every* $y_0 \in A$, *where* $A \subset \mathscr{Y}$, tancone $(A;y_0)$ *is a closed, pointed cone, and* tancone $(A;y_0) \subset \overline{\text{cone}}\,(A - y_0)$.

27 LEMMA. *If A is a subset of* $\mathscr{Y}$ *which is finitely open in itself* [*in particular, if A is convex—see* (1.40)], *and* $y_0 \in A$, *then* $\overline{\text{cone}}\,(A - y_0) = \text{tancone}\,(A;y_0) = \overline{\text{co}}\,\{\text{tancone}\,(A;y_0)\}$.

28 LEMMA. *If W is an open convex cone in* $\mathscr{Y}$, *then* $\alpha W + \beta \bar{W} = W$ *whenever* $\alpha > 0$ *and* $\beta \geq 0$; *in particular* $W + \bar{W} = W$.

Proof. Let $\lambda = \alpha/(\alpha + \beta)$ and $\mu = \beta/(\alpha + \beta)$, so that $0 < \lambda \leq 1$ and $\lambda + \mu = 1$. It now follows from Lemma (19) that $\lambda W + \mu \bar{W} = W$. Since W and $\bar{W}$ are cones [see Lemma (23)], $\lambda W = \alpha W = W$ and $\mu \bar{W} = \beta \bar{W} = \bar{W}$ if $\mu > 0$, and our desired conclusion follows at once.

29 LEMMA. *If A and B are compact sets in* $\mathscr{Y}$, *then* $\alpha A + \beta B$ *is compact for all* $\alpha, \beta \in R$.

Proof. The set $A \times B$ is compact in $\mathscr{Y}^2$ [see (3.35)], and the map $(y_1,y_2) \to \alpha y_1 + \beta y_2$ from $\mathscr{Y}^2$ into $\mathscr{Y}$ is continuous by Lemma (2), so that our conclusion follows at once from Theorem (3.26).

30 A linear topological space $\mathscr{Y}$ will be said to be *locally convex* if there is a local base [see (9)] for $\mathscr{Y}$ each of whose members is convex.

A particularly important class of locally convex linear topological spaces is the class of so-called normed spaces, which are defined as follows:

31 A *norm* is a functional defined on a linear vector space, with its value at any $y \in \mathscr{Y}$ ordinarily denoted by $\|y\|$, with the following properties:

(a) $\|y\| \geq 0$ for all $y \in \mathscr{Y}$,
(b) $\|y\| = 0$ if and only if $y = 0$,
(c) $\|\alpha y\| = |\alpha|\,\|y\|$ for all $\alpha \in R$ and $y \in \mathscr{Y}$,
(d) $\|y_1 + y_2\| \leq \|y_1\| + \|y_2\|$ for all $y_1, y_2 \in \mathscr{Y}$.

32 If $\mathscr{Y}$ is a space with a norm $\|\cdot\|$, we may define a topology on $\mathscr{Y}$ [under which $\mathscr{Y}$ becomes a linear topological space, by virtue of Theorem (11)] by choosing as a local base all sets of the form $\{y: \|y\| < \alpha\}$ where $\alpha > 0$. This topology will be referred to as the *norm topology* for $\mathscr{Y}$, and the resulting linear topological space will be referred to as a *normed linear vector space*. Note that every normed linear vector space is Hausdorff [see (3.29)].

33 If $\|\cdot\|_1$ and $\|\cdot\|_2$ are two norms defined on the same linear space $\mathscr{Y}$, then the norm topologies induced by these two norms coincide [see (10)] whenever there are constants $\lambda_1 > 0$ and $\lambda_2 > 0$ such that

$$\lambda_1\|y\|_1 \leq \|y\|_2 \leq \lambda_2\|y\|_1 \quad \text{for all} \quad y \in \mathscr{Y},$$

In this case, the two norms $\|\cdot\|_1$ and $\|\cdot\|_2$ will be said to be *equivalent*. It is evident that if there are two equivalent norms on a linear vector space $\mathscr{Y}$, then the two linear topological spaces corresponding to the two norm topologies on $\mathscr{Y}$ are topologically isomorphic [in the sense of (12)].

It is useful to notice that the norm is itself a continuous functional in the norm topology of a normed linear vector space.

34 Two normed linear vector spaces $\mathscr{Y}$ and $\mathscr{Z}$ will be said to be *isometrically isomorphic* if the two spaces are isomorphic [see (1.4)] as linear spaces (with correspondence denoted by $\leftrightarrow$) such that $\|y\| = \|z\|$ whenever $y \leftrightarrow z$. It is evident that if $\mathscr{Y}$ and $\mathscr{Z}$ are isometrically isomorphic, then they are also topologically isomorphic (in the norm topologies).

35 A sequence $y_1, y_2, \ldots,$ in a normed linear vector space will be called a *Cauchy sequence* if $\lim_{m,n\to\infty} \|y_m - y_n\| = 0$. A normed linear vector space $\mathscr{Y}$ will be said to be *complete* if, whenever $y_1, y_2, \ldots,$ is a Cauchy sequence in $\mathscr{Y}$, there is a (necessarily unique) element $y \in \mathscr{Y}$ such that $\lim_{n\to\infty} \|y - y_n\| = 0$. Complete normed linear vector spaces will also be referred to as *Banach spaces*.

36 It is not hard to see that the functional $\xi \to |\xi|$ [see (2.13)] is a norm on R^n under which R^n becomes a Banach space. This norm will be called the *Euclidean norm* on R^n. It is easily seen to be equivalent to the two norms $\|(\xi^1, \ldots, \xi^n)\|_1 = \sum_1^n |\xi^i|$ and $\|(\xi^1, \ldots, \xi^n)\|_2 = \max\{|\xi^i|: i = 1, \ldots, n\}$. The corresponding norm topology is easily seen to coincide with the Euclidean topology defined in (3.40). Other important examples of Banach spaces will be discussed in Section 6.

37 The sets $\{y: \|y\| < \alpha\}$ and $\{y: \|y\| \leq \alpha\}$, where $\alpha > 0$, will be called, respectively, open and closed *balls in* $\mathscr{Y}$. It is not hard to see that every open (respectively, closed) ball in $\mathscr{Y}$ is an open (respectively, closed) set in the norm topology. Also, it is easily verified that every ball (whether open or closed) is convex, so

that a normed linear vector space is locally convex. A set in a normed linear vector space will be said to be *bounded* if it is contained in some ball.

38 If $\mathscr{Y}_1, \ldots, \mathscr{Y}_m$ are normed linear vector spaces, we may define a norm on $\mathscr{Y}_1 \times \cdots \times \mathscr{Y}_m$ as follows:

39 $$\|(y_1, \ldots, y_m)\| = \max_{1 \leq i \leq m} \|y_i\|_i,$$

where $\|\cdot\|_i$ denotes the norm in $\mathscr{Y}_i$ for each $i = 1, \ldots, m$. [It is straightforward to verify that this norm on $\mathscr{Y}_1 \times \cdots \times \mathscr{Y}_m$ satisfies conditions (a)–(d) of (31).] The normed linear vector space $\mathscr{Y}_1 \times \cdots \times \mathscr{Y}_m$, with norm defined by (39), will be called the *direct product* of the normed linear vector spaces $\mathscr{Y}_1, \ldots, \mathscr{Y}_m$, and the norm given by (39) will be called the *product* of the norms $\|\cdot\|_1, \ldots, \|\cdot\|_m$.

40 It is not hard to see that the norm on a product space given by (39) is equivalent [in the sense of (33)] to either of the two norms $(\sum_1^m \|y_i\|_i^2)^{1/2}$ or $\sum_{i=1}^m \|y_i\|_i$.

41 If $\mathscr{Y}_1, \ldots, \mathscr{Y}_m$ are complete, then it is easily seen that $\mathscr{Y}_1 \times \cdots \times \mathscr{Y}_m$ also is, so that a direct product of Banach spaces is itself a Banach space.

42 It should be noted that if $\mathscr{Y}_1, \ldots, \mathscr{Y}_m$ are normed linear vector spaces, then we have indicated two ways of defining a topology on the product space $\mathscr{Y}_1 \times \cdots \times \mathscr{Y}_m$: the norm topology induced by the product norm as defined in (38), and the product topology obtained from considering $\mathscr{Y}_1, \ldots, \mathscr{Y}_m$ as linear topological spaces, as was described in (17). On the basis of (10), however, it is easy to see that these two topologies are identical.

We close this section with the following useful characterization of conditionally compact sets in normed linear vector spaces.

43 THEOREM. *A set in a normed linear vector space $\mathscr{Y}$ is conditionally compact if and only if it is sequentially compact* [*see* (3.14)].

Proof. See [4, Theorem 15, p. 22].

5. Functions from One Linear Topological Space into Another

In this section we shall consider properties of functions which are defined on a linear topological space (or on a subset of such a space) and which take on their values in another (or possibly the same) linear topological space. Primarily, we shall confine ourselves to continuous functions.

We first point out that if $\mathscr{Y}$ and $\mathscr{Z}$ are linear topological spaces, $A \subset \mathscr{Y}$, $y_0 \in A$, $f: A \to \mathscr{Z}$, $g: A \to \mathscr{Z}$, and if f and g are continuous at y_0, then $(\alpha f + \beta g)$ is also continuous at y_0, for all $\alpha, \beta \in R$.

1 If $\mathscr{Y}$ and $\mathscr{Z}$ are linear topological spaces, then we shall denote by $\mathbf{B}(\mathscr{Y},\mathscr{Z})$ the set of all continuous linear operators from $\mathscr{Y}$ into $\mathscr{Z}$. It follows from what was just said and from (2.10) that $\mathbf{B}(\mathscr{Y},\mathscr{Z})$ may be considered to be a linear vector space (and we shall so consider it). Note that a linear operator from $\mathscr{Y}$ into $\mathscr{Z}$ is continuous if and only if it is continuous at $0 \in \mathscr{Y}$. Also note that the identity operator on $\mathscr{Y}$ belongs to $\mathbf{B}(\mathscr{Y},\mathscr{Y})$.

2 If $\mathscr{Y}$ and $\mathscr{Z}$ are normed linear vector spaces, then a linear operator $f: \mathscr{Y} \to \mathscr{Z}$ is continuous if and only if $\sup_{\|y\|=1} \|f(y)\| = \sup_{\|y\| \le 1} \|f(y)\| < \infty$ (see [4, Lemma 4, p. 59]). In this case, we may define a norm on $\mathbf{B}(\mathscr{Y},\mathscr{Z})$ through the relation

$$\|f\| = \sup_{\|y\|=1} \|f(y)\|, \quad f \in \mathbf{B}(\mathscr{Y},\mathscr{Z}),$$

so that $\mathbf{B}(\mathscr{Y},\mathscr{Z})$ thereby itself becomes a normed linear vector space. It is evident that

$$\|f(y)\| \le \|f\| \, \|y\| \quad \text{for all} \quad y \in \mathscr{Y} \quad \text{and} \quad f \in \mathbf{B}(\mathscr{Y},\mathscr{Z}).$$

If $\mathscr{X}$ is also a normed linear vector space, $f \in \mathbf{B}(\mathscr{X},\mathscr{Y})$, and $g \in \mathbf{B}(\mathscr{Y},\mathscr{Z})$, then $\|g \circ f\| \le \|g\| \, \|f\|$. If $\mathscr{Z}$ is a Banach space, then $\mathbf{B}(\mathscr{Y},\mathscr{Z})$ also is (see [3, Theorem 5.7.3, p. 103]).

4 If $\mathscr{Y}$ is a linear topological space, then we shall denote $\mathbf{B}(\mathscr{Y},R)$ by $\mathscr{Y}^*$; i.e., $\mathscr{Y}^*$ consists of all continuous linear functionals [see (2.8)] defined on $\mathscr{Y}$. Evidently, $\mathscr{Y}^* \subset \mathscr{Y}^+$ [see (2.10)]. We shall refer to $\mathscr{Y}^*$ as the *conjugate space* (or *dual space*) of $\mathscr{Y}$. If $\mathscr{Y}$ is normed, we may of course define a norm (and the associated norm topology) on $\mathscr{Y}^*$, as indicated in (2).

5 We can define another very useful topology on $\mathscr{Y}^*$, whether or not $\mathscr{Y}$ is normed (so long as $\mathscr{Y}$ is a linear topological space): the so-called *weak* topology* (or $\mathscr{Y}$ topology). This topology, under which $\mathscr{Y}^*$ becomes a locally convex [see (4.30)] linear topological space (see [4, pp. 419–420]) is defined by taking as local base all sets of the form

$$\{\ell : \ell \in \mathscr{Y}^*, \ \ell(y_j) < \varepsilon \quad \text{for} \quad j = 1, \ldots, \nu\}$$

where $\varepsilon > 0$ and $\{y_1, \ldots, y_\nu\}$ is some finite subset of $\mathscr{Y}$.

6   It follows from (4.10) and Theorem (3.25B) that if $\mathscr{Y}$ is a normed linear vector space, then the norm topology on $\mathscr{Y}^*$ is stronger than the weak* topology on $\mathscr{Y}^*$.

7   We may also define a topology on any locally convex linear topological space $\mathscr{Y}$ by means of $\mathscr{Y}^*$. Namely, we define the so-called *weak topology* (or $\mathscr{Y}^*$ topology) on $\mathscr{Y}$ (under which $\mathscr{Y}$ becomes a locally convex linear topological space—see [4, p. 419]) by taking as local base all sets of the form

$$\{y : y \in \mathscr{Y}, \ \ell_j(y) < \varepsilon \quad \text{for} \quad j = 1, \ldots, v\},$$

where $\varepsilon > 0$ and $\{\ell_1, \ldots, \ell_v\}$ is some finite subset of $\mathscr{Y}^*$.

8   LEMMA. *Let $\mathscr{Y}$ be a linear topological space. Then the linear functionals in $(\mathscr{Y}^*)^+$ which are continuous in the weak* topology of $\mathscr{Y}^*$ are precisely those functionals $\ell^+ \in (\mathscr{Y}^*)^+$ which may be represented in the form $\ell^+(\ell) = \ell(y_0)$ for all $\ell \in \mathscr{Y}^*$ (for some fixed $y_0 \in \mathscr{Y}$, which in general depends on ℓ^+). If also $\mathscr{Y}$ is locally convex, then the set of all linear functionals in $\mathscr{Y}^+$ which are continuous in the weak topology of $\mathscr{Y}$ coincides with $\mathscr{Y}^*$.*

Proof. See [4, Theorem 9, p. 421].

9   If $\mathscr{Y}$ is a locally convex linear topological vector space (e.g., if $\mathscr{Y}$ is normed), it follows from (4.10) and Theorem (3.25B) that the original topology on $\mathscr{Y}$ is stronger than the weak topology (hence the term weak topology). In particular, the norm topology of a normed linear vector space is stronger than its weak topology.

10   If $\mathscr{Y} = R^n$, then we have already seen that $(R^n)^+$ is isomorphic to R^n [see (2.12)], and it is easily seen that $(R^n)^+ = (R^n)^*$. Further, the norm and weak topologies on R^n coincide, as do the norm and weak* topologies on $(R^n)^*$. In addition, $(R^n)^*$ and R^n are topologically isomorphic. For these reasons, one commonly says (as we ourselves shall) that $(R^n)^* = R^n$.

11   THEOREM. *If f is a linear operator from R^n into some linear topological space $\mathscr{Z}$, then f is continuous.*

Proof. See [4, Corollary 4, p. 245].

The following three lemmas turn out to be useful.

12   LEMMA. *If $\mathscr{Y}$ is a linear topological space, $A \subset \mathscr{Y}$, $B \subset \mathscr{Y}$, and $\ell \in \mathscr{Y}^*$ separates A and B [see (2.19)], then ℓ also separates $\bar{A}$ and $\bar{B}$.*

Proof. Suppose that $\ell(y_1) \leq \gamma \leq \ell(y_2)$ for all $y_1 \in A$, $y_2 \in B$, and some $\gamma \in R$. But, by Theorem (3.25C), the sets $A_1 = \{y:$

$\ell(y) \leq \gamma\}$ and $B_1 = \{y:\ell(y) \geq \gamma\}$ are closed in $\mathscr{Y}$. Hence, since $A_1 \supset A$ and $B_1 \supset B$, $A_1 \supset \bar{A}$ and $B_1 \supset \bar{B}$, i.e., ℓ separates $\bar{A}$ and $\bar{B}$. |||

13 LEMMA. *If a linear functional ℓ on a linear topological space separates two sets, one of which has an interior point, then ℓ is continuous.*

Proof. See [4, Lemma 7, p. 417].

14 LEMMA. *If $\mathscr{Y}$ is a linear topological space, $y_0 \in \mathscr{Y}$, $\ell \in \mathscr{Y}^*$, and $\ell \neq 0$, then, for every neighborhood N of 0 in $\mathscr{Y}$, there are points y_1 and y_2 in $y_0 + N$ such that $\ell(y_1) < \ell(y_0) < \ell(y_2)$.*

Proof. Since $\ell \neq 0$, there is a $y_3 \in \mathscr{Y}$ such that $\ell(y_3) > 0$. Let $\varepsilon > 0$ be such that $\pm\varepsilon y_3 \in N$ [see Corollary (4.3)]. Then, if we set $y_1 = y_0 - \varepsilon y_3$ and $y_2 = y_0 + \varepsilon y_3$, so that $y_1, y_2 \in y_0 + N$, $\ell(y_1) < \ell(y_0) < \ell(y_2)$. |||

If $\mathscr{Y}$ is a linear topological space, then we have a number of separation theorems [see (2.19)] involving functionals in $\mathscr{Y}^*$. These theorems, which we shall now state, play a very important role in the theory which we shall develop.

15 THEOREM. *Let A and B be disjoint convex sets in a linear topological space $\mathscr{Y}$, and suppose that A has an interior point. Then there is a nonzero continuous linear functional $\ell \in \mathscr{Y}^*$ that separates $\bar{A}$ and $\bar{B}$.*

Theorem (15) follows at once from Theorem (2.21) and Lemmas (4.20), (12), and (13).

16 COROLLARY. *If B is a closed convex set in a locally convex linear topological space $\mathscr{Y}$, $y_0 \in \mathscr{Y}$, and $y_0 \notin B$, then there is a linear functional $\ell \in \mathscr{Y}^*$ such that*

$$\ell(y_0) > \sup_{y \in B} \ell(y).$$

Proof. Let N be a convex neighborhood of 0 in $\mathscr{Y}$ such that $y_0 + N$ does not meet B. (Such a neighborhood exists by our hypotheses.) Then our result follows at once from Theorem (15) and Lemma (14). |||

17 COROLLARY. *If y_1 and y_2 are distinct points in a locally convex linear topological space $\mathscr{Y}$, then there is a linear functional $\ell \in \mathscr{Y}^*$ such that $\ell(y_1) \neq \ell(y_2)$.*

18 COROLLARY. *Let A and B be disjoint convex sets in a linear topological space $\mathscr{Y}$ such that A has an interior point and such that*

there is a point $y_0 \in \bar{A} \cap \bar{B}$. *Then there is a nonzero, continuous linear functional* $\ell \in \mathscr{Y}^*$ *such that*

$$\ell(y_1) \leq \ell(y_0) \leq \ell(y_2) \quad \textit{for all} \quad y_1 \in \bar{A} \quad \textit{and} \quad y_2 \in \bar{B}.$$

19 THEOREM. *Let* A *be a convex set in* R^m *such that* $0 \notin \operatorname{int} A$. *Then there is a nonzero vector* $b \in R^m$ *such that* $b \cdot \xi \leq 0$ *for all* $\xi \in A$.

Proof. If $0 \notin \bar{A}$, then our result follows at once from Corollary (16), (10), and (2.13). For a proof of the case where $0 \in \bar{A}$, so that 0 is a boundary point of A, see [5, Theorem 8, p. 20]. |||

The following lemma will prove to be useful.

20 LEMMA. *Let* $\mathscr{Y}$ *be a linear topological space, and let* $\mathscr{L}$ *be an arbitrary set in* $\mathscr{Y}^*$. *Then if* $\ell_0 \in \mathscr{Y}^*$ *is not in the weak* closure of* co co $\mathscr{L}$, *there is a* $y_0 \in \mathscr{Y}$ *such that*

21 $$\ell_0(y_0) > 0 \geq \ell(y_0) \quad \textit{for all} \quad \ell \in \mathscr{L}.$$

Proof. By Corollary (16) and Lemma (8), there is an element $y_0 \in \mathscr{Y}$ such that

22 $$\ell_0(y_0) > \sup_{\ell \in \text{co co } \mathscr{L}} \ell(y_0).$$

But since co co $\mathscr{L}$ is a cone, the sup in the righthand side of (22) must be zero, which implies that (21) holds. |||

23 If K is a cone in a linear topological space $\mathscr{Y}$, then we shall call the set

$$\{\ell : \ell \in \mathscr{Y}^*, \ell(y) \leq 0 \quad \text{for all} \quad y \in K\}$$

the *dual* (or *conjugate*) of the cone K, and shall denote it by K^*. It is easily seen that K^* is a convex cone which is closed in the weak* topology of $\mathscr{Y}^*$.

24 LEMMA. *Let* K *be an open cone in a linear topological space* $\mathscr{Y}$, *and let* $\ell \in K^*$, *where* $\ell \neq 0$. *Then* $\ell(y) < 0$ *for all* $y \in K$.

Proof. This lemma follows at once from Lemma (14). |||

25 Suppose that $\mathscr{Y}_1, \ldots, \mathscr{Y}_m$ are linear topological spaces. It follows from (2.11) that $\ell \in (\mathscr{Y}_1 \times \cdots \times \mathscr{Y}_m)^*$ if and only if there are functionals $\ell_i \in \mathscr{Y}_i^*$, for $i = 1, \ldots, m$, such that $\ell(y_1, \ldots, y_m) = \ell_1(y_1) + \cdots + \ell_m(y_m)$ for every $(y_1, \ldots, y_m) \in \mathscr{Y}_1 \times \cdots \times \mathscr{Y}_m$. From this it follows at once $(\mathscr{Y}_1 \times \cdots \times \mathscr{Y}_m)^*$ is isomorphic to $(\mathscr{Y}_1^* \times \cdots \times \mathscr{Y}_m^*)$. An almost immediate consequence of this is

the assertion that if $K_1, \ldots, K_m$ are cones in linear topological spaces $\mathscr{Y}_1, \ldots, \mathscr{Y}_m$, and $K = K_1 \times \cdots \times K_m$, then $\ell \in K^*$ if and only if there are $\ell_i \in K_i^*$ for $i = 1, \ldots, m$ such that $\ell(y_1, \ldots, y_m) = \sum_{i=1}^m \ell_i(y_i)$ for all $(y_1, \ldots, y_m) \in \mathscr{Y}_1 \times \cdots \times \mathscr{Y}_m$.

26 A function $f: A \to \mathscr{Z}$, where A is a set in a linear vector space $\mathscr{Y}$ which is finitely open in itself and $\mathscr{Z}$ is a topological space, will be said to be *finitely continuous* if, for every $y_0 \in A$ and each finite subset $\{y_1, \ldots, y_m\}$ of A,

$$f\left(y_0 + \sum_{j=1}^{m} \lambda^j (y_j - y_0)\right) \xrightarrow[\substack{(\lambda^1, \ldots, \lambda^m) \to (0, \ldots, 0) \\ (\lambda^1, \ldots, \lambda^m) \in \bar{R}_+^m}]{} f(y_0),$$

i.e., if the function $(\lambda^1, \ldots, \lambda^m) \to f(y_0 + \sum_{j=1}^m \lambda^j (y_j - y_0))$ from $\bar{R}_+^m$ into $\mathscr{Z}$ is continuous at $0 \in R^m$. If $A = \mathscr{Y}$, $\mathscr{Z}$ is a linear topological space, and f is affine (and, in particular, linear), then f is finitely continuous. [This is a direct consequence of Theorem (11).] Alternately, if $\mathscr{Y}$ is a linear topological space and $f: A \to \mathscr{Z}$ is continuous, then f is finitely continuous. Finally, if $f: A \to \mathscr{Z}$ is finitely continuous, A' is a subset of $\mathscr{Z}$ such that $f(A) \subset A'$, g is a function from A' into a topological space $\mathscr{Z}$, and g is continuous on $f(A)$, then $g \circ f$ is a finitely continuous function from A into $\mathscr{Z}'$.

27 A function $f: A \to \mathscr{Z}$, where A is a subset of $\mathscr{Y}$, and $\mathscr{Y}$ and $\mathscr{Z}$ are linear topological spaces, will be said to be *uniformly continuous* on a subset A_1 of A if, for every neighborhood N' of 0 in $\mathscr{Z}$, there is a neighborhood N of 0 in $\mathscr{Y}$ such that $f(y) \in f(y_1) + N'$ whenever $y_1 \in A_1$ and $y \in (y_1 + N) \cap A$.

Note that Definition (27) is slightly different from the usual definition of uniform continuity. [Ordinarily, the last relation in the definition is replaced by "$y \in (y_1 + N) \cap A_1$."]

28 A family $\mathscr{F}$ of functions $f: A \to \mathscr{Z}$, where A is a subset of $\mathscr{Y}$, and $\mathscr{Y}$ and $\mathscr{Z}$ are linear topological spaces, will be said to be *equicontinuous at a point* $y_0 \in A$ if, for every neighborhood N' of 0 in $\mathscr{Z}$, there is a neighborhood N of 0 in $\mathscr{Y}$ such that $f(y) \in f(y_0) + N'$ for all $f \in \mathscr{F}$, so long as $y \in (y_0 + N) \cap A$. If the family $\mathscr{F}$ is equicontinuous at each point of a subset A_1 of A, then we shall say that $\mathscr{F}$ is *equicontinuous on* A_1. If $\mathscr{F}$ is equicontinuous on A, then we shall simply say that $\mathscr{F}$ is *equicontinuous.*

29 Note that if $\mathscr{F}$ is a finite collection of functions and if each of these functions is continuous at a point $y_0 \in A$ (respectively,

continuous), then $\mathscr{F}$ is equicontinuous at y_0 (respectively, equicontinuous). More generally, any finite union of equicontinuous families is equicontinuous (whether at a point, or in general).

30 A family $\mathscr{F}$ of functions $f: A \to \mathscr{Z}$, where A is a subset of $\mathscr{Y}$, and $\mathscr{Y}$ and $\mathscr{Z}$ are linear topological spaces, will be said to be *uniformly equicontinuous on* a subset A_1 of A if, for every neighborhood N' of 0 in $\mathscr{Z}$, there is a neighborhood N of 0 in $\mathscr{Y}$ such that $f(y) \in f(y_1) + N'$ for all $f \in \mathscr{F}$ whenever $y_1 \in A_1$ and $y \in (y_1 + N) \cap A$. Clearly, if $\mathscr{F}$ is uniformly equicontinuous on A_1, then $\mathscr{F}$ is equicontinuous on A_1.

31 THEOREM. *If $\mathscr{F}$ is a family of functions $f: A \to \mathscr{Z}$, where A is a subset of $\mathscr{Y}$, and $\mathscr{Y}$ and $\mathscr{Z}$ are linear topological spaces, and if A_1 is a subset of A which is compact in $\mathscr{Y}$ and on which $\mathscr{F}$ is equicontinuous, then $\mathscr{F}$ is uniformly equicontinuous on A_1.*

Proof. Let N' be an arbitrary neighborhood of 0 in $\mathscr{Z}$. Let N'' be a neighborhood of 0 in $\mathscr{Z}$ such that $N'' + N'' \subset N'$ [see Corollary (4.4)], and let $N''' = N'' \cap (-N'')$ [see Corollary (4.6)]. Since $\mathscr{F}$ is equicontinuous on A_1, for every $y' \in A_1$ there is a neighborhood $N_{y'}$ of 0 in $\mathscr{Y}$ such that $f(y) \in f(y') + N'''$ for all $f \in \mathscr{F}$ whenever $y \in (y' + N_{y'}) \cap A$. For each $y' \in A_1$, let $\tilde{N}_{y'}$ be a neighborhood 0 in $\mathscr{Y}$ such that $\tilde{N}_{y'} + \tilde{N}_{y'} \subset N_{y'}$. The family $\{y' + \tilde{N}_{y'}: y' \in A_1\}$ evidently forms a covering of A_1, so that, since A_1 is compact, there are points $y'_1, \ldots, y'_\rho$ such that $\bigcup_{i=1}^{\rho} (y'_i + \tilde{N}_{y'_i}) \supset A_1$. Let $N = \bigcap_{i=1}^{\rho} \tilde{N}_{y'_i}$.

Now suppose that $y_1 \in A_1$ and that $y \in (y_1 + N) \cap A$. Then $y_1 \in y'_j + \tilde{N}_{y'_j} \subset y'_j + N_{y'_j}$ for some $j = 1, \ldots, \rho$, and $y \in y'_j + \tilde{N}_{y'_j} + N \subset y'_j + \tilde{N}_{y'_j} + \tilde{N}_{y'_j} \subset y'_j + N_{y'_j}$. Consequently, for every $f \in \mathscr{F}$, $f(y) \in f(y'_j) + N'''$ and $f(y_1) \in f(y'_j) + N'''$, so that $f(y) - f(y_1) \in N''' - N''' \subset N'' + N'' \subset N'$. |||

32 COROLLARY. *If f is a continuous function from A into $\mathscr{Z}$, where A is a subset of $\mathscr{Y}$, and $\mathscr{Y}$ and $\mathscr{Z}$ are linear topological spaces, and if A_1 is a subset of A which is compact in $\mathscr{Y}$, then f is uniformly continuous on A_1.*

33 It is useful to extend the concepts of convergence which we introduced in (3.16–23) to the case of linear topological spaces. Indeed, suppose that $\mathscr{X}$ and $\mathscr{Z}$ are linear topological spaces, that $x_0 \in \mathscr{X}$, $A_1 \subset A \subset \mathscr{X}$, and $A' \subset \mathscr{Z}$, that f is a function from A into $\mathscr{Z}$, and that f_1 is the restriction of f to A_1. Then we shall say

that $f(x)$ *tends* (or *converges*) *to* A' *as* x *tends to* x_0, written as

$$f(x) \to A' \quad \text{as} \quad x \to x_0, \quad \text{or} \quad f(x) \xrightarrow[x \to x_0]{} A',$$

if, for every neighborhood N' of 0 in $\mathscr{Y}$, there is a neighborhood N of 0 in $\mathscr{X}$ such that $f(x) \in A' + N'$ whenever $x \in (x_0 + N) \cap A$. If $f_1(x)$ tends to A' as x tends to x_0 [even though $f(x)$ may not], then we shall write

$$f(x) \to A' \quad \text{as} \quad x \to x_0 \quad \text{and} \quad x \in A_1, \quad \text{or} \quad f(x) \xrightarrow[\substack{x \to x_0 \\ x \in A_1}]{} A'.$$

If $\mathscr{X} = R$, we shall use the notation $x \to x_0^+$ (respectively, $x \to x_0^-$) to mean $x \to x_0$ and $x \in (x_0 + R_+)$ [respectively, $x \in (x_0 + R_-)$].

34 If $\mathscr{X}$, $\mathscr{Y}$, A', and x_0 are as in the second sentence of (33), Y is an arbitrary set, $A_1 \subset A \subset \mathscr{X} \times Y$, $Y_1 \subset Y$, and f is a function from A into $\mathscr{Y}$, then we shall say that $f(x,y)$ *tends* (or *converges*) *to* A' *as* x *tends to* x_0 *uniformly with respect to* $y \in Y_1$, written as

$$f(x,y) \xrightarrow[x \to x_0]{} A' \quad \text{uniformly w.r. to} \quad y \in Y_1,$$

if, for every neighborhood N' of 0 in $\mathscr{Y}$, there is a neighborhood N of 0 in $\mathscr{X}$ such that $f(x,y) \in A' + N'$ whenever $(x,y) \in [(x_0 + N) \times Y_1] \cap A$. We shall also write

35 $$f(x,y) \xrightarrow[\substack{x \to x_0 \\ x \in A_1}]{} A' \quad \text{uniformly w.r. to} \quad y \in Y_1$$

with a meaning that should now be clear, and shall write $x \to x_0^+$ or $x \to x_0^-$ in (35) in place of $x \to x_0$ and $x \in A_1$ when $\mathscr{X} = R$ and A_1 is as previously described.

It is easy to see that if A' consists of a single point $x' \in \mathscr{Y}$, then all of the definitions of convergence to A' introduced above coincide with the corresponding convergence definitions of (3.16–23).

36 Note that if $\mathscr{X}$, $\mathscr{Y}$, x_0, A_1, and A are as in the second sentence of (33), f and g are functions from A into $\mathscr{Y}$, $A' \subset \mathscr{Y}$, and $A'' \subset \mathscr{Y}$, then, if $f(x) \to A'$ as $x \to x_0$ and $g(x) \to A''$ as $x \to x_0$, we also have that $\alpha f(x) + \beta g(x) \to \alpha A' + \beta A''$ as $x \to x_0$ (for all α, $\beta \in R$), and similarly if we adjoin "$x \in A_1$" to "$x \to x_0$." Further, an analogous statement can be made for uniform convergence. Finally, with respect to direct products, an assertion similar to that in (3.37) can be made.

37 If A is a subset of an arbitrary set X, and f is a map from A into X, then any $x \in A$ that satisfies the equation $x = f(x)$ will be called a *fixed point* of f.

The following so-called fixed point theorem will be very important in the sequel.

38 THEOREM. *Let C be a compact, convex set in a locally convex linear topological space, and let f be a continuous function from C into C. Then f has at least one fixed point.*

Proof. See [4, Theorem 5, p. 456].

39 Given two sets K and K_1 in R^m, we shall say that K_1 is a *simplicial linearization* of K if, for every pair (S,η), where S is a simplex in R^m such that $0 \in S \subset K_1$ and $\eta \in R_+$, there exist a continuous function $\chi_0 : S \to R^m$ and a linear function $\chi_1 : R^m \to R^m$ (both possibly depending on S and η) such that

40 $$|\chi_0(\xi) - \xi| < \eta \quad \text{for every} \quad \xi \in S,$$

41 $$\chi_1 \circ \chi_0(S) \subset K.$$

Loosely speaking, K_1 is a simplicial linearization of K if any simplex $S \subset K_1$ and containing 0 can be mapped into a subset of K by means of a "slight distortion" (the continuous map χ_0) followed by a linear transformation (the map χ_1).

42 We shall adopt the convention that if $0 \notin K_1$, where $K_1 \subset R^m$, then K_1 is a simplicial linearization of every set $K \subset R^m$. In particular, the empty set is a simplicial linearization of every set $K \subset R^m$.

We now prove the following lemma.

43 LEMMA. *If $0 \notin K \subset R^m$ and $K_1 \subset R^m$ is a simplicial linearization of K, then $0 \notin \text{int } K_1$. If, in addition, K_1 is convex, then there is a nonzero vector $b \in R^m$ such that $b \cdot \xi \leq 0$ for all $\xi \in K_1$.*

Proof. Let us argue by contradiction, and suppose that $0 \in \text{int } K_1$. Then, as is easily seen, there is a simplex S of dimension m such that $S \subset K_1$ and $0 \in \text{int } S$. Let $\eta > 0$ be such that $\xi \in R^m$ and $|\xi| < \eta$ imply that $\xi \in S$. By hypothesis, there exist a continuous function $\chi_0 : S \to R^m$ and a linear function $\chi_1 : R^m \to R^m$ such that (40) and (41) hold. Let χ_2 be the continuous map from S into R^m defined by $\chi_2(\xi) = \xi - \chi_0(\xi)$, $\xi \in S$. Now (40) implies that $\chi_2(S) \subset S$, and it then follows from Theorems (38) and (4.21) that, for some $\xi_1 \in S$, $\chi_2(\xi_1) = \xi_1$, or $\chi_0(\xi_1) = 0$. The linearity of χ_1 together with (41) now imply that $0 = \chi_1 \circ \chi_0(\xi_1) \in$

K, contradicting our hypothesis. The second conclusion of the lemma is an immediate consequence of Theorem (19). |||

44 We make the following definition. If $\mathscr{Y}$ and $\mathscr{Z}$ are linear topological spaces, and $A \subset \mathscr{Y}$, then a function $f: A \to \mathscr{Z}$ will be said to be a *continuous linear* function if it is the restriction to A of a function in $\mathbf{B}(\mathscr{Y},\mathscr{Z})$. Clearly every continuous linear function $f: A \to \mathscr{Z}$ is linear in the sense of (2.7), and continuous in the sense of (3.24). As we agreed to do in (2.7), we shall sometimes use the same symbol to denote a linear continuous function and its (linear) extension in $\mathbf{B}(\mathscr{Y},\mathscr{Z})$, even though this extension may not be unique.

We close this section with the following theorem.

45 THEOREM. *Let* $\mathbf{G}$ *be an open convex set in* R^m *and let* f *be a convex functional defined on* $\mathbf{G}$. *Then* f *is continuous.*

Proof. See [1, Theorem 7, p. 193].

6. Some Special Linear Topological Spaces

In this section we shall discuss some particular linear topological spaces and their conjugates which we shall make particular use of in the sequel.

We have already discussed R^n and $(R^n)^*$ in (1.3), (2.12 and 13), (3.40), (4.13 and 36), and (5.10 and 11).

1 It is often advantageous to consider an element of R^n as either a row matrix or as a column matrix (in which case we shall use the terms *row vector* or *column vector*), and thereby take advantage of the formalism of matrix algebra. For example, it is easily seen that $f \in \mathbf{B}(R^\mu,R^m)$ [see (5.1)], for arbitrary positive integers μ and m, if and only if there are real numbers α_{ij}, $i = 1, \ldots, m$, $j = 1, \ldots, \mu$, such that

2
$$f(\xi^1,\ldots,\xi^\mu) = \left(\sum_{j=1}^{\mu} \alpha_{1j}\xi^j, \ldots, \sum_{j=1}^{\mu} \alpha_{mj}\xi^j\right)$$
$$\text{for all} \quad (\xi^1,\ldots,\xi^\mu) \in R^\mu.$$

Relation (2) can, of course, be compactly written in matrix notation as

$$f(\xi) = \xi' = \Phi\xi,$$

where ξ and ξ' are column vectors of dimensions μ and m, respectively, and Φ is the $(m \times \mu)$ matrix whose i, j-th element is

α_{ij}. Further, if ξ and η are vectors in R^m, then we can (and sometimes shall) write the dot product $\xi \cdot \eta$ as the matrix product of the row vector ξ with the column vector η.

3 On the other hand, an $(m \times \mu)$ matrix Φ may be considered to be an element of $R^{m\mu}$ in an obvious way, and we shall sometimes find it advantageous to so consider it. In particular, when discussing properties of matrices which require topological concepts, such as continuity or differentiability of matrix-valued functions, we shall consider an $(m \times \mu)$ matrix to be an element of $R^{m\mu}$ in this way, with the Euclidean topology [see (3.40)] on $R^{m\mu}$. However, we shall also on occasion talk about the *norm* of an $(m \times \mu)$ matrix Φ, which we shall write as $|\Phi|$, and thereby mean the norm of the corresponding linear operator in $\mathbf{B}(R^\mu,R^m)$ [see (5.2)]; i.e.,

4 $$|\Phi| = \sup_{|\xi|=1} |\Phi\xi|,$$

where $|\Phi\xi|$ denotes the Euclidean norm [see (4.36)] of the element $\Phi\xi$ in R^m, and $|\xi|$ denotes the Euclidean norm of ξ in R^μ. By (5.3), we have that $|\Phi\xi| \le |\Phi|\,|\xi|$ for all $\xi \in R^\mu$. Also, it is worth noting that the norm of a matrix Φ defined by (4) is equivalent [in the sense of (4.33)] to the Euclidean norm of Φ, when Φ is considered to be an element of $R^{m\mu}$.

Let us now turn to some function spaces, i.e., linear topological spaces whose elements are functions.

5 If I is a compact interval $[t_1,t_2]$ and n is a positive integer, then we shall denote by $\mathscr{C}^n(I)$, or simply $\mathscr{C}^n$ if the choice of I is clear, the normed linear vector space consisting of all continuous functions x from I into R^n, with norm defined by

6 $$\|x\| = \max_{t\in I} |x(t)|$$

[see (3.46)]. We shall denote $\mathscr{C}^1(I)$ simply by $\mathscr{C}(I)$ or by $\mathscr{C}$.

7 For each $j = 1, \ldots, n$, let e_j denote the vector in R^n whose j-th component is one and whose remaining components vanish. Further, for each function $x:I \to R^n$, let x^j denote the real-valued function $t \to e_j \cdot x(t)$, $t \in I$. It is easily seen that $x \in \mathscr{C}^n(I)$ if and only if $x^j \in \mathscr{C}(I)$ for each j. Consequently, the correspondence $x \to (x^1, \ldots, x^n)$ between $\mathscr{C}^n$ and $\mathscr{C} \times \cdots \times \mathscr{C}$ (n times) defines an isomorphism between these two spaces (as linear vector spaces).

8 Let us show that the norm on $\mathscr{C}^n$ defined by (6) is equivalent [in the sense of (4.33)] to the product norm defined in (4.38). But the product norm [which we shall denote by $\|\cdot\|_p$ to distinguish it from the norm defined by (6)] is given by

$$\|x\|_p = \max_{1\le j\le n} \max_{t\in I} |x^j(t)| = \max_{t\in I}\left[\max_{1\le j\le n} |x^j(t)|\right],$$

and, for every $\xi = (\xi^1,\ldots,\xi^n) \in R^n$, $\max\{|\xi^j|: j = 1,\ldots,n\} \le |\xi| \le \sqrt{n}\max\{|\xi^j|: j = 1,\ldots,n\}$. From this it follows that $\|x\|_p \le \|x\| \le \sqrt{n}\|x\|_p$ for all $x \in \mathscr{C}^n$, which implies, by virtue of (4.33), that the norms $\|\cdot\|_p$ and $\|\cdot\|$ are equivalent. Hence, $\mathscr{C}^n(I)$ is topologically isomorphic [see (4.12)] to $(\mathscr{C}(I))^n$.

9 It is not hard to prove that $\mathscr{C}^n$ (for each $n \ge 1$) is complete, i.e., that $\mathscr{C}^n$ is a Banach space.

10 If I is a compact interval $[t_1,t_2]$, then we shall denote by $NBV(I)$ the normed linear vector space of all functions λ from I into R which (i) are of bounded variation on I, (ii) are continuous from the right in (t_1,t_2) [i.e., which have the property that $\lambda(t+\varepsilon)\to\lambda(t)$ as $\varepsilon\to 0^+$ for every t, $t_1 < t < t_2$], and (iii) satisfy $\lambda(t_2) = 0$, with $\|\lambda\|$ equal to the total variation of λ, written as $TV\,\lambda$ and defined as

11 $$TV\,\lambda = \sup \sum_{i=1}^{\nu} |\lambda(s_i) - \lambda(s_{i-1})|,$$

with the supremum taken over all subdivisions $t_1 = s_0 < s_1 < \cdots < s_\nu = t_2$ of I.

There is an intimate relationship between the spaces $(\mathscr{C}^n(I))^*$ and $(NBV(I))^n$, as is demonstrated by the following two theorems.

12 THEOREM. *The spaces* $(\mathscr{C}(I))^*$ [*as a normed linear vector space—see* (5.4)] *and* $NBV(I)$ *are isometrically isomorphic* [*see* (4.34)] *with the correspondence between two elements* $\ell \in (\mathscr{C}(I))^*$ *and* $\lambda \in NBV(I)$ *defined by the relation*

13 $$\ell(x) = \int_I x(t)\,d\lambda(t) \quad \text{for all} \quad x \in \mathscr{C}(I).$$

Proof. See [8, Theorem 4.32–C, p. 200].

14 THEOREM. *The spaces* $(\mathscr{C}^n(I))^*$ *and* $(NBV(I))^n$ *are isomorphic, with the correspondence between two elements* $\ell \in (\mathscr{C}^n(I))^*$ *and* $\lambda = (\lambda^1,\ldots,\lambda^n) \in (NBV(I))^n$ *defined by the relation*

15 $$\ell(x) = \int_I \sum_{j=1}^{n} x^j(t)\,d\lambda^j(t) \quad \text{for every} \quad x = (x^1,\ldots,x^n) \in \mathscr{C}^n(I).$$

Proof. Theorem (14) follows at once from Theorem (12), the last sentence of (8), and (5.25). |||

16 COROLLARY. *Let* $f \in \mathbf{B}(\mathscr{C}^n(I), R^n)$. *Then there are functions* $\lambda^{ij} \in NBV(I)$ (*for* $i, j = 1, \ldots, n$) *such that* [*setting* $f = (f^1, \ldots, f^n)$]

17 $$f^i(x) = \int_I \sum_{j=1}^{n} x^j(t)\, d\lambda^{ij}(t) \quad \textit{for every}$$
$$x = (x^1, \ldots, x^n) \in \mathscr{C}^n(I) \quad \textit{and each} \quad i = 1, \ldots, n.$$

Further, $TV\, \lambda^{ij} \leq \|f\|$ *for every* i *and* j.

Proof. The existence of functions $\lambda^{ij} \in NBV(I)$ such that (17) holds follows at once from Theorem (14) because $f^i \in (\mathscr{C}^n(I))^*$ for each $i = 1, \ldots, n$. Furthermore, for each i and j,

18 $$\|f\| = \sup_{\|x\|=1} |f(x)| \geq \sup_{\|x\|=1} |f^i(x)|$$
$$\geq \sup_{\|x^j\|=1} \left| \int_I x^j(t)\, d\lambda^{ij}(t) \right| = TV\, \lambda^{ij},$$

the last inequality in (18) being a consequence of Theorem (12). |||

19 In accordance with our remarks in (1), we point out that relations (15) and (17) may be rewritten in matrix notation in the forms

20 $$\ell(x) = \int_I d\lambda(t) x(t),$$

21 $$f(x) = \int_I d\Lambda(t) x(t),$$

where λ is the row-vector-valued function with elements $\lambda^1, \ldots, \lambda^n$, x is the column-vector-valued function with elements $x^1, \ldots, x^n$, Λ is the $(n \times n)$-matrix-valued function with elements λ^{ij} ($i, j = 1, \ldots, n$), and f is a column vector. We may regard (20) and (21) either as formal equivalents of (15) and (17), respectively, or, alternatively and equivalently, we may regard (20) and (21) as formulas involving Stieltjes integrals defined (in an obvious manner) in terms of matrices.

The following theorem on the conjugate cone of a certain cone in $\mathscr{C}(I)$ will prove to be useful.

22 THEOREM. *Let* K *denote the closed convex cone in* $\mathscr{C}(I)$ *defined as follows*:

$$K = \{x \colon x(t) \leq 0 \quad \textit{for all} \quad t \in I_1\},$$

where I_1 *is some given closed subset of* I. *Then* K^* [*see* (5.23)] *consists of all linear functionals* ℓ *in* $(\mathscr{C}(I))^*$ *with the representation*

(13) *with* $\lambda \in NBV(I)$, *where* λ *is nondecreasing on* I *and is constant on any subinterval of* I *which does not meet* I_1.

Proof. The theorem may be proved by a contradiction argument and the usage of specific suitable elements of K. The details are left as an exercise. |||

23 For each compact interval $I = [t_1,t_2]$ and each $p \in [1,\infty)$, let $L_p(I)$ denote the normed linear vector space consisting of all measurable functions $z: I \to R$ such that $|z(\cdot)|^p$ is integrable over I, with the norm defined by the relation

24 $$\|z\| = \left(\int_I |z(t)|^p \, dt \right)^{1/p}.$$

Further, let $L_\infty(I)$ denote the normed linear vector space consisting of all functions $z: I \to R$ which are essentially bounded, with the norm defined by

25 $$\|z\| = \operatorname*{ess\,sup}_{t \in I} |z(t)|.$$

26 In the spaces $L_p(I)$ $(1 \leq p \leq \infty)$ we shall, as is customary, not distinguish between functions which differ only on a set of measure zero. It is then easily seen that (25) defines a norm on L_∞; for a proof that (24) defines a norm on L_p, for $1 \leq p < \infty$, see [8, Example 5, p. 90]. The spaces $L_p(I)$ $(1 \leq p \leq \infty)$ are complete, and hence are Banach spaces (see [4, Theorem 6, p. 146 and Corollary 14, p. 150]).

27 It turns out that $(L_p(I))^*$ is isometrically isomorphic to $L_q(I)$ for every $p \in (1,\infty)$, where $q = p/(1 - p)$, and that $(L_1(I))^*$ is isometrically isomorphic to $L_\infty(I)$ (see [4, Theorem 1, p. 286, and Theorem 5, p. 289]). However, we shall be primarily interested in $L_\infty(I)$, whose dual is more difficult to specify.

28 THEOREM. *For every* $\in (L_\infty(I))^*$, *there is a finitely additive, real-valued measure (which is not necessarily countably additive)* $\mathbf{v}$, *defined on all Lebesgue measurable subsets of* I, *such that* $\mathbf{v}(I_0) = 0$ *for all subsets* I_0 *of* I *of Lebesgue measure zero, and such that*

29 $$\ell(z) = \int_I z(s) \, d\mathbf{v}(s) \quad \textit{for all} \quad z \in L_\infty(I).$$

Proof. See [4, Theorem 16 and Definition 15, p. 296].

30 For every positive integer n, we define $L_\infty^n(I)$ (where I is the usual compact interval $[t_1,t_2]$) as the normed linear vector space consisting of all functions $z: I \to R^n$ such that $|z(\cdot)|$ is essentially

bounded, with the norm defined by

$$\|z\| = \operatorname*{ess\,sup}_{t \in I} |z(t)|.$$

31 Arguing as in (7) and (8), we can show that $L_\infty^n(I)$ is topologically isomorphic to $(L_\infty(I))^n$.

The following theorem is analogous to Theorem (22):

32 THEOREM. *Let K denote the closed convex cone in $L_\infty(I)$ defined as follows:*

$$K = \{z : z(t) \le 0 \textit{ for almost all } t \in I'\},$$

where I' is some given Lebesgue measurable subset of I. Then K^* [see (5.23)] consists of all linear functionals $\ell \in (L_\infty(I))^*$ with the representation (29) for some finitely additive, nonnegative real-valued measure $\mathbf{v}$, *defined on all Lebesgue measurable subsets of I, such that $\mathbf{v}(\hat{I}) = 0$ for every Lebesgue measurable subset $\hat{I}$ of I whose intersection with I' has Lebesgue measure zero.*

Theorem (32) may be proved much in the same way as Theorem (22).

33 Now suppose that A is a subset of some topological space X (which is typically a linear topological space), that $\mathscr{X}$ is some normed linear vector space, and that $\mathscr{Y}$ is a subspace of the linear vector consisting of all continuous functions from A into $\mathscr{X}$. Then we shall define a topology on $\mathscr{Y}$, which we shall refer to as the *topology of pointwise convergence*, under which $\mathscr{Y}$ becomes a locally convex linear topological space, by taking as local base all sets of the form

$$\left\{T : T \in \mathscr{Y},\ \max_{1 \le j \le \nu} \|Tx_j\| < \varepsilon\right\}$$

where $\{x_1, \ldots, x_\nu\}$ is some finite subset of A, and $\varepsilon > 0$ [see Theorem (4.11)].

7. A General Theory of Differentiation

In this, the final section of this chapter, we shall consider various types of differentiability properties of functions from one linear vector space into a linear topological space. These properties will be exploited to a great extent in the subsequent chapters of this book.

Throughout this section, $\mathscr{Y}$ will denote a linear vector space and $\mathscr{X}$ a linear topological space, and we shall consider functions defined on subsets of $\mathscr{Y}$ and taking on their values in $\mathscr{X}$. For some of our definitions, we shall have to assume that $\mathscr{Y}$ is also a linear topological space, and for others, that $\mathscr{Y}$ and $\mathscr{X}$ are both normed linear vector spaces. We shall also suppose, throughout this section, that $\mathscr{E}$ is a subset of $\mathscr{Y}$ which is finitely open in itself [so that, by Lemma (1.41), cone $(\mathscr{E} - e_0)$ is convex for every $e_0 \in \mathscr{E}$].

1 We shall say that a function $\varphi : \mathscr{E} \to \mathscr{X}$ is *Gâteaux directionally differentiable at* a point $e_0 \in \mathscr{E}$ if there is a (necessarily unique) function $D\varphi$: cone $(\mathscr{E} - e_0) \to \mathscr{X}$ such that

2 $$\frac{\varphi(e_0 + \varepsilon y) - \varphi(e_0)}{\varepsilon} - D\varphi(y) \xrightarrow[\varepsilon \to 0^+]{} 0 \quad \text{for all} \quad y \in \text{cone}\,(\mathscr{E} - e_0).$$

[Note that, because $\mathscr{E}$ is finitely open in itself, for all $y \in$ cone $(\mathscr{E} - e_0)$, $e_0 + \varepsilon y \in \mathscr{E}$ for all $\varepsilon > 0$ sufficiently small.] We shall say that φ is *finitely directionally differentiable at* $e_0 \in \mathscr{E}$ if, for every convex polyhedron [see (1.35)] $P \subset$ cone $(\mathscr{E} - e_0)$, the convergence in (2) is uniform w.r. to $y \in P$ [see (3.20–23)]. In either of the two preceding cases, we shall refer to $D\varphi$ as the *directional differential* of φ at e_0. (The adjectives "Gâteaux" or "finite" will often be omitted when we refer to directional differentials.)

3 If φ is Gâteaux (respectively, finitely) directionally differentiable at each point of $\mathscr{E}$, then we shall simply say that φ is *Gâteaux* (respectively, *finitely*) *directionally differentiable.*

4 Note that we have imposed no linearity requirement on $D\varphi$ in the preceding definitions. However, if φ is Gâteaux (respectively, finitely) directionally differentiable at $e_0 \in \mathscr{E}$, and if the directional differential $D\varphi$ of φ at e_0 is *linear* [see (2.7)], then we shall say that φ is *Gâteaux* (respectively, *finitely*) *differentiable at* e_0, and we shall then refer to $D\varphi$ as the *Gâteaux* (respectively, *finite*) *differential* of φ at e_0. If φ is Gâteaux (respectively, finitely) differentiable at each point of $\mathscr{E}$, then we shall simply say that φ is *Gâteaux* (respectively, *finitely*) *differentiable.*

5 Clearly, differentiability implies directional differentiability (whether Gâteaux or finite), and finite differentiability implies Gâteaux differentiability (whether directional or "plain").

6 THEOREM. *If a function φ from a convex set $\mathscr{E}$ in $\mathscr{Y}$ into $\mathscr{Z}$ is W-convex—where W is a closed, convex cone in $\mathscr{Z}$—and if φ is also Gâteaux directionally differentiable at $e_0 \in \mathscr{E}$, then the directional differentional $D\varphi$ of φ at e_0 is also W-convex.*

Proof. Let $y_1, y_2 \in \text{cone}(\mathscr{E} - e_0)$. Because $\mathscr{E}$ is convex, $e_0 + \varepsilon\lambda y_1 + \varepsilon(1 - \lambda)y_2 \in \mathscr{E}$ for all $\lambda \in [0,1]$ whenever $\varepsilon > 0$ is sufficiently small. Hence, for all such λ and ε,

$$\begin{aligned}\varphi(e_0 + \varepsilon\lambda y_1 + \varepsilon(1 - \lambda)y_2) &= \varphi(\lambda(e_0 + \varepsilon y_1) + (1 - \lambda)(e_0 + \varepsilon y_2)) \\ &\leq_W \lambda\varphi(e_0 + \varepsilon y_1) + (1 - \lambda)\varphi(e_0 + \varepsilon y_2),\end{aligned}$$

which means that

$$\frac{\varphi(e_0 + \varepsilon\lambda y_1 + \varepsilon(1 - \lambda)y_2) - \varphi(e_0)}{\varepsilon} - \lambda\frac{\varphi(e_0 + \varepsilon y_1) - \varphi(e_0)}{\varepsilon} - (1 - \lambda)\frac{\varphi(e_0 + \varepsilon y_2) - \varphi(e_0)}{\varepsilon} \in \frac{1}{\varepsilon}W = W.$$

Passing to the limit as $\varepsilon \to 0^+$, we conclude that

$$D\varphi(\lambda y_1 + (1 - \lambda)y_2) - \lambda D\varphi(y_1) - (1 - \lambda)D\varphi(y_2) \in \bar{W} = W,$$

which at once implies our desired conclusion. |||

7 LEMMA. *If φ is a convex functional defined on a convex set $\mathscr{E} \subset \mathscr{Y}$, and e_0 is an internal point of $\mathscr{E}$, then φ is Gâteaux directionally differentiable at e_0, and the directional differential $D\varphi$ of φ at e_0 is convex on $\mathscr{Y}$.*

Proof. Let $y \in \mathscr{Y}$ be arbitrary. Then $e_0 \pm \varepsilon y \in \mathscr{E}$ whenever $\varepsilon > 0$ is sufficiently small. Then, for all such $\varepsilon > 0$ and all $\lambda \in [0,1]$,

$$\begin{aligned}\frac{\varphi(e_0 + \varepsilon\lambda y) - \varphi(e_0)}{\varepsilon} &= \frac{\varphi((1 - \lambda)e_0 + \lambda(e_0 + \varepsilon y)) - \varphi(e_0)}{\varepsilon} \\ &\leq \lambda\frac{\varphi(e_0 + \varepsilon y) - \varphi(e_0)}{\varepsilon},\end{aligned}$$

which means that the difference quotient $[\varphi(e_0 + \varepsilon y) - \varphi(e_0)]/\varepsilon$ is monotonically nonincreasing as a function of ε, as $\varepsilon \to 0^+$. If we can show that this difference quotient is bounded from below for $\varepsilon > 0$ sufficiently small, it will follow that this quotient tends to a limit as $\varepsilon \to 0^+$, which means that φ is Gâteaux directionally differentiable at e_0. But if $\varepsilon_0 > 0$ is such that $e_0 \pm \varepsilon_0 y \in \mathscr{E}$, then, for all ε with $0 < \varepsilon < \varepsilon_0$,

$$0 = \varphi\left(\frac{\varepsilon_0}{\varepsilon + \varepsilon_0}(e_0 + \varepsilon y) + \frac{\varepsilon}{\varepsilon + \varepsilon_0}(e_0 - \varepsilon_0 y)\right) - \varphi(e_0)$$

$$\leq \frac{\varepsilon_0}{\varepsilon + \varepsilon_0}[\varphi(e_0 + \varepsilon y) - \varphi(e_0)] + \frac{\varepsilon}{\varepsilon + \varepsilon_0}[\varphi(e_0 - \varepsilon_0 y) - \varphi(e_0)]$$

so that

$$\frac{\varphi(e_0 + \varepsilon y) - \varphi(e_0)}{\varepsilon} \geq -\frac{\varphi(e_0 - \varepsilon_0 y) - \varphi(e_0)}{\varepsilon_0}$$

$$\text{for all} \quad \varepsilon \in (0,\varepsilon_0).$$

The convexity of $D\varphi$ follows at once from Theorem (6) and (2.33). |||

8 Corollary. *If φ is a function from a convex set $\mathscr{E}$ in $\mathscr{Y}$ into R^μ which is $\bar{R}^\mu$-convex, and e_0 is an internal point of $\mathscr{E}$, then φ is Gâteaux directionally differentiable at e_0, and the directional differential $D\varphi$ of φ at e_0 is $\bar{R}^\mu$-convex on $\mathscr{Y}$.*

Proof. Corollary (8) follows at once from Lemma (7) and (2.33). |||

The following lemma is a direct consequence of the definition of an affine function [see (2.16)].

9 Lemma. *If $\varphi:\mathscr{Y} \to \mathscr{Z}$ is affine, then φ is finitely differentiable, and the finite differential $D\varphi$ of φ at any point in $\mathscr{Y}$ coincides with the linear part of φ.*

10 Theorem. *If $\varphi:\mathscr{E} \to \mathscr{Z}$ is finitely differentiable, then φ is finitely continuous [see (5.26)].*

Proof. Let $\{y_0,y_1,\ldots,y_m\} \subset \mathscr{E}$, and let $P = \text{co}\,\{y_1 - y_0,\ldots,y_m - y_0\}$, so that P is a convex polyhedron and $P \subset \text{cone}\,(\mathscr{E} - e_0)$ [see Lemma (1.41)]. Let $D\varphi$ be the finite differential of φ at y_0. Then, if $\lambda^1, \ldots, \lambda^m$ are sufficiently small nonnegative numbers, not all zero,

11 $$\varphi\left(y_0 + \sum_{j=1}^m \lambda^j(y_j - y_0)\right) - \varphi(y_0) - \sum_{j=1}^m \lambda^j D\varphi(y_j - y_0)$$

$$= \left(\sum_{i=1}^m \lambda^i\right)\left[\frac{\varphi\left(y_0 + \left(\sum_{i=1}^m \lambda^i\right)\sum_{j=1}^m \frac{\lambda^j}{\sum_k \lambda^k}(y_j - y_0)\right) - \varphi(y_0)}{\sum_{i=1}^m \lambda^i}\right.$$

$$\left. - D\varphi\left(\sum_{j=1}^m \frac{\lambda^j}{\sum_k \lambda^k}(y_j - y_0)\right)\right].$$

But the term in the brackets in (11) tends to 0 as $(\lambda^1, \ldots, \lambda^m) \to 0$ by the definition of a finite differential, so that (11) implies, by virtue of Corollary (4.3), that

$$\varphi\left(y_0 + \sum_{j=1}^{m} \lambda^j(y_j - y_0)\right) \xrightarrow[\substack{(\lambda^1, \ldots, \lambda^m) \to 0 \\ (\lambda^1, \ldots, \lambda^m) \in \bar{R}^m_+}]{} \varphi(y_0),$$

i.e., φ is finitely continuous. |||

12 It is worth noting, and quite straightforward to verify, that a function $\varphi: \mathscr{E} \to \mathscr{Z}$ is finitely directionally differentiable (respectively, finitely differentiable) at a point $e_0 \in \mathscr{E}$, if and only if, for each finite subset $\{y_1, \ldots, y_m\}$ of $\mathscr{E}$, the function

$$(\lambda^1, \ldots, \lambda^m) \to \varphi\left(e_0 + \sum_{i=1}^{m} \lambda^i(y_i - y_0)\right)$$

from $\bar{R}^m_+ \cap N$ (where N is some sufficiently small neighborhood of 0 in R^m) into $\mathscr{Z}$ is finitely directionally differentiable (respectively, finitely differentiable) at $0 \in R^m$. In (46), we shall discuss the relation of finite differentiability to the usual notion of differentiability when $\mathscr{Y} = R^m$ and $\mathscr{Z} = R$.

We shall now introduce a weaker differentiability condition, which we shall refer to as semidifferentiability. In what follows, we shall suppose that W is a closed, convex cone in $\mathscr{Z}$ [which must be pointed—see Lemma (4.23)].

13 We shall say that a function $\phi: \mathscr{E} \to \mathscr{Z}$ is *Gâteaux W-semidifferentiable at* a point $e_0 \in \mathscr{E}$ if there is a W-convex function [see (2.32)] $D\phi$: cone $(\mathscr{E} - e_0) \to \mathscr{Z}$, with $D\phi(0) = 0$, such that [see (5.33)]

14 $$\frac{\phi(e_0 + \varepsilon y) - \phi(e_0)}{\varepsilon} - D\phi(y) \xrightarrow[\varepsilon \to 0^+]{} W$$

$$\text{for all } y \in \text{cone}\,(\mathscr{E} - e_0).$$

We shall say that ϕ is *finitely W-semidifferentiable at* $e_0 \in \mathscr{E}$ if, for every convex polyhedron $P \subset$ cone $(\mathscr{E} - e_0)$, the convergence in (14) is uniform w.r. to $y \in P$ [see (5.34)]. In either of the two preceding cases, we shall refer to $D\phi$ as a *W-semidifferential* of ϕ at e_0. (The adjectives "Gâteaux" or "finite" will often be omitted when we refer to semidifferentials.) Note that a W-semidifferential is, in general, *not* unique.

15 If ϕ is Gâteaux (respectively, finitely) W-semidifferentiable at each point of $\mathscr{E}$, then we shall simply say that ϕ is *Gâteaux* (respectively, *finitely*) *W-semidifferentiable*.

16 If $\mathscr{X} = R$ and $W = \bar{R}_-$ (and this particular case is especially important), then (as is easily seen) the functional ϕ is Gâteaux W-semidifferentiable at $e_0 \in \mathscr{E}$ if and only if there is a convex functional $D\phi$ defined on cone $(\mathscr{E} - e_0)$, with $D\phi(0) = 0$, such that†

17
$$\limsup_{\varepsilon \to 0^+} \frac{\phi(e_0 + \varepsilon y) - \phi(e_0)}{\varepsilon} \leq D\phi(y)$$

$$\text{for all } y \in \text{cone } (\mathscr{E} - e_0).$$

In this case, we shall simply say that ϕ is *Gâteaux semidifferentiable at* e_0. Further, ϕ is finitely W-semidifferentiable at e_0 if and only if the convergence in (17) is uniform (in an evident sense) w.r. to $y \in P$, for every convex polyhedron $P \subset$ cone $(\mathscr{E} - e_0)$. In this latter case, we shall simply say that ϕ is *finitely semidifferentiable at* e_0. In either case, we shall simply refer to $D\phi$ as a *semidifferential* (the adjectives Gâteaux or finite will usually be omitted) of ϕ at e_0.

18 If $\mathscr{X} = R^\mu$ and $W = \overline{R^\mu_-}$ (for some positive integer μ), then it is easily seen that a function $\phi = (\phi^1, \ldots, \phi^\mu)\colon \mathscr{E} \to \mathscr{X}$ is Gâteaux (respectively, finitely) W-semidifferentiable at $e_0 \in \mathscr{E}$ if and only if ϕ^i is Gâteaux (respectively, finitely) semidifferentiable at e_0 for each $i = 1, \ldots, \mu$.

19 If a function $\phi\colon \mathscr{E} \to R$ is Gâteaux (respectively, finitely) semidifferentiable at each point of $\mathscr{E}$, then we shall simply say that ϕ is *Gâteaux* (respectively, *finitely*) *semidifferentiable*.

Clearly, finite W-semidifferentiability implies Gâteaux W-semidifferentiability.

20 Note that the bigger the closed, convex cone W, the weaker is the requirement of (either Gâteaux or finite) W-semidifferentiability [see (2.36)]. Indeed, if $W = \mathscr{X}$, then every function

† For a functional f defined on a subset of R, we say that $\limsup_{\varepsilon \to 0^+} f(\varepsilon) = a \in R$ if $f(\varepsilon) \to (-\infty, a]$ as $\varepsilon \to 0^+$ in the sense of (5.33), and if it is not true that $f(\varepsilon) \to (-\infty, a']$ as $\varepsilon \to 0^+$ for any $a' < a$. If $f(\varepsilon) \to (-\infty, a]$ as $\varepsilon \to 0^+$ for every $a \in R$ (respectively, for no $a \in R$), then we shall say that $\limsup_{\varepsilon \to 0^+} f(\varepsilon) = -\infty$ (respectively, $= \infty$).

$\phi:\mathscr{E} \to \mathscr{Z}$ is finitely W-semidifferentiable. On the other hand, if $W = \{0\}$, then $\phi:\mathscr{E} \to \mathscr{Z}$ is Gâteaux (respectively, finitely) W-semidifferentiable at $e_0 \in \mathscr{E}$ if and only if ϕ is Gâteaux (respectively, finitely) differentiable at e_0 [see (2.34) and Lemma (2.37)]. Thus, every function $\phi:\mathscr{E} \to \mathscr{Z}$ which is Gâteaux (respectively, finitely) differentiable at $e_0 \in \mathscr{E}$ is also Gâteaux (respectively, finitely) W-semidifferentiable at e_0 (no matter what the closed, convex cone $W \subset \mathscr{Z}$). Further, if $\phi:\mathscr{E} \to \mathscr{Z}$ is only Gâteaux (respectively, finitely) *directionally* differentiable at $e_0 \in \mathscr{E}$, and if the directional differential $D\phi$ of ϕ at e_0 is W-convex [as must be the case, by Theorem (6), if $\mathscr{E}$ is convex and ϕ is W-convex], then ϕ is also Gâteaux (respectively, finitely) W-semidifferentiable at e_0, with $D\phi$ as a W-semidifferential.

21 THEOREM. *If $\phi:\mathscr{Y} \to \mathscr{Z}$ is a W-convex function, then ϕ is finitely W-semidifferentiable at each point $e_0 \in \mathscr{Y}$, and the function $y \to \phi(e_0 + y) - \phi(e_0):\mathscr{Y} \to \mathscr{Z}$ is a finite W-semidifferential of ϕ at e_0.*

Proof. Let us denote the function $y \to \phi(e_0 + y) - \phi(e_0)$ by $D\phi$. Obviously, $D\phi$ is W-convex on $\mathscr{Y}$ and $D\phi(0) = 0$. Also, since ϕ is W-convex, whenever $0 < \varepsilon \leq 1$,

$$\begin{aligned}\phi(e_0 + \varepsilon y) &= \phi((1 - \varepsilon)e_0 + \varepsilon(e_0 + y)) \\ &\leq_W (1 - \varepsilon)\phi(e_0) + \varepsilon\phi(e_0 + y) \quad \text{for all} \quad y \in \mathscr{Y},\end{aligned}$$

i.e.,

$$\frac{\phi(e_0 + \varepsilon y) - \phi(e_0)}{\varepsilon} - D\phi(y) \in \frac{1}{\varepsilon} W = W$$

$$\text{whenever} \quad 0 < \varepsilon \leq 1, \quad \text{for all} \quad y \in \mathscr{Y},$$

which immediately implies that our desired conclusion holds. |||

22 COROLLARY. *If ϕ is a convex functional defined on $\mathscr{Y}$, then ϕ is finitely semidifferentiable at each point $e_0 \in \mathscr{Y}$, and the function $y \to \phi(e_0 + y) - \phi(e_0):\mathscr{Y} \to R$ is a finite semidifferential of ϕ at e_0.*

We shall now introduce a stronger concept of differentiability, so-called dual differentiability. Henceforth in this section (unless we specifically specify the contrary) we shall suppose that $\mathscr{Y}$ is a linear topological space and that A is some arbitrary subset of $\mathscr{Y}$.

23 A function $\varphi:A \to \mathscr{Z}$ will be said to be *dually differentiable at* a point $y_0 \in A$ if there is a (necessarily unique) continuous linear

function [see (5.44)] $D\varphi\colon \overline{\text{co}}$ tancone $(A;y_0) \to \mathscr{X}$ [see (4.24–27)] such that [see (3.16–19)]

24 $$\frac{\varphi(y_0 + \varepsilon(\tilde{y} + y)) - \varphi(y_0)}{\varepsilon} - D\varphi(\tilde{y}) \xrightarrow[\substack{(y,\varepsilon)\to(0,0) \\ \varepsilon>0 \\ y_0+\varepsilon(\tilde{y}+y)\in A}]{} 0$$

for all $\tilde{y} \in$ tancone $(A;y_0)$.

[The convergence $(y,\varepsilon) \to (0,0)$ is of course to be understood in the product topology for $\mathscr{Y} \times R$ [see (4.17)], i.e., we have a "dual limit" in (24)—whence the term "dual differential".] In this case, we shall refer to $D\varphi$ as the *dual differential* of φ at y_0. If φ is dually differentiable at each point of A, then we shall simply say that φ is *dually differentiable.*

25 A function $\varphi\colon A \to \mathscr{X}$ will be said to be *dually W-semidifferentiable at* a point $y_0 \in A$ (where, as before, W is a closed, convex cone in $\mathscr{X}$) if there is a continuous, W-convex function $D\varphi$: $\overline{\text{co}}$ tancone $(A;y_0) \to \mathscr{X}$, with $D\varphi(0) = 0$, such that [see (5.33)]

$$\frac{\varphi(y_0 + \varepsilon(\tilde{y} + y)) - \varphi(y_0)}{\varepsilon} - D\varphi(\tilde{y}) \xrightarrow[\substack{(y,\varepsilon)\to(0,0) \\ \varepsilon>0 \\ y_0+\varepsilon(\tilde{y}+y)\in A}]{} W$$

for all $\tilde{y} \in$ tancone $(A;y_0)$.

In this case, we shall refer to $D\varphi$ as a *dual W-semidifferential of* φ at y_0. Note that a dual W-semidifferential is generally *not* unique. If φ is dually W-semidifferentiable at each point of A, then we shall simply say that φ is *dually W-semidifferentiable.*

26 If $\mathscr{X} = R$, then we shall simply say *dually semidifferentiable* in place of dually $\bar{R}_-$-semidifferentiable, and *dual semidifferential* in place of dual $\bar{R}_-$-semidifferential.

27 If $\mathscr{X} = R^\mu$ and $W = \bar{R}^\mu_-$ (for some positive integer μ), then it is easily seen that a function $\varphi = (\varphi^1, \ldots, \varphi^\mu)$: $A \to \mathscr{X}$ is dually W-semidifferentiable at $y_0 \in A$ if and only if φ^i is dually semidifferentiable at y_0 for each $i = 1, \ldots, \mu$. [Compare with (18).]

28 The observations in the first, second, and fourth sentences in (20) remain in effect if we replace "finite(ly)" or "Gâteaux" by "dual(ly)". In particular, dual W-semidifferentiability is implied by dual differentiability (for any closed, convex cone $W \subset \mathscr{X}$,

including $W = \{0\}$). As we shall see in the next theorem, dual differentiability (respectively, dual W-semidifferentiability) implies finite differentiability (respectively, finite W-semidifferentiability).

29 THEOREM. *Let $\mathscr{E}$ be a set in a linear topological space $\mathscr{Y}$ which is finitely open in itself, and let $\varphi:\mathscr{E} \to \mathscr{Z}$ be a function which is dually differentiable (respectively, dually W-semidifferentiable—where W is a closed, convex cone in the linear topological space $\mathscr{Z}$) at $e_0 \in \mathscr{E}$ with dual differential (respectively, dual W-semidifferential) $D\varphi$. Then φ is finitely differentiable (respectively, finitely W-semidifferentiable) at e_0 with finite differential (respectively, finite W-differential) $D\varphi$.*

Proof. The theorem follows directly from the following lemma, Lemma (4.27), Theorem (4.21), the second sentence of (28), and the third sentence in (20). |||

30 LEMMA. *Let φ be a function from a set A in a linear topological space $\mathscr{Y}$ into a linear topological space $\mathscr{Z}$. Suppose that φ is dually W-semidifferentiable at some point $y_0 \in A$, with dual W-semidifferential $D\varphi$. Then, for any compact set $C \subset$ tancone $(A;y_0)$,*

$$\frac{\varphi(y_0 + \varepsilon(\tilde{y} + y)) - \varphi(y_0)}{\varepsilon} - D\varphi(\tilde{y}) \xrightarrow[\substack{(y,\varepsilon)\to(0,0)\\ \varepsilon>0\\ y_0+\varepsilon(\tilde{y}+y)\in A}]{} W$$

uniformly w.r. to $\tilde{y} \in C$.

Proof. Let N' be an arbitrary neighborhood of 0 in $\mathscr{Z}$, and let N'_1 be a neighborhood of 0 in $\mathscr{Z}$ such that $N'_1 + N'_1 \subset N'$ [see Corollary (4.4)]. By hypothesis, for each $\tilde{y} \in C$, there are a neighborhood $N_{\tilde{y}}$ of 0 in $\mathscr{Y}$ and a number $\varepsilon_{\tilde{y}} > 0$ such that

31 $$\frac{\varphi(y_0 + \varepsilon(\tilde{y} + y)) - \varphi(y_0)}{\varepsilon} - D\varphi(\tilde{y}) \in W + N'_1$$

$$\text{and} \quad D\varphi(\tilde{y}) - D\varphi(\hat{y}) \in N'_1$$

whenever $y \in N_{\tilde{y}}$, $\hat{y} - \tilde{y} \in N_{\tilde{y}}$, $\hat{y} \in C$, $0 < \varepsilon < \varepsilon_{\tilde{y}}$, and $(y_0 + \varepsilon(\tilde{y} + y)) \in A$. For each $\tilde{y} \in C$, let $\hat{N}_{\tilde{y}}$ be a neighborhood of 0 in $\mathscr{Y}$ such that $\hat{N}_{\tilde{y}} + \hat{N}_{\tilde{y}} \subset N_{\tilde{y}}$. Since C is compact, there is a subset $\{y_1, \ldots, y_m\}$ of C such that $\bigcup_{i=1}^m (y_i + \hat{N}_{y_i}) \supset C$. Let $N = \bigcap_{i=1}^m \hat{N}_{y_i}$ and $\varepsilon_0 = \min\{\varepsilon_{y_i}: i = 1, \ldots, m\}$. Then the relations $\hat{y} \in C$, $y \in N$, $0 < \varepsilon < \varepsilon_0$, and $(y_0 + \varepsilon(\hat{y} + y)) \in A$ imply that, for some $j = 1, \ldots, m$, $\hat{y} \in y_j + \hat{N}_{y_j}$, $\hat{y} + y \in \hat{y} + \hat{N}_{y_j} \supset$

$y_j + N_{y_j}$, and $0 < \varepsilon < \varepsilon_{y_j}$, and thence, by (31)—with $\tilde{y}$ replaced by y_j, and y by $(\hat{y} + y - y_j)$—that

$$\frac{\varphi(y_0 + \varepsilon(\hat{y} + y)) - \varphi(y_0)}{\varepsilon} - D\varphi(\hat{y})$$
$$= \left[\frac{\varphi(y_0 + \varepsilon(y_j + (\hat{y} + y - y_j))) - \varphi(y_0)}{\varepsilon} - D\varphi(y_j)\right]$$
$$+ [D\varphi(y_j) - D\varphi(\hat{y})] \in W + N_1' + N_1' \subset W + N',$$

which completes our proof. |||

The following theorem is analogous to Theorem (21).

32 THEOREM. *If $\phi:\mathscr{Y} \to \mathscr{X}$ is a continuous W-convex function, then ϕ is dually W-semidifferentiable at each point $y_0 \in \mathscr{Y}$, and the function $y \to \phi(y_0 + y) - \phi(y_0):\mathscr{Y} \to \mathscr{X}$ is a dual W-semidifferential of ϕ at y_0.*

Proof. Let us denote the function $y \to \phi(y_0 + y) - \phi(y_0)$ by $D\phi$. Arguing as in the proof of Theorem (21), we can show that

$$\frac{\phi(y_0 + \varepsilon y) - \phi(y_0)}{\varepsilon} - D\phi(\tilde{y}) \in [D\phi(y) - D\phi(\tilde{y})] + W$$
$$\text{whenever} \quad 0 < \varepsilon \leq 1, \quad \text{for all} \quad y, \tilde{y} \in \mathscr{Y},$$

from which our desired conclusion follows at once, because of the continuity of ϕ. |||

33 COROLLARY. *If ϕ is a continuous convex functional defined on $\mathscr{Y}$, then ϕ is dually semidifferentiable at each point $y_0 \in \mathscr{Y}$, and the function $y \to \phi(y_0 + y) - \phi(y_0):\mathscr{Y} \to R$ is a dual semidifferential of ϕ at y_0.*

We now turn to the following "chain rules" of differentiation.

34 THEOREM. *Let $\mathscr{Y}$ be a linear vector space and $\mathscr{E}$ a set in $\mathscr{Y}$ which is finitely open in itself, let $\mathscr{X}$ and $\mathscr{X}'$ be linear topological spaces and A a subset of $\mathscr{X}'$, and let $\phi:\mathscr{E} \to \mathscr{X}'$ and $\varphi: A \to \mathscr{X}$ be functions such that $A \supset \phi(\mathscr{E})$. Then if ϕ is Gâteaux (respectively, finitely) differentiable at $e_0 \in \mathscr{E}$ with Gâteaux (respectively, finite) differential $D\phi$, and φ is dually differentiable (respectively, dually W-semidifferentiable) at $\phi(e_0)$, with dual differential (respectively, dual W-semidifferential) $D\varphi$, then $\varphi \circ \phi$ is Gâteaux (respectively, finitely) differentiable (respectively, W-semidifferentiable) at e_0 with Gâteaux (respectively, finite) differential (respectively, W-semidifferential) $D\varphi \circ D\phi$.*

35 THEOREM. *Let* $\mathscr{Y}$, $\mathscr{Z}$, *and* $\mathscr{Z}'$ *be linear topological spaces and* A *and* A' *subsets of* $\mathscr{Y}$ *and* $\mathscr{Z}'$, *respectively, and let* $\varphi_1 : A \to \mathscr{Z}'$ *and* $\varphi_2 : A' \to \mathscr{Z}$ *be functions such that* $A' \supset \varphi_1(A)$. *Suppose that* φ_1 *is dually differentiable at* $y_0 \in A$ *with dual differential* $D\varphi_1$, *and that* φ_2 *is dually differentiable (respectively, dually W-semidifferentiable) at* $\varphi_1(y_0)$ *with dual differential (respectively, dual W-semidifferential)* $D\varphi_2$. *Then* $\varphi_2 \circ \varphi_1$ *is dually differentiable (respectively, dually W-semidifferentiable) at* y_0 *with dual differential (respectively, dual W-semidifferential)* $D\varphi_2 \circ D\varphi_1$.

36 The proofs of Theorems (34) and (35), the former of the two being based in part on Lemma (30), are quite straightforward, and will be left as an exercise to the reader. It should be noted that if, in Theorem (34), ϕ is only assumed to be Gâteaux directionally differentiable at e_0, and φ is dually differentiable at $\phi(e_0)$, then $\varphi \circ \phi$ is Gâteaux directionally differentiable at e_0 with directional differential $D\varphi \circ D\phi$.

In order to define our last (and strongest) differential, we shall suppose that $\mathscr{Y}$ and $\mathscr{Z}$ are Banach spaces, and that A is a set in $\mathscr{Y}$ with a nonempty interior.

37 A function $\varphi : A \to \mathscr{Z}$ will be said to be *Fréchet differentiable at* a point $y_0 \in \operatorname{int} A$ if there is a (necessarily unique) function $D\varphi \in \mathbf{B}(\mathscr{Y},\mathscr{Z})$ such that

$$\frac{\varphi(y_0 + y) - \varphi(y_0) - D\varphi(y)}{\|y\|} \xrightarrow[\substack{y \to 0 \\ y \neq 0 \\ y_0 + y \in A}]{} 0.$$

In this case, we shall refer to $D\varphi$ as the *Fréchet differential* (or *Fréchet derivative*) of φ at y_0. If A is open and φ is Fréchet differentiable at each point of A, then we shall simply say that φ is *Fréchet differentiable.*

The following lemma is obvious [compare with Theorem (10)].

38 LEMMA. *If* $\varphi : A \to \mathscr{Z}$ *is Fréchet differentiable at* $y_0 \in \operatorname{int} A$, *then* φ *is continuous at* y_0.

39 COROLLARY. *If* $\varphi : A \to \mathscr{Z}$ *is Fréchet differentiable, then* φ *is continuous.*

The following lemma, whose proof is immediate, is analogous to Lemma (9).

40 LEMMA. *If* $\varphi : \mathscr{Y} \to \mathscr{Z}$ *is affine and continuous, then* φ *is Fréchet differentiable and its Fréchet differential (at any point in* $\mathscr{Y}$*) coincides with its linear part.*

We also have the following (easily verified) chain rule for Fréchet differentials.

41   THEOREM. *Let $\mathscr{Y}$, $\mathscr{Z}$, and $\mathscr{Z}'$ be Banach spaces, let A and A' be sets in $\mathscr{Y}$ and $\mathscr{Z}'$, respectively, with nonempty interiors, and let $\varphi_1 : A \to \mathscr{Z}'$ and $\varphi_2 : A' \to \mathscr{Z}$ be functions such that $A' \supset \varphi_1(A)$. Then, if φ_1 and φ_2 are Fréchet differentiable at a point $y_0 \in$ int A and at $\varphi_1(y_0) \in$ int A', respectively, with respective Fréchet differentials $D\varphi_1$ and $D\varphi_2$, $\varphi_2 \circ \varphi_1$ is also Fréchet differentiable (at y_0) with Fréchet differential $D\varphi_2 \circ D\varphi_1$.*

There will be occasions when we wish to emphasize that $D\varphi$ denotes a differential (or W-semidifferential)—whether finite, dual, etc.—of φ at a particular y_0 in the domain of φ. In this case, we shall write $D\varphi(y_0; \cdot)$ instead of $D\varphi(\cdot)$ or $D\varphi$.

The following theorem in essence generalizes the conventional mean-value theorem of the differential calculus.

42   THEOREM. *Suppose that $\mathscr{Y}$ and $\mathscr{Z}$ are Banach spaces, that A is some set in $\mathscr{Y}$, that y' and y'' are points of A such that* co $\{y', y''\} \subset$ int A, *and that φ is a function from A into $\mathscr{Z}$ which is Fréchet differentiable at each point of* co $\{y', y''\}$. *Then*

$$\|\varphi(y') - \varphi(y'')\| \le \|y' - y''\| \cdot \sup_{y \in \text{co}\,\{y', y''\}} \|D\varphi(y; \cdot)\|.$$

Proof. For each $\lambda \in [0,1]$, let y_λ denote $(\lambda y' + (1 - \lambda)y'')$, so that co $\{y', y''\} = \{y_\lambda : 0 \le \lambda \le 1\}$. Let $M = \sup_{0 \le \lambda \le 1} \|D\varphi(y_\lambda; \cdot)\|$, and let $\varepsilon > 0$ be fixed but arbitrary. For each $\lambda \in [0,1]$, there is a $\delta_\lambda > 0$ such that

$$\frac{\|\varphi(y_{\lambda'}) - \varphi(y_\lambda) - D\varphi(y_\lambda; y_{\lambda'} - y_\lambda)\|}{\|y_{\lambda'} - y_\lambda\|} < \varepsilon$$

whenever $0 < |\lambda' - \lambda| < 2\delta_\lambda$.

Since $y_{\lambda'} - y_\lambda = (\lambda' - \lambda)(y' - y'')$, this means that, whenever $|\lambda' - \lambda| < 2\delta_\lambda$, then [see (5.3)]

$$\begin{aligned}
\|\varphi(y_{\lambda'}) - \varphi(y_\lambda)\| &\le \|\varphi(y_{\lambda'}) - \varphi(y_\lambda) - D\varphi(y_\lambda; y_{\lambda'} - y_\lambda)\| \\
&\quad + \|D\varphi(y_\lambda; y_{\lambda'} - y_\lambda)\| \\
&\le \varepsilon \|y_{\lambda'} - y_\lambda\| + \|D\varphi(y_\lambda; \cdot)\| \, \|y_{\lambda'} - y_\lambda\| \\
&\le (M + \varepsilon)|\lambda' - \lambda| \, \|y' - y''\|.
\end{aligned}$$

Since $[0,1]$ is compact in R, there are numbers $\lambda_1, \ldots, \lambda_\nu$, with $0 = \lambda_1 < \lambda_2 < \cdots < \lambda_\nu = 1$, such that $\lambda_{i+1} - \lambda_i < \delta_{\lambda_{i+1}} + \delta_{\lambda_i}$

for each $i = 1, \ldots, \nu - 1$, which implies that

$$\|\varphi(y_{\lambda_{i+1}}) - \varphi(y_{\lambda_i})\| \leq (M + \varepsilon)(\lambda_{i+1} - \lambda_i)\|y' - y''\| \quad \text{for} \quad i = 1, \ldots, \nu - 1.$$

But

$$\begin{aligned}\|\varphi(y') - \varphi(y'')\| &= \|[\varphi(y_{\lambda_\nu}) - \varphi(y_{\lambda_{\nu-1}})] + [\varphi(y_{\lambda_{\nu-1}}) - \varphi(y_{\lambda_{\nu-2}})] \\ &\quad + \cdots + [\varphi(y_{\lambda_2}) - \varphi(y_{\lambda_1})]\| \\ &\leq \sum_{i=1}^{\nu-1} \|\varphi(y_{\lambda_{i+1}}) - \varphi(y_{\lambda_i})\| \\ &\leq (M + \varepsilon)\|y' - y''\| \sum_{i=1}^{\nu-1} (\lambda_{i+1} - \lambda_i) \\ &= (M + \varepsilon)\|y' - y''\|.\end{aligned}$$

Since $\varepsilon > 0$ was arbitrary, $\|\varphi(y') - \varphi(y'')\| \leq M\|y' - y''\|$. |||

In case $\mathscr{X} = R$, Theorem (42) has the following stronger form.

43 THEOREM. *Let A be a set in a Banach space $\mathscr{Y}$ and let y' and y'' be points of A such that* co $\{y', y''\} \subset$ int A. *Then if φ is a functional on A which is Fréchet differentiable at each point of* co $\{y', y''\}$, *there is a point $y''' \in$* co $\{y', y''\}$ *such that*

$$\varphi(y') - \varphi(y'') = D\varphi(y'''; y' - y'').$$

Theorem (43) follows easily from the conventional mean-value theorem of the differential calculus.

The following theorem essentially states that Fréchet differentiability is a stronger requirement than dual differentiability.

44 THEOREM. *Suppose that $\varphi: A \to \mathscr{X}$ is Fréchet differentiable at $y_0 \in$ int A. Then φ is also dually differentiable at y_0, and the Fréchet and dual differentials of φ at y_0 coincide.*

Proof. Let $D\varphi$ denote the Fréchet differential of φ at y_0, and let $\tilde{y} \in \mathscr{Y}$ be fixed but arbitrary. Hence, if $\varepsilon > 0$ and $y \in \mathscr{Y}$ are such that ε and $\|y\|$ are sufficiently small, then $y_0 + \varepsilon(\tilde{y} + y) \in A$ and

$$\frac{\varphi(y_0 + \varepsilon(\tilde{y} + y)) - \varphi(y_0)}{\varepsilon} - D\varphi(\tilde{y})$$

$$= \|\tilde{y} + y\| \frac{\varphi(y_0 + \varepsilon(\tilde{y} + y)) - \varphi(y_0) - D\varphi(\varepsilon(\tilde{y} + y))}{\|\varepsilon(\tilde{y} + y)\|} + D\varphi(y).$$

The expression in the right-hand side of the preceding equation tends to 0 as $(y, \varepsilon) \to (0,0)$ by Definition (37), as is to be shown. |||

We may summarize the relative strength of the various differentiability conditions that we have introduced [see also Theorem (29)] with the following diagram:

Fréchet $\Rightarrow$ dual $\Rightarrow$ finitely $\Rightarrow$ Gâteaux $\Rightarrow$ Gâteaux directionally.

45 If we exclude the first and last items from the preceding diagram, then this diagram also applies to W-semidifferentiability. Finally [see (20) and (28)], differentiability (whether dual, finite, or Gâteaux) implies the corresponding W-semidifferentiability (no matter what the closed, convex cone $W \subset \mathscr{X}$, including $\{0\}$).

46 If $\mathscr{Y} = R^m$ for some positive integer m, and if $\mathscr{X} = R$, then Fréchet differentiability is equivalent to the usual notion of the existence of a total derivative, and the Fréchet differential coincides with the total differential. Further, if $\mathscr{Y} = R^m$ (and $\mathscr{X}$ is an arbitrary Banach space), then it is not hard to show that finite differentiability at a point $y_0 \in \operatorname{int} A$ implies Fréchet differentiability at y_0. [The argument, which is based on the fact that every ball [see (4.37)] in R^m is contained in a convex polyhedron and on Theorem (5.11) is left as an exercise to the reader.] Thus, in R^m, the concepts of Fréchet, dual, and finite differentiability are essentially equivalent. Consequently, if a function φ has as domain a set $A \subset R^m$ and if $y_0 \in \operatorname{int} A$, then we shall simply say that φ is *differentiable* at y_0 instead of φ is Fréchet differentiable at y_0. Of course, when $\mathscr{Y}$ is infinite-dimensional, then the preceding remarks fail to carry over.

47 If the functions $\varphi_1 : A \to \mathscr{X}$ and $\varphi_2 : A \to \mathscr{X}$ (where A is a set in $\mathscr{Y}$, and $\mathscr{Y}$ and $\mathscr{X}$ are Banach spaces) are Fréchet differentiable at a point $y_0 \in \operatorname{int} A$ with differentials $D\varphi_1$ and $D\varphi_2$, respectively, then it is easy to show [see (5.36)] that $(\alpha\varphi_1 + \beta\varphi_2)$, for any α, $\beta \in R$, is also Fréchet differentiable at y_0, with Fréchet differential $(\alpha D\varphi_1 + \beta D\varphi_2)$. Analogous statements hold for dual differentials, finite and Gâteaux differentials, and finite and Gâteaux directional differentials. For W-semidifferentials (whether dual, finite, or Gâteaux), an analogous conclusion also holds, provided we confine ourselves to *nonnegative* coefficients α and β [see (1.13) and (2.34)].

48 If a function $\varphi : A \to \mathscr{X}$ (where A is an open set in $\mathscr{Y}$, and $\mathscr{Y}$ and $\mathscr{X}$ are Banach spaces) is Fréchet differentiable, and the mapping $y \to D\varphi(y;\cdot)$: $A \to \mathbf{B}(\mathscr{Y},\mathscr{X})$ is continuous with respect to the

norm topology of $\mathbf{B}(\mathcal{Y}, \mathcal{Z})$ [see (5.2)], then we shall say that φ is *continuously differentiable*. Note that if $\varphi_1: A \to \mathcal{Z}$ and $\varphi_2: A \to \mathcal{Z}$ are continuously differentiable, then $(\alpha\varphi_1 + \beta\varphi_2)$ also is, for any $\alpha, \beta \in R$.

49   Let $\mathcal{Y}, \mathcal{Z}_1, \ldots, \mathcal{Z}_m$ be Banach spaces, let $A \subset \mathcal{Y}$, and, for each $i = 1, \ldots, m$, let φ_i be a function from A into $\mathcal{Z}_i$. Then it is easily shown that the function [see (2.5)] $\varphi = (\varphi_1, \ldots, \varphi_m): A \to (\mathcal{Z}_1 \times \cdots \times \mathcal{Z}_m)$ is Fréchet differentiable at a point $y_0 \in \text{int } A$ [with respect to the product topology in $(\mathcal{Z}_1 \times \cdots \times \mathcal{Z}_m)$—see (4.38–42)] if and only if each φ_i $(i = 1, \ldots, m)$ is Fréchet differentiable at y_0. In this case, the Fréchet differential $D\varphi$ of φ at y_0 is given by $D\varphi = (D\varphi_1, \ldots, D\varphi_m)$, where, for each i, $D\varphi_i$ is the Fréchet differential of φ_i at y_0. Similar assertions hold for finite and Gâteaux differentials (and directional differentials), and for dual differentials [see (3.37 and 38)].

50   Further, if $\mathcal{Y}, \mathcal{Z}_1, \ldots, \mathcal{Z}_m$ are linear topological spaces, A is a subset of $\mathcal{Y}$, and, for each $i = 1, \ldots, m$, W_i is a closed, convex cone in $\mathcal{Z}_i$ and ϕ_i is a function from A into $\mathcal{Z}_i$, then $\phi = (\phi_1, \ldots, \phi_m): A \to (\mathcal{Z}_1 \times \cdots \times \mathcal{Z}_m)$ is dually W-semidifferentiable [where $W = (W_1 \times \cdots \times W_m)$] at a point $y_0 \in A$ if and only if each ϕ_i is dually W_i-semidifferentiable at y_0 [see (2.35)]. In this case, $(D\phi_1, \ldots, D\phi_m)$ is a dual W-semidifferential of ϕ at y_0 if and only if, for each i, $D\phi_i$ is a dual W_i-semidifferential of ϕ_i at y_0. Analogous assertions hold for Gâteaux and finite W-semidifferentials.

51   Let $\mathcal{Y}_1, \ldots, \mathcal{Y}_m$ be topological spaces, let $\mathcal{Z}$ be a Banach space, let A be a subset of $(\mathcal{Y}_1 \times \cdots \times \mathcal{Y}_m)$, and let φ be a function from A into $\mathcal{Z}$. For each $i = 1, \ldots, m$ and every $\tilde{y} = (\tilde{y}_1, \ldots, \tilde{y}_m) \in A$, let

52
$$A_i(\tilde{y}) = \{y_i : y_i \in \mathcal{Y}_i, (\tilde{y}_1, \ldots, \tilde{y}_{i-1}, y_i, \tilde{y}_{i+1}, \ldots, \tilde{y}_m) \in A\},$$
$$i = 1, \ldots, m,$$

so that $A_i(\tilde{y}) \subset \mathcal{Y}_i$. If, for some $j = 1, \ldots, m$ and some $\tilde{y} = (\tilde{y}_1, \ldots, \tilde{y}_m) \in A$, $\tilde{y}_j \in \text{int } A_i(\tilde{y})$ (as must be the case if $\tilde{y} \in \text{int } A$), $\mathcal{Y}_j$ is a Banach space, and the function

$$y_j \to \varphi(\tilde{y}_1, \ldots, \tilde{y}_{j-1}, y_j, \tilde{y}_{j+1}, \ldots, \tilde{y}_m): A_j(\tilde{y}) \to \mathcal{Z}$$

is Fréchet differentiable at $\tilde{y}_j$, then we shall say that φ is *Fréchet differentiable with respect to its j-th argument at $\tilde{y}$*. In this case, we shall refer to the corresponding Fréchet differential—which will be denoted by $D_j\varphi$ [or by $D_j\varphi(\tilde{y};\cdot)$, if we wish to emphasize that it is at $\tilde{y}$]—as the *Fréchet partial differential* (or *Fréchet partial derivative*) of φ w.r. to its j-th argument at $\tilde{y}$. Note that $D_j\varphi \in \mathbf{B}(\mathscr{Y}_j,\mathscr{Z})$. If $A_j(\tilde{y})$ is open for all $\tilde{y} \in A$ and if φ is Fréchet differentiable w.r. to its j-th argument at every $\tilde{y} \in A$, then we shall simply say that φ is *Fréchet differentiable w.r. to its j-th argument*; if, in addition, the mapping $\tilde{y} \to D_j\varphi(\tilde{y};\cdot): A \to \mathbf{B}(\mathscr{Y}_j,\mathscr{Z})$ is continuous [with respect to the norm topology of $\mathbf{B}(\mathscr{Y}_j,\mathscr{Z})$], then we shall say that φ is *continuously differentiable w.r. to its j-th argument*. If $\mathscr{Y}_1, \ldots, \mathscr{Y}_m$ are all Banach spaces, and if φ is Fréchet differentiable at some point $\tilde{y} \in \operatorname{int} A$, then it is easy to see that φ is Fréchet differentiable w.r. to each of its arguments at $\tilde{y}$ (i.e., w.r. to its j-th argument for each $j = 1, \ldots, m$), and that

53 $$D\varphi(\tilde{y};y_1,\ldots,y_m) = \sum_{j=1}^{m} D_j\varphi(\tilde{y};y_j)$$

$$\text{for all} \quad (y_1,\ldots,y_m) \in (\mathscr{Y}_1 \times \cdots \times \mathscr{Y}_m) \quad \text{and} \quad \tilde{y} \in A.$$

The converse is not generally true; i.e., if φ is Fréchet differentiable w.r. to each of its arguments at $\tilde{y} \in A$, then it does not necessarily follow that φ is Fréchet differentiable at $\tilde{y}$. However, we do have the following theorem.

54 THEOREM. *Let $\mathscr{Y}_1, \ldots, \mathscr{Y}_m$ and $\mathscr{Z}$ be Banach spaces and A an open set in $(\mathscr{Y}_1 \times \cdots \times \mathscr{Y}_m)$. Then a continuous function $\varphi: A \to \mathscr{Z}$ is continuously differentiable if and only if it is continuously differentiable w.r. to each of its arguments, in which case* (53) *holds.*

Proof. See [3, pp. 167–168].

The following is an implicit function theorem which we shall require in the sequel.

55 THEOREM. *Let $\mathscr{X}$ and $\mathscr{Z}$ be Banach spaces, let O be an open set in $\mathscr{Z} \times \mathscr{X}$, and let F be a continuously differentiable function from O into $\mathscr{X}$. Let $(z_0,x_0) \in O$ be such that $F(z_0,x_0) = 0$ and suppose that $D_2F(z_0,x_0;\cdot)$ is a homeomorphism* [see (3.28)] *of $\mathscr{X}$ onto itself. Then there are a number $\varepsilon_1 > 0$ and a continuously differentiable function q: $\{z: z \in \mathscr{Z}, \|z - z_0\| < \varepsilon_1\} \to \mathscr{X}$ such that* (i) $q(z_0) = x_0$, (ii) $(z,q(z)) \in O$ *and* $F(z,q(z)) = 0$ *whenever* $\|z - z_0\| < \varepsilon_1$, *and*

$$\text{(iii)} \quad Dq(z_0;\cdot) = -[D_2F(z_0,x_0;\cdot)]^{-1} \circ [D_1F(z_0,x_0;\cdot)].$$

Proof. For the proof of a more general implicit function theorem, see [3, Theorem 10.2.1, p. 265].

56 Suppose that $\mathscr{Y}_1, \ldots, \mathscr{Y}_m$, and $\mathscr{Z}$ are linear topological spaces, that A is a subset of $(\mathscr{Y}_1 \times \cdots \times \mathscr{Y}_m)$, and that $\varphi: A \to \mathscr{Z}$ is a function which is dually differentiable at a point $\tilde{y} = (\tilde{y}_1, \ldots, \tilde{y}_m) \in A$—with dual differential $D\varphi(\tilde{y}; y_1, \ldots, y_m)$, so that $D\varphi$ is a linear operator from $(\mathscr{Y}_1 \times \cdots \times \mathscr{Y}_m)$ into $\mathscr{Z}$ [see (23) and (5.44)]. Then, for each $j = 1, \ldots, m$, we shall refer to the function $y_j \to D\varphi(\tilde{y}; 0, \ldots, 0, y_j, 0, \ldots, 0)$ from $\mathscr{Y}_j$ into $\mathscr{Z}$ as a *formal dual partial differential* of φ with respect to its j-th argument at $\tilde{y}$, and shall denote it by $D_j\varphi$ or by $D_j\varphi(\tilde{y}; \cdot)$. Note that here, formula (53) evidently remains in force. We point out that if $\overline{\text{co}}$ tancone $(A; \tilde{y}) = (\mathscr{Y}_1 \times \cdots \times \mathscr{Y}_m)$, then the formal dual partial differentials are actual dual partial differentials in a sense similar to that of (51).

57 In a completely analogous manner, if $\mathscr{Y}_1, \ldots, \mathscr{Y}_m$, $\mathscr{Z}$ and A are as in (56) and $\varphi: A \to \mathscr{Z}$ is a function which is dually W-semidifferentiable at a point $\tilde{y} \in A$ with a *linear* (as well as continuous) dual W-semidifferential $D\varphi$, then we can define *formal dual partial W-semidifferentials* of φ with respect to each of its arguments at $\tilde{y}$, which will also be denoted by $D_j\varphi$ or by $D_j\varphi(\tilde{y}; \cdot)$. Here again, (53) remains in force.

We now present a brief discussion of Fréchet differentials for the special case where $\mathscr{Y} = R^\mu$ and $\mathscr{Z} = R^m$ (for some positive integers μ and m), using the matrix viewpoint which was introduced in (6.1).

58 Thus, suppose that $A \subset R^\mu$, that $\xi_0 \in \text{int}\, A$, and that χ is a function from A into R^m which is Fréchet differentiable at ξ_0 [see (46)]. Then the Fréchet differential $D\chi(\xi_0; \cdot) \in \mathbf{B}(R^\mu, R^m)$, and [see (6.1)] there is an $(m \times \mu)$ matrix, which we shall denote by $\mathscr{D}\chi$ [or by $\mathscr{D}\chi(\xi_0)$ if we wish to emphasize the fact that the differential is at ξ_0], and which we shall refer to as *the Jacobian matrix of* χ *at* ξ_0, such that

59 $$D\chi(\xi_0; \xi) = \mathscr{D}\chi(\xi_0)\xi \quad \text{for all} \quad \xi \in R^\mu$$

[of course, ξ in (59) must be considered to be a column vector]. In this case, the i, j-th element of $\mathscr{D}\chi(\xi_0)$ is nothing more than the conventional partial derivative of the i-th "component" of χ with respect to the j-th "component" of $\xi \in R^\mu$. If $m = 1$, then $\mathscr{D}\chi(\xi_0)$ is a row vector which coincides with the ordinary gradient of χ at ξ_0. Note that $\|D\chi(\xi_0; \cdot)\|$ [defined as in (5.2)] coincides with

$|\mathscr{D}\chi(\xi_0)|$ [defined as in (6.4)]. If A is open and χ is Fréchet differentiable, then it is evident that χ is continuously differentiable if and only if the mapping $\xi \to \mathscr{D}\chi(\xi)$ from A into $R^{m\mu}$ [see (6.3)] is continuous.

60 Now suppose that $\mathscr{Y}_1, \ldots, \mathscr{Y}_m$ are topological spaces, with $\mathscr{Y}_j = R^\mu$, for some $j = 1, \ldots, m$ (and some positive integer μ), that $A \subset (\mathscr{Y}_1 \times \cdots \times \mathscr{Y}_m)$, that $\tilde{y} = (\tilde{y}_1, \ldots, \tilde{y}_m) \in A$ is such that $\tilde{y}_j \in \operatorname{int} A_j(\tilde{y})$ [see (52)], and that φ is a function from A into R^n (for some positive integer n) which is Fréchet differentiable with respect to its j-th argument at $\tilde{y}$. Then the Fréchet partial differential $D_j\varphi(\tilde{y};\cdot) \in \mathbf{B}(R^\mu,R^n)$, and there is an $(n \times \mu)$ matrix, which we shall in this case denote by $\mathscr{D}_j\varphi$ [or by $\mathscr{D}_j\varphi(\tilde{y})$ or $\mathscr{D}_j\varphi(\tilde{y}_1, \ldots, \tilde{y}_m)$, if we wish to emphasize that the partial differential is at $\tilde{y} = (\tilde{y}_1, \ldots, \tilde{y}_m)$], and which we shall refer to as a *Jacobian matrix* of φ at $\tilde{y} = (\tilde{y}_1, \ldots, \tilde{y}_m)$ (if necessary, for clarity, we may say "the Jacobian matrix of φ w.r. to its j-th argument") such that

61 $$D_j\varphi(\tilde{y};y_j) = \mathscr{D}_j\varphi(\tilde{y})y_j \quad \text{for all} \quad y_j \in R^\mu.$$

As in (58), $\|D_j\varphi(\tilde{y};\cdot)\| = |\mathscr{D}_j\varphi(\tilde{y})|$. If $A_j(\tilde{y})$ is open for all $\tilde{y} \in A$ and φ is Fréchet differentiable w.r. to its j-th argument, then φ is continuously differentiable w.r. to its j-th argument if and only if the mapping $\tilde{y} \to \mathscr{D}_j\varphi(\tilde{y})\colon A \to R^{n\mu}$ is continuous. We shall on occasion also concern ourselves with differentiability properties of the map $\tilde{y} \to \mathscr{D}_j\varphi(\tilde{y})$, in which case we shall simply say that $\mathscr{D}_j\varphi$ has this or that differentiability property.

We close this section and this chapter with the following lemma regarding subgradients [see (2.29)] and Fréchet differentials of convex functionals.

62 LEMMA. *Let χ be a convex functional on R^μ. Then if χ is Fréchet differentiable at some $\xi_0 \in R^\mu$, $\mathscr{D}\chi(\xi_0)$ is the unique subgradient of χ at ξ_0.*

Proof. See [10, Theorem 25.1, p. 242].

References

1. C. Berge, *Topological Spaces*, Macmillan, New York, 1963.
2. N. Bourbaki, *Eléments de Mathématique*, Livre V, *Espaces Vectoriels Topologiques*, Hermann et Cie., Paris, 1953.
3. J. Dieudonné, *Foundations of Modern Analysis*, Academic Press, New York, 1960.

4. N. Dunford and J. T. Schwartz, *Linear Operators*, Part I, Interscience, New York, 1958.
5. H. G. Eggleston, *Convexity*, Cambridge University Press, Cambridge, 1958.
6. J. L. Kelley, I. Namioka et al., *Linear Topological Spaces*, D. Van Nostrand, Princeton, New Jersey, 1963.
7. E. J. McShane and T. A. Botts, *Real Analysis*, D. Van Nostrand, Princeton, New Jersey, 1959.
8. A. E. Taylor, *Introduction to Functional Analysis*, John Wiley, New York, 1958.
9. F. A. Valentine, *Convex Sets*, McGraw-Hill, New York, 1964.
10. R. T. Rockafellar, *Convex Analysis*, Princeton University Press, Princeton, New Jersey, 1970.

CHAPTER II

A Basic Optimization Problem in Simplified Form

THIS CHAPTER IS in many ways a capsule version of the entire book. We shall here investigate the optimization problem which consists in finding the minimum of a function ϕ^0 on a given set $\mathscr{E}$, subject to equality constraints $\varphi^i(e) = 0$ for $i = 1, \ldots, m$ and inequality constraints $\phi^i(e) \leq 0$ for $i = 1, \ldots, \mu$, where $\varphi^1, \ldots, \varphi^m$, ϕ^0, $\phi^1, \ldots, \phi^\mu$ are given, real-valued functions defined on $\mathscr{E}$. Under the assumptions that $\mathscr{E}$ is a subset of a linear vector space which is finitely open in itself [see (I.1.39)], that $\varphi^1, \ldots, \varphi^m$ are finitely differentiable [see (I.7.4)], and that ϕ^0, $\phi^1, \ldots, \phi^\mu$ are finitely semidifferentiable [see (I.7.13, 16, and 19)], we shall show in Section 1 that every solution of our problem satisfies a generalized Lagrange multiplier rule.

In Section 2, we shall apply our multiplier rule to a particular class of optimization problems defined in terms of operator equations in a Banach space, in which it is desired to find the "best" operator from a given class of admissible operators, subject to certain given equality and inequality constraints on the fixed points of these operators. For this problem, as we shall see, under suitable hypotheses, the multiplier rule takes the form of a maximum principle which every solution of the problem must satisfy. We shall then specialize this result further to obtain a maximum principle for optimal control problems with ordinary differential equation constraints.

In Section 3, we shall investigate sufficient conditions for a point $e_0 \in \mathscr{E}$ to be a solution of our basic problem, and shall show that, under suitable additional hypotheses, dealing primarily with convexity conditions on the problem data, the necessary conditions of the multiplier rule are also sufficient.

In Section 4, we shall apply one of our sufficiency theorems to the particular class of optimization problems discussed in Section 2 (defined in terms of operator equations), to show that the maximum principle, which was found to be a necessary condition, is also (under suitable assumptions on the problem data) sufficient, so long as one additional technical condition is satisfied. This result will then be specialized to optimal control problems with constraints in the form of ordinary differential equations.

We shall also briefly touch on problems which differ from the one discussed in the first sentence of this chapter in that there are either no equality constraints or no inequality constraints. As is to be expected, all our results carry over (with certain evident simplifications) to these problems.

All of the results of this chapter will be obtained under the assumption that the underlying set $\mathscr{E}$ for the optimization problem is finitely open in itself. Unfortunately, in many important applications, particularly in those from nonlinear optimal control theory, the set $\mathscr{E}$ is *not* finitely open in itself. However, it turns out that, in most of these applications, $\mathscr{E}$ is "almost" finitely open in itself, and, in fact, is sufficiently close to being finitely open in itself that the basic multiplier rule derived in this chapter remains in force as a necessary condition for optimality, so long as $\varphi^1, \ldots, \varphi^m$, ϕ^0, $\phi^1, \ldots, \phi^\mu$ satisfy hypotheses which are only slightly stronger than those which are imposed in this chapter. Unfortunately, it is an extremely delicate matter to give a suitable, precise definition for the concept of "almost" finitely open in itself. This is because the class of all sets with this "almost" f.o.i.i. property must, on the one hand, be sufficiently small that the multiplier rule remains in effect, yet must, on the other hand, be sufficiently large that it will include the sets that arise in the optimal control problems which we wish to investigate. Indeed, much of Chapters III and IV is devoted to an elaboration of the preceding sentence. Nevertheless, the essence of Chapters III and IV is contained in Sections 1 and 2 of this chapter, and for this reason the reader is advised to carefully digest this material before proceeding to the remainder of the book.

We should also point out that, in Chapters III and IV, we investigate problems containing inequality constraints which are considerably more general than those of this chapter, and also investigate other optimization problems, such as minimax problems.

1. A Simplified Multiplier Rule

In this section we shall develop a necessary condition in the form of a multiplier rule for the following optimization problem.

1 *Problem.* Given a set $\mathscr{E}$ in a linear vector space $\mathscr{Y}$ and real-valued functions $\varphi^1, \ldots, \varphi^m, \phi^0, \phi^1, \ldots, \phi^\mu$ defined on $\mathscr{E}$, find an element $e_0 \in \mathscr{E}$ that achieves a minimum for ϕ^0 (on $\mathscr{E}$) subject to the constraints $\varphi^i(e) = 0$ for $i = 1, \ldots, m$ and $\phi^i(e) \leq 0$ for $i = 1, \ldots, \mu$.

To derive our multiplier rule, we shall suppose that the data of Problem (1) satisfy the following hypotheses:

2 HYPOTHESIS. *The set $\mathscr{E}$ is finitely open in itself* [*see* (I.1.39)].

3 HYPOTHESIS. *The functions $\varphi^1, \ldots, \varphi^m$ are finitely differentiable* [*see* (I.7.4)], and the *functions $\phi^0, \phi^1, \ldots, \phi^\mu$ are finitely semidifferentiable* [*see* (I.7.13, 16, *and* 19)].

Our multiplier rule has the following form.

4 THEOREM. *Let e_0 be a solution of Problem* (1), *and suppose that Hypotheses* (2) *and* (3) *hold. Then, for every*† *collection* ($i = 0, \ldots, \mu$) *of finite semidifferentials $D\phi^i(e_0;\cdot)$ of the ϕ^i at e_0, there exist vectors $\alpha = (\alpha^1, \ldots, \alpha^m) \in R^m$ and $\beta = (\beta^0, \beta^1, \ldots, \beta^\mu) \in \bar{R}_-^{\mu+1}$* [*see* (I.1.20)], *not both zero, such that*

5 $$\sum_{i=1}^{m} \alpha^i D\varphi^i(e_0;y) + \sum_{i=0}^{\mu} \beta^i D\phi^i(e_0;y) \leq 0 \quad \text{for all} \quad y \in \operatorname{cone}(\mathscr{E} - e_0),$$

6 $$\sum_{i=1}^{\mu} \beta^i \phi^i(e_0) = 0,$$

where, for each $i = 1, \ldots, m$, $D\varphi^i(e_0;\cdot)$ denotes the finite differential of φ^i at e_0.

The principal content of Theorem (4) lies in the inequality (5) which involves the "multipliers" $\alpha^1, \ldots, \alpha^m, \beta^0, \beta^1, \ldots, \beta^\mu$. Note that (6)—because β^i, as well as $\phi^i(e_0)$, is nonpositive for each $i = 1, \ldots, \mu$—is equivalent to the statement $\beta^i\phi^i(e_0) = 0$ for each $i = 1, \ldots, \mu$.

Proof of Theorem (4). Let e_0 be a solution of Problem (1), let $D\varphi^i(e_0;\cdot)$ denote the finite differential of φ^i at e_0, and let $D\phi^i(e_0;\cdot)$ be a finite semidifferential of ϕ^i at e_0 (for each i).

† Recall that a finite semidifferential is generally not unique [see (I.7.14)].

For each i, we shall denote $D\varphi^i(e_0;\cdot)$ and $D\phi^i(e_0;\cdot)$ simply by $D\varphi^i(\cdot)$ and $D\phi^i(\cdot)$, respectively. Further, let $\varphi = (\varphi^1, \ldots, \varphi^m)$, let $D\varphi = (D\varphi^1, \ldots, D\varphi^m)$, and let

$$\mathscr{J} = \{e: e \in \mathscr{E}, \phi^i(e) < 0 \text{ for } i = 1, \ldots, \mu, \phi^0(e) < \phi^0(e_0)\},$$

7 $$\mathscr{K} = \{y: y \in \text{cone}(\mathscr{E} - e_0), \phi^i(e_0) + D\phi^i(y) < 0 \text{ for } i = 1, \ldots, \mu, D\phi^0(y) < 0\}.$$

It follows at once that $0 \notin \varphi(\mathscr{J})$. Further, it is a direct consequence of the convexity of the functions $D\phi^0, \ldots, D\phi^\mu$ and the convexity of the set cone $(\mathscr{E} - e_0)$ [see Lemma (I.1.41)] that $\mathscr{K}$ is a convex set in $\mathscr{Y}$ (if it is not empty). Loosely speaking, $(\mathscr{K} + e_0)$ may be viewed as a "first-order" convex approximation to $\mathscr{J}$ near e_0. More precisely, we shall show that $D\varphi(\mathscr{K})$ is a simplicial linearization [see (I.5.39)] of $\varphi(\mathscr{J})$. Note that, since $D\varphi$ is linear, $D\varphi(\mathscr{K})$ is convex, by (I.2.18).

Thus, let S be an arbitrary simplex in R^m such that $0 \in S \subset D\varphi(\mathscr{K})$, and let $\eta \in R_+$ be arbitrary. Let the vertices of S be $D\varphi(y_1), \ldots, D\varphi(y_k)$, where $\{y_1, \ldots, y_k\} \subset \mathscr{K}$, and let $\mathscr{S}$ denote the simplex in $\mathscr{Y}$ whose vertices are $y_1, \ldots, y_k$. [Note that $y_1, \ldots, y_k$ are in general position (see (I.1.22)) because $D\varphi(y_1), \ldots, D\varphi(y_k)$ are (see (I.1.31)).] Since $D\varphi$ is linear, $D\varphi(\mathscr{S}) = S$. Also, since $\mathscr{K}$ is convex, $\mathscr{S} \subset \mathscr{K} \subset$ cone $(\mathscr{E} - e_0)$. For each $j = 1, \ldots, k$, let $D\phi^0(y_j) = \eta_j^0$ and let $\phi^i(e_0) + D\phi^i(y_j) = \eta_j^i$ for each $i = 1, \ldots, \mu$, so that $\eta_j^i < 0$ for all $i = 0, 1, \ldots, \mu$ and each $j = 1, \ldots, k$. Let $\tilde{\eta} = \max\{\eta_j^i: i = 0, 1, \ldots, \mu, j = 1, \ldots, k\}$, so that $\tilde{\eta} < 0$. It follows at once from the convexity of the functions $D\phi^0, \ldots, D\phi^\mu$ that

$$\phi^i(e_0) + D\phi^i(y) \le \tilde{\eta} < 0 \quad \text{for} \quad i = 1, \ldots, \mu \quad \text{and}$$
$$D\phi^0(y) \le \tilde{\eta} < 0 \quad \text{for all} \quad y \in \mathscr{S}.$$

Further, inasmuch as

8 $$\phi^i(e_0) \le 0 \quad \text{for} \quad i = 1, \ldots, \mu,$$

this means that

9 $$\phi^i(e_0) + \varepsilon D\phi^i(y) \le \varepsilon\tilde{\eta} < 0 \quad \text{for} \quad i = 1, \ldots, \mu$$
$$\text{and} \quad \varepsilon D\phi^0(y) \le \varepsilon\tilde{\eta} < 0 \quad \text{for all} \quad y \in \mathscr{S} \quad \text{and all} \quad \varepsilon \in (0,1).$$

Let us show that, for some $\varepsilon_0 > 0$, $(e_0 + \varepsilon\mathscr{S}) \subset \mathscr{J}$ whenever $0 < \varepsilon < \varepsilon_0$. Since $\mathscr{S} \subset \mathscr{K} \subset \text{cone}\,(\mathscr{E} - e_0)$, and $\mathscr{E}$ is finitely open in itself, there is a number $\varepsilon_1 \in (0,1)$ such that $(e_0 + \varepsilon\mathscr{S}) \subset \mathscr{E}$ for all $\varepsilon \in (0,\varepsilon_1)$. Further [see (I.7.16)], there is a number $\varepsilon_0 \in (0,\varepsilon_1)$ such that

$$\frac{\phi^i(e_0 + \varepsilon y) - \phi^i(e_0)}{\varepsilon} \leq D\phi^i(y) - \frac{\tilde{\eta}}{2}$$

$$\text{for all} \quad y \in \mathscr{S} \quad \text{and each} \quad i = 0, 1, \ldots, \mu$$

whenever $0 < \varepsilon < \varepsilon_0 < 1$. But this, by virtue of (9), directly implies that $(e_0 + \varepsilon\mathscr{S}) \subset \mathscr{J}$ whenever $0 < \varepsilon < \varepsilon_0$.

Further, since $\varphi(e_0) = 0$, there is a number $\tilde{\varepsilon} \in (0,\varepsilon_0)$ such that

10 $$\left|\frac{\varphi(e_0 + \tilde{\varepsilon}y)}{\tilde{\varepsilon}} - D\varphi(y)\right| < \eta \quad \text{for all} \quad y \in \mathscr{S}.$$

For each $\xi \in S$, let $y(\xi)$ denote the point in $\mathscr{S}$ whose barycentric coordinates [see (I.1.33)] coincide with those of ξ (in S). Evidently,

11 $$D\varphi(y(\xi)) = \xi \quad \text{for all} \quad \xi \in S.$$

Let us define the mapping $\chi_0 : S \to R^m$ and the linear operator $\chi_1 : R^m \to R^m$ as follows:

$$\chi_0(\xi) = \frac{\varphi(e_0 + \tilde{\varepsilon}y(\xi))}{\tilde{\varepsilon}}, \quad \chi_1(\xi) = \tilde{\varepsilon}\xi.$$

Since φ is finitely continuous [see Theorem (I.7.10)], we easily conclude, using Theorem (I.4.21), that χ_0 is continuous, and it follows from (10) and (11) that $|\chi_0(\xi) - \xi| < \eta$ for all $\xi \in S$. Also,

$$\chi_1 \circ \chi_0(S) = \varphi(e_0 + \tilde{\varepsilon}\mathscr{S}) \subset \varphi(\mathscr{J}),$$

which means that $D\varphi(\mathscr{K})$ is indeed a simplicial linearization of $\varphi(\mathscr{J})$.

Since $0 \notin \varphi(\mathscr{J})$, Lemma (I.5.43) now implies that there is a nonzero vector $\tilde{\alpha} \in R^m$ such that $\tilde{\alpha} \cdot D\varphi(y) \leq 0$ for all $y \in \mathscr{K}$. Consequently, the open convex cone $R_-^{\mu+2}$ [see (I.1.19)] in $R^{\mu+2}$ does not meet the following subset B of $R^{\mu+2}$:

$$B = \{(\gamma^0, \gamma^1, \ldots, \gamma^\mu, -\tilde{\alpha} \cdot D\varphi(y)) : y \in \text{cone}\,(\mathscr{E} - e_0),$$

$$\phi^i(e_0) + D\phi^i(y) \leq \gamma^i \quad \text{for} \quad i = 1, \ldots, \mu, D\phi^0(y) \leq \gamma^0\}.$$

It follows at once from the linearity of $D\varphi$, the convexity of the functions $D\phi^i$ and of the set cone $(\mathscr{E} - e_0)$ [see Lemma (I.1.41)] that B is a convex set. Also, because $D\phi^i(0) = 0$ for each i [see (I.7.16)], and since (8) holds, the point $(0,\phi^1(e_0),\ldots,\phi^\mu(e_0),0) \in B \cap \bar{R}_-^{\mu+2}$. Hence, by Corollary (I.5.18) [also see (I.5.10)], there is a nonzero vector $\hat{\beta} = (\hat{\beta}^0, \hat{\beta}^1, \ldots, \hat{\beta}^{\mu+1}) \in R^{\mu+2}$ such that

12 $$\hat{\beta}\cdot\zeta_1 \leq \sum_{i=1}^{\mu} \hat{\beta}^i\phi^i(e_0) \leq \hat{\beta}\cdot\zeta_2 \quad \text{for all} \quad \zeta_1 \in \bar{R}_-^{\mu+2} \quad \text{and} \quad \zeta_2 \in B.$$

But this implies, by virtue of (8), that $\hat{\beta}^i \geq 0$ for all $i = 0, \ldots, \mu + 1$, and that $\hat{\beta}^i\phi^i(e_0) = 0$ for $i = 1, \ldots, \mu$. Further, (12) and the definition of B imply that

$$\sum_{i=0}^{\mu} \hat{\beta}^i D\phi^i(y) - \hat{\beta}^{\mu+1}\tilde{\alpha}\cdot D\varphi(y) \geq 0 \quad \text{for all} \quad y \in \text{cone}\,(\mathscr{E} - e_0).$$

If we set $\beta = -(\hat{\beta}^0, \ldots, \hat{\beta}^\mu)$ and $\alpha = \hat{\beta}^{\mu+1}\tilde{\alpha}$, so that $|\alpha| + |\beta| > 0$, we immediately conclude that $\beta \in \bar{R}_-^{\mu+1}$ and that (5) and (6) hold. |||

13 Note that if the $D\phi^i(e_0;\cdot)$ for $i = 0, \ldots, \mu$ are linear [as must be the case if the ϕ^i are finitely differentiable and (for each i) $D\phi^i(e_0;\cdot)$ is the finite differentiable of ϕ^i at e_0—see (I.7.4)], then (5) implies that

14 $$\sum_{i=1}^{m} \alpha^i D\varphi^i(e_0;e) + \sum_{i=0}^{\mu} \beta^i D\phi^i(e_0;e) \leq \sum_{i=1}^{m} \alpha^i D\varphi^i(e_0;e_0) + \sum_{i=0}^{\mu} \beta^i D\phi^i(e_0;e_0) \quad \text{for all} \quad e \in \mathscr{E}.$$

If also e_0 is an internal point [see (I.1.42)] of $\mathscr{E}$, then (14) implies, by virtue of Lemma (I.2.39), that

$$\sum_{i=1}^{m} \alpha^i D\varphi^i(e_0;\cdot) + \sum_{i=0}^{\mu} \beta^i D\phi^i(e_0;\cdot) = 0.$$

15 If the inequality constraints $\phi^i(e) \leq 0$ for $i = 1, \ldots, \mu$ are deleted in Problem (1), and Hypothesis (3) is correspondingly "thinned out," then the multiplier rule of Theorem (4) remains in force, with obvious modifications [including the omission of Eq. (6)]. Indeed, it is only necessary to make some obvious omissions in the proof of Theorem (4) presented previously.

A similar conclusion holds if the equality constraints $\varphi^i(e) = 0$ for $i = 1, \ldots, m$ are deleted in Problem (1). Since the proof of

Theorem (4) in this case must be somewhat modified, let us dwell on this case in a bit more detail. To be precise, we consider the following problem.

16 *Problem.* Given a set $\mathscr{E}$ in a linear vector space $\mathscr{Y}$ and real-valued functions $\phi^0, \phi^1, \ldots, \phi^\mu$ defined on $\mathscr{E}$, find an element $e_0 \in \mathscr{E}$ that achieves a minimum for ϕ^0 (on $\mathscr{E}$) subject to the constraints $\phi^i(e) \leq 0$ for $i = 1, \ldots, \mu$.

In this case, Hypothesis (3) is replaced by the following one.

17 HYPOTHESIS. *The functions $\phi^0, \phi^1, \ldots, \phi^\mu$ are Gâteaux semi-differentiable* [*see* (I.7.13, 16, *and* 19)].

Note the weaker differentiability requirement on the ϕ^i in Hypothesis (17) in comparison to Hypothesis (3) (Gâteaux vs. finite semdifferentiability).

Our multiplier rule then takes the following form.

18 THEOREM. *Let e_0 be a solution of Problem* (16) *and suppose that Hypotheses* (2) *and* (17) *hold. Then, for every collection* ($i = 0, \ldots, \mu$) *of Gâteau semidifferentials $D\phi^i(e_0;\cdot)$ of the ϕ^i at e_0, there is a nonzero vector $\beta = (\beta^0, \beta^1, \ldots, \beta^\mu) \in \bar{R}_-^{\mu+1}$ such that*

19 $$\sum_{i=0}^{\mu} \beta^i D\phi^i(e_0;y) \leq 0 \quad \textit{for all} \quad y \in \operatorname{cone}(\mathscr{E} - e_0),$$

20 $$\sum_{i=1}^{\mu} \beta^i \phi^i(e_0) = 0.$$

Proof. Let e_0 be a solution of Problem (16), and let (for each $i = 0, \ldots, \mu$) $D\phi^i(e_0;\cdot)$—which we shall simply denote by $D\phi^i(\cdot)$—be a Gâteaux semidifferential of ϕ^i at e_0. Further, let $\mathscr{K}$ be defined as in the proof of Theorem (4) [see (7)]. We shall show that $\mathscr{K}$ is empty.

Suppose the contrary; i.e., let $\tilde{y} \in \mathscr{K}$. Let $\eta = \max\{\phi^1(e_0) + D\phi^1(\tilde{y}), \ldots, \phi^\mu(e_0) + D\phi^\mu(\tilde{y}), D\phi^0(\tilde{y})\}$, so that $\eta < 0$. By (I.7.16), and because $\mathscr{E}$ is finitely open in itself, there is a number $\varepsilon_0 \in (0,1)$ such that

$$\frac{\phi^i(e_0 + \varepsilon_0 \tilde{y}) - \phi^i(e_0)}{\varepsilon_0} \leq D\phi^i(\tilde{y}) - \frac{\eta}{2} \quad \text{for each} \quad i = 0, 1, \ldots, \mu,$$

$$e_0 + \varepsilon_0 \tilde{y} \in \mathscr{E}.$$

If we set $\tilde{e} = e_0 + \varepsilon_0 \tilde{y}$, so that $\tilde{e} \in \mathscr{E}$, we at once conclude—since (8) and (9) with $y = \tilde{y}$ hold here too—that $\phi^i(\tilde{e}) < 0$ for $i = 1, \ldots, \mu$

and that $\phi^0(\tilde{e}) < \phi^0(e_0)$, contradicting the hypothesis that e_0 is a solution of Problem (16).

Hence, $\mathscr{K}$ is empty, which means that the set

$$B = \{(D\phi^0(y) + \gamma^0, \phi^1(e_0) + D\phi^1(y) + \gamma^1, \ldots, \phi^\mu(e_0) + D\phi^\mu(y) + \gamma^\mu) : y \in \operatorname{cone}(\mathscr{E} - e_0), \gamma^i \geq 0 \quad \text{for} \quad i = 0, \ldots, \mu\}$$

in $R^{\mu+1}$ does not meet the open convex cone $R^{\mu+1}_-$. It follows from the convexity of the functions $D\phi^i$ and of the set cone $(\mathscr{E} - e_0)$ [see Lemma (I.1.41)] that B is a convex set. Also, since $D\phi^i(0) = 0$ for each i [see (I.7.16)] and (8) holds, $(0, \phi^1(e_0), \ldots, \phi^\mu(e_0)) \in B \cap \bar{R}^{\mu+1}_-$. Hence, by Corollary (I.5.18) [also see (I.5.10)], there is a nonzero vector $\beta = (\beta^0, \beta^1, \ldots, \beta^\mu) \in R^{\mu+1}$ such that

$$\beta \cdot \zeta_1 \leq \sum_{i=1}^{\mu} \beta^i \phi^i(e_0) \leq \beta \cdot \zeta_2 \quad \text{for all} \quad \zeta_1 \in B \quad \text{and} \quad \zeta_2 \in \bar{R}^{\mu+1}_-.$$

But this is easily seen to imply that $\beta \in \bar{R}^{\mu+1}_-$ and that (19) and (20) hold. |||

The observations of (13) have obvious analogs here.

2. Optimization Problems with Operator Equation Constraints

In this section we shall apply the multiplier rule of Section 1 to obtain necessary conditions in the form of a maximum principle for optimization problems defined with the aid of operator equations in Banach spaces, and we shall specialize the general maximum principle to obtain a special maximum principle for optimal control problems containing constraints in the form of ordinary differential equations.

We first consider the following problem.

1 *Problem.* Let $\mathscr{X}$ be a Banach space [see (I.4.35)], let A be an open set in $\mathscr{X}$, let $\mathscr{W}$ be a given set of continuous operators from A into $\mathscr{X}$, and let $\hat{\varphi}^1, \ldots, \hat{\varphi}^m, \hat{\phi}^0, \hat{\phi}^1, \ldots, \hat{\phi}^\mu$ be given functionals defined on A. Then we wish to find a pair $(x_0, T_0) \in A \times \mathscr{W}$ such that

$$x_0 = T_0 x_0,$$

$$\hat{\varphi}^i(x_0) = 0 \quad \text{for} \quad i = 1, \ldots, m,$$

$$\hat{\phi}^i(x_0) \leq 0 \quad \text{for} \quad i = 1, \ldots, \mu,$$

and such that $\hat{\phi}^0(x_0) \leq \hat{\phi}^0(x)$ for all $x \in A$ that satisfy the relations

$$\text{(a)}\ \hat{\varphi}^i(x) = 0 \quad \text{for} \quad i = 1, \ldots, m,$$

$$\text{(b)}\ \hat{\phi}^i(x) \leq 0 \quad \text{for} \quad i = 1, \ldots, \mu,$$

$$\text{(c)}\ x = Tx \quad \text{for some} \quad T \in \mathscr{W}.$$

Problem (1) includes, as special cases, a large number of so-called optimal control problems, of which the most celebrated is perhaps the following one.

2 *Problem.* Let there be given a compact interval $I = [t_1,t_2]$, an open set $\mathbf{G}$ in R^n, arbitrary sets V and U in R^r such that $V \supset \bar{U}$, a continuous function $f:\mathbf{G} \times V \times I \to R^n$, a set $\mathscr{U}$ of functions $u:I \to U$ all of whose members are measurable and essentially bounded, and functions $\chi^0, \ldots, \chi^{\mu+m}$ from $\mathbf{G} \times \mathbf{G}$ into R. Then we wish to find an absolutely continuous function $x_0:I \to \mathbf{G}$ which satisfies the relations

3 $$\dot{x}_0(t) = f(x_0(t), u_0(t), t) \quad \text{for almost all} \quad t \in I \quad \text{and some} \quad u_0 \in \mathscr{U},$$

4 $$\chi^i(x_0(t_1),x_0(t_2)) \leq 0 \quad \text{for} \quad i = 1, \ldots, \mu,$$

5 $$\chi^i(x_0(t_1),x_0(t_2)) = 0 \quad \text{for} \quad i = \mu + 1, \ldots, \mu + m,$$

and such that $\chi^0(x_0(t_1),x_0(t_2)) \leq \chi^0(x(t_1),x(t_2))$ for all absolutely continuous functions $x:I \to \mathbf{G}$ which satisfy

$$\text{(a)}\ \dot{x}(t) = f(x(t), u(t), t) \quad \text{a.e. on } I \quad \text{for some} \quad u \in \mathscr{U},$$

$$\text{(b)}\ \chi^i(x(t_1),x(t_2)) \leq 0 \quad \text{for} \quad i = 1, \ldots, \mu,$$

$$\text{(c)}\ \chi^i(x(t_1),x(t_2)) = 0 \quad \text{for} \quad i = \mu + 1, \ldots, \mu + m.$$

[In the preceding, we have written $\dot{x}(t)$ to mean the ordinary derivative of x at t, which, had we used the notation of Section 7 of Chapter I, would instead have been written as $\mathscr{D}x(t)$. Conforming with the usual notation, we shall throughout this book write $\dot{x}(t)$ instead of $\mathscr{D}x(t)$ in this situation.]

In different terms, Problem (2) consists of finding an optimal "control" function u_0 from the class $\mathscr{U}$ of "admissible controls" such that there is a corresponding "phase" trajectory x_0 [corresponding in the sense that the ordinary differential equation (3) is satisfied] which satisfies given constraints on the initial and final values [i.e., satisfies relations (4) and (5)] and, in so doing, achieves a minimum for the function χ^0 on the initial and final values of the trajectory.

6 We can identify Problem (2) as a special case of Problem (1) if we identify $\mathscr{X}$ with $\mathscr{C}^n(I)$ [see (I.6.5)], A with the set of all functions in $\mathscr{X}$ whose range is contained in $\mathbf{G}$, the functionals $\hat{\varphi}^i$ $(i = 1, \ldots, m)$ with the functions $x \to \chi^{\mu+i}(x(t_1),x(t_2))$, the functionals $\hat{\phi}^i$ $(i = 0, \ldots, \mu)$ with the functions $x \to \chi^i(x(t_1),x(t_2))$, and $\mathscr{W}$ with the set of all integral operators $T: A \to \mathscr{X}$ which have the form

7 $$(Tx)(t) = \xi + \int_{t_1}^{t} f(x(s),u(s),s)\, ds, \quad t \in I,$$

for some $\xi \in R^n$ and $u \in \mathscr{U}$. [It is easily seen that every operator of the form of (7) is continuous.]

Let us now return to Problem (1) and apply the multiplier rule developed in Section 1 to obtain necessary conditions that solutions of this problem must satisfy.

In order to assure that Hypotheses (2) and (3), respectively, are satisfied, we must impose some requirements on the set $\mathscr{W}$ and on the functions $\hat{\varphi}^i$, $\hat{\phi}^i$. Thus, we shall make the following hypotheses:

8 HYPOTHESIS. *Each $T \in \mathscr{W}$ has at most one fixed point [see (I.5.37)].*

9 HYPOTHESIS. *The set $\mathscr{W}$ is convex, every $T \in \mathscr{W}$ is continuously differentiable [see (I.7.48)], and, for each $x \in A$ and $T \in \mathscr{W}$, $[E - DT(x;\cdot)]$—where E denotes the identity operator on $\mathscr{X}$ [see (I.2.9)]—is a (linear) homeomorphism [see (I.3.28)] of $\mathscr{X}$ onto itself.*

10 HYPOTHESIS. *The functions $\hat{\varphi}^1, \ldots, \hat{\varphi}^m$, and $\hat{\phi}^0, \ldots, \hat{\phi}^\mu$ are dually differentiable [see (I.7.23)].*

11 To set up a correspondence between Problem (1) and Problem (1.1), we let $\mathscr{Y}$ denote the linear vector space of all continuously differentiable operators from A into $\mathscr{X}$, and denote by $\mathscr{Y}_1$ the set of all $T \in \mathscr{Y}$ which have exactly one fixed point. Let $\tilde{\varphi}_0$ denote the function from $\mathscr{Y}_1$ into A which assigns to each T its fixed point. Now, under the supposition that Hypothesis (9) is satisfied, if we denote $\mathscr{W} \cap \mathscr{Y}_1$ by $\mathscr{E}$, denote the restriction of $\tilde{\varphi}_0$ to $\mathscr{E}$ by φ_0, and set $\varphi^i = \hat{\varphi}^i \circ \varphi_0$ for $i = 1, \ldots, m$ and $\phi^i = \hat{\phi}^i \circ \varphi_0$ for $i = 0, \ldots, \mu$, then Problem (1) clearly may be identified with Problem (1.1). Let us further show that if Hypotheses (8)–(10) are satisfied, then Hypotheses (1.2 and 3), hold.

To verify that Hypothesis (1.2) holds, we shall make use of the implicit function theorem proved in Chapter I [Theorem (I.7.55)].

12 Thus, let $T_0 \in \mathscr{E} = \mathscr{W} \cap \mathscr{Y}_1$ be arbitrary, and let $\{T_1, \ldots, T_\rho\}$ be an arbitrary finite subset of $\mathscr{W}$. Let us denote $\varphi_0(T_0)$ by x_0. Define the function $F: R^\rho \times A \to \mathscr{X}$ as follows:

$$F(\lambda,x) = F(\lambda^1, \ldots, \lambda^\rho, x) = x - T_0 x - \sum_{i=1}^{\rho} \lambda^i (T_i x - T_0 x).$$

Evidently, $F(0,x_0) = x_0 - T_0 x_0 = 0$, and, by Hypothesis (9), $[E - DT_0(x_0;\cdot)] = D_2 F(0,x_0;\cdot)$ is a homeomorphism of $\mathscr{X}$ onto itself. Further, F is obviously continuously differentiable. Hence, by the implicit function Theorem (I.7.55), there are a number $\varepsilon_1 > 0$ and a continuously differentiable function q: $\{\lambda: \lambda \in R^\rho, |\lambda| < \varepsilon_1\} \to A$ such that (i) $q(0) = x_0$, (ii) $F(\lambda, q(\lambda)) = 0$ whenever $|\lambda| < \varepsilon_1$, and (iii)

$$Dq(0;\lambda) = [E - DT_0(x_0;\cdot)]^{-1} \circ \left[\sum_{i=1}^{\rho} \lambda^i (T_i x_0 - T_0 x_0) \right]$$

$$\text{for all} \quad \lambda \in R^\rho.$$

Since $\mathscr{W}$ is convex [by Hypothesis (9)], this at once implies that $\mathscr{E}$ is finitely open in itself, that $(\mathscr{W} - T_0) \subset \text{cone}\,(\mathscr{E} - T_0)$, and that φ_0 is finitely differentiable at T_0 with

$$D\varphi_0(T_0;T) = [E - DT_0(x_0;\cdot)]^{-1} T x_0 \quad \text{for all} \quad T \in \mathscr{Y}.$$

Inasmuch as $T_0 \in \mathscr{E}$ was arbitrary, we have shown that φ_0 is finitely differentiable.

13 It now follows from Hypothesis (10) and Theorem (I.7.34) that, for each i, the functions $\varphi^i = \hat{\varphi}^i \circ \varphi_0$ and $\phi^i = \hat{\phi}^i \circ \varphi_0$ are finitely differentiable with finite differentials given by the formulas

$$D\varphi^i(T_0;T) = D\hat{\varphi}^i(x_0;[E - DT_0(x_0;\cdot)]^{-1} \circ T x_0), \quad i = 1, \ldots, m,$$

$$D\phi^i(T_0;T) = D\hat{\phi}^i(x_0;[E - DT_0(x_0;\cdot)]^{-1} \circ T x_0), \quad i = 0, \ldots, \mu,$$

for all $T_0 \in \mathscr{E}$ and $T \in \mathscr{Y}$, where $x_0 = \varphi_0(T_0)$ and $D\varphi^i$, etc., have the obvious meanings.

Consequently, Hypotheses (1.2 and 3) are satisfied if Hypotheses (8)–(10) are, and we can appeal to Theorem (1.4) [also see (1.13)] to obtain the following result.

14 THEOREM. *Let* (x_0,T_0) *be a solution of Problem* (1) *where Hypotheses* (8)–(10) *are satisfied. Then, there exist vectors* $\alpha = (\alpha^1,\ldots,\alpha^m) \in R^m$ *and* $\beta = (\beta^0,\beta^1,\ldots,\beta^\mu) \in \bar{R}_-^{\mu+1}$, *not both zero, such that*

15 $$\left(\sum_{i=1}^{m} \alpha^i D\hat{\varphi}^i(x_0;\cdot) + \sum_{i=0}^{\mu} \beta^i D\hat{\phi}^i(x_0;\cdot)\right) \circ [E - DT_0(x_0;\cdot)]^{-1}(Tx_0)$$
$$\leq \left(\sum_{i=1}^{m} \alpha^i D\hat{\varphi}^i(x_0;\cdot) + \sum_{i=0}^{\mu} \beta^i D\hat{\phi}^i(x_0;\cdot)\right) \circ [E - DT_0(x_0;\cdot)]^{-1}(x_0)$$
$$\text{for all} \quad T \in \mathscr{W},$$

16 $$\sum_{i=1}^{\mu} \beta^i \hat{\phi}^i(x_0) = 0,$$

where, for each $i = 1,\ldots,m$, $D\hat{\varphi}^i(x_0;\cdot)$ *denotes the dual differential of* $\hat{\varphi}^i$ *at* x_0, *and similarly for* $D\hat{\phi}^i(x_0;\cdot)$.

17 If the inequality constraints $\hat{\phi}^i(x) \leq 0$ for $i = 1,\ldots,\mu$, or the equality constraints $\hat{\varphi}^i(x) = 0$ for $i = 1,\ldots,m$, in Problem (1) are deleted, and if Hypothesis (10) is correspondingly simplified, then Theorem (14) remains in force with some obvious modifications [see (1.15) and Theorem (1.18)].

18 Inasmuch as the main content of Theorem (14) is the inequality (15) which asserts that a certain linear functional on $\mathscr{W}$ attains its maximum at T_0, we shall refer to Theorem (14) as a *maximum principle.* As we shall see in Chapter IV, this theorem remains in force if we considerably relax the convexity requirement on $\mathscr{W}$ in Hypothesis (9), provided that we somewhat strengthen Hypothesis (10).

Let us specialize Theorem (14) to Problem (2). In order that Hypotheses (8)–(10) hold, we must impose some conditions on the data of this problem. Thus, we make the following hypotheses.

19 HYPOTHESIS. *The function* $f:\mathbf{G} \times V \times I \to R^n$ *is continuously differentiable with respect to its first argument* [*see* (I.7.52)]. *Further, for each pair* u_1, u_2 *in* $\mathscr{U}$ *and every* $\lambda \in [0,1]$, *there is a function* $u_3 \in \mathscr{U}$ (*possibly depending on* u_1, u_2, *and* λ) *such that*

$$\lambda f(\xi,u_1(t),t) + (1-\lambda)f(\xi,u_2(t),t) = f(\xi,u_3(t),t)$$
$$\text{for all} \quad \xi \in \mathbf{G} \quad \text{and} \quad t \in I.$$

20 HYPOTHESIS. *The functions* $\chi^i:\mathbf{G} \times \mathbf{G} \to R$, *for each* $i = 0,\ldots,\mu + m$, *are continuously differentiable* [*see* (I.7.48)].

Let us show that Hypotheses (8)–(10) are satisfied whenever Hypotheses (19) and (20) are, if we make the identification between Problems (1) and (2) indicated in (6). First, it follows directly from Theorems (I.7.41 and 54) and Lemma (I.7.40) that [under Hypothesis (20)] the functions $\hat{\varphi}^i$ and $\hat{\phi}^i$ (for each i) are Fréchet differentiable with [see (I.7.60) for an explanation of the notation]

$$D\hat{\phi}^i(x;\delta x) = \sum_{j=1}^{2} \mathscr{D}_j\chi^i(x(t_1),x(t_2))\delta x(t_j), \quad i = 0, \ldots, \mu,$$

$$D\hat{\varphi}^i(x;\delta x) = \sum_{j=1}^{2} \mathscr{D}_j\chi^{\mu+i}(x(t_1),x(t_2))\delta x(t_j), \quad i = 1, \ldots, m,$$

for all $x \in A$ and $\delta x \in \mathscr{C}^n$, so that, by Theorem (I.7.44), Hypothesis (10) holds. Under Hypothesis (19), Hypothesis (8) holds by Theorem (A.31) of the Appendix, and it follows from Theorem (A.5) of the Appendix that the operator T defined by (7)—with $u \in \mathscr{U}$—is continuously differentiable with

$$(DT(x;\delta x))(t) = \int_{t_1}^{t} \mathscr{D}_1 f(x(s),u(s),s)\delta x(s)\, ds, \quad t \in I,$$

for all $x \in A$ and $\delta x \in \mathscr{C}^n$. It follows at once from Hypothesis (19) that $\mathscr{W}$ is convex, and it is a consequence of Corollary (A.28) and Theorem (A.5) in the Appendix [also see (IV.3.3)] that $[E - DT(x;\cdot)]$ is a homeomorphism of $\mathscr{C}^n$ onto itself for all $T \in \mathscr{W}$ and $x \in A$, so that also Hypothesis (9) holds.

Consequently, we can appeal to Theorem (14) and conclude that if x_0 is a solution of Problem (2) satisfying Eq. (3), and if Hypotheses (19) and (20) hold, then there is a nonzero vector $\alpha = (\alpha^0, \alpha^1, \ldots, \alpha^{\mu+m}) \in R^{\mu+m+1}$ such that

21 $$\sum_{i=0}^{\mu+m} \sum_{j=1}^{2} \alpha^i \mathscr{D}_j\chi^i(x_0(t_1),x_0(t_2))\delta x(t_j) \leq 0$$

for all functions $\delta x \in \mathscr{C}^n$ that satisfy an equation of the form

22 $$\delta x(t) = \xi + \int_{t_1}^{t} \mathscr{D}_1 f(x_0(s),u_0(s),s)\delta x(s)\, ds + \int_{t_1}^{t} [f(x_0(s),u(s),s) - f(x_0(s),u_0(s),s)]\, ds, \quad t \in I,$$

for some $u \in \mathscr{U}$ and some $\xi \in R^n$, and such that

23 $$\alpha^i \leq 0 \quad \text{for} \quad i = 0, \ldots, \mu,$$

24 $$\sum_{i=1}^{\mu} \alpha^i\chi^i(x_0(t_1),x_0(t_2)) = 0.$$

Let us introduce the following notations:

$$\mathscr{D}_j^0\chi^i = \mathscr{D}_j\chi^i(x_0(t_1),x_0(t_2))$$

$$\text{for} \quad i = 0, \ldots, \mu + m \quad \text{and} \quad j = 1 \text{ or } 2,$$

$$\mathscr{D}^0 f(t) = \mathscr{D}_1 f(x_0(t),u_0(t),t) \quad \text{for all} \quad t \in I,$$

$$\delta f_u(t) = f(x_0(t),u(t),t) - f(x_0(t),u_0(t),t)$$

$$\text{for all} \quad u \in \mathscr{U} \quad \text{and} \quad t \in I.$$

Equation (22) is obviously equivalent to the linear inhomogeneous ordinary differential equation

25 $$\delta\dot{x}(t) = \mathscr{D}^0 f(t)\delta x(t) + \delta f_u(t) \quad \text{a.e. on } I,$$

with initial value

26 $$\delta x(t_1) = \xi.$$

Using the variation of parameters formula [see Theorem (A.51) in the Appendix], we can write the solution of (25) and (26) in the form

27 $$\delta x(t) = \Phi(t)\xi + \Phi(t)\int_{t_1}^{t} \Phi^{-1}(s)\delta f_u(s)\,ds, \quad t \in I,$$

where Φ is the absolutely containuous $(n \times n)$-matrix-valued function defined on I that satisfies

28 $$\dot{\Phi}(t) = \mathscr{D}^0 f(t)\Phi(t) \quad \text{a.e. on } I, \quad \Phi(t_1) = \text{the identity matrix}.$$

Because (21) holds for all $\delta x \in \mathscr{C}^n$ that satisfy (22) (for some $u \in \mathscr{U}$ and $\xi \in R^n$), we can now conclude that

29 $$\sum_{i=0}^{\mu+m} \alpha^i \left\{[\mathscr{D}_1^0\chi^i + \mathscr{D}_2^0\chi^i\Phi(t_2)]\xi + \mathscr{D}_2^0\chi^i\Phi(t_2)\int_{t_1}^{t_2} \Phi^{-1}(t)\delta f_u(t)\,dt\right\} \leq 0$$

$$\text{for all} \quad \xi \in R^n \quad \text{and} \quad u \in \mathscr{U}.$$

Since $u_0 \in \mathscr{U}$ and $\delta f_{u_0} = 0$, (29) implies, in particular, that

$$\sum_{i=0}^{\mu+m} \alpha^i[\mathscr{D}_1^0\chi^i + \mathscr{D}_2^0\chi^i\Phi(t_2)]\xi \leq 0 \quad \text{for all} \quad \xi \in R^n,$$

which is possible only if

30 $$\sum_{i=0}^{\mu+m} \alpha^i[\mathscr{D}_1^0\chi^i + \mathscr{D}_2^0\chi^i\Phi(t_2)] = 0.$$

Since (29) must also hold for $\xi = 0$, we further conclude that

31 $$\int_{t_1}^{t_2} \psi(t)f(x_0(t),u(t),t)\,dt \le \int_{t_1}^{t_2} \psi(t)f(x_0(t),u_0(t),t)\,dt \quad \text{for all} \quad u \in \mathscr{U},$$

where $\psi(\cdot)$ is the absolutely continuous n (row)-vector-valued function defined on I given by

32 $$\psi(t) = \sum_{i=0}^{\mu+m} \alpha^i \mathscr{D}_2^0 \chi^i \Phi(t_2)\Phi^{-1}(t).$$

But [see Theorem (A.47) in the Appendix]

$$\frac{d}{dt}[\Phi^{-1}(t)] = -\Phi^{-1}(t)\mathscr{D}^0 f(t) \quad \text{a.e. on } I,$$

so that, by (32),

33 $$\dot{\psi}(t) = -\psi(t)\mathscr{D}_1 f(x_0(t),u_0(t),t) \quad \text{a.e. on } I.$$

Also, (32) and (30) imply [inasmuch as $\Phi(t_1) = \Phi^{-1}(t_1) =$ the identity matrix] that

34 $$\psi(t_2) = \sum_{i=0}^{\mu+m} \alpha^i \mathscr{D}_2 \chi^i(x_0(t_1),x_0(t_2)),$$

35 $$\psi(t_1) = -\sum_{i=0}^{\mu+m} \alpha^i \mathscr{D}_1 \chi^i(x_0(t_1),x_0(t_2)).$$

Thus we have proved the following theorem.

36 THEOREM. *Let x_0 be a solution of Problem* (2) *satisfying Eq.* (3), *and suppose that Hypotheses* (19) *and* (20) *hold. Then there exists an absolutely continuous n (row)-vector-valued function ψ defined on I that satisfies the linear ordinary differential equation* (33), *the maximum principle* (31), *and the boundary conditions* (34) *and* (35), *where $\alpha = (\alpha^0, \alpha^1, \ldots, \alpha^{\mu+m})$ is some nonzero vector in $R^{\mu+m+1}$ that satisfies* (23) *and* (24).

37 The function ψ in the preceding theorem is commonly referred to as an *adjoint variable* because Eq. (33) is the adjoint of the homogeneous part of Eq. (25), which is itself a "linearization" of the basic equation, Eq. (3). The fundamental inequality (31) is referred to as a *maximum principle in integral form* because it is evidently equivalent to the relation

38 $$\int_{t_1}^{t_2} \psi(t)f(x_0(t),u_0(t),t)\,dt = \max_{u\in\mathscr{U}} \int_{t_1}^{t_2} \psi(t)f(x_0(t),u(t),t)\,dt.$$

Equations (34) and (35) are commonly referred to as transversality conditions because they generalize the transversality conditions of the classical calculus of variations.

The necessary conditions of Theorem (36) are clearly trivial if $\psi(t) \equiv 0$ on I, and it is, therefore, of interest to investigate conditions under which $\psi(t) \not\equiv 0$. By Corollary (A.49), either $\psi \equiv 0$ on I, or $\psi(t) \neq 0$ for all $t \in I$. Thus, if we can find sufficient conditions so that either $\psi(t_1) \neq 0$ or $\psi(t_2) \neq 0$, or, equivalently, so that $(\psi(t_1),\psi(t_2)) \neq 0$ (in R^{2n}), then these conditions are also sufficient that $\psi(t) \neq 0$ for all $t \in I$. But, by (34) and (35), $(\psi(t_1),\psi(t_2))$ cannot vanish if there is *no* nonzero vector $\tilde{\alpha} = (\tilde{\alpha}^0, \tilde{\alpha}^1, \ldots, \tilde{\alpha}^{\mu+m}) \in R^{\mu+m+1}$ satisfying (23), (24) (with $\alpha = \tilde{\alpha}$), and the equality

39 $$\sum_{i=0}^{\mu+m} \tilde{\alpha}^i(\mathscr{D}_1^0\chi^i,\mathscr{D}_2^0\chi^i) = 0 \quad (\text{in } R^{2n}).$$

With this in mind, let us impose the following condition.

40 CONDITION. *There is at least one vector* $(\xi_1,\xi_2) \in R^n \times R^n$ *such that*

$$\mathscr{D}_1^0\chi^i \cdot \xi_1 + \mathscr{D}_2^0\chi^i \cdot \xi_2 < 0 \quad \textit{for } i = 0$$

and each $i = 1, \ldots, \mu$ *such that* $\chi^i(x_0(t_1),x_0(t_2)) = 0$;

$$\mathscr{D}_1^0\chi^i \cdot \xi_1 + \mathscr{D}_2^0\chi^i \cdot \xi_2 = 0 \quad \textit{for} \quad i = \mu + 1, \ldots, \mu + m;$$

and the m vectors $(\mathscr{D}_1^0\chi^i,\mathscr{D}_2^0\chi^i)$, $i = \mu + 1, \ldots, \mu + m$, *are linearly independent in* $(R^n)^2$.

Condition (40) in essence asserts that Problem (2) is well-posed in the sense that the initial and final value constraints on the trajectory x are at x_0, to "first order," compatible with one another and with the functional being minimized.

It is easy to see that if Condition (40) holds, then there is no nonzero vector $\tilde{\alpha} = (\tilde{\alpha}^0, \tilde{\alpha}^1, \ldots, \tilde{\alpha}^{\mu+m}) \in R^{\mu+m+1}$ satisfying (23), (24) (with $\alpha = \tilde{\alpha}$), and (39), so that $\psi(t) \neq 0$ for all $t \in I$.

We point out that a sufficient condition for Condition (40) to hold is that the collection of vectors $(\mathscr{D}_1^0\chi^i,\mathscr{D}_2^0\chi^i)$ for those $i = 0, \ldots, \mu + m$ such that $\chi^i(x_0(t_1),x_0(t_2)) = 0$ is linearly independent.

Finally, we note that (38), or, equivalently, (31), take on a much simpler form if the set $\mathscr{U}$ consists of all essentially bounded, measurable functions $u: I \to U$. This case is particularly important in applications. Indeed, in this case, (38) is equivalent to the "pointwise" maximum principle

41 $$\psi(t)f(x_0(t),u_0(t),t) = \max_{v\in U} \psi(t)f(x_0(t),v,t) \quad \text{for almost all} \quad t \in I.$$

To show this, we first remark that (41) obviously implies (38). We prove the converse by means of a contradiction argument. Thus, suppose that the set

$$\hat{I} = \{t\colon t \in I,\ \psi(t)f(x_0(t),u_0(t),t) < \sup_{v\in U} \psi(t)f(x_0(t),v,t)\}$$

has positive measure. Let us denote the function

$$t \to \psi(t)f(x_0(t),u_0(t),t)\colon I \to R$$

by H_0, so that H_0 is measurable on I. Let $\hat{t} \in \hat{I}$ be a regular point for H_0, i.e., suppose that

42 $$\lim_{\text{meas } J\to 0} \frac{\text{meas }(H_0^{-1}(O) \cap J)}{\text{meas } J} = 1,$$

where J is an arbitrary subinterval of I that contains $\hat{t}$, for every neighborhood O of $H_0(\hat{t})$. Such a point $\hat{t}$ exists because almost every point of I is a regular point for H_0. (This is a property of measurable functions.) Since $\hat{t} \in \hat{I}$, there is a point $\hat{v} \in U$ such that

43 $$H_0(\hat{t}) = \psi(\hat{t})f(x_0(\hat{t}),u_0(\hat{t}),\hat{t}) < \psi(\hat{t})f(x_0(\hat{t}),\hat{v},\hat{t}).$$

Relations (42) and (43), together with the continuity of the functions ψ, f, and x_0, imply that the (measurable) set

$$I^* = \{t\colon t \in I,\ H_0(t) = \psi(t)f(x_0(t),u_0(t),t) < \psi(t)f(x_0(t),\hat{v},t)\}$$

has positive measure. Let us define $\hat{u} \in \mathscr{U}$ as follows:

$$\hat{u}(t) = \begin{cases} u_0(t) & \text{if} \quad t \in I,\ t \notin I^*, \\ \hat{v} & \text{if} \quad t \in I^*. \end{cases}$$

But then

$$\int_{t_2}^{t_1} \psi(t)f(x_0(t),u_0(t),t)\,dt < \int_{t_1}^{t_2} \psi(t)f(x_0(t),\hat{u}(t),t)\,dt,$$

which contradicts (38), and thereby proves that (under the stated hypothesis) (41) holds.

44 Relations (38) and (41) are sometimes written compactly with the aid of an auxiliary function—the so-called *Hamiltonian.* Namely, let H, the Hamiltonian function, be the function from

$R^n \times \mathbf{G} \times V \times I$ into R defined by

$$H(\eta,\xi,v,t) = \eta \cdot f(\xi,v,t).$$

Then (38) and (41) may, respectively, be rewritten in the forms

45 $$\int_{t_1}^{t_2} H(\psi(t),x_0(t),u_0(t),t)\,dt = \max_{u\in\mathscr{U}} \int_{t_1}^{t_2} H(\psi(t),x_0(t),u(t),t)\,dt,$$

46 $$H(\psi(t),x_0(t),u_0(t),t) = \max_{v\in U} H(\psi(t),x_0(t),v,t) \quad \text{for almost all} \quad t \in I.$$

47 Of the hypotheses that we made for Problem (1) [Hypotheses (8)–(10)], the one that is the most severe is the requirement [in Hypothesis (9)] that the set $\mathscr{W}$ is convex. Without this hypothesis, the set $\mathscr{E} = \mathscr{W} \cap \mathscr{Y}_1$ will not in general be finitely open in itself, and it is not possible to make use of the multiplier rule developed in Section 1 in order to obtain necessary conditions for our problem. Since the set $\mathscr{W}$ is *not* convex in many problems of interest, it is necessary to develop a multiplier rule without the hypothesis that $\mathscr{E}$ is finitely open in itself. As we discussed in the introduction to this chapter, this will be done in Chapter III.

48 On the other hand, theorems on the existence of solutions for Problem (1) are difficult to obtain unless the functions $\hat{\varphi}^i$ and $\hat{\phi}^i$ are continuous (which they fortunately almost always are), and unless the set $\mathscr{W}$ is convex. For many problems of interest, although $\mathscr{W}$ is not convex, its closure $\overline{\mathscr{W}}$ in the space $\mathscr{Y}$ [as defined in (11)]—with respect to the topology of pointwise convergence [see (I.6.33)]—is convex (and it is precisely this fact which we exploit in extending our multiplier rule in Chapter III). Further, in these problems, one often can prove the existence of solutions if one replaces $\mathscr{W}$ by $\overline{\mathscr{W}}$, and, if, in addition, φ_0 is continuous on $\overline{\mathscr{W}} \cap \mathscr{Y}_1$ [where $\mathscr{Y}_1$ and φ_0 are as defined in (11)] with respect to the topology of pointwise convergence. But, under these circumstances, we have the following situation: (i) The original problem (1) may not have a solution. (ii) The "relaxed version" of Problem (1), obtained by enlarging $\mathscr{W}$ to $\overline{\mathscr{W}}$, does have a solution, and also satisfies Hypotheses (8)–(10). (iii) If $(\bar{x}_0, \bar{T}_0) \in A \times \overline{\mathscr{W}}$ is any solution of the "relaxed" problem, N_1 is any neighborhood of 0 in $\mathscr{C}^n$, and N_2 is any neighborhood of 0 in $\mathscr{Y}$, there is a pair $(x_0,T_0) \in A \times \mathscr{W}$ such that $x_0 = T_0 x_0$, $x_0 \in \bar{x}_0 + N_1$, $T_0 \in \bar{T}_0 + N_2$; i.e., any solution of the relaxed problem can be approximated "to any degree of accuracy" by a pair $(x_0,T_0) \in$

$A \times \mathscr{W}$ which is "admissible" in the sense of the original problem. Further (assuming that the $\hat{\varphi}^i$ and $\hat{\phi}^i$ are continuous), this means that, since $\bar{x}_0$ satisfies the given equality and inequality constraints, the approximating point x_0 "almost" satisfies them, and $\hat{\phi}^0(x_0)$ is almost the same as the minimum "payoff" $\hat{\phi}^0(\bar{x}_0)$ of the relaxed problem. Consequently, under the preceding circumstances, there is much to be said for trying to solve the relaxed problem rather than the original one—which may, in fact, have no solution—and, in order to solve the relaxed problem, we can make use of the necessary conditions developed in this section.

The convexity hypothesis for $\mathscr{W}$ is reflected in Problem (2) in the convexity hypothesis for the function f and the set $\mathscr{U}$ in Hypothesis (19). This latter hypothesis will be satisfied, for example, when (a) f is of the form

$$f(\xi,v,t) = f_1(\xi,t) + f_2(\xi,t)v, \quad (\xi,v,t) \in \mathbf{G} \times V \times I,$$

where f_1 and f_2 are continuous functions from $\mathbf{G} \times I$ into R^n and into the set of all $(n \times r)$ matrices, respectively, which are also continuously differentiable w.r. to their first argument, and (b) $\mathscr{U}$ is a convex subset of $L^r_\infty(I)$ [see (I.6.30)], e.g., if $\mathscr{U}$ is the set of all essentially bounded, measurable functions from I into U, and U is a convex set in R^r.

3. Convex Problems and Sufficient Conditions

In this section we shall depart somewhat from the principal theme of this book—the investigation of necessary conditions—and shall instead examine sufficient conditions for e_0 to be a solution of Problem (1.1), or of Problem (1.16). We shall consider in some detail so-called convex problems which are characterized by the fact that the set $\mathscr{E}$ and the functions ϕ^i are convex, and that the functions φ^i are affine. We shall see that, when Problem (1.1) is convex, the multiplier rule of Section 1 can be put in a particularly simple form. Further, we shall see that, for such problems, under some suitable, additional—but relatively mild—hypotheses, the multiplier rule yields conditions which are not only necessary but also sufficient for optimality.

1 Problem (1.1) will be said to be *convex* if (i) $\mathscr{E}$ is a convex set in $\mathscr{Y}$, (ii) the functions ϕ^i, $i = 0, \ldots, \mu$, are defined and convex

on $\mathscr{Y}$ [see (I.2.22)], and (iii) the functions φ^i, $i = 1, \ldots, m$, are defined and affine on $\mathscr{Y}$ [see (I.2.16)].

2 Problem (1.1), if it is convex, will be said to be *well-posed* if (a) there is a vector $e^* \in \mathscr{E}$ such that $\varphi^i(e^*) = 0$ for $i = 1, \ldots, m$ and such that $\phi^i(e^*) < 0$ for $i = 1, \ldots, \mu$, and (b) $0 \in \operatorname{int} \varphi(\mathscr{E})$, where $\varphi = (\varphi^1, \ldots, \varphi^m)$.†

Loosely speaking, a convex problem is well-posed if the equality and inequality constraints are consistent on the constraint set $\mathscr{E}$, and if none of the equality constraints is redundant.

For convex problems, the multiplier rule takes the following form:

3 THEOREM. *Let $e_0 \in \mathscr{E}$ be a solution of Problem* (1.1), *and suppose that this problem is convex. Then there exist vectors $\alpha = (\alpha^1, \ldots, \alpha^m) \in R^m$ and $\beta = (\beta^0, \beta^1, \ldots, \beta^\mu) \in \bar{R}_-^{\mu+1}$* [*see* (I.1.20)], *not both zero, such that*

4 $$\left(\sum_{i=1}^{m} \alpha^i\varphi^i + \sum_{i=0}^{\mu} \beta^i\phi^i\right)(e) \leq \left(\sum_{i=1}^{m} \alpha^i\varphi^i + \sum_{i=0}^{\mu} \beta^i\phi^i\right)(e_0) = \beta^0\phi^0(e_0) \quad \textit{for all} \quad e \in \mathscr{E}.$$

Proof. Since every convex set is finitely open in itself [see (I.1.40)], Hypothesis (1.2) is satisfied. Further, it follows from Lemma (I.7.9) that the φ^i are finitely differentiable, with finite differential at e_0 given by $D\varphi^i = \varphi_L^i$, the linear part of φ^i. Finally, by Corollary (I.7.22), we conclude that the ϕ^i are finitely semidifferentiable, with semidifferentials at e_0 given by $D\phi^i(e_0;\cdot) = \phi^i(e_0 + \cdot) - \phi^i(e_0)$. Hence, Hypothesis (1.3) is satisfied, and we can appeal to Theorem (1.4) and conclude that there exist vectors $\alpha = (\alpha^1, \ldots, \alpha^m) \in R^m$ and $\beta = (\beta^0, \beta^1, \ldots, \beta^\mu) \in \bar{R}_-^{\mu+1}$, not both

† We may assume, without loss of generality, that $\varphi(\mathscr{E})$ has dimension m (i.e., has interior points)—see (I.2.18) and (I.1.34). Indeed, if $\varphi(\mathscr{E})$ has dimension less than m, then there is a nonzero vector $\tilde{\alpha} = (\tilde{\alpha}^1, \ldots, \tilde{\alpha}^m) \in R^m$ such that $\tilde{\alpha} \cdot \varphi(e) = 0$ for all $e \in \mathscr{E}$ [note that $0 \in \varphi(\mathscr{E})$ by (a)]. For the sake of definiteness, say $\tilde{\alpha}^1 \neq 0$. Then we have $\varphi^1(e) = -\sum_{i=2}^{m} (\tilde{\alpha}^i/\tilde{\alpha}^1)\varphi^i(e)$ for all $e \in \mathscr{E}$, so that the problem constraint $\varphi^1(e) = 0$ is redundant and may be discarded. If $(\varphi^2, \ldots, \varphi^m)(e)$ has dimension $(m - 1)$, we simply reformulate our problem, omitting the unnecessary constraint $\varphi^1(e) = 0$. If $(\varphi^2, \ldots, \varphi^m)(e)$ has dimension less than $(m - 1)$, we successively repeat the procedure just described until we finally arrive at the desired form for our problem—unless all of the φ^i vanish identically, in which case we discard all of the equality constraints and obtain a problem of the form of Problem (1.16).

zero, such that

5 $$\sum_{i=1}^{m} \alpha^i \varphi_L^i(e - e_0) + \sum_{i=0}^{\mu} \beta^i[\phi^i(e) - \phi^i(e_0)] \leq 0 \quad \text{for all} \quad e \in \mathscr{E},$$

and such that (1.6) holds. Inasmuch as (for each $i = 1, \ldots, m$) $\varphi^i(e_0) = 0$, φ_L^i is linear, and $\varphi^i(y) = \varphi_L^i(y) + \varphi^i(0)$ for all $y \in \mathscr{Y}$, (4) follows at once from (5) and (1.6). |||

It is interesting to note that the necessary conditions of Theorem (3) are sufficient for optimality, whether or not Problem (1.1) is convex, and even if Hypotheses (1.2 and 3) do not hold, so long as $\beta^0 \neq 0$ and e_0 satisfies the problem equality and inequality constraints. Indeed, we have the following theorem.

6 THEOREM. *If, for Problem* (1.1) (*whether or not it is convex*), *the vectors* $e_0 \in \mathscr{E}$, $\alpha = (\alpha^1, \ldots, \alpha^m) \in R^m$, *and* $\beta = (\beta^0, \beta^1, \ldots, \beta^\mu) \in \bar{R}_-^{\mu+1}$ *satisfy* (4) *as well as the relations*

7 $$\beta^0 \neq 0, \varphi^i(e_0) = 0 \quad \textit{for} \quad i = 1, \ldots, m,$$
$$\textit{and} \quad \phi^i(e_0) \leq 0 \quad \textit{for} \quad i = 1, \ldots, \mu,$$

then e_0 *is a solution of the problem.*

Proof. Without loss of generality, we shall suppose that $\beta^0 = -1$, so that (4) implies that

8 $$\phi^0(e_0) \leq \phi^0(e) - \sum_{i=1}^{m} \alpha^i \varphi^i(e) - \sum_{i=1}^{\mu} \beta^i \phi^i(e) \quad \text{for all} \quad e \in \mathscr{E}.$$

Suppose that $e' \in \mathscr{E}$ satisfies the relations $\varphi^i(e') = 0$ for $i = 1, \ldots, m$ and $\phi^i(e') \leq 0$ for $i = 1, \ldots, \mu$. Then it follows from (8), since $\beta^i \leq 0$ for each i, that

$$\phi^0(e_0) \leq \phi^0(e') - \sum_{i=1}^{\mu} \beta^i \phi^i(e') \leq \phi^0(e'),$$

i.e., e_0 is a solution of the problem. |||

It is now evidently of interest to find an answer to the following question: If Problem (1.1) is convex, what are sufficient conditions on the problem data to ensure that Theorem (3) holds with $\beta^0 \neq 0$? Indeed, under these supplemental conditions, Theorem (3) [by virtue of Theorem (6)] will yield conditions which are both necessary and sufficient for e_0 to be a solution of Problem (1.1). It turns out that one sufficient condition is that Problem (1.1) is well-posed.

9 THEOREM. *Suppose that Problem* (1.1) *is convex and well-posed. Then* $e_0 \in \mathscr{E}$ *is a solution of this problem if and only if there exist vectors* $\alpha_0 = (\alpha_0^1, \ldots, \alpha_0^m) \in R^m$ *and* $\beta_0 = (\beta_0^1, \ldots, \beta_0^\mu) \in \bar{R}_+^\mu$ *such that*

10 $$\varphi^i(e_0) = 0 \quad \text{for} \quad i = 1, \ldots, m, \quad \text{and} \quad \phi^i(e_0) \le 0 \quad \text{for} \quad i = 1, \ldots, \mu,$$

11 $$\left(\phi^0 + \sum_{i=1}^{m} \alpha_0^i \varphi^i + \sum_{i=1}^{\mu} \beta_0^i \phi^i\right)(e) \geq \left(\phi^0 + \sum_{i=1}^{m} \alpha_0^i \varphi^i + \sum_{i=1}^{\mu} \beta_0^i \phi^i\right)(e_0) = \phi^0(e_0) \quad \text{for all} \quad e \in \mathscr{E}.$$

Proof. We begin with the necessity. Let e_0 be a solution of Problem (1.1), so that (10) is evidently satisfied. By Theorem (3), there are vectors $\alpha = (\alpha^1, \ldots, \alpha^m) \in R^m$ and $\beta = (\beta^0, \beta^1, \ldots, \beta^\mu) \in \bar{R}_-^{\mu+1}$, not both zero, such that (4) holds. Let us show that $\beta^0 \neq 0$.

Let $e^* \in \mathscr{E}$ be the vector whose existence is asserted in Definition (2). Then, if $\beta^0 = 0$, (4) would imply that $\sum_{i=1}^{\mu} \beta^i \phi^i(e^*) \le 0$, which, since $\beta^i \le 0$ for each i, is possible only if $\beta^i = 0$ for $i = 0, \ldots, \mu$. This in turn would imply that $\alpha \neq 0$ and that (4) takes the form

$$\sum_{i=1}^{m} \alpha^i \varphi^i(e) \le 0 \quad \text{for all} \quad e \in \mathscr{E},$$

contradicting [by virtue of Lemma (I.5.14)], the hypothesis that $0 \in \operatorname{int} \varphi(\mathscr{E})$. Hence, $\beta^0 < 0$.

If we now denote α^i/β^0 by α_0^i for each $i = 1, \ldots, m$ and β^i/β^0 by β_0^i for each $i = 1, \ldots, \mu$, then evidently $\beta_0 = (\beta_0^1, \ldots, \beta_0^\mu) \in \bar{R}_+^\mu$, and (11) is an immediate consequence of (4).

The sufficiency follows at once from Theorem (6). |||

Relations (10) and (11), which, for a well-posed, convex problem, are necessary and sufficient for $e_0 \in \mathscr{E}$ to be a solution of Problem (1.1), can be written compactly in the form of a so-called saddlepoint condition. Let us define the real-valued function Λ on $\mathscr{E} \times R^m \times \bar{R}_+^\mu$ as follows:

12 $$\Lambda(e,\alpha,\beta) = \left(\phi^0 + \sum_{i=1}^{m} \alpha^i \varphi^i + \sum_{i=1}^{\mu} \beta^i \phi^i\right)(e), \quad (e,\alpha,\beta) \in \mathscr{E} \times R^m \times \bar{R}_+^\mu.$$

We shall refer to the function Λ as the *Lagrangian* for Problem (1.1). Then it is easily verified that $(e_0,\alpha_0,\beta_0) \in \mathscr{E} \times R^m \times \bar{R}^{\mu}_{+}$ satisfies (10) and (11) if and only if

13 $$\min_{e \in \mathscr{E}} \Lambda(e,\alpha_0,\beta_0) = \Lambda(e_0,\alpha_0,\beta_0) = \max_{\substack{\alpha \in R^m \\ \beta \in \bar{R}^{\mu}_{+}}} \Lambda(e_0,\alpha,\beta).$$

For obvious reasons, relation (13) is known as a saddlepoint condition.

We can now restate Theorem (9) as follows in terms of the Lagrangian.

14 THEOREM. *Suppose that Problem* (1.1) *is convex and well-posed. Then $e_0 \in \mathscr{E}$ is a solution of this problem if and only if there exist vectors $\alpha_0 \in R^m$ and $\beta_0 \in \bar{R}^{\mu}_{+}$ such that* (13) *holds, where the functional Λ is defined by* (12).

Note that if Problem (1.1) is convex, then the functions ϕ^i are Gâteaux directionally differentiable, and the directional differentials $D\phi^i(y;\cdot)$, for every $y \in \mathscr{Y}$, are themselves convex functionals [see Lemma (I.7.7)]. If the Gâteaux directional differentials of the ϕ^i, at a solution point e_0 of Problem (1.1), are at the same time finite semidifferentials [see (I.7.16)], then we may appeal to Theorem (1.4) and obtain necessary conditions which, at first sight, differ from those of Theorem (3). Namely, we can obtain the result of Theorem (3) but with (4) replaced by the relations

15 $$\sum_{i=1}^{m} \alpha^i \varphi^i_{\mathrm{L}}(e - e_0) + \sum_{i=0}^{\mu} \beta^i D\phi^i(e_0; e - e_0) \leq 0 \quad \text{for all} \quad e \in \mathscr{E},$$

16 $$\sum_{i=1}^{\mu} \beta^i \phi^i(e_0) = 0,$$

where, in (15), for each i, $D\phi^i(e_0;\cdot)$ denotes the directional differential of ϕ^i at e_0, and φ^i_{L} the linear part of φ^i.

It turns out that the two sets of necessary conditions are actually equivalent, as follows from the following lemma.

17 LEMMA. *Suppose that Problem* (1.1) *is convex. Further, let $e_0 \in \mathscr{E}$, $\alpha = (\alpha^1, \ldots, \alpha^m) \in R^m$ and $\beta = (\beta^0, \beta^1, \ldots, \beta^{\mu}) \in \bar{R}^{\mu+1}_{-}$ be such that*

18 $$\varphi^i(e_0) = 0 \quad \textit{for} \quad i = 1, \ldots, m.$$

Then (4) *holds if and only if* (15) *and* (16) *hold, where, for each i,*

$D\phi^i(e_0;\cdot)$ denotes the Gâteaux directional differential of ϕ^i at e_0, and ${\varphi^i}_{\mathrm{L}}$ the linear part of φ^i.

Proof. First suppose that $\alpha \in R^m$, $\beta \in \bar{R}_-^{\mu+1}$, and $e_0 \in \mathscr{E}$ satisfy (18) and (4), so that they certainly also satisfy (16). Since $\mathscr{E}$ is convex, $e_0 + \varepsilon y = (1 - \varepsilon)e_0 + \varepsilon(e_0 + y) \in \mathscr{E}$ whenever $y \in \mathscr{E} - e_0$ and $\varepsilon \in (0,1)$, so that, by virtue of (4),

19 $$\sum_{i=1}^{m} \alpha^i \frac{\varphi^i(e_0 + \varepsilon y) - \varphi^i(e_0)}{\varepsilon} + \sum_{i=0}^{\mu} \beta^i \frac{\phi^i(e_0 + \varepsilon y) - \phi^i(e_0)}{\varepsilon} \leq 0$$

$$\text{for all} \quad y \in \mathscr{E} - e_0 \quad \text{and} \quad \varepsilon \in (0,1).$$

Passing to the limit in (19) as $\varepsilon \to 0^+$, we at once arrive at (15).

Conversely, suppose that $e_0 \in \mathscr{E}$, $\alpha \in R^m$, and $\beta \in \bar{R}_-^{\mu+1}$ satisfy (15), (16), and (18). The equality in (4) then follows at once, and, since ϕ^i (for each i) is convex,

$$\frac{\phi^i(e_0 + \varepsilon(e - e_0)) - \phi^i(e_0)}{\varepsilon} \leq \phi^i(e) - \phi^i(e_0)$$

$$\text{for all } e \in \mathscr{Y} \quad \text{and} \quad \varepsilon \in (0,1) \quad \text{for each } i.$$

Passing to the limit in this relation as $\varepsilon \to 0^+$, we conclude that

$$D\phi^i(e_0; e - e_0) \leq \phi^i(e) - \phi^i(e_0)$$

$$\text{for all} \quad e \in \mathscr{Y} \quad \text{and each} \quad i = 0, 1, \ldots, \mu,$$

which, by virtue of (15), at once implies that the inequality in (4) also holds. |||

20 Note that the convexity of the set $\mathscr{E}$ was not used in the second part of the proof of Lemma (17).

It follows from Lemma (17) that, whenever Problem (1.1) is convex, relations (10) and (11) hold (for some fixed $e_0 \in \mathscr{E}, \alpha_0 \in R^m$, and $\beta_0 \in \bar{R}_+^{\mu}$) if and only if (10) and the relations

21 $$D\phi^0(e_0; e - e_0) + \sum_{i=1}^{m} \alpha_0^i \varphi_{\mathrm{L}}^i(e - e_0) + \sum_{i=1}^{\mu} \beta_0^i D\phi^i(e_0; e - e_0) \geq 0$$

$$\text{for all} \quad e \in \mathscr{E},$$

22 $$\sum_{i=1}^{\mu} \beta_0^i \phi^i(e_0) = 0$$

hold, where $D\phi^i(e_0;\cdot)$ and φ_{L}^i are as in the lemma statement.

Theorems (3) and (9) may now be reformulated as follows in "differential form."

23 THEOREM. *Let $e_0 \in \mathscr{E}$ be a solution of Problem* (1.1) *and suppose that this problem is convex. Then there are vectors $\alpha = (\alpha^1, \ldots, \alpha^m) \in R^m$ and $\beta = (\beta^0, \beta^1, \ldots, \beta^\mu) \in \bar{R}_-^{\mu+1}$, not both zero, such that* (15) *and* (16) *hold, where, for each i, $D\phi^i(e_0;\cdot)$ denotes the Gâteaux directional differential of ϕ^i at e_0, and φ_L^i the linear part of φ^i.*

24 THEOREM. *Suppose that Problem* (1.1) *is convex and well-posed. Then $e_0 \in \mathscr{E}$ is a solution of this problem if and only if there exist vectors $\alpha_0 = (\alpha_0^1, \ldots, \alpha_0^m) \in R^m$ and $\beta_0 = (\beta_0^1, \ldots, \beta_0^\mu) \in \bar{R}_+^\mu$ such that* (10), (21), *and* (22) *hold where (for each i) $D\phi^i(e_0;\cdot)$ and φ_L^i are as in Theorem* (23).

Note that, in Theorems (23) and (24), we do not impose the requirement of *finite* directional differentiability on the ϕ^i [compare with Theorem (1.4)], but only require *Gâteaux* directional differentiability, which, by Lemma (I.7.7), follows automatically from the hypothesized convexity of the functions ϕ^i.

Unfortunately, the "differential form" of the sufficiency theorem (6) is weaker than Theorem (6) itself, and we must content ourselves with the following result, which is a direct consequence of what has already been shown.

25 THEOREM. *Suppose that, for Problem* (1.1), *the functions φ^i (for $i = 1, \ldots, m$) are defined and affine on $\mathscr{Y}$, and that the functions ϕ^i (for $i = 0, \ldots, \mu$) are defined and convex on $\mathscr{Y}$. Further, suppose that the vectors $e_0 \in \mathscr{E}$, $\alpha = (\alpha^1, \ldots, \alpha^m) \in R^m$, and $\beta = (\beta^0, \beta^1, \ldots, \beta^\mu) \in \bar{R}_-^{\mu+1}$ satisfy* (7), (15), *and* (16), *where, for each i, $D\phi^i(e_0;\cdot)$ denotes the Gâteaux directional differential of ϕ^i at e_0, and φ_L^i the linear part of φ^i. Then e_0 is a solution of the problem.*

Theorem (25) is weaker than Theorem (6) in that no restrictions are placed on the problem data in Theorem (6), whereas in Theorem (25) hypotheses are on the functions φ^i and ϕ^i [*but not on the set $\mathscr{E}$*—see (20)].

Let us now turn to convex problems in the absence of equality constraints, i.e., to Problem (1.16). We begin with two definitions which are obvious analogs of Definitions (1) and (2).

26 Problem (1.16) will be said to be *convex* if (i) $\mathscr{E}$ is a convex set in $\mathscr{Y}$, and (ii) the functions ϕ^i, $i = 0, \ldots, \mu$, are defined and convex on $\mathscr{Y}$.

27 Problem (1.16), if it is convex, will be said to be *well-posed* if there is a vector $e^* \in \mathscr{E}$ such that $\phi^i(e^*) < 0$ for $i = 1, \ldots, \mu$.

We can then prove, in much the same fashion that we proved Theorems (3), (6), (9), (14), (23), (24), and (25), the following theorems, which are the obvious analogs of these preceding theorems.

28 THEOREM. *Let* $e_0 \in \mathscr{E}$ *be a solution of Problem* (1.16), *and suppose that this problem is convex. Then there exists a nonzero vector* $\beta = (\beta^0, \beta^1, \ldots, \beta^\mu) \in \bar{R}^{\mu+1}_{-}$ *such that*

29 $$\sum_{i=0}^{\mu} \beta^i \phi^i(e) \leq \sum_{i=0}^{\mu} \beta^i \phi^i(e_0) = \beta^0 \phi^0(e_0) \quad \text{for all} \quad e \in \mathscr{E}.$$

30 THEOREM. *If, for Problem* (1.16) (*whether or not it is convex*), *the vectors* $e_0 \in \mathscr{E}$ *and* $\beta = (\beta^0, \beta^1, \ldots, \beta^\mu) \in \bar{R}^{\mu+1}_{-}$ *satisfy* (29) *as well as the relations*

31 $$\beta^0 \neq 0 \quad \text{and} \quad \phi^i(e_0) \leq 0 \quad \text{for} \quad i = 1, \ldots, \mu,$$

then e_0 *is a solution of the problem.*

32 THEOREM. *Suppose that Problem* (1.16) *is convex and well-posed. Then* $e_0 \in \mathscr{E}$ *is a solution of this problem if and only if there exists a vector* $\beta_0 = (\beta_0^1, \ldots, \beta_0^\mu) \in \bar{R}^{\mu}_{+}$ *such that*

33 $$\phi^i(e_0) \leq 0 \quad \text{for} \quad i = 1, \ldots, \mu,$$

34 $$\left(\phi^0 + \sum_{i=1}^{\mu} \beta_0^i \phi^i\right)(e) \geq \left(\phi^0 + \sum_{i=1}^{\mu} \beta_0^i \phi^i\right)(e_0) = \phi^0(e_0) \quad \text{for all} \quad e \in \mathscr{E}.$$

Equivalently, $e_0 \in \mathscr{E}$ *is a solution of this problem if and only if there exists a vector* $\beta_0 \in \bar{R}^{\mu}_{+}$ *such that the saddlepoint condition*

35 $$\min_{e \in \mathscr{E}} \Lambda(e, \beta_0) = \Lambda(e_0, \beta_0) = \max_{\beta \in \bar{R}^{\mu}_{+}} \Lambda(e_0, \beta)$$

holds, where the Lagrangian function $\Lambda : \mathscr{E} \times \bar{R}^{\mu}_{+} \to R$ *is defined by the formula*

36 $$\Lambda(e, \beta) = \phi^0(e) + \sum_{i=1}^{\mu} \beta^i \phi^i(e), \; e \in \mathscr{E}, \; \beta \in \bar{R}^{\mu}_{+}.$$

37 THEOREM. *Let* $e_0 \in \mathscr{E}$ *be a solution of Problem* (1.16) *and suppose that the problem is convex. Then there is a nonzero vector* $\beta = (\beta^0, \beta^1, \ldots, \beta^\mu) \in \bar{R}^{\mu+1}_{-}$ *such that*

38 $$\sum_{i=0}^{\mu} \beta^i D\phi^i(e_0; e - e_0) \leq 0 \quad \text{for all} \quad e \in \mathscr{E},$$

39 $$\sum_{i=1}^{\mu} \beta^i \phi^i(e_0) = 0,$$

where, for each i, $D\phi^i(e_0;\cdot)$ *denotes the Gâteaux directional differential of* ϕ^i *at* e_0.

40 THEOREM. *Suppose that Problem* (1.16) *is convex and well-posed. Then* $e_0 \in \mathscr{E}$ *is a solution of this problem if and only if there exists a vector* $\beta_0 = (\beta_0^1, \ldots, \beta_0^\mu) \in \bar{R}_+^\mu$ *such that* (33) *holds and such that*

$$D\phi^0(e_0; e - e_0) + \sum_{i=1}^{\mu} \beta_0^i D\phi^i(e_0; e - e_0) \geq 0 \quad \text{for all} \quad e \in \mathscr{E},$$

$$\sum_{i=1}^{\mu} \beta_0^i \phi^i(e_0) = 0,$$

where, for each i, $D\phi^i(e_0;\cdot)$ *is as in Theorem* (37).

41 THEOREM. *Suppose that, for Problem* (1.16), *the functions* ϕ^0, $\phi^1, \ldots, \phi^\mu$ *are defined and convex on* $\mathscr{Y}$. *Further, suppose that the vectors* $e_0 \in \mathscr{E}$ *and* $\beta = (\beta^0, \beta^1, \ldots, \beta^\mu) \in \bar{R}_-^{\mu+1}$ *satisfy* (31), (38), *and* (39), *where, for each i,* $D\phi^i(e_0;\cdot)$ *is as in Theorem* (37). *Then* e_0 *is a solution of the problem.*

In an obvious manner (and we shall leave the details to the reader), we can derive results which are analogous to those already obtained in this section for problems that differ from Problems (1.1 and 16), in that the inequality constraints $\phi^i(e) \leq 0$ for $i = 1, \ldots, \mu$ are deleted.

4. Sufficient Conditions for Problems with Operator Equation Constraints

In this section, we shall apply the theory developed in the preceding section—in particular, Theorem (3.25)—to Problem (2.1).

Let us suppose that the data of Problem (2.1) satisfy the following hypotheses.

1 HYPOTHESIS. *The set* $A = \mathscr{X}$.

2 HYPOTHESIS. *The functions* $\hat{\varphi}^i$ *(for* $i = 1, \ldots, m$*) are affine, and the functions* $\hat{\phi}^i$ *(for* $i = 0, \ldots, \mu$*) are convex.*

3 HYPOTHESIS. *The set* $\mathscr{W}$ *consists of all operators* $T:\mathscr{X} \to \mathscr{X}$ *of the form*

4 $$Tx = Lx + \tilde{x},\ x \in \mathscr{X},$$

where L is some fixed operator in $\mathbf{B}(\mathscr{X},\mathscr{X})$ [*see* (I.5.1)], *and* $\tilde{x}$ *belongs to some given subset* $\tilde{\mathscr{X}}$ *of* $\mathscr{X}$. *We shall also suppose that L is such that* $(E - L)$*—where E denotes the identity operator on* $\mathscr{X}$*—is a (linear) homeomorphism of* $\mathscr{X}$ *onto itself.*

If we suppose that the data of Problem (2.1) satisfy Hypotheses (1) and (3), then, in the notation of (2.11), $\mathscr{W} \subset \mathscr{Y}_1$ so that $\mathscr{E} = \mathscr{W}$, and

5 $$\varphi_0(T) = (E - L)^{-1}T(0)$$

for all $T \in \mathscr{W}$. Let us extend φ_0 to all of $\mathscr{Y}$ by using formula (5) for all $T \in \mathscr{Y}$, and let us now use the symbol φ_0 to denote this extension. Clearly, φ_0 is a linear function from $\mathscr{Y}$ into $\mathscr{X}$, and, under Hypothesis (2), the mappings $\varphi^i = \hat{\varphi}^i \circ \varphi_0$ (for $i = 1, \ldots, m$) are affine [see (I.2.17)] and the mappings $\phi^i = \hat{\phi}^i \circ \varphi_0$ (for $i = 0, \ldots, \mu$) are convex [see (I.2.23)], and hence Gâteaux directionally differentiable [see Lemma (I.7.7)]. Making the identification between Problems (2.1) and (1.1) indicated in (2.11), we may now appeal to Theorem (3.25) to obtain the following result.

6 THEOREM. *Suppose that, for Problem* (2.1), *Hypotheses* (1)–(3) *are satisfied. Further, suppose that* $x_0 \in \mathscr{X}$, $\tilde{x}_0 \in \tilde{\mathscr{X}}$, $\alpha = (\alpha^1, \ldots, \alpha^m) \in R^m$, *and* $\beta = (\beta^0, \beta^1, \ldots, \beta^\mu) \in \bar{R}_-^{\mu+1}$ *satisfy the relations*

7 $$x_0 = Lx_0 + \tilde{x}_0,$$

8 $$\hat{\varphi}^i(x_0) = 0 \quad \textit{for} \quad i = 1, \ldots, m \quad \textit{and} \quad \hat{\phi}^i(x_0) \leq 0 \quad \textit{for} \quad i = 1, \ldots, \mu,$$

9 $$\beta^0 \neq 0,$$

10 $$\sum_{i=1}^{\mu} \beta^i \hat{\phi}^i(x_0) = 0,$$

11 $$\sum_{i=1}^{m} \alpha^i \hat{\varphi}^i_{\mathrm{L}}(x - x_0) + \sum_{i=0}^{\mu} \beta^i D\hat{\phi}^i(x_0; x - x_0) \leq 0 \quad \textit{whenever} \quad x = Lx + \tilde{x} \quad \textit{for some} \quad \tilde{x} \in \tilde{\mathscr{X}},$$

where, for each i, $D\hat{\phi}^i(x_0;\cdot)$ *denotes the Gâteaux directional differential of* $\hat{\phi}^i$ *at* x_0, *and* $\hat{\varphi}^i_{\mathrm{L}}$ *the linear part of* $\hat{\varphi}^i$. *Then, if* $T_0: \mathscr{X} \to \mathscr{X}$ *is the operator defined by*

12 $$T_0 x = Lx + \tilde{x}_0, \; x \in \mathscr{X},$$

(x_0, T_0) *is a solution of Problem* (2.1).

13 We point out that if not only Hypotheses (1)–(3) are satisfied, but also $\tilde{\mathscr{X}}$ is convex (so that $\mathscr{W}$ is convex), then, by Theorem (3.23), the sufficient conditions of Theorem (6)—but with (9)

replaced by the weaker condition $|\alpha| + |\beta| > 0$—are necessary for (x_0,T_0) to be a solution of Problem (2.1). Of course, we can arrive at the same conclusion on the basis of Theorem (2.14), if the Gâteaux directional differentials of $\hat{\phi}^0, \ldots, \hat{\phi}^\mu$ are at the same time dual differentials and the $\hat{\varphi}^i$ are continuous [see Lemma (I.7.40)]. Thus, under the indicated hypotheses, the necessary conditions of the maximum principle [Theorem (2.14)] are also sufficient for optimality if, in addition, (9) holds. As we shall see in Chapter IV, the necessary conditions hold even if we considerably relax the requirement that $\tilde{\mathscr{X}}$ is a convex set [based on Theorem (1.4), we can already weaken the requirement to $\tilde{\mathscr{X}}$ being finitely open in itself]. The sufficient conditions, of course, hold for an arbitrary subset $\tilde{\mathscr{X}}$ of $\mathscr{X}$.

Let us specialize the results we have just obtained to a problem which is very closely related to Problem (2.2).

14 *Problem.* Let there be given a compact interval $I = [t_1,t_2]$, an integrable $(n \times n)$ matrix-valued function $\mathbf{A}$ defined on I, a set $\mathscr{U}$ of functions $u: I \to R^r$ all of whose members are measurable and essentially bounded, a continuous function $h: R^r \times I \to R^n$, a continuous function $f^0: R^n \times I \to R$ which is continuously differentiable with respect to its first argument [see (I.7.51)] and which has the property that the function $\xi \to f^0(\xi,t): R^n \to R$ is convex for every $t \in I$, vectors $\zeta_0, \zeta_1, \ldots, \zeta_m$ and $\eta_0, \eta_1, \ldots, \eta_m$ in R^n, numbers $v_1, \ldots, v_m$, and a convex, continuously differentiable functional χ defined on $R^n \times R^n$. Then we wish to find an absolutely continuous function $x_0: I \to R^n$ which satisfies the relations

$$\dot{x}_0(t) = \mathbf{A}(t)x_0(t) + h(u_0(t),t) \quad \text{for almost all} \quad t \in I \quad \text{and some} \quad u_0 \in \mathscr{U},$$

$$\zeta_i \cdot x_0(t_1) + \eta_i \cdot x_0(t_2) = v_i \quad \text{for} \quad i = 1, \ldots, m,$$

$$\chi(x_0(t_1),x_0(t_2)) \leq 0,$$

and such that

$$\zeta_0 \cdot x_0(t_1) + \eta_0 \cdot x_0(t_2) + \int_{t_1}^{t_2} f^0(x_0(t),t)\,dt \leq \zeta_0 \cdot x(t_1) + \eta_0 \cdot x(t_2) + \int_{t_1}^{t_2} f^0(x(t),t)\,dt$$

for all absolutely continuous functions $x: I \to R^n$ which satisfy

(a) $\dot{x}(t) = \mathbf{A}(t)x(t) + h(u(t),t)$ a.e. on I for some $u \in \mathscr{U}$, (b) $\zeta_i \cdot x(t_1) + \eta_i \cdot x(t_2) = v_i$ for $i = 1, \ldots, m$, and (c) $\chi(x(t_1),x(t_2)) \leq 0$.

We may identify the preceding problem with Problem (2.1) by identifying both $\mathscr{X}$ and A with $\mathscr{C}^n(I)$, μ with 1, by defining the functionals $\hat{\varphi}^1, \ldots, \hat{\varphi}^m, \hat{\phi}^0$, and $\hat{\phi}^1$ on $\mathscr{X}$ through the formulas

$$\hat{\varphi}^i(x) = \zeta_i \cdot x(t_1) + \eta_i \cdot x(t_2) - v_i \quad \text{for} \quad i = 1, \ldots, m,$$

$$\hat{\phi}^1(x) = \chi(x(t_1),x(t_2)),$$

$$\hat{\phi}^0(x) = \zeta_0 \cdot x(t_1) + \eta_0 \cdot x(t_2) + \int_{t_1}^{t_2} f^0(x(t),t)\, dt,$$

and by identifying $\mathscr{W}$ with the set of all integral operators $T:\mathscr{X} \to \mathscr{X}$ which have the form

$$(Tx)(t) = \xi + \int_{t_1}^{t} \mathbf{A}(s)x(s)\, ds + \int_{t_1}^{t} h(u(s),s)\, ds, \; t \in I,$$

for some $\xi \in R^n$ and some $u \in \mathscr{U}$. Having made the indicated identifications, it is obvious that Hypotheses (1) and (2) are satisfied, and, if we identify $\tilde{\mathscr{X}}$ with the set of all functions $\tilde{x} \in \mathscr{X}$ of the form

15 $$\tilde{x}(t) = \xi + \int_{t_1}^{t} h(u(s),s)\, ds, \; t \in I,$$

where $\xi \in R^n$ and $u \in \mathscr{U}$, and define the (linear) operator $L:\mathscr{X} \to \mathscr{X}$ through the formula

16 $$(Lx)(t) = \int_{t_1}^{t} \mathbf{A}(s)x(s)\, ds, \; t \in I, \; x \in \mathscr{X},$$

then we easily see, by virtue of Corollary (A.28) in the Appendix, that Hypothesis (3) is also satisfied.

Further, it follows from Theorem (I.7.41) and Lemma (I.7.40) that $\hat{\phi}^0$ and $\hat{\phi}^1$ are Fréchet differentiable, with Fréchet differentials [which are then certainly also Gâteaux directional differentials—see Theorems (I.7.44 and 29) and (I.7.5)] given by the formulas

$$D\hat{\phi}^0(x;\delta x) = \zeta_0 \cdot \delta x(t_1) + \eta_0 \cdot \delta x(t_2) + \int_{t_1}^{t_2} \mathscr{D}_1 f^0(x(t),t)\delta x(t)\, dt,$$

$$D\hat{\phi}^1(x;\delta x) = \sum_{j=1}^{2} \mathscr{D}_j \chi(x(t_1),x(t_2))\delta x(t_j),$$

for all $x \in \mathscr{X}$ and $\delta x \in \mathscr{X}$.

Using the variation of parameters formula [see Theorem (A.51)], we can write the solution of the equation $x = Lx + \tilde{x}$—where L is given by (16) and $\tilde{x}$ by (15) (for some $u \in \mathscr{U}$)—which, of course, is equivalent to the linear inhomogeneous ordinary differential equation $\dot{x}(t) = \mathbf{A}(t)x(t) + h(u(t),t)$, $t \in I$, with initial condition $x(t_1) = \xi$, in the form

$$x(t) = \Phi(t)\xi + \Phi(t)\int_{t_1}^{t} \Phi^{-1}(s)h(u(s),s)\,ds, \quad t \in I,$$

where Φ is the absolutely continuous $(n \times n)$-matrix-valued function defined on I that satisfies

$$\dot{\Phi}(t) = \mathbf{A}(t)\Phi(t) \quad \text{a.e. on } I, \quad \Phi(t_1) = \text{the identity matrix.}$$

Note that, by Theorem (A.47),

17 $$\frac{d}{dt}[\Phi^{-1}(t)] = -\Phi^{-1}(t)\mathbf{A}(t)$$

$$\text{a.e. on } I, \quad \Phi^{-1}(t_1) = \text{identity matrix.}$$

We may now appeal to Theorem (6) and thereby conclude that if, in Problem (14), the absolutely continuous function $x_0 \in \mathscr{C}^n$, the function $u_0 \in \mathscr{U}$, and the vector $(\alpha^{-1},\alpha^0,\alpha^1, \ldots ,\alpha^m) \in R^{m+2}$ satisfying the relations

18 $$\alpha^0 < 0$$

19 $$\alpha^{-1} \leq 0 \quad \text{and} \quad \alpha^{-1} = 0 \quad \text{if} \quad \chi(x_0(t_1),x_0(t_2)) < 0,$$

20 $$\dot{x}_0(t) = \mathbf{A}(t)x_0(t) + h(u_0(t),t) \quad \text{a.e. on } I,$$

21 $$\chi(x_0(t_1),x_0(t_2)) \leq 0,$$

22 $$\zeta_i \cdot x_0(t_1) + \eta_i \cdot x_0(t_2) = v_i \quad \text{for} \quad i = 1, \ldots, m,$$

as well as the inequality

23 $$\begin{aligned}\sum_{i=-1}^{m} \alpha^i(\zeta_i + \eta_i\Phi(t_2)) \cdot \xi &+ \sum_{i=-1}^{m} \alpha^i\eta_i\Phi(t_2)\int_{t_1}^{t_2} \Phi^{-1}(s)[h(u(s),s) \\ &- h(u_0(s),s)]\,ds + \alpha^0 \int_{t_1}^{t_2} \mathscr{D}_1 f^0(x_0(t),t)\Phi(t)\,dt\,\xi \\ &+ \alpha^0 \int_{t_1}^{t_2} \mathscr{D}_1 f^0(x_0(t),t)\Phi(t)\int_{t_1}^{t} \Phi^{-1}(s)[h(u(s),s) \\ &- h(u_0(s),s)]\,ds\,dt \leq 0 \quad \text{for all} \quad u \in \mathscr{U} \quad \text{and} \quad \xi \in R^n\end{aligned}$$

[where, for convenience of notation, we have set

24 $$\mathscr{D}_1\chi(x_0(t_1),x_0(t_2)) = \zeta_{-1} \quad \text{and} \quad \mathscr{D}_2\chi(x_0(t_1),x_0(t_2)) = \eta_{-1},$$

and the η_i, the ζ_i, and $\mathscr{D}_1 f^0$ are to be considered row vectors, and h and ξ column vectors], then x_0 is a solution of the problem.

If we interchange the order of integration in the double integral in (23) and take into account that $u_0 \in \mathscr{U}$, we easily can convince ourselves that (23) can hold if and only if

25 $$\sum_{i=-1}^{m} \alpha^i(\zeta_i + \eta_i \Phi(t_2)) + \alpha^0 \int_{t_1}^{t_2} \mathscr{D}_1 f^0(x_0(t),t)\Phi(t)\,dt = 0, \quad \text{and}$$

26 $$\int_{t_1}^{t_2} \left[\sum_{i=-1}^{m} \alpha^i \eta_i \Phi(t_2) + \alpha^0 \int_{s}^{t_2} \mathscr{D}_1 f^0(x_0(t),t)\Phi(t)\,dt \right] \Phi^{-1}(s)[h(u(s),s) - h(u_0(s),s)]\,ds \leq 0 \quad \text{for all} \quad u \in \mathscr{U}.$$

Let us set

27 $$\psi(s) = \left[\sum_{i=-1}^{m} \alpha^i \eta_i \Phi(t_2) + \alpha^0 \int_{s}^{t_2} \mathscr{D}_1 f^0(x_0(t),t)\Phi(t)\,dt \right] \Phi^{-1}(s) \quad \text{for} \quad s \in I,$$

so that ψ is an absolutely continuous n-row-vector-valued function defined on I. It is easily seen, by virtue of (17), that (25) and (27) are satisfied if and only if ψ is the [unique—see Theorem (A.51)] solution of the linear inhomogeneous ordinary differential equation

28 $$\dot{\psi}(t) = -\psi(t)\mathbf{A}(t) - \alpha^0 \mathscr{D}_1 f^0(x_0(t),t) \quad \text{a.e. on } I$$

satisfying the boundary conditions

29 $$\psi(t_1) = -\sum_{i=-1}^{m} \alpha^i \zeta_i,$$

30 $$\psi(t_2) = \sum_{i=-1}^{m} \alpha^i \eta_i.$$

Now, by (27), (26) can be rewritten in the equivalent form

31 $$\int_{t_1}^{t_2} \psi(s)h(u(s),s)\,ds \leq \int_{t_1}^{t_2} \psi(s)h(u_0(s),s)\,ds \quad \text{for all} \quad u \in \mathscr{U}.$$

Thus, we have proved the following result.

32 THEOREM. *Suppose that, for Problem* (14), *the absolutely continuous functions* x_0 *and* ψ *in* $\mathscr{C}^n(I)$ (*the former of which is to be considered a column vector, and the latter a row vector*), *the function* $u_0 \in \mathscr{U}$, *and the vector* $\alpha = (\alpha^{-1}, \alpha^0, \alpha^1, \ldots, \alpha^m) \in R^{m+2}$ *satisfy*

the relations (18)–(22) *and* (28)–(31), *where* η_{-1} *and* ζ_{-1} *are defined by* (24). *Then* x_0 *is a solution of the problem.*

As we pointed out in (13), if the set of functions

$$\left\{t \to \int_{t_1}^{t} h(u(s),s)\, ds,\ t \in I : u \in \mathscr{U}\right\}$$

is convex—so that the set $\tilde{\mathscr{X}}$ of all functions $\tilde{x}$ of the form of (15) with $\xi \in R^n$ and $u \in \mathscr{U}$ is also convex—and if x_0 is a solution of Problem (14) satisfying (20) (with $u_0 \in \mathscr{U}$), then there is a nonzero vector $(\alpha^{-1},\alpha^0,\alpha^1,\ldots,\alpha^m) \in R^{m+2}$ such that (19)–(24) hold. Further, as we just showed, if (23) holds, then there is an absolutely continuous n-row-vector-valued function ψ defined on I satisfying (28)–(31). Thus, in this case, as usual, only (18) stands in the way of having conditions which are both necessary and sufficient for optimality.

If, in Problem (14), the functions $u \in \mathscr{U}$ all take on their values in some set U in R^r, then (31) can evidently be replaced as a sufficient condition in Theorem (32) by the "pointwise" maximum condition

33 $$\psi(t)h(u_0(t),t) = \max_{v \in U} \psi(t)h(v,t) \quad \text{a.e. on } I.$$

On the other hand, if $\mathscr{U}$ consists of *all* essentially bounded, measurable functions from I into U, then [arguing exactly as in the proof of (2.41)] we can show that (33) and (31) are equivalent.

In closing, we remark that the results of this section can be modified in an obvious way if we replace the given problems by ones in which the equality and/or inequality constraints have been omitted.

CHAPTER III

A General Multiplier Rule

OUR AIM IN this chapter is to generalize the multiplier rules [Theorems (II.1.4 and 18)] developed in Section 1 of the preceding chapter. Our generalizations will be in a number of directions. First, we shall weaken the requirement that the underlying set $\mathscr{E}$ is finitely open in itself, in order to be able to apply our results to more general optimization problems with operator equation constraints than those considered in Chapter II [as was discussed in the introduction to Chapter II and in (II.2.47)]. Second, we shall allow optimization problems with inequality constraints that are considerably more general than those of Problems (II.1.1 and 16). More precisely, we shall replace the inequality constraint $(\phi^1, \ldots, \phi^\mu)(e) \in \bar{R}^\mu_-$ of Chapter II by the more general constraint $\phi_1(e) \in Z_1$, where Z_1 is a given convex body [see (I.4.18)] in a linear topological space $\mathscr{Z}_1$, and ϕ_1 is some given function from $\mathscr{E}$ into $\mathscr{Z}_1$. Such a generalization will turn out to be important when we wish to investigate optimal control problems with so-called restricted phase co-ordinates, as we shall see in Chapter IV.

In addition, we shall consider other types of optimization problems, specifically minimax problems and so-called vector-valued criterion function optimization problems. In order to obtain a multiplier rule which is valid for these two classes of problems, as well as for the problems of the type of Chapter II, we shall introduce the concept of a certain type of extremal. The multiplier rule which we shall develop will be satisfied by every extremal of this type, and since the solution of any of the aforementioned problems will also be such an extremal, we shall in a single theorem obtain necessary conditions for all three types of problems.

Interestingly enough, the multiplier rule developed in Chapter II can, under certain hypotheses, be somewhat sharpened, and we shall indicate how this can be done in Section 4 of this chapter.

In the last section of this chapter, we shall extend the results of Section 3 of Chapter II on convex problems and sufficient conditions to problems with the more general type of inequality constraints mentioned earlier, and to minimax problems and optimization problems with vector-valued criterion functions as well.

1. The Concept of an Extremal

Let us turn back to Problem (II.1.1). If we denote the function $(\varphi^1, \ldots, \varphi^m)$ from $\mathscr{E}$ into R^m by φ, and the function $(\phi^1, \ldots, \phi^\mu)$ from $\mathscr{E}$ into R^μ by ϕ_1, then this problem is clearly equivalent to the following one:

1 *Problem.* Given a set $\mathscr{E}$ in a linear vector space $\mathscr{Y}$ and functions $\varphi: \mathscr{E} \to R^m$, $\phi^0: \mathscr{E} \to R$, and $\phi_1: \mathscr{E} \to R^\mu$, find an element $e_0 \in \mathscr{E}$ that achieves a minimum for ϕ^0 (on $\mathscr{E}$) subject to the constraints $\varphi(e) = 0$ and $\phi_1(e) \in \bar{R}^\mu_-$.

The following problem is evidently a generalization of Problem (1).

2 *Problem.* Given arbitrary sets $\mathscr{E}$ and $\mathscr{E}'$ with $\mathscr{E} \subset \mathscr{E}'$, a linear topological space $\mathscr{Z}_1$ and a convex body [see (I.4.18)] Z_1 in $\mathscr{Z}_1$, and functions $\varphi: \mathscr{E}' \to R^m$, $\phi^0: \mathscr{E}' \to R$, and $\phi_1: \mathscr{E}' \to \mathscr{Z}_1$, find an element $e_0 \in \mathscr{E}$ that achieves a minimum for ϕ^0 (on $\mathscr{E}$) subject to the constraints $\varphi(e) = 0$ and $\phi_1(e) \in Z_1$.

The following problem, which slightly generalizes Problem (II.1.1), is also an important particular case of Problem (2).

3 *Problem.* Given arbitrary sets $\mathscr{E}$ and $\mathscr{E}'$ with $\mathscr{E} \subset \mathscr{E}'$ and real-valued functions $\varphi^1, \ldots, \varphi^m, \phi^0, \phi^1, \ldots, \phi^\mu$ defined on $\mathscr{E}'$, find an element $e_0 \in \mathscr{E}$ that achieves a minimum for ϕ^0 (on $\mathscr{E}$) subject to the constraints $\varphi^i(e) = 0$ for $i = 1, \ldots, m$ and $\phi^i(e) \leq 0$ for $i = 1, \ldots, \mu$.

In Problem (2), it often turns out [e.g., in the special case of Problem (3)] that the convex body Z_1 is a closed convex cone in $\mathscr{Z}_1$ which is defined in terms of a given subset $\mathscr{L}$ of $\mathscr{Z}_1^*$ [see (I.5.4)] through the formula

4 $$Z_1 = \{\zeta : \zeta \in \mathscr{Z}_1, \ell(\zeta) \leq 0 \text{ for all } \ell \in \mathscr{L}\}.$$

Of course, if $\mathscr{L}$ is an arbitrary subset of $\mathscr{Z}_1^*$, then the set defined by (4) will not necessarily have a nonempty interior (i.e., Z_1 may not be a convex body), and indeed this set may even be empty [although, if it is not empty, it must evidently be a closed, convex cone—see (I.1.13), (I.3.4), and Theorem (I.3.25C)]. However, if $\mathscr{L}$ is a finite subset of $\mathscr{Z}_1^*$ whose elements are linearly independent, then it follows from Corollary (I.2.15) and Theorem (I.3.25B) that the set Z_1 given by (4) must have a nonempty interior, and is thus a convex body.

5   It is often the case in Problem (2) [e.g., in the special case of Problem (3)] that $\mathscr{Z}_1 = R^\mu$ for some positive integer μ. On the other hand, as we shall see in Chapters IV and V, there are problems in optimal control which are most conveniently cast in the form of Problem (2) with $\mathscr{Z}_1 = R^\mu \times \mathscr{C}^\nu(I)$ (for some compact interval I and some positive integer ν). For example, if we modify Problem (II.2.2) by adjoining an inequality constraint of the form

$$\tilde{\chi}^i(x(t),t) \leq 0 \quad \text{for all} \quad t \in I \quad \text{and each} \quad i = 1, \ldots, \nu,$$

where $\tilde{\chi}^1, \ldots, \tilde{\chi}^\nu$ are given continuous functions from $\mathbf{G} \times I$ into R, then precisely this type of situation arises.

Thus, we shall, in this chapter, investigate Problem (2) without assuming that $\mathscr{Z}_1$ is necessarily finite-dimensional.

6   Of course, we may generalize Problem (2) even further, by removing the restriction that the range of φ is finite-dimensional, i.e., by permitting "infinite-dimensional" equality constraints. However, a satisfactory theory for problems with such restrictions is still in the formative stage, and consequently we shall have to content ourselves in this book to the investigation of problems with "finite-dimensional" equality constraints. Fortunately, it also turns out that the equality constraints that arise in most important applications are finite-dimensional.

7   There are two other important classes of optimization problems which are often encountered in applications: problems with a vector-valued criterion function, and minimax problems; and we shall also be interested in investigating them. Instead of investigating each of these three classes of problems individually, we shall use a more efficient approach. Namely, we shall find the essential common feature which solutions of each of these problems possess, and we shall investigate the properties of this

common feature. This common feature will be referred to as a certain type of *extremality*.

Before introducing this concept of extremality, let us precisely state the minimax and vector-valued criterion function problems.

8 *Problem* (Minimax). Given arbitrary sets $\mathscr{E}$ and $\mathscr{E}'$ with $\mathscr{E} \subset \mathscr{E}'$, linear topological spaces $\mathscr{X}_1$ and $\mathscr{X}_2$ and a convex body Z_1 in $\mathscr{X}_1$, a nonempty set $\mathscr{L} \subset \mathscr{X}_2^*$, and functions $\varphi : \mathscr{E}' \to R^m$, $\phi_1 : \mathscr{E}' \to \mathscr{X}_1$, and $\phi_2 : \mathscr{E}' \to \mathscr{X}_2$, find an element $e_0 \in \mathscr{E}$ that achieves a minimum (on $\mathscr{E}$) for $\sup_{\ell \in \mathscr{L}} \ell \circ \phi_2(e)$ subject to the constraints $\varphi(e) = 0$ and $\phi_1(e) \in Z_1$.

An important special case of Problem (8) which, for the sake of completeness, we shall state separately, is the following one:

9 *Problem.* Given arbitrary sets $\mathscr{E}$ and $\mathscr{E}'$ with $\mathscr{E} \subset \mathscr{E}'$ and real-valued functions $\varphi^1, \ldots, \varphi^m$, $\phi^1, \ldots, \phi^\mu$, $\phi^{\mu+1}, \ldots, \phi^\nu$ defined on $\mathscr{E}'$, find an element $e_0 \in \mathscr{E}$ that achieves a minimum (on $\mathscr{E}$) for $\max_{\mu+1 \leq i \leq \nu} \phi^i(e)$ subject to the constraints $\varphi^i(e) = 0$ for $i = 1, \ldots, m$ and $\phi^i(e) \leq 0$ for $i = 1, \ldots, \mu$.

10 *Problem* (Vector Valued Criterion Function). Given arbitrary sets $\mathscr{E}$ and $\mathscr{E}'$ with $\mathscr{E} \subset \mathscr{E}'$, linear topological spaces $\mathscr{X}_1$ and $\mathscr{X}_2$ and a convex body Z_1 in $\mathscr{X}_1$, a nonempty set $\mathscr{L} \subset \mathscr{X}_2^*$, and functions $\varphi : \mathscr{E}' \to R^m$, $\phi_1 : \mathscr{E}' \to \mathscr{X}_1$ and $\phi_2 : \mathscr{E}' \to \mathscr{X}_2$, find an element $e_0 \in \mathscr{E}$ that satisfies the relations $\varphi(e_0) = 0$ and $\phi_1(e_0) \in Z_1$ such that there is no element $e_1 \in \mathscr{E}$ satisfying the conditions (i) $\varphi(e_1) = 0$ and $\phi_1(e_1) \in Z_1$, (ii) $\ell \circ \phi_2(e_1) \leq \ell \circ \phi_2(e_0)$ for every $\ell \in \mathscr{L}$, and (iii) $\tilde{\ell} \circ \phi_2(e_1) < \tilde{\ell} \circ \phi_2(e_0)$ for some $\tilde{\ell} \in \mathscr{L}$.

11 Problem (10) is usually referred to as an optimization problem with a vector-valued criterion function. Loosely speaking, such a problem is characterized by the presence of a number of "cost" functionals: the functionals $\ell \circ \phi_2$, for $\ell \in \mathscr{L}$. Since it is in general not possible to simultaneously achieve a minimum for more than one functional on a given set, it is not unreasonable in this situation to try to find elements $e_0 \in \mathscr{E}$ which (a) satisfy the problem constraints and (b) have no better alternative in the sense that there is no element which (i) satisfies the problem constraints, (ii) is no "worse" than e_0 with respect to any cost functional, and (iii) is "better" than e_0 with respect to at least one cost functional. Indeed, this is precisely the problem formulation which is found in Problem (10).

It is clear that Problem (2) may be considered to be a particular case either of Problem (8) or of Problem (10). However, it will not be convenient for us to adopt this point of view, and we shall not do so.

An important special case of Problem (10), which we shall state separately for the sake of completeness, is the following [compare with Problem (9)]:

12 *Problem.* Given arbitrary sets $\mathscr{E}$ *and* $\mathscr{E}'$ *with* $\mathscr{E} \subset \mathscr{E}'$ *and real-valued functions* $\varphi^1, \ldots, \varphi^m, \phi^1, \ldots, \phi^\mu, \phi^{\mu+1}, \ldots, \phi^\nu$ *defined on* $\mathscr{E}'$, find an element $e_0 \in \mathscr{E}$ that satisfies the relations $\varphi^i(e_0) = 0$ for $i = 1, \ldots, m$ and $\phi^i(e_0) \leq 0$ for $i = 1, \ldots, \mu$ such that there is no element $e_1 \in \mathscr{E}$ satisfying the conditions (i) $\varphi^i(e_1) = 0$ for $i = 1, \ldots, m$ and $\phi^i(e_1) \leq 0$ for $i = 1, \ldots, \mu$, (ii) $\phi^i(e_1) \leq \phi^i(e_0)$ for every $i = \mu + 1, \ldots, \nu$, and (iii) $\phi^j(e_1) < \phi^j(e_0)$ for some $j = \mu + 1, \ldots, \nu$.

Problem (3) is clearly a special case either of Problem (9) or of Problem (12), but again we shall, for convenience, not adopt this viewpoint.

13 Let us now turn to the common feature that solutions of Problems (2), (8), and (10) possess, which we discussed in (7). In order to define an extremal of the kind which will be referred to as (φ,ϕ,Z)-extremal, there must be given

(i) arbitrary sets $\mathscr{E}$ and $\mathscr{E}'$ with $\mathscr{E} \subset \mathscr{E}'$,
(ii) a linear topological space $\mathscr{Z}$,
(iii) a convex body Z in $\mathscr{Z}$,
(iv) a function $\phi: \mathscr{E}' \to \mathscr{Z}$, and
(v) a function $\varphi: \mathscr{E}' \to R^m$ (for some positive integer m).

14 Definition. *If* $\mathscr{E}'$, $\mathscr{E}$, $\mathscr{Z}$, Z, ϕ, *and* φ *are as indicated in* (13), *then an element* $e_0 \in \mathscr{E}$ *will be called a* (φ,ϕ,Z)*-extremal if* (i) $\varphi(e_0) = 0$, (ii) $\phi(e_0) \in Z$, *and* (iii) *the set* $\{e\colon e \in \mathscr{E}, \varphi(e) = 0$, *and* $\phi(e) \in \operatorname{int} Z\}$ *is empty.*

15 As we shall now show, a solution of Problem (2), (8), or (10) is always a (φ,ϕ,Z)-extremal, for suitably chosen φ, ϕ, and Z. [In Problems (8) and (10), this will actually not be true unless we make some technical assumptions regarding the set $\mathscr{L}$.] In the next section we shall show that, under suitable hypotheses on $\mathscr{E}, \mathscr{E}', Z, \phi$, and φ, a (φ,ϕ,Z)-extremal must satisfy a multiplier rule which is very analogous to Theorem (II.1.4). This multiplier rule, in turn, can (and will) be specialized to obtain individual multiplier

rules for Problems (2), (8), and (10), the first of which will be seen to include Theorem (II.1.4) as a special case.

We begin with Problem (2).

16 THEOREM. *If e_0 is a solution of Problem (2), then e_0 is a (φ,ϕ,Z)-extremal for $\mathscr{Z} = R \times \mathscr{Z}_1$, $Z = \bar{R}_- \times Z_1$, and $\phi = (\phi^0 - \phi^0(e_0), \phi_1)$.*

Proof. By Lemma (I.4.22), Z is a convex body in $\mathscr{Z}$, so that conditions (i)–(v) of (13) hold. Then, using (I.3.34), we can arrive at our desired conclusion by direct calculation. |||

For Problems (8) and (10), we shall have to make hypotheses on the sets

17 $$Z_2^\kappa = \{\zeta : \zeta \in \mathscr{Z}_2, \ell(\zeta) \leq \kappa \text{ for all } \ell \in \mathscr{L}\},$$

defined by (17) for each $\kappa \in R$.

Note that, for each $\kappa \in R$, Z_2^κ (if it is not empty—and $0 \in Z_2^\kappa$ for every $\kappa \geq 0$) is a closed convex set [see (I.3.4) and Theorem (I.3.25C)], and that Z_2^0 is a closed, convex cone.

Let us also consider the sets (defined for each $\kappa \in R$)

18 $$V_2^\kappa = \left\{\zeta : \zeta \in \mathscr{Z}_2, \sup_{\ell \in \mathscr{L}} \ell(\zeta) < \kappa\right\}.$$

Evidently, $0 \in V_2^\kappa$ for each $\kappa > 0$, but the sets V_2^κ for $\kappa \leq 0$ may be empty. However, if $\sup_{\ell \in \mathscr{L}} \ell(\zeta) < 0$ for some $\zeta \in \mathscr{Z}_2$, then clearly none of the sets V_2^κ ($\kappa \in R$) is empty. If V_2^κ is not empty, then it is convex. Further, as is easily verified,

19 $$V_2^\kappa + Z_2^0 = V_2^\kappa \quad \text{for every} \quad \kappa \in R.$$

Now let $e_0 \in \mathscr{E}$ be a solution of Problem (8). We set

20 $$\kappa_0 = \sup_{\ell \in \mathscr{L}} \ell \circ \phi_2(e_0),$$

and make the following two hypotheses.

21 HYPOTHESIS. *Either $\kappa_0 \neq 0$, or V_2^0 is not empty* [*see* (18) *and* (20)].

22 HYPOTHESIS. $V_2^{\kappa_0} = \operatorname{int} Z_2^{\kappa_0}$ [*see* (17), (18), *and* (20)].

23 Note that, by virtue of Theorem (I.3.25B) and Lemma (I.5.14), Hypothesis (22) holds whenever $\mathscr{L}$ is a finite subset of $\mathscr{Z}_2^*$ with $0 \notin \mathscr{L}$. If also the elements of $\mathscr{L}$ are linearly independent, then, by Corollary (I.2.15), V_2^0 is not empty [so that Hypothesis (21) also holds].

24 THEOREM. *Let* $e_0 \in \mathscr{E}$ *be a solution of Problem* (8) *subject to Hypotheses* (21) *and* (22). *Then* e_0 *is a* (φ,ϕ,Z)*-extremal for* $\mathscr{Z} = \mathscr{Z}_1 \times \mathscr{Z}_2$, $Z = Z_1 \times Z_2^{\kappa_0}$, *and* $\phi = (\phi_1,\phi_2)$.

Proof. It follows from Hypotheses (21) and (22) and our discussion of the sets Z_2^{κ} and V_2^{κ} that $Z_2^{\kappa_0}$ is a convex body whose interior is $V_2^{\kappa_0}$. Hence, by Lemma (I.4.22), Z is a convex body, and (i)–(v) of (13) hold. By (I.3.34), int Z = (int Z_1) $\times$ $V_2^{\kappa_0}$. Evidently, $\varphi(e_0) = 0$ and $\phi(e_0) \in Z$. If $e_1 \in \mathscr{E}$ satisfies $\varphi(e_1) = 0$ and $\phi(e_1) \in$ int Z, then $\phi_1(e_1) \in (\text{int } Z_1) \subset Z_1$, and $\phi_2(e_1) \in V_2^{\kappa_0}$, or

$$\sup_{\ell \in \mathscr{L}} \ell \circ \phi_2(e_1) < \kappa_0 = \sup_{\ell \in \mathscr{L}} \ell \circ \phi_2(e_0),$$

contradicting that e_0 is a solution of Problem (8). |||

For Problem (10), Hypotheses (21) and (22) are replaced by the following one.

25 HYPOTHESIS. *The set* V_2^0 [*see* (18)] *is not empty, and* V_2^0 = int Z_2^0 [*see* (17)].

26 THEOREM. *Let* $e_0 \in \mathscr{E}$ *be a solution of Problem* (10) *subject to Hypothesis* (25). *Then* e_0 *is a* (φ,ϕ,Z)*-extremal for* $\mathscr{Z} = \mathscr{Z}_1 \times \mathscr{Z}_2$, $Z = Z_1 \times Z_2^0$, *and* $\phi = (\phi_1, \phi_2 - \phi_2(e_0))$.

Theorem (26) follows in much the same way as Theorems (16) and (24).

27 We point out that Hypothesis (25) holds whenever $\mathscr{L}$ is a finite subset of $\mathscr{Z}_2^*$ whose elements are linearly independent [see (23)].

28 We may also consider variants of Problems (2), (8), and (10) in which the equality constraint and/or the generalized inequality (i.e., convex body) constraint is absent. In case only the generalized inequality constraint is omitted, we may still identify solutions of these problems with (φ,ϕ,Z)-extremals, provided that we redefine $\mathscr{Z}$, Z, and ϕ in evident ways. In the case of no equality constraints, we must introduce a new type of extremal, the so-called (ϕ,Z)-extremal.

29 DEFINITION. *If* $\mathscr{E}'$, $\mathscr{E}$, $\mathscr{Z}$, Z, *and* ϕ *are as in* (13)—*but with* (v) *omitted—then an element* $e_0 \in \mathscr{E}$ *will be called a* (ϕ,Z)*-extremal if* $\phi(e_0) \in Z$ *and if the set* $\{e: e \in \mathscr{E}, \phi(e) \in \text{int } Z\}$ *is empty.*

30 We leave it to the reader to show that solutions of Problems (2), (8), and (10)—but with the equality constraints omitted—subject to the same hypotheses as before, are at the same time (ϕ,Z)-extremals, where $\mathscr{Z}$, Z, and ϕ are defined precisely as before.

2. Abstract Multiplier Rules

In this section we shall obtain generalized multiplier rules [which are analogous to, and generalizations of, Theorems (II.1.4 and 18)] representing necessary conditions which (φ,ϕ,Z)-extremals, or (ϕ,Z)-extremals, must satisfy, under suitable hypotheses for $\mathscr{E}$, $\mathscr{E}'$, etc.

We first shall consider (φ,ϕ,Z)-extremals. Thus, suppose that e_0 is a (φ,ϕ,Z)-extremal and that $\mathscr{E}$, φ, ϕ satisfy the following hypotheses, which are natural generalizations of Hypotheses (II.1.2 and 3).

1 HYPOTHESIS. *The set $\mathscr{E}'$ is a set in a linear vector space $\mathscr{Y}$, and both $\mathscr{E}$ and $\mathscr{E}'$ are finitely open in themselves* [*see* (I.1.39)].

2 HYPOTHESIS. *The function φ is finitely differentiable at e_0* [*see* (I.7.4)], *and is finitely continuous* [*see* (I.5.26)].

3 HYPOTHESIS. *There is a closed, convex cone $W \subset \mathscr{Z}$ such that* (i) (int Z) + W = int Z *and* (ii) *the function ϕ is finitely W-semidifferentiable at e_0* [*see* (I.7.13)].

4 Note that, under Hypotheses (1)–(3), the finite differential of φ and a finite W-semidifferential of ϕ at e_0 are defined on [cone $(\mathscr{E}' - e_0)$] $\supset$ [co co $(\mathscr{E} - e_0)$] [see Lemma (I.1.41)].

5 Also note that if ϕ is finitely differentiable at e_0, then Hypothesis (3) is satisfied because we can choose $W = \{0\}$ [see (I.7.20)]. If $Z = Z' + z_0$, where $z_0 \in \mathscr{Z}$ and Z' is a (convex) cone, then Hypothesis (3) is satisfied whenever ϕ is finitely $\bar{Z}'$-semidifferentiable at e_0, inasmuch as we can then choose $W = \bar{Z}'$ [see Lemmas (I.4.19, 23, and 28)].

6 We also point out that if, for some sets Z_1 and W in $\mathscr{Z}$, $0 \in W$ and $Z_1 + W = Z$, then, as is easily seen, (int Z_1) + W = int Z_1.

We can show [and shall in Theorem (18)] that, if e_0 is a (φ,ϕ,Z)-extremal and Hypotheses (1)–(3) hold, then, for every finite W-semidifferential $D\phi(e_0;\cdot)$ of ϕ at e_0, there exist a vector $\alpha \in R^m$ and a linear functional $\ell \in \mathscr{Z}^*$, not both zero, such that

7 $$\alpha \cdot D\varphi(e_0;y) + \ell \circ D\phi(e_0;y) \leq 0 \quad \text{for all} \quad y \in \text{co co}\,(\mathscr{E} - e_0),$$

8 $$\ell(z) \geq \ell \circ \phi(e_0) \quad \text{for all} \quad z \in \bar{Z},$$

where $D\varphi(e_0;\cdot)$ denotes the finite differential of φ at e_0.

As we shall see in Section 3, this result includes Theorem (II.1.4) as a special case.

As we noted in Chapter II, in many applications—particularly those in the theory of optimal control—Hypothesis (1) is violated because the set $\mathscr{E}$ is not finitely open in itself. Consequently, we must look for alternate hypotheses under which the just-indicated multiplier rule is valid.

9 It turns out that, in the applications in which we are interested, although $\mathscr{E}$ is not finitely open in itself, $\mathscr{E}'$ is a subset (which *is* finitely open in itself) of a linear *topological* space $\mathscr{Y}$, and $\mathscr{E}$ is "almost" finitely open in itself in the sense that, for every finite subset $\{e_1, \ldots, e_\rho\}$ of $\mathscr{E}$ and each neighborhood N of 0 in $\mathscr{Y}$, there exist a number $\varepsilon_0 > 0$ and a *continuous* function $\gamma: \{\lambda: \lambda = (\lambda^1, \ldots, \lambda^\rho) \in R^\rho,\ 0 \leq \lambda^i \leq \varepsilon_0$ for all $i\} \to N$ such that $e_0 + \sum_{i=1}^{\rho} \lambda^i(e_i - e_0) + \gamma(\lambda^1, \ldots, \lambda^\rho) \in \mathscr{E}$ whenever $0 \leq \lambda^i \leq \varepsilon_0$ for each $i = 1, \ldots, \rho$. The set $\mathscr{E}$ is evidently finitely open in itself if and only if we can choose (for some $\varepsilon_0 > 0$) γ to vanish identically. Thus, loosely speaking, γ is a measure of how far the set $\mathscr{E}$ is from being finitely open in itself. Since γ may be chosen to be "arbitrarily small," $\mathscr{E}$ is "almost" finitely open in itself.

Having relaxed the requirement on $\mathscr{E}$ from that of being finitely open in itself to that of being "almost" so, it is only natural to ask: Can one tighten the requirements on the set $\mathscr{E}'$ and on the functions φ and ϕ in such a way that the previously indicated necessary conditions remain in force? The answer is in the affimative; indeed, it is sufficient to require that $\mathscr{E}'$ is open and that the functions φ and ϕ, in addition to satisfying Hypotheses (2) and (3), are continuous on $\mathscr{E}'$.

10 Unfortunately, in the applications to optimal control theory which are one of our principal goals, the set $\mathscr{E}'$ is ordinarily *not* open (although it is usually finitely open), and the functions φ and ϕ are often *not* continuous on $\mathscr{E}'$. In fact, φ and ϕ are often not even continuous on $\mathscr{E}$, and are continuous only on certain special subsets of $\mathscr{E}'$. With $\mathscr{E}$ only "almost" finitely open in itself, with $\mathscr{E}'$ only finitely open, and with φ and ϕ continuous only on suitable subsets of $\mathscr{E}'$, the previously indicated necessary conditions generally do not hold. However, it is true that, for the applications that interest us, the set $\mathscr{E}$ is not only "almost" finitely open in itself, but has an additional property (which is tied to the sets on which φ and ϕ are continuous) under which the necessary conditions do hold.

We shall now make some hypotheses which meet the conflicting requirements which we have just described. That is, on the one hand, these hypotheses are sufficiently strong that, with them, our desired multiplier rule does hold as a necessary condition, while, on the other hand, they are sufficiently weak that the problem data in the applications of interest do satisfy them.

11 HYPOTHESIS. *The set* $\mathscr{E}'$ *is a set in a linear topological space* $\mathscr{Y}$ *and is finitely open in itself.*

12 HYPOTHESIS. *The function* φ *is finitely differentiable at* e_0.

13 HYPOTHESIS. *There is a closed, convex cone* $W \subset \mathscr{X}$ *such that* (i) $(\text{int } Z) + W = \text{int } Z$, *and* (ii) *the function* ϕ *is finitely* W-*semidifferentiable at* e_0.

14 HYPOTHESIS. *For every finite subset* $\{e_1, \ldots, e_\rho\}$ *of* $\mathscr{E}$, there *exists a subset* $\mathscr{E}_0$ *of* $\mathscr{E}$ *and a number* $\varepsilon_0 > 0$ *with the following properties*:

A. *The set* $\mathscr{R}$ *defined by*

15 $$\mathscr{R} = \left\{ e \colon e = e_0 + \sum_{i=1}^{\rho} \lambda^i (e_i - e_0), 0 \leq \lambda^i \leq \varepsilon_0 \text{ for } i = 1, \ldots, \rho \right\}$$

is contained in $\mathscr{E}'$.

B. *The functions* φ *and* ϕ *are continuous on* $\mathscr{R} \cup \mathscr{E}_0$ [*see* (I.3.24)].

C. *For every neighborhood* N *of* 0 *in* $\mathscr{Y}$, *there exists a continuous function* γ: $\{\lambda\colon \lambda = (\lambda^1, \ldots, \lambda^\rho) \in R^\rho,\ 0 \leq \lambda^i \leq \varepsilon_0$ *for* $i = 1, \ldots, \rho\} \to N$ *such that*

16 $$e_0 + \sum_{i=1}^{\rho} \lambda^i (e_i - e_0) + \gamma(\lambda^1, \ldots, \lambda^\rho) \in \mathscr{E}_0$$

$$\text{whenever} \quad 0 \leq \lambda^i \leq \varepsilon_0 \quad \text{for each} \quad i.$$

17 Note that Hypotheses (1) and (11) are essentially the same with regard to $\mathscr{E}'$, although in the latter we require $\mathscr{Y}$ to be a linear *topological* space; Hypothesis (12) is slightly weaker than Hypothesis (2), and Hypotheses (13) and (3) coincide, so that remarks (4)–(6) are pertinent here too. It is clear that Hypothesis (14) implies that $\mathscr{E}$ is "almost" finitely open in itself in the sense previously indicated, but now, in addition, we require that (16) holds, where $\mathscr{E}_0$ is a subset of $\mathscr{E}$ on which φ and ϕ are continuous. In fact, φ and ϕ must be continuous on $\mathscr{E}_0 \cup \mathscr{R}$, where $\mathscr{R}$ is given by (15).

We can now state the promised necessary conditions, which we shall refer to as an *abstract multiplier rule.*

18 THEOREM. *Let e_0 be a (φ,ϕ,Z)-extremal, and suppose that either Hypotheses* (1)–(3) *or Hypotheses* (11)–(14) *hold. Then, for every*† *finite W-semidifferential $D\phi(e_0;\cdot)$ of ϕ at e_0, there exist a vector $\alpha \in R^m$ and a linear functional $\ell \in \mathscr{Z}^*$, not both zero, such that*

19 $$\alpha \cdot D\varphi(e_0; y) + \ell \circ D\phi(e_0; y) \leq 0 \quad \textit{for all} \quad y \in \text{co co}(\mathscr{E} - e_0),$$

20 $$\ell(z) \geq \ell \circ \phi(e_0) \quad \textit{for all} \quad z \in \bar{Z},$$

where $D\varphi(e_0;\cdot)$ denotes the finite differential of φ at e_0.

Proof. In essence, the proof of this theorem is very similar to that of Theorem (II.1.4), and the reader who has fully understood the latter proof may find it convenient to refer back to Chapter II as he reads the present proof.

Thus, suppose that e_0 is a (φ,ϕ,Z)-extremal satisfying one of the indicated sets of hypotheses, and let $D\phi(e_0;\cdot)$ be a finite W-semidifferential of ϕ at e_0. We first introduce some notation. Let us denote $D\varphi(e_0;\cdot)$ (the finite differential of φ at e_0) and $D\phi(e_0;\cdot)$ simply by $D\varphi(\cdot)$ and $D\phi(\cdot)$, respectively, and let

$$\mathscr{J} = \{e: e \in \mathscr{E}, \phi(e) \in \text{int } Z\},$$

21 $$\mathscr{K} = \{y: y \in \text{co co}(\mathscr{E} - e_0), \phi(e_0) + D\phi(y) \in \text{int } Z\}.$$

It follows at once that $0 \notin \varphi(\mathscr{J})$. Further, it is a direct consequence of the convexity of the sets [co co $(\mathscr{E} - e_0)$] and int Z [see Lemma I.4.19)], the W-convexity of the function $D\phi$, and the relation (int Z) + W = int Z that $\mathscr{K}$ is a convex set in $\mathscr{Y}$ (if it is not empty). Loosely speaking, $(\mathscr{K} + e_0)$ may be looked upon as a "first-order" convex approximation to $\mathscr{J}$ near e_0. More precisely, we shall show that $D\varphi(\mathscr{K})$ is a simplicial linearization [see (I.5.39)] of $\varphi(\mathscr{J})$. Note that, since $D\varphi$ is linear, $D\varphi(\mathscr{K})$ is convex, by (I.2.18).

Thus, let S be an arbitrary simplex in R^m such that $0 \in S \subset D\varphi(\mathscr{K})$, and let $\eta \in R_+$ be arbitrary. Let the vertices of S be $D\varphi(y_1), \ldots, D\varphi(y_k)$, where $\{y_1, \ldots, y_k\} \subset \mathscr{K}$, and let $\mathscr{S}$ denote the simplex in $\mathscr{Y}$ whose vertices are $y_1,\ldots,y_k$. [Note that $y_1,\ldots,y_k$ are in general position (see (I.1.22)) because $D\varphi(y_1), \ldots, D\varphi(y_k)$ are (see (I.1.31)).] Since $D\varphi$ is linear, $D\varphi(\mathscr{S}) = S$. Also, because

† Recall that a W-semidifferential is generally not unique [see (I.7.14)].

$\mathscr{K}$ is convex, $\mathscr{E} \subset \mathscr{E}'$, and $\mathscr{E}'$ is finitely open in itself, [see Lemma (I.1.41)] $\mathscr{S} \subset \mathscr{K} \subset$ co co $(\mathscr{E}' - e_0) =$ cone $(\mathscr{E}' - e_0)$. Further, $[\phi(e_0) + D\phi(y_j)] \in \text{int } Z$ for $j = 1, \ldots, k$, so that there are neighborhoods $N'_1, \ldots, N'_k$ of 0 in $\mathscr{Z}$ such that

22 $$\phi(e_0) + D\phi(y_j) + N'_j \subset \text{int } Z \quad \text{for} \quad j = 1, \ldots, k.$$

Let

23 $$N' = \bigcap_{j=1}^{k} N'_j.$$

We shall show that

24 $$\phi(e_0) + \varepsilon[D\phi(y) + N'] \subset \text{int } Z$$
$$\text{for all} \quad y \in \mathscr{S} \quad \text{and every} \quad \varepsilon \in (0,1).$$

Indeed, if $y \in \mathscr{S}$, then there are numbers $\beta^1, \ldots, \beta^k$ in $\bar{R}_+$ such that $y = \sum_{j=1}^k \beta^j y_j$ and $\sum_{j=1}^k \beta^j = 1$. It follows from the W-convexity of $D\phi$ that

$$D\phi(y) \in \sum_{j=1}^{k} \beta^j D\phi(y_j) + W.$$

Hence, making use of (22) and (23), we conclude, by virtue of the convexity of int Z, that

$$\phi(e_0) + D\phi(y) + N' \subset \sum_{j=1}^{k} \beta^j[\phi(e_0) + D\phi(y_j) + N'_j] + W$$
$$\subset \sum_{j=1}^{k} \beta^j(\text{int } Z) + W \subset (\text{int } Z) + W = \text{int } Z,$$

i.e.,

25 $$\phi(e_0) + D\phi(y) + N' \subset \text{int } Z \quad \text{for all} \quad y \in \mathscr{S}.$$

Now $\phi(e_0) \in Z$ [see Definition (1.14)], and

$$\phi(e_0) + \varepsilon[D\phi(y) + N'] = (1 - \varepsilon)\phi(e_0) + \varepsilon[\phi(e_0) + D\phi(y) + N']$$
$$\text{for every} \quad y \in \mathscr{S} \quad \text{and} \quad \varepsilon \in R,$$

so that, if we take into account (25) and Lemma (I.4.19), (24) follows at once.

Now let us suppose that Hypotheses (1)–(3) hold. Here it follows from Lemma (I.1.41) that cone $(\mathscr{E} - e_0) =$ co co $(\mathscr{E} - e_0)$. Consequently, $y_j \in$ cone $(\mathscr{E} - e_0)$ for each j, and, because $\mathscr{E}$ is finitely open in itself, there is an $\varepsilon_1 \in (0,1)$ such that $e_0 + \varepsilon\mathscr{S} \subset \mathscr{E}$ for all $\varepsilon \in [0,\varepsilon_1)$. By definition of $D\phi$, there is an $\varepsilon_0 \in (0,\varepsilon_1)$ such that

$$\frac{\phi(e_0 + \varepsilon y) - \phi(e_0)}{\varepsilon} \in D\phi(y) + N' + W \quad \text{for all} \quad y \in \mathscr{S}$$

whenever $0 < \varepsilon < \varepsilon_0 < 1$. From this, it follows at once, by virtue of (24) and because $(\text{int } Z) + \varepsilon W = (\text{int } Z) + W = \text{int } Z$ for all $\varepsilon > 0$, that $(e_0 + \varepsilon\mathscr{S}) \subset \mathscr{J}$ whenever $0 < \varepsilon < \varepsilon_0$. Arguing as in the proof of Theorem (II.1.4), we can now show that $D\varphi(\mathscr{K})$ is indeed a simplicial linearization of $\varphi(\mathscr{J})$.

Now let us turn to the case where Hypotheses (11)–(14) hold. Here, we cannot conclude that $y_j \in \text{cone}\,(\mathscr{E} - e_0)$ for each j. However, since $y_j \in \text{co co}\,(\mathscr{E} - e_0)$ for each j, there are elements $e_1, \ldots, e_\rho$ in $\mathscr{E}$ and numbers β_j^i ($i = 1, \ldots, \rho$ and $j = 1, \ldots, k$) in $\bar{R}_+$ such that [see (I.1.29)]

$$y_j = \sum_{i=1}^{\rho} \beta_j^i(e_i - e_0) \quad \text{for} \quad j = 1, \ldots, k.$$

Let

$$\beta_0 = 1 + \max\,\{\beta_j^i \colon i = 1, \ldots, \rho; j = 1, \ldots, k\}.$$

By Hypothesis (14), there exist a subset $\mathscr{E}_0$ of $\mathscr{E}$ and a number $\varepsilon_0 > 0$ satisfying Conditions A, B, and C of the hypothesis. We now prove the following lemma.

26 LEMMA. *There is a number $\bar{\varepsilon} \in (0,1)$ such that, for every $\tilde{\varepsilon} \in (0,\bar{\varepsilon})$ and every neighborhood N of 0 in $\mathscr{Y}$, there are maps Θ and $\tilde{\gamma}$ from $\mathscr{S}$ into $\mathscr{Y}$ (possibly depending on $\tilde{\varepsilon}$ and N), with Θ continuous, such that*

27 $$\Theta(\mathscr{S}) \subset \mathscr{E}_0,$$

28 $$\Theta(y) = e_0 + \tilde{\varepsilon}y + \tilde{\gamma}(y) \quad \textit{for all} \quad y \in \mathscr{S},$$

29 $$\tilde{\gamma}(\mathscr{S}) \subset N,$$

30 $$e_0 + \tilde{\varepsilon}\mathscr{S} \subset \mathscr{R}\ [\textit{see } (15)].$$

Roughly speaking, Lemma (26) asserts that $\mathscr{S}$ can be mapped into $\mathscr{E}_0$ by means of a "severe" contraction (multiplication by $\tilde{\varepsilon}$), followed by a translation through e_0 and a "small" continuous distortion (the addition of $\tilde{\gamma}$).

Proof. For every $y \in \mathscr{S}$, let $\beta^1(y), \ldots, \beta^k(y)$ denote the barycentric coordinates [see (I.1.33))] of y in $\mathscr{S}$, and let

31 $$\tilde{\beta}^i(y) = \sum_{j=1}^{k} \beta^j(y)\beta_j^i \quad \text{for} \quad i = 1, \ldots, \rho.$$

It follows by a trivial calculation that, for every $y \in \mathscr{S}$,

$$y = \sum_{i=1}^{\rho} \tilde{\beta}^i(y)(e_i - e_0),$$

$$0 \leq \tilde{\beta}^i(y) \leq \beta_0 \quad \text{for each} \quad i = 1, \ldots, \rho.$$

Let us set $\bar{\varepsilon} = \varepsilon_0/(\varepsilon_0 + \beta_0)$, so that $0 < \bar{\varepsilon} \leq \varepsilon_0$ and $\bar{\varepsilon} < 1$, and let us arbitrarily choose a number $\tilde{\varepsilon} \in (0,\bar{\varepsilon})$ and a neighborhood N of 0 in $\mathscr{Y}$. Evidently,

32 $$e_0 + \tilde{\varepsilon} y = e_0 + \sum_{i=1}^{\rho} \tilde{\varepsilon}\tilde{\beta}^i(y)(e_i - e_0) \quad \text{for every} \quad y \in \mathscr{S},$$

$$0 \leq \tilde{\varepsilon}\tilde{\beta}^i(y) < \varepsilon_0 \quad \text{for each} \quad i = 1, \ldots, \rho \quad \text{and every} \quad y \in \mathscr{S},$$

so that (30) holds. Let γ be a continuous map from $\{\lambda : \lambda = (\lambda^1, \ldots, \lambda^\rho) \in R^\rho, 0 \leq \lambda^i \leq \varepsilon_0 \text{ for } i = 1, \ldots, \rho\}$ into N such that (16) holds, and let $\tilde{\gamma} : \mathscr{S} \to N$ be the map defined by the formula

$$\tilde{\gamma}(y) = \gamma(\tilde{\varepsilon}\tilde{\beta}^1(y), \ldots, \tilde{\varepsilon}\tilde{\beta}^\rho(y)), \quad y \in \mathscr{S}.$$

Now (29) evidently holds and it follows from (31) and Theorems (I.4.21) and (I.3.27) that $\tilde{\gamma}$ is continuous. If we define the map $\Theta : \mathscr{S} \to \mathscr{Y}$ through (28), so that Θ is evidently continuous, then (27) is an immediate consequence of (16) and (32), completing the proof of the lemma. |||

We shall now show that if the number $\tilde{\varepsilon} \in (0,\bar{\varepsilon})$ and the neighborhood N of 0 in $\mathscr{Y}$ are suitably chosen, then the map Θ correspondingly determined by Lemma (26) satisfies the following relations:

33 $$\Theta(\mathscr{S}) \subset \mathscr{J},$$

34 $$\left| \frac{\varphi \circ \Theta(y)}{\tilde{\varepsilon}} - D\varphi(y) \right| < \eta \quad \text{for all} \quad y \in \mathscr{S}.$$

This will at once imply our desired conclusion that $D\varphi(\mathscr{K})$ is a simplicial linearization of $\varphi(\mathscr{J})$. Indeed, we then can define the maps $\chi_0 : S \to R^m$ and $\chi_1 : R^m \to R^m$ as follows:

35 $$\chi_0(\xi) = \frac{\varphi \circ \Theta(y(\xi))}{\tilde{\varepsilon}}, \quad \chi_1(\xi) = \tilde{\varepsilon}\xi,$$

where, for each $\xi \in S$, $y(\xi)$ is the point in $\mathscr{S}$ whose barycentric coordinates (in $\mathscr{S}$) coincide with those of ξ (in S), so that $D\varphi(y(\xi)) = \xi$ for all $\xi \in S$. Evidently, χ_1 is linear, and it follows from the continuity of Θ, Condition B of Hypothesis (14), (27), and Theorems (I.3.27) and (I.4.21) that χ_0 is continuous. Further, if (33) and (34) hold, then $|\chi_0(\xi) - \xi| < \eta$ for all $\xi \in S$, and $\chi_1 \circ \chi_0(S) \subset \varphi(\mathscr{J})$; i.e., $D\varphi(\mathscr{K})$ is a simplicial linearization of $\varphi(\mathscr{J})$.

Thus it only remains to show that (33) and (34) hold, for judicious choices of $\tilde{\varepsilon}$ and of N.

Let N'' be a neighborhood of 0 in $\mathscr{Z}$ such that $N'' + N'' \subset N'$ [see Corollary (I.4.4)], where N' is defined by (23) and (22). By Hypotheses (11)–(13), there is an $\varepsilon_3 > 0$ such that, for all $y \in \rho\beta_0[\text{co}\,\{e_1 - e_0, \ldots, e_\rho - e_0, 0\}]$ and *a fortiori* for all $y \in \mathscr{S}$,

36 $$\frac{\phi(e_0 + \varepsilon y) - \phi(e_0)}{\varepsilon} \in D\phi(y) + N'' + W,$$

37 $$\left|\frac{\varphi(e_0 + \varepsilon y) - \varphi(e_0)}{\varepsilon} - D\varphi(y)\right| < \frac{\eta}{2}$$

whenever $0 < \varepsilon < \varepsilon_3$. Let $\tilde{\varepsilon}$ be an arbitrary fixed number such that $0 < \tilde{\varepsilon} < \max\,\{\varepsilon_3, \bar{\varepsilon}\}$.

It follows from (I.2.18) that the set $\mathscr{R}$ [see (15)] is a convex polyhedron, and, by Theorem (I.4.21), is therefore compact. By Hypothesis (14), Condition B, φ and ϕ are continuous on $\mathscr{R} \cup \mathscr{E}_0$, so that, by Corollary (I.5.32), the restrictions to $\mathscr{R} \cup \mathscr{E}_0$ of φ and ϕ are uniformly continuous on $\mathscr{R}$ [see (I.5.27)]. Consequently, there is a neighborhood N of 0 in $\mathscr{Y}$ such that

38 $$|\varphi(e + e') - \varphi(e)| < \tilde{\varepsilon}\eta/2,$$

39 $$\phi(e + e') - \phi(e) \in \tilde{\varepsilon}N''$$

whenever $e \in \mathscr{R}$, $e' \in N$, and $e + e' \in \mathscr{R} \cup \mathscr{E}_0$ [see Corollary (I.4.6)]. With this choice of $\tilde{\varepsilon}$ and N, let us consider the corresponding maps Θ and $\tilde{\gamma}$ [as specified in Lemma (26)]. We have, for all $y \in \mathscr{S}$, by virtue of Hypothesis (13), (24), (27)–(30), (36), and (39) that

$$\begin{aligned}
\phi \circ \Theta(y) &= [\phi(e_0 + \tilde{\varepsilon}y + \tilde{\gamma}(y)) - \phi(e_0 + \tilde{\varepsilon}y)] \\
&\quad + \tilde{\varepsilon}\frac{\phi(e_0 + \tilde{\varepsilon}y) - \phi(e_0)}{\tilde{\varepsilon}} + \phi(e_0) \\
&\in \tilde{\varepsilon}N'' + \tilde{\varepsilon}[D\phi(y) + N'' + W] + \phi(e_0) \\
&\in \phi(e_0) + \tilde{\varepsilon}[D\phi(y) + N'] + \tilde{\varepsilon}W \\
&\subset (\text{int } Z) + W = \text{int } Z,
\end{aligned}$$

i.e., $\phi \circ \Theta(\mathscr{S}) \subset \text{int } Z$. But this relation, together with (27) (since $\mathscr{E}_0 \subset \mathscr{E}$), implies that (33) holds.

Also—see (37) and (38) and note that $\varphi(e_0) = 0$ by Definition (1.14)—

$$\left| \frac{\varphi \circ \Theta(y)}{\tilde{\varepsilon}} - D\varphi(y) \right| \leq \frac{1}{\tilde{\varepsilon}} |\varphi(e_0 + \tilde{\varepsilon}y + \tilde{\gamma}(y)) - \varphi(e_0 + \tilde{\varepsilon}y)|$$
$$+ \left| \frac{\varphi(e_0 + \tilde{\varepsilon}y) - \varphi(e_0)}{\tilde{\varepsilon}} - D\varphi(y) \right| < \frac{\eta}{2} + \frac{\eta}{2} = \eta$$

for all $y \in \mathscr{S}$; i.e., (34) holds.

Thus, we have shown that, either under Hypotheses (1)–(3) or under Hypotheses (11)–(14), $D\varphi(\mathscr{K})$ is a simplicial linearization of $\varphi(\mathscr{J})$.

Since $0 \notin \varphi(\mathscr{J})$, Lemma (I.5.43) now implies that there is a nonzero vector $\tilde{\alpha} \in R^m$ such that $\tilde{\alpha} \cdot D\varphi(y) \leq 0$ for all $y \in \mathscr{K}$. Consequently, the open, convex set [see (I.1.16) and (I.3.35)] $(\text{int } Z) \times R_-$ in $\mathscr{Z} \times R$ does not meet the subset B of $\mathscr{Z} \times R$ defined as follows:

40 $$B = \{(z, -\tilde{\alpha} \cdot D\varphi(y)) : y \in \text{co co } (\mathscr{E} - e_0), z \in \mathscr{Z}, \phi(e_0) + D\phi(y) \in z + W\}.$$

It follows at once from the linearity of $D\varphi$, the W-convexity of $D\phi$, and the convexity of co co $(\mathscr{E} - e_0)$ that B is a convex set. Also, because $D\phi(0) = 0$ [see (I.7.13)] and $0 \in W$ by Lemma (I.4.23), $(\phi(e_0),0) \in B$, and, since $\phi(e_0) \in Z$, $(\phi(e_0),0) \in \bar{B} \cap \overline{(\text{int } Z) \times R_-}$ [see (I.3.34) and Lemma (I.4.19)]. Hence, by Corollary (I.5.18), there is a nonzero linear functional $\tilde{\ell} \in (\mathscr{Z} \times R)^*$ such that

$$\tilde{\ell}((z_1,\xi_1)) \leq \tilde{\ell}((\phi(e_0),0)) \leq \tilde{\ell}((z_2,\xi_2)) \quad \text{for all} \quad (z_1,\xi_1) \in B,$$
$$z_2 \in \bar{Z} \quad \text{and} \quad \xi_2 \in \bar{R}_-.$$

But, by (I.5.25), this means that there are a linear functional $\ell \in \mathscr{Z}^*$ and a real number η_1, not both zero, such that

41 $$\ell \circ \phi(e_0) + \ell \circ D\phi(y) - \eta_1 \tilde{\alpha} \cdot D\varphi(y) \leq \ell \circ \phi(e_0) \quad \text{for all} \quad y \in \text{co co } (\mathscr{E} - e_0),$$

42 $$\ell(z) \geq \ell \circ \phi(e_0) \quad \text{for all} \quad z \in \bar{Z}.$$

If we set $\alpha = -\eta_1 \tilde{\alpha}$, so that α and ℓ cannot both vanish, then we see that inequalities (19) and (20) precisely coincide with (41) and (42). |||

The abstract multiplier rule takes on special forms if the problem data satisfy suitable hypotheses, as is indicated in the following corollaries.

43 COROLLARY. *If, in Theorem* (18), $\phi(e_0) \in \text{int } Z$, *then* $\ell = 0$ *and* $\alpha \neq 0$.

Proof. This corollary follows at once from (20) and Lemma (I.5.14). |||

44 COROLLARY. *If, in Theorem* (18), *Z is a cone, then* (20) *holds if and only if* (i) $\ell \in -(\bar{Z})^*$ [*see* (I.5.23)] *and* (ii) $\ell \circ \phi(e_0) = 0$.

Proof. If Z is a cone and e_0 is a (φ,ϕ,Z)-extremal, so that $\phi(e_0) \in Z$, then it follows from Lemma (I.4.23) that 0 and $2\phi(e_0)$ both belong to $\bar{Z}$. But then (20) at once implies $\ell \circ \phi(e_0) = 0$, which means that $\ell \in -(\bar{Z})^*$. The converse is obvious. |||

45 COROLLARY. *If, in Theorem* (18), $D\phi$ *is linear* [*see* (I.2.7)], *then* (19) *is equivalent to the maximum principle*

$$(\alpha \cdot D\varphi + \ell \circ D\phi)(e) \leq (\alpha \cdot D\varphi + \ell \circ D\phi)(e_0) \quad \textit{for all} \quad e \in \mathscr{E}.$$

If also e_0 *is an internal point* [*see* (I.1.42)] *of* $\mathscr{E}$, *then*

$$\alpha \cdot D\varphi + \ell \circ D\phi = 0.$$

Proof. The first conclusion of the corollary is obvious; the second conclusion follows at once from Lemma (I.2.39). |||

46 COROLLARY. *Let* e_0 *be a* (φ,ϕ,Z)*-extremal, and suppose that either Hypotheses* (1), (2) *or Hypotheses* (11), (12), (14) *hold. Further, suppose that* ϕ *is defined and W-convex on* $\mathscr{Y}$, *for some closed, convex cone* $W \subset \mathscr{X}$ *such that* $(\text{int } Z) + W = \text{int } Z$. *Then there exist a vector* $\alpha \in R^m$ *and a linear functional* $\ell \in \mathscr{X}^*$, *not both zero, such that*

47 $$(\alpha \cdot D\varphi + \ell \circ \phi)(e) \leq (\alpha \cdot D\varphi + \ell \circ \phi)(e_0) \quad \textit{for all} \quad e \in e_0 + \text{co co}(\mathscr{E} - e_0),$$

48 $$\ell(z) \geq \ell \circ \phi(e_0) \quad \textit{for all} \quad z \in \bar{Z},$$

where $D\varphi$ *denotes the finite differential of* φ *at* e_0.

Proof. If ϕ is defined and W-convex on $\mathscr{Y}$, then, by Theorem (I.7.21), ϕ is finitely W-semidifferentiable, with the function $y \to \phi(e_0 + y) - \phi(e_0)$ as a finite W-semidifferential at e_0. Hence,

Hypothesis (3) [or (13)] is satisfied, and our conclusion follows at once from Theorem (18). |||

49 Note that the inequality in (47) holds, in particular, for all $e \in \mathscr{E}$.

50 If the hypotheses of Corollary (46) are satisfied, and if, in addition, ϕ is finitely directionally differentiable at e_0, so that, by the last sentence in (I.7.20), ϕ is also finitely W-semidifferentiable at e_0, then it appears that Theorem (18) will yield two sets of necessary conditions: one in which the basic multiplier rule has the form of (19), where $D\phi(e_0;\cdot)$ is the directional differential of ϕ at e_0; and a second in which (19) has been replaced by (47), although the "multipliers" α and ℓ in (19) need not be the same as those in (47). However, for a given α and ℓ, one of these multiplier rules holds if and only if the other one does (with the same α and ℓ), so that the two rules are essentially equivalent. Further, we may even relax the requirement on ϕ from finite to Gâteaux directional differentiability (at e_0)—and note that, if $\mathscr{X} = R^\mu$ and $W = \bar{R}^\mu_-$, then, by Corollary (I.7.8), ϕ must be Gâteaux directionally differentiable. The preceding conclusions follow at once from the following Lemma, which is closely related to Lemma (II.3.17).

51 LEMMA. *Let e_0 be a (φ,ϕ,Z)-extremal, and suppose that Hypotheses* (11) *and* (12) *hold, except that in the former it suffices to assume that $\mathscr{Y}$ is a linear vector space. Further, suppose that ϕ is defined and W-convex on $\mathscr{Y}$, for some closed convex cone $W \subset \mathscr{X}$ such that* (int Z) + W = int Z, *and that ϕ is Gâteaux directionally differentiable at e_0. Let $\ell \in \mathscr{X}^*$ satisfy* (48). *Then* (19) *holds (for some $\alpha \in R^m$)—with $D\phi(e_0;\cdot)$ denoting the Gâteaux directional differential of ϕ at e_0, and $D\varphi(e_0;\cdot)$ the finite differential of φ at e_0—if and only if* (47) *holds—where $D\varphi(\cdot) = D\varphi(e_0;\cdot)$.*

Proof. Suppose that, under the indicated hypotheses, $\alpha \in R^m$ and $\ell \in \mathscr{X}^*$ satisfying (47) and (48). It then follows at once that

$$\alpha \cdot D\varphi(y) + \ell\left(\frac{\phi(e_0 + \varepsilon y) - \phi(e_0)}{\varepsilon}\right) \leq 0$$

$$\text{for all} \quad y \in \text{co co}(\mathscr{E} - e_0) \quad \text{and} \quad \varepsilon > 0.$$

Passing to the limit as $\varepsilon \to 0^+$, and taking into account the continuity of ℓ, we conclude that (19) holds.

Conversely, suppose that (19) and (48) hold. Since ϕ is defined and W-convex on $\mathscr{Y}$, we have that $\phi(e_0 + \varepsilon y) = \phi((1 - \varepsilon)e_0 + \varepsilon(e_0 + y)) \leq_W (1 - \varepsilon)\phi(e_0) + \varepsilon\phi(e_0 + y)$ for all $y \in \mathscr{Y}$ and $\varepsilon \in (0,1)$, or that

$$\frac{\phi(e_0 + \varepsilon y) - \phi(e_0)}{\varepsilon} - [\phi(e_0 + y) - \phi(e_0)] \in W$$

$$\text{for all} \quad y \in \mathscr{Y} \quad \text{and} \quad \varepsilon \in (0,1).$$

Passing to the limit as $\varepsilon \to 0^+$, we conclude that (recall that W is closed)

52 $$D\phi(y) - [\phi(e_0 + y) - \phi(e_0)] \in W \quad \text{for all} \quad y \in \mathscr{Y}.$$

Since $\phi(e_0) \in Z$ and (int Z) + W = int Z, it directly follows, by virtue of Lemma (I.4.19), that $\phi(e_0) + W \subset \bar{Z}$. Because of (48), this implies that $\ell \in -(W^*)$, as a result of which (47) is a direct consequence of (19) and (52). |||

Let us consider the following supplemental hypothesis in Theorem (18).

53 HYPOTHESIS. *The space $\mathscr{Y}$ is a linear topological space, $D\phi(e_0;\cdot)$ [which is defined on* cone $(\mathscr{E}' - e_0)$] *is continuous, and either* (i) $\overline{\text{co co}}\,(\mathscr{E} - e_0) \subset \text{cone}\,(\mathscr{E}' - e_0)$ *or* (ii) *$D\phi(e_0;\cdot)$ can be extended to $\overline{\text{co co}}\,(\mathscr{E} - e_0)$ in such a way that it remains a continuous function with values in $\mathscr{Z}$.*

54 COROLLARY. *If, in Theorem* (18), *Hypothesis* (53) *also holds for some finite W-semidifferential $D\phi(e_0;\cdot)$ of ϕ at e_0, and if $D\varphi(e_0;\cdot)$ is a continuous linear function* [*see* (I.5.44)], *then relation* (19) *takes the stronger form*

55 $$\alpha \cdot D\varphi(e_0;y) + \ell \circ D\phi(e_0;y) \leq 0 \quad \textit{for all} \quad y \in \overline{\text{co co}}\,(\mathscr{E} - e_0).$$

Proof. The corollary follows at once from Theorems (18) and (I.3.25 and 27). |||

Now let us turn to (ϕ,Z)-extremals. Corresponding to Hypotheses (1) and (3) [note that Hypothesis (2) is here meaningless], we shall make the following hypotheses:

56 HYPOTHESIS. *The set $\mathscr{E}'$ is a set in a linear vector space $\mathscr{Y}$, and both $\mathscr{E}$ and $\mathscr{E}'$ are finitely open in themselves.*

57 HYPOTHESIS. *There is a closed, convex cone* $W \subset \mathscr{Z}$ *such that* (i) (int Z) + W = int Z, *and* (ii) *the function* ϕ *is Gâteaux W-semidifferentiable at* e_0 [*see* (I.7.13)].

Hypothesis (56) coincides with Hypothesis (1), but Hypothesis (57) is weaker than Hypothesis (3) in that we require Gâteaux, rather than finite, W-semidifferentiability for ϕ at e_0.

Corresponding to Hypotheses (11) and (14), we shall make the following hypotheses:

58 HYPOTHESIS. *The set* $\mathscr{E}'$ *is a set in a linear topological space* $\mathscr{Y}$ *and is finitely open in itself.*

59 HYPOTHESIS. *For every finite subset* $\{e_1, \ldots, e_\rho\}$ *of* $\mathscr{E}$, *there exist a subset* $\mathscr{E}_0$ *of* $\mathscr{E}$ *and a number* $\varepsilon_0 > 0$ *with the following properties*:

A. *The set* $\mathscr{R}$ *defined by*

$$\mathscr{R} = \left\{ e \colon e = e_0 + \sum_{i=1}^{\rho} \lambda^i (e_i - e_0), 0 \leq \lambda^i \leq \varepsilon_0 \text{ for } i = 1, \ldots, \rho \right\}$$

is contained in $\mathscr{E}'$.

B. *The function* ϕ *is continuous on* $\mathscr{R} \cup \mathscr{E}_0$.

C. *For every neighborhood N of* 0 *in* $\mathscr{Y}$ *and every* $e \in \mathscr{R}$, *there is an element* $e' \in N$ *(possibly depending on N and on e) such that* $(e + e') \in \mathscr{E}_0$.

Hypothesis (58) coincides with Hypothesis (11), and Hypothesis (59) differs from Hypothesis (14) primarily in that Condition C of the former is somewhat weaker than Condition C of the latter.

The remarks (4)–(6) and (17), with obvious modifications in (4), (5), and (17), are pertinent here also.

The abstract multiplier rule for (ϕ, Z)-extremals then takes the following form.

60 THEOREM. *Let* e_0 *be a* (ϕ, Z)*-extremal, and suppose that either Hypotheses* (56) *and* (57) *or Hypotheses* (58), (57), *and* (59) *hold. Then, for every Gâteaux W-semidifferential* $D\phi(e_0;\cdot)$ *of* ϕ *at* e_0, *there exists a nonzero linear functional* $\ell \in \mathscr{Z}^*$ *such that*

61 $$\ell \circ D\phi(e_0;y) \leq 0 \quad \text{for all} \quad y \in \text{co co}(\mathscr{E} - e_0),$$

62 $$\ell(z) \geq \ell \circ \phi(e_0) \quad \text{for all} \quad z \in \bar{Z}.$$

Proof. In essence, the proof of this theorem is very similar to that of Theorem (II.1.18).

Thus, suppose that e_0 is a (ϕ,Z)-extremal satisfying one of the indicated sets of hypotheses, and let $D\phi(e_0;\cdot)$ be a Gâteaux W-semidifferential of ϕ at e_0. Let us denote $D\phi(e_0;\cdot)$ simply by $D\phi(\cdot)$, and define $\mathscr{K}$ as in the proof of Theorem (18), by means of (21). We shall show that $\mathscr{K}$ is empty.

Suppose the contrary; i.e., let $\tilde{y} \in \mathscr{K}$, so that $\phi(e_0) + D\phi(\tilde{y}) \in \operatorname{int} Z$, which means that there is a neighborhood N' of 0 in $\mathscr{Z}$ such that $\phi(e_0) + D\phi(\tilde{y}) + N' \subset \operatorname{int} Z$. Because $\phi(e_0) \in Z$ [see Definition (1.29)], it now follows from Lemma (I.4.19) that

63
$$\begin{aligned}\phi(e_0) &+ \varepsilon[D\phi(\tilde{y}) + N'] \\ &= (1 - \varepsilon)\phi(e_0) + \varepsilon[\phi(e_0) + D\phi(\tilde{y}) + N'] \subset \operatorname{int} Z \\ &\qquad \text{for all} \quad \varepsilon \in (0,1).\end{aligned}$$

First let us suppose that Hypotheses (56) and (57) hold. Here it follows from Lemma (I.1.41) that $\operatorname{cone}(\mathscr{E} - e_0) = \operatorname{co}\operatorname{co}(\mathscr{E} - e_0)$. Consequently, $\tilde{y} \in \operatorname{cone}(\mathscr{E} - e_0)$, and, because $\mathscr{E}$ is finitely open in itself, there is an $\varepsilon_1 \in (0,1)$ such that $e_0 + \varepsilon\tilde{y} \in \mathscr{E}$ for all $\varepsilon \in [0,\varepsilon_1)$. By definition of $D\phi$, there is an $\varepsilon_0 \in (0,\varepsilon_1)$ such that

64
$$\frac{\phi(e_0 + \varepsilon_0\tilde{y}) - \phi(e_0)}{\varepsilon_0} \in D\phi(\tilde{y}) + N' + W.$$

If we set $\tilde{e} = e_0 + \varepsilon_0\tilde{y}$, so that $\tilde{e} \in \mathscr{E}$, we at once conclude, on the basis of (63) and (64) and the relation $(\operatorname{int} Z) + \varepsilon_0 W = (\operatorname{int} Z) + W = \operatorname{int} Z$, that $\phi(\tilde{e}) \in \operatorname{int} Z$, contradicting the hypothesis that e_0 is a (ϕ,Z)-extremal. Hence, $\mathscr{K}$ must be empty.

Now let us turn to the case where Hypotheses (58), (57), and (59) hold. Here we cannot conclude that $\tilde{y} \in \operatorname{cone}(\mathscr{E} - e_0)$. However, since $\tilde{y} \in \operatorname{co}\operatorname{co}(\mathscr{E} - e_0)$, there are elements $e_1, \ldots, e_\rho$ in $\mathscr{E}$ and numbers $\beta^1, \ldots, \beta^\rho$ in R_+ such that $\tilde{y} = \sum_{i=1}^{\rho} \beta^i(e_i - e_0)$. By Hypothesis (59), there exist a subset $\mathscr{E}_0$ of $\mathscr{E}$ and a number $\varepsilon_0 > 0$ satisfying Conditions A, B, and C of the hypothesis.

Let N'' be a neighborhood of 0 in $\mathscr{Z}$ such that $N'' + N'' \subset N'$ [see Corollary (I.4.4)]. By definition of $D\phi$, we can choose a number $\tilde{\varepsilon} \in (0,1)$ such that $\tilde{\varepsilon}\beta^i \le \varepsilon_0$ for $i = 1, \ldots, \rho$ and such that

65
$$\frac{\phi(e_0 + \tilde{\varepsilon}\tilde{y}) - \phi(e_0)}{\tilde{\varepsilon}} \in D\phi(\tilde{y}) + N'' + W.$$

By Hypothesis (59), Condition B, there is a neighborhood N of

0 in $\mathscr{Y}$ such that

66 $$\phi(e_0 + \tilde{\varepsilon}\tilde{y} + y') - \phi(e_0 + \tilde{\varepsilon}\tilde{y}) \in \tilde{\varepsilon}N'' \quad \text{whenever} \quad y' \in N$$
$$\text{and} \quad (e_0 + \tilde{\varepsilon}\tilde{y} + y') \in \mathscr{R} \cup \mathscr{E}_0$$

[see Corollary (I.4.6)], and, by Condition C of this hypothesis, there is an $e' \in N$ such that $e_0 + \tilde{\varepsilon}\tilde{y} + e' \in \mathscr{E}_0$. Let $\tilde{e} = e_0 + \tilde{\varepsilon}\tilde{y} + e'$. It then follows from (63), (65), and (66) that

$$\begin{aligned}
&\phi(\tilde{e}) \\
&= \phi(e_0 + \tilde{\varepsilon}\tilde{y} + e') - \phi(e_0 + \tilde{\varepsilon}\tilde{y}) + \tilde{\varepsilon}\frac{\phi(e_0 + \tilde{\varepsilon}\tilde{y}) - \phi(e_0)}{\tilde{\varepsilon}} + \phi(e_0) \\
&\quad \in \tilde{\varepsilon}N'' + \tilde{\varepsilon}[D\phi(\tilde{y}) + N'' + W] + \phi(e_0) \\
&\quad \subset \phi(e_0) + \tilde{\varepsilon}[D\phi(\tilde{y}) + N'] + \tilde{\varepsilon}W \\
&\quad \subset (\text{int } Z) + W = \text{int } Z,
\end{aligned}$$

which, since $\tilde{e} \in \mathscr{E}_0 \subset \mathscr{E}$, contradicts the hypothesis that e_0 is a (ϕ,Z)-extremal.

Thus, we have shown that, either under Hypotheses (56) and (57) or under Hypotheses (58), (57), and (59), the set $\mathscr{K}$ is empty. Consequently, the open, convex set (int Z) in $\mathscr{Z}$ does not meet the set $B = [\phi(e_0) + D\phi(\text{co co } (\mathscr{E} - e_0)) - W]$ in $\mathscr{Z}$. It follows at once from the W-convexity of $D\phi$ and the convexity of co co $(\mathscr{E} - e_0)$ that B is a convex set. Also, because $D\phi(0) = 0$ [see (I.7.13)] and $0 \in W$ [see Lemma (I.4.23)], $\phi(e_0) \in B \cap Z$, i.e., [see Lemma (I.4.19)] $\phi(e_0) \in \bar{B} \cap \overline{(\text{int } Z)}$. Hence, by Corollary (I.5.18), there is a nonzero linear function $\ell \in \mathscr{Z}^*$ such that

$$\ell(z_1) \leq \ell \circ \phi(e_0) \leq \ell(z_2) \quad \text{for all} \quad z_1 \in B \quad \text{and} \quad z_2 \in \bar{Z},$$

which at once implies that (61) and (62) hold. |||

67 We point out (as is easily verified) that Theorem (60) remains in force if we weaken Hypothesis (59) by replacing Condition B by the following:

B′. *The restriction to* $\mathscr{R} \cup \mathscr{E}_0$ *of the function* ϕ *is continuous at each point of* $\mathscr{R}$.

Just as was true in the case of Theorem (18), Theorem (60) takes on special forms if the problem data satisfy suitable hypotheses, as is indicated in the following corollaries, which may be proved in essentially the same way as Corollaries (44)–(46) and (54).

68 COROLLARY. *If, in Theorem* (60), *Z is a cone, then* (62) *holds if and only if* $\ell \in -(\bar{Z})^*$ *and* $\ell \circ \phi(e_0) = 0$.

69 COROLLARY. *If, in Theorem* (60), $D\phi$ *is linear, then* (61) *is equivalent to the maximum principle*

$$\ell \circ D\phi(e) \leq \ell \circ D\phi(e_0) \quad \text{for all} \quad e \in \mathscr{E}.$$

If also e_0 *is an internal point of* $\mathscr{E}$, *then* $\ell \circ D\phi = 0$.

70 COROLLARY. *Let* e_0 *be a* (ϕ, Z)*-extremal, and suppose that either Hypothesis* (56) *or Hypotheses* (58) *and* (59) *hold. Further, suppose that* ϕ *is defined and W-convex on* $\mathscr{Y}$, *for some closed, convex cone* $W \subset \mathscr{Z}$ *such that* $(\text{int } Z) + W = \text{int } Z$. *Then there exists a nonzero linear functional* $\ell \in \mathscr{Z}^*$ *such that*

71 $$\ell \circ \phi(e) \leq \ell \circ \phi(e_0) \quad \textit{for all} \quad e \in e_0 + \text{co co}(\mathscr{E} - e_0),$$
$$\ell(z) \geq \ell \circ \phi(e_0) \quad \textit{for all} \quad z \in \bar{Z}.$$

Note that the inequality (71) holds, in particular, for all $e \in \mathscr{E}$.

72 If the hypotheses of Corollary (70) are satisfied, and if, in addition, ϕ is Gâteaux directionally differentiable at e_0 [as must be the case, by Corollary (I.7.8), if $\mathscr{Z} = R^\mu$ and $W = \bar{R}^\mu_-$], then Theorem (60) will yield two equivalent sets of necessary conditions: one in which the basic multiplier rule has the form of (61), where $D\phi(e_0;\cdot)$ is the Gâteaux directional differential of ϕ at e_0; and a second in which (61) has been replaced by (71). This follows from an obvious analog of Lemma (51).

73 COROLLARY. *If, in Theorem* (60), *Hypothesis* (53) *also holds for some Gâteaux W-semidifferential* $D\phi(e_0;\cdot)$ *of* ϕ *at* e_0, *then relation* (61) *takes the stronger form*

74 $$\ell \circ D\phi(e_0; y) \leq 0 \quad \textit{for all} \quad y \in \overline{\text{co co}}(\mathscr{E} - e_0).$$

3. Specializations of the Abstract Multiplier Rules

In this section we shall apply the abstract multiplier rules derived in the preceding section in order to obtain special multiplier rules that solutions of Problems (1.2), (1.8) (minimax), and (1.10) (vector-valued criterion) must satisfy. We begin with Problem (1.2).

1 THEOREM. *Let* e_0 *be a solution of Problem* (1.2). *Suppose that* ϕ^0 *is finitely semidifferentiable* [*see* (I.7.16)] *at* e_0, *and that either*

Hypotheses (2.1–3) *or Hypotheses* (2.11–14) *hold, but with* ϕ, $\mathscr{Z}$, *Z, and W replaced by* ϕ_1, $\mathscr{Z}_1$, Z_1, *and* W_1, *respectively, and, in the case of Hypothesis* (2.14), *also suppose in Condition* B *that* ϕ^0 *is continuous on* $\mathscr{R} \cup \mathscr{E}_0$. *Then, for every finite semidifferential* $D\phi^0(e_0;\cdot)$ *of* ϕ^0 *at* e_0 *and every finite* W_1*-semidifferential* $D\phi_1(e_0;\cdot)$ *of* ϕ_1 *at* e_0, *there exist a vector* $\alpha \in R^m$, *a number* $\beta^0 \leq 0$, *and a linear functional* $\ell_1 \in \mathscr{Z}_1^*$, *not all zero, such that*

2
$$\alpha \cdot D\varphi(e_0; y) + \beta^0 D\phi^0(e_0; y) + \ell_1 \circ D\phi_1(e_0; y) \leq 0 \quad \textit{for all} \quad y \in \text{co co}(\mathscr{E} - e_0),$$
$$\ell_1(z_1) \geq \ell_1 \circ \phi_1(e_0) \quad \textit{for all} \quad z_1 \in \bar{Z}_1,$$

where $D\varphi(e_0;\cdot)$ *denotes the finite differential of* φ *at* e_0.

Proof. By Theorem (1.16), e_0 is a (φ,ϕ,Z)-extremal with $\mathscr{Z} = R \times \mathscr{Z}_1$, $Z = \bar{R}_- \times Z_1$, and $\phi = (\phi^0 - \phi^0(e_0), \phi_1)$. If we set $W = \bar{R}_- \times W_1$, so that W is a closed, convex cone in $\mathscr{Z}$ [see (I.1.16) and (I.3.35)], then (I.7.50) and (I.3.34) imply that Hypothesis (2.3) [or, equivalently, (2.13)] holds and that $(D\phi^0(e_0;\cdot), D\phi_1(e_0;\cdot))$ is a finite W-semidifferential of ϕ at e_0. Finally, if Hypothesis (2.14) is satisfied with ϕ replaced by ϕ_1, and if ϕ^0 is continuous on $\mathscr{R} \cup \mathscr{E}_0$, then, by (I.3.38), ϕ is continuous on $\mathscr{R} \cup \mathscr{E}_0$. Hence, we may appeal to Theorem (2.18), and, making use of (I.5.25) and (I.3.34), directly obtain our desired conclusions. |||

The following corollaries follow directly from Corollaries (2.44–46 and 54).

4 Corollary. *If, in Theorem* (1), Z_1 *is a cone, then* (3) *holds if and only if* $\ell_1 \in -(\bar{Z}_1)^*$ *and* $\ell_1 \circ \phi_1(e_0) = 0$.

5 Corollary. *If, in Theorem* (1), $D\phi^0$ *and* $D\phi_1$ *are linear, then* (2) *is equivalent to the maximum principle*

$$(\alpha \cdot D\varphi + \beta^0 D\phi^0 + \ell_1 \circ D\phi_1)(e) \leq (\alpha \cdot D\varphi + \beta^0 D\phi^0 + \ell_1 \circ D\phi_1)(e_0) \quad \textit{for all} \quad e \in \mathscr{E}.$$

If also e_0 *is an internal point of* $\mathscr{E}$, *then*

$$\alpha \cdot D\varphi + \beta^0 D\phi^0 + \ell_1 \circ D\phi_1 = 0.$$

6 Corollary. *Let* e_0 *be a solution of Problem* (1.2) *such that* ϕ^0 *is defined and convex on* $\mathscr{Y}$, *and* ϕ_1 *is defined and* W_1*-convex on*

$\mathscr{Y}$, *for some closed, convex cone* $W_1 \subset \mathscr{Z}_1$ *such that* $(\text{int } Z_1) + W_1 = \text{int } Z_1$. *Further, suppose that either Hypotheses* (2.1 *and* 2) *or Hypotheses* (2.11, 12, *and* 14)—*with* $\phi = (\phi^0, \phi_1)$ *in Condition* B *of* (2.14)—*hold. Then there exists a vector* $\alpha \in R^m$, *a number* $\beta^0 \leq 0$, *and a linear functional* $\ell_1 \in \mathscr{Z}_1^*$, *not all zero, such that*

7 $$(\alpha \cdot D\varphi + \beta^0\phi^0 + \ell_1 \circ \phi_1)(e) \leq (\alpha \cdot D\varphi + \beta^0\phi^0 + \ell_1 \circ \phi_1)(e_0) \quad \textit{for all} \quad e \in e_0 + \text{co co}\,(\mathscr{E} - e_0),$$

8 $$\ell_1(z_1) \geq \ell_1 \circ \phi_1(e_0) \quad \textit{for all} \quad z_1 \in \bar{Z}_1,$$

where $D\varphi$ *denotes the finite differential of* φ *at* e_0.

9 On the basis of (2.50) and (I.7.49), we can conclude that if the hypotheses of Corollary (6) are satisfied, and if, in addition, ϕ_1 is Gâteaux directionally differentiable at e_0 [as must be the case, by Corollary (I.7.8), if $\mathscr{Z}_1 = R^\mu$ and $W_1 = \bar{R}^\mu_-$], then Theorem (1) yields two *equivalent* sets of necessary conditions: one in which the basic multiplier rule has the form of (2), where $D\varphi(e_0;\cdot)$, $D\phi^0(e_0;\cdot)$, and $D\phi_1(e_0;\cdot)$ denote, respectively, the finite differential of φ, the Gâteaux directional differential of ϕ^0 [which exists by Lemma (I.7.7.)], and the Gâteaux directional differential of ϕ_1, all at e_0, and a second in which (2) is replaced by (7), with $D\varphi = D\varphi(e_0;\cdot)$.

10 Corollary. *If, in Theorem* (1), *Hypothesis* (2.53) *also holds with* $D\phi = (D\phi^0, D\phi_1)$, *for some finite semidifferential* $D\phi^0$ *of* ϕ^0 *at* e_0 *and some finite* W_1-*semidifferential* $D\phi_1$ *of* ϕ_1 *at* e_0, *and if* $D\varphi(e_0;\cdot)$ *is a continuous linear function* [*see* (I.5.44)], *then relation* (2) *takes the stronger form*

11 $$\alpha \cdot D\varphi(e_0; y) + \beta^0 D\phi^0(e_0; y) + \ell_1 \circ D\phi_1(e_0; y) \leq 0 \quad \textit{for all} \quad y \in \overline{\text{co co}}\,(\mathscr{E} - e_0).$$

Now let us turn to Problems (1.8 and 10).

12 Theorem. *Let* e_0 *be a solution of Problem* (1.8) [*respectively,* (1.10)], *and suppose that Hypotheses* (1.21 *and* 22) [*respectively, Hypothesis* (1.25)] *hold. Further, suppose that either Hypotheses* (2.1 *and* 2) *or Hypotheses* (2.11, 12, *and* 14)—*with* $\phi = (\phi_1, \phi_2)$ *hold. Finally, suppose that* ϕ_i (*for* $i = 1, 2$) *is finitely* W_i-*semidifferentiable at* e_0—*for some closed, convex cone* $W_1 \subset \mathscr{Z}_1$ *such that* $(\text{int } Z_1) + W_1 = \text{int } Z_1$, *and with* $W_2 = Z_2^0$ [*see* (1.17)]. *Then, for every finite* W_i-*semidifferentials* $D\phi_i(e_0;\cdot)$ *of* ϕ_i *at* e_0

$(i = 1, 2)$, *there exist a vector* $\alpha \in R^m$ *and linear functionals* $\ell_i \in \mathscr{X}_i^*$ $(i = 1, 2)$, *not all zero, such that*

13 $$\alpha \cdot D\varphi(e_0; y) + \ell_1 \circ D\phi_1(e_0; y) + \ell_2 \circ D\phi_2(e_0; y) \leq 0 \quad \textit{for all} \quad y \in \operatorname{co}\operatorname{co}(\mathscr{E} - e_0),$$

14 $$\ell_1(z_1) \geq \ell_1 \circ \phi_1(e_0) \quad \textit{for all} \quad z_1 \in \bar{Z}_1,$$

15 $$\ell_2(z_2) \geq \ell_2 \circ \phi_2(e_0) \quad \textit{for all} \quad z_2 \in Z_2^{\kappa o} \quad [\textit{see}\ (1.17\ \textit{and}\ 20)]$$

(*respectively,*

16 $$\ell_2(z_2) \geq 0 \textit{ for all } z_2 \in Z_2^0),$$

where $D\varphi(e_0;\cdot)$ *denotes the finite differential of* φ *at* e_0.

Proof. By Theorem (1.24) [respectively, (1.26)], e_0 is a (φ,ϕ,Z)-extremal with $\mathscr{X} = \mathscr{X}_1 \times \mathscr{X}_2$, $Z = Z_1 \times Z_2^{\kappa o}$ [respectively, $Z = Z_1 \times Z_2^0$], and $\phi = (\phi_1, \phi_2)$ [respectively, $\phi = (\phi_1, \phi_2 - \phi_2(e_0))$]. If we set $W = W_1 \times Z_2^0$, so that W is a closed convex cone in $\mathscr{X}$ [see (I.1.16) and (I.3.35)], then the theorem hypotheses together with (I.7.50), (I.3.34), and (1.19), imply that Hypothesis (2.3) [or, equivalently, (2.13)] holds and that $(D\phi_1(e_0;\cdot), D\phi_2(e_0;\cdot))$ is a finite W-semidifferential of ϕ at e_0. Hence, we may appeal to Theorem (2.18), and, by virtue of Theorem (I.5.25) and (I.3.34), conclude that there exist a vector $\alpha \in R^m$ and linear functionals $\ell_i \in \mathscr{X}_i^*$ $(i = 1, 2)$, not all zero, such that (13) holds and such that $\ell_1(z_1) + \ell_2(z_2) \geq \ell_1 \circ \phi_1(e_0) + \ell_2 \circ \phi_2(e_0)$ for all $z_1 \in \bar{Z}_1$ and $z_2 \in Z_2^{\kappa o}$ [respectively, $\ell_1(z_1) + \ell_2(z_2) \geq \ell_1 \circ \phi_1(e_0)$ for all $z_1 \in \bar{Z}_1$ and $z_2 \in Z_2^0$]. Since $\phi_1(e_0) \in Z_1$ and $\phi_2(e_0) \in Z_2^{\kappa o}$ [respectively, and $0 \in Z_2^0$], this implies that (14) and (15) [respectively, (14) and (16)] hold. |||

If we replace Hypothesis (1.21) in Theorem (12) by a stronger hypothesis (which turns out to be satisfied in many cases of interest), then conclusion (15) takes on a special form. Thus, consider the following hypothesis for Problem (1.8).

17 HYPOTHESIS. *There is an element* $\zeta_0 \in \mathscr{X}_2$ *such that* $\ell(\zeta_0) = 1$ *for all* $\ell \in \mathscr{L}$.

18 Note that if Hypothesis (17) holds, then, as is easily verified [see (1.17 and 18)],

19 $$V_2^\kappa = V_2^0 + \kappa\zeta_0 \quad \text{and} \quad Z_2^\kappa = Z_2^0 + \kappa\zeta_0 \quad \text{for all} \quad \kappa \in R.$$

Since $0 \in V_2^\kappa$ for every $\kappa > 0$, (19) implies that V_2^0 is not empty. Hence, if Hypothesis (17) holds, then Hypothesis (1.21) also does.

Further, by (19), if Hypothesis (17) holds, then for Hypothesis (1.22) to hold, it is sufficient that the second part of Hypothesis (1.25) hold.

20 We also remark that if $\mathscr{L}$ is a finite subset of $\mathscr{Z}_2^*$ whose elements are linearly independent, then, by virtue of Corollary (I.2.15), Hypothesis (17) holds [compare with (1.23 and 27)].

21 COROLLARY. (to Theorem 12). *If, in Theorem* (12), *for Problem* (1.8), *we replace the assumption that Hypothesis* (1.21) *holds by the assumption that Hypothesis* (17) *holds, then all of the conclusions of the theorem remain in effect, and* (15) *is equivalent to the conditions* [*see* (1.17 *and* 20) *and* (I.5.23)]

22 $$\ell_2 \in -(Z_2^0)^* \quad \textit{and} \quad \ell_2 \circ \phi_2(e_0) = \kappa_0 \ell_2(\zeta_0).$$

Proof. We have already seen in (18) that, under the hypotheses of the corollary, (19) holds and Hypothesis (1.21) holds, so that all of the conclusions of Theorem (12) carry over. Because of (19), (15) takes the form

23 $$\ell_2(z_2) \geq \ell_2 \circ \phi_2(e_0) - \kappa_0 \ell_2(\zeta_0) \quad \text{for all} \quad z_2 \in Z_2^0.$$

But $\zeta_1 = [\phi_2(e_0) - \kappa_0\zeta_0] \in Z_2^0$ and, since Z_2^0 is a cone, $\frac{1}{2}\zeta_1$ and $2\zeta_1$ belong to Z_2^0 as well, from which it follows, by virtue of (23), that $\ell_2(\zeta_1) = 0$ and $\ell_2(z_2) \geq 0$ for all $z_2 \in Z_2^0$, i.e., that (22) holds.

Conversely, (22) obviously implies (23) and hence (15). |||

Note that if, in Problems (1.8) or (1.10), ϕ_1 is finitely differentiable at e_0, then ϕ_1 satisfies the requirements of Theorem (12), because we can then choose $W_1 = \{0\}$ [see (I.7.20)]. Alternately, if the convex body Z_1 is also a cone in $\mathscr{Z}_1$, and if ϕ_1 is finitely $\bar{Z}_1$-semidifferentiable at e_0, then ϕ_1 also satisfies the requirements of Theorem (12), because we can then choose $W_1 = Z_1$ [see Lemmas (I.4.19, 23, and 28)].

The following corollaries to Theorem (12) follow directly from Corollaries (2.44–46 and 54).

24 COROLLARY. *If, in Theorem* (12), Z_1 *is a cone, then* (14) *holds if and only if* $\ell_1 \in (-\bar{Z}_1)^*$ *and* $\ell_1 \circ \phi_1(e_0) = 0$.

25 COROLLARY. *If, in Theorem* (12), $D\phi_1$ *and* $D\phi_2$ *are linear, then* (13) *is equivalent to the maximum principle*

$$(\alpha \cdot D\varphi + \ell_1 \circ D\phi_1 + \ell_2 \circ D\phi_2)(e) \leq (\alpha \cdot D\varphi + \ell_1 \circ D\phi_1 + \ell_2 \circ D\phi_2)(e_0)$$
$$\textit{for all} \quad e \in \mathscr{E}.$$

If also e_0 *is an internal point of* $\mathscr{E}$, *then*

$$\alpha \cdot D\varphi + \ell_1 \circ D\phi_1 + \ell_2 \circ D\phi_2 = 0.$$

26 COROLLARY. *Let* e_0 *be a solution of Problem* (1.8) [*respectively,* (1.10)], *and suppose that Hypotheses* (1.21 *and* 22) [*respectively,* (1.25)] *hold. Further, suppose that either Hypotheses* (2.1 *and* 2) *or Hypotheses* (2.11, 12, *and* 14)—*with* $\phi = (\phi_1, \phi_2)$—*hold. Finally, suppose that* ϕ_i (*for* $i = 1, 2$) *is defined and* W_i*-convex on* $\mathscr{Y}$—*for some closed, convex cone* $W_1 \subset \mathscr{Z}_1$ *such that* $(\text{int } Z_1) + W_1 = \text{int } Z_1$, *and with* $W_2 = Z_2^0$ [*see* (1.17)]. *Then there exist a vector* $\alpha \in R^m$ *and linear functionals* $\ell_i \in \mathscr{Z}_i^*$, $i = 1, 2$, *not all zero, such that* (14) *and* (15) [*respectively,* (14) *and* (16)] *hold, and such that*

27
$$(\alpha \cdot D\varphi + \ell_1 \circ \phi_1 + \ell_2 \circ \phi_2)(e) \leq (\alpha \cdot D\varphi + \ell_1 \circ \phi_1 + \ell_2 \circ \phi_2)(e_0)$$
$$\text{for all} \quad e \in e_0 + \text{co co}\,(\mathscr{E} - e_0),$$

where $D\varphi$ *denotes the finite differential of* φ *at* e_0. *If, in the case of Problem* (1.8), *Hypothesis* (17) *holds, then* (15) *is equivalent to* (22).

28 On the basis of (2.50) and (I.7.49), we can conclude that if the hypotheses of Corollary (26) are satisfied, and if, in addition, ϕ_1 and ϕ_2 are Gâteaux directionally differentiable at e_0, then we can obtain two equivalent sets of necessary conditions, much as was described in (9).

29 COROLLARY. *If, in Theorem* (12), *Hypothesis* (2.53) *also holds with* $D\phi = (D\phi_1, D\phi_2)$, *for some finite* W_i*-semidifferentials* $D\phi_i(e_0;\cdot)$ *of* ϕ_i *at* e_0 $(i = 1, 2)$, *and if* $D\varphi(e_0;\cdot)$ *is a continuous linear function* [*see* (I.5.44)], *then relation* (13) *takes the stronger form*

30
$$\alpha \cdot D\varphi(e_0; y) + \ell_1 \circ D\phi_1(e_0; y) + \ell_2 \circ D\phi_2(e_0; y) \leq 0$$
$$\text{for all} \quad y \in \overline{\text{co co}}\,(\mathscr{E} - e_0).$$

31 Note that, if $\phi_1(e_0) \in (\text{int } Z_1)$, then relation (3) [or, equivalently, (8) or (14)] implies, by Lemma (I.5.14), that $\ell_1 \equiv 0$.

32 If Problems (1.2, 8, and 10) are modified by omitting the generalized inequality constraint $\phi_1(e) \in Z_1$, then it is easy to see that all of the preceding results of this section remain in force, provided, of course, that we omit all of the hypotheses and

conclusions regarding ϕ_1, and delete terms involving ϕ_1 in expressions such as (2).

33 On the other hand, if we modify Problems (1.2, 8, and 10) by omitting the equality constraint $\varphi(e) = 0$, then we can appeal to Theorem (2.60) [also see (1.30)], and, arguing essentially as in the proofs of Theorems (1) and (12), obtain the following multiplier rules [which are the obvious analogs of Theorems (1) and (12)].

34 THEOREM. *Let e_0 be a solution of Problem* (1.2), *but with the equality constraint $\varphi(e) = 0$ deleted. Suppose that ϕ^0 is Gâteaux semidifferentiable at e_0, and that either Hypotheses* (2.56 *and* 57) *or Hypotheses* (2.57–59) *hold, but with ϕ, $\mathscr{Z}$, Z, and W replaced by ϕ_1, $\mathscr{Z}_1$, Z_1, and W_1, respectively, and, in the case of Hypothesis* (2.59), *also suppose in Condition B that ϕ^0 is continuous on $\mathscr{R} \cup \mathscr{E}_0$. Then, for every Gâteaux semifferential $D\phi^0(e_0;\cdot)$ of ϕ^0 at e_0 and every Gâteaux W_1-semidifferential $D\phi_1(e_0;\cdot)$ of ϕ_1 at e_0, there exist a number $\beta^0 \leq 0$ and a linear functional $\ell_1 \in \mathscr{Z}_1^*$, not both zero, such that*

35 $$\beta^0 D\phi^0(e_0; y) + \ell_1 \circ D\phi_1(e_0; y) \leq 0 \quad \textit{for all} \quad y \in \text{co}\, co\, (\mathscr{E} - e_0),$$

36 $$\ell_1(z_1) \geq \ell_1 \circ \phi_1(e_0) \quad \textit{for all} \quad z_1 \in \bar{Z}_1.$$

37 THEOREM. *Let e_0 be a solution of Problem* (1.8) [*respectively,* (1.10)]—*but with the equality constraint $\varphi(e) = 0$ deleted—and suppose that Hypotheses* (1.21 *and* 22) [*respectively,* (1.25)] *hold. Further, suppose that either Hypothesis* (2.56) *or Hypotheses* (2.58 *and* 59)—*with $\phi = (\phi_1, \phi_2)$—hold. Finally, suppose that ϕ_i (for $i = 1, 2$) is Gâteaux W_i-semidifferentiable at e_0—for some closed, convex cone $W_1 \subset \mathscr{Z}_1$ such that* $(\text{int}\, Z_1) + W_1 = \text{int}\, Z_1$, *and with $W_2 = Z_2^0$* [*see* (1.17)]. *Then, for every Gâteaux W_i-semidifferentials $D\phi_i(e_0;\cdot)$ of ϕ_i at e_0 ($i = 1, 2$), there exist linear functionals $\ell_i \in \mathscr{Z}_i^*$ ($i = 1, 2$), not both zero, such that*

38 $$\ell_1 \circ D\phi_1(e_0; y) + \ell_2 \circ D\phi_2(e_0; y) \leq 0 \quad \textit{for all} \quad y \in \text{co}\,\text{co}\,(\mathscr{E} - e_0),$$

39 $$\ell_1(z_1) \geq \ell_1 \circ \phi_1(e_0) \quad \textit{for all} \quad z_1 \in \bar{Z}_1,$$

40 $$\ell_2(z_2) \geq \ell_2 \circ \phi_2(e_0) \quad \textit{for all} \quad z_2 \in Z_2^{\kappa 0}$$

(*respectively,*

41 $$\ell_2(z_2) \geq 0 \textit{ for all } z_2 \in Z_2^0).$$

42 We shall leave it to the reader to obtain corollaries to Theorems (34) and (37) which correspond to Corollaries (4)–(6), (10), (21), (24)–(26), and (29). Further, remarks which are the obvious analogs of (9), (28), (31), and (32) also can be made here.

4. An Extension of the Abstract Multiplier Rules

If e_0 is a (φ,ϕ,Z)-extremal, or a (ϕ,Z)-extremal, in the special case where $\mathscr{X} = R^\nu$ and $Z = \bar{R}^\nu_- + z_0$ (for some positive integer ν and some fixed $z_0 \in R^\nu$), then the multiplier rules [Theorems (2.18 and 60)] can be somewhat simplified, and, in addition, somewhat sharpened. We shall obtain these modified multiplier rules in this section, and shall apply them to obtain particular multiplier rules for Problems (1.3), (1.9) (minimax), and (1.12) (vector-valued criterion).

Corresponding to Hypothesis (2.3) [or (2.13)], we shall make the following hypothesis for the function $\phi:\mathscr{E}' \to R^\nu$ [see (I.7.18)].

1 HYPOTHESIS. *For each i, ϕ^i is finitely semidifferentiable at e_0, and at least one finite semidifferential of ϕ^i at e_0 can be extended onto $\mathscr{Y}$ in such a way that it remains a convex functional.*

We then have the following multiplier rule, corresponding to Theorem (2.18) and Corollary (2.46).

2 THEOREM. *Let e_0 be a (φ,ϕ,Z)-extremal for the special case where $\mathscr{X} = R^\nu$ and $Z = \bar{R}^\nu_- + z_0$, for some fixed $z_0 \in R^\nu$, and suppose that either Hypotheses* (2.1 *and* 2) *or Hypotheses* (2.11, 12, *and* 14) *hold. Further, suppose that Hypothesis* (1) *holds. Then, for every set of finite semidifferentials $D\phi^i(e_0;\cdot)$, $i = 1, \ldots, \nu$, of the ϕ^i at e_0 of the type described in Hypothesis* (1), *there exist vectors $\alpha \in R^m$ and $\beta \in \bar{R}^\nu_-$, not both zero, and a linear function $\Lambda = (\Lambda^1, \ldots, \Lambda^\nu): \mathscr{Y} \to R^\nu$ such that*

3 $$\alpha \cdot D\varphi(e_0; y) + \beta \cdot D\phi(e_0; y) \leq 0 \quad \text{for all} \quad y \in \operatorname{co}\operatorname{co}(\mathscr{E} - e_0),$$

4 $$(\alpha \cdot D\varphi + \beta \cdot \Lambda)(e) \leq (\alpha \cdot D\varphi + \beta \cdot \Lambda)(e_0) \quad \text{for all} \quad e \in \mathscr{E},$$

5 $$\Lambda^i(y) \leq D\phi^i(e_0; y) \quad \text{for all} \quad y \in \mathscr{Y} \quad \text{and each} \quad i,$$

6 $$\beta \cdot \phi(e_0) = \beta \cdot z_0,$$

where $D\varphi(e_0;\cdot) = D\varphi(\cdot)$ is the finite differential of φ at e_0 and $D\phi = (D\phi^1, \ldots, D\phi^\nu)$.

If we replace Hypothesis (1) *by the assumption that* $\phi^1, \ldots, \phi^\nu$ *are all defined and convex on* $\mathscr{Y}$, *then the preceding conclusions remain in effect, with* $D\phi^i(e_0;\cdot)$ (*for each* i) *denoting the Gâteaux directional differential of* ϕ^i *at* e_0, *and, in addition, relations* (3) *and* (5) *are, respectively, equivalent to the relations*

7
$$(\alpha \cdot D\varphi + \beta \cdot \phi)(e) \leq (\alpha \cdot D\varphi + \beta \cdot \phi)(e_0) \quad \textit{for all} \quad e \in e_0 + \operatorname{co}\operatorname{co}(\mathscr{E} - e_0),$$

8
$$\Lambda^i(y) \leq \phi^i(e_0 + y) - \phi^i(e_0) \quad \textit{for all} \quad y \in \mathscr{Y} \quad \textit{and each} \quad i.$$

Let us defer the proof of Theorem (2) until later in this section.

9 The principal novel feature of the preceding multiplier rule is the inequality (4) which is in terms of the *linear* function Λ which [see (5)] has the property that, for each i, Λ^i is a subgradient of $D\phi^i(e_0;\cdot)$ at 0 [see (I.2.29)] and, if $\phi^1, \ldots, \phi^\nu$ are all defined and convex on $\mathscr{Y}$, a subgradient of ϕ^i at e_0 as well [see (8)]. It is trivially verified that (3) is a direct consequence of (4) and (5) when $\beta \in \bar{R}^\nu_-$, so that relations (4)–(6) are stronger necessary conditions than relations (3) and (6). If each $D\phi^i$ is linear, then Λ must coincide with $D\phi$ if (5) is to hold, in which case (3) and (4) are equivalent. However, if $D\phi^i$ is not linear for some i (although it must be convex), then (4) is often a more useful relation than (3). This is because inequality (4) contains only linear terms, whereas (3) contains the convex functions $D\phi^i$, and an inequality involving only linear terms is usually much easier to manipulate and reformulate into an easily interpretable form than an equality containing nonlinear terms.

Let us now specialize Theorem (2) to obtain multiplier rules for Problems (1.3, 9, and 12). For Problem (1.3), we have the following theorem.

10 THEOREM. *Let* e_0 *be a solution of Problem* (1.3), *and suppose that either Hypotheses* (2.1 *and* 2) *or Hypotheses* (2.11, 12, *and* 14) *hold, where* $\varphi = (\varphi^1, \ldots, \varphi^m)$ *and* $\phi = (\phi^0, \phi^1, \ldots, \phi^\mu)$. *Further, suppose that Hypothesis* (1) *holds. Then, for every set of finite semi-differentials* $D\phi^i(e_0;\cdot)$, $i = 0, \ldots, \mu$, *of the* ϕ^i *at* e_0 *of the type described in Hypothesis* (1), *there exist vectors* $\alpha \in R^m$ *and* $\beta = (\beta^0, \beta^1, \ldots, \beta^\mu) \in \bar{R}^{\mu+1}_-$, *not both zero, and a linear function* $\Lambda = (\Lambda^0, \ldots, \Lambda^\mu)\colon \mathscr{Y} \to R^{\mu+1}$ *such that* (3)–(5) *hold and such that*

11
$$\sum_{i=1}^{\mu} \beta^i \phi^i(e_0) = 0,$$

where $D\varphi(e_0;\cdot) = D\varphi(\cdot)$ *is the finite differential of* φ *at* e_0 *and* $D\phi = (D\phi^0, D\phi^1, \ldots, D\phi^\mu)$.

If we replace Hypothesis (1) *by the assumption that* $\phi^0, \phi^1, \ldots, \phi^\mu$ *are all defined and convex on* $\mathscr{Y}$, *then the preceding conclusions remain in effect, with* $D\phi^i$ (*for each* i) *denoting the Gâteaux directional differential of* ϕ^i *at* e_0, *and, in addition, relations* (3) *and* (5) *are, respectively, equivalent to* (7) *and* (8).

Proof. By Theorem (1.16), e_0 is a (φ,ϕ,Z)-extremal with $\mathscr{Z} = R^{\mu+1}$, $Z = \bar{R}^{\mu+1}_-$, and $\phi = (\phi^0 - \phi^0(e_0), \phi^1, \ldots, \phi^\mu)$. Theorem (10) now follows at once from Theorem (2). |||

For Problems (1.9 and 12), we have the following theorem.

12 THEOREM. *Let* e_0 *be a solution either of Problem* (1.9) *or of Problem* (1.12), *and suppose that either Hypotheses* (2.1 *and* 2) *or Hypotheses* (2.11, 12, *and* 14) *hold, where* $\varphi = (\varphi^1, \ldots, \varphi^m)$ *and* $\phi = (\phi^1, \ldots, \phi^\nu)$. *Further, suppose that Hypothesis* (1) *holds. Then, for every set of finite semidifferentials* $D\phi^i(e_0;\cdot)$, $i = 1, \ldots, \nu$, *of the* ϕ^i *at* e_0 *of the type described in Hypothesis* (1), *there exist vectors* $\alpha \in R^m$ *and* $\beta = (\beta^1, \ldots, \beta^\nu) \in \bar{R}^\nu_-$, *not both zero, and a linear function* $\Lambda = (\Lambda^1, \ldots, \Lambda^\nu)\colon \mathscr{Y} \to R^\nu$ *such that* (3)–(5) *and* (11) *hold, and, in the case of Problem* (1.9), *such that also*

13 $$\sum_{i=\mu+1}^{\nu} \beta^i\phi^i(e_0) = (\max\{\phi^j(e_0)\colon j = \mu+1, \ldots, \nu\}) \sum_{i=\mu+1}^{\nu} \beta^i,$$

where $D\varphi(e_0;\cdot)$, $D\varphi$, *and* $D\phi$ *are as in Theorem* (2).

If we replace Hypothesis (1) *by the assumption that* $\phi^1, \ldots, \phi^\nu$ *are all defined and convex on* $\mathscr{Y}$, *then the preceding conclusions remain in effect, with* $D\phi^i$ (*for each* i) *denoting the Gâteaux directional differential of* ϕ^i *at* e_0, *and, in addition, relations* (3) *and* (5) *are, respectively, equivalent to* (7) *and* (8).

Proof. By the remarks in (1.23 and 27), Hypotheses (1.21, 22, and 25) hold. Therefore, by Theorems (1.24 and 26), e_0 is a (φ,ϕ,Z)-extremal, with $\mathscr{Z} = R^\nu$; $Z = \bar{R}^\mu_- \times (\bar{R}^{\nu-\mu}_- + \kappa_0(1, \ldots, 1))$ in the case of Problem (1.9), where $\kappa_0 = \max\{\phi^j(e_0)\colon j = \mu + 1, \ldots, \nu\}$, and $Z = \bar{R}^\nu_-$ in the case of Problem (1.12); and $\phi = (\phi^1, \ldots, \phi^\nu)$ in the case of Problem (1.9), and

$$\begin{aligned}\phi = (\phi^1, \ldots, \phi^\mu, \phi^{\mu+1}, \ldots, \phi^\nu)\\ - (0, \ldots, 0, \phi^{\mu+1}(e_0), \ldots, \phi^\nu(e_0))\end{aligned}$$

in the case of Problem (1.12). Except for (11) and (13) in the case of Problem (1.9), our desired conclusions now follow at once from Theorem (2).

For Problem (1.9), (6) implies that

14 $$\sum_{i=1}^{\mu} \beta^i \phi^i(e_0) + \sum_{i=\mu+1}^{\nu} \beta^i[\phi^i(e_0) - \kappa_0] = 0.$$

But, since e_0 is a solution of the problem and $\beta \in \bar{R}^{\nu}_{-}$, each of the two sums in the left-hand side of (14) must be non-negative, i.e., must vanish, which means that (11) and (13) hold. |||

15 Because $\beta^i \leq 0$ and $\phi^i(e_0) \leq 0$ for $i = 1, \ldots, \mu$, relation (11) in Theorems (10) and (12) implies that $\beta^i \phi^i(e_0) = 0$ for each $i = 1, \ldots, \mu$. In Theorem (12), relation (13) [which is valid for Problem (1.9)], evidently implies that

$$\beta^i[\phi^i(e_0) - \max\{\phi^j(e_0) : j = \mu + 1, \ldots, \nu\}] = 0 \quad \text{for} \quad i = \mu + 1, \ldots, \nu.$$

Relation (13) could also have been obtained in Theorem (12) by appealing to Corollary (2.21).

Corresponding to Corollaries (2.45 and 54), we have the following easily verified corollaries of Theorems (2), (10), and (12).

16 COROLLARY *If, in Theorems* (2), (10), *or* (12), $D\phi$ *is linear, then* (3) *is equivalent to the maximum principle*

$$(\alpha \cdot D\varphi + \beta \cdot D\phi)(e) \leq (\alpha \cdot D\varphi + \beta \cdot D\phi)(e_0) \quad \textit{for all} \quad e \in \mathscr{E}.$$

If also e_0 *is an internal point of* $\mathscr{E}$, *then* $\alpha \cdot D\varphi + \beta \cdot D\phi = 0$.

17 COROLLARY *If, in Theorems* (2), (10), *or* (12), $\mathscr{Y}$ *is a linear topological space,* $D\phi(e_0;\cdot)$ *is continuous on* $\mathscr{Y}$, *and* $D\varphi(e_0;\cdot)$ *is a continuous linear function* [*see* (I.5.44)], *then* Λ *is continuous* (i.e., $\Lambda^i \in \mathscr{Y}^*$ *for each* i) *and relations* (3) *and* (4) *take the stronger forms*

18 $$\alpha \cdot D\varphi(e_0; y) + \beta \cdot D\phi(e_0; y) \leq 0 \quad \textit{for all} \quad y \in \overline{\text{co}}\,\overline{\text{co}}\,(\mathscr{E} - e_0),$$

19 $$(\alpha \cdot D\varphi + \beta \cdot \Lambda)(e) \leq (\alpha \cdot D\varphi + \beta \cdot \Lambda)(e_0) \quad \textit{for all} \quad e \in \bar{\mathscr{E}}.$$

The continuity of Λ in Corollary (17) follows from (5), Theorem (I.3.25), and Lemma (I.5.13).

20 If Problems (1.3, 9, and 12) are modified by omitting the inequality constraints $\phi^i(e) \leq 0$ for $i = 1, \ldots, \mu$, then it is easy to see that Theorems (10) and (12), as well as Corollaries (16) and

(17), remain in force, provided, of course, that we omit all terms with indices $1, \ldots, \mu$ or set $\mu = 0$ in the hypotheses and conclusions, and, in particular, omit (11).

On the other hand, if we modify Problems (1.3, 9, and 12) by omitting the equality constraints $\varphi^i(e) = 0$ for $i = 1, \ldots, m$, then we can obtain necessary conditions only if we modify the multiplier rule (2) to obtain a rule which corresponds to Theorem (2.60) and Corollary (2.70). This modified rule is as follows.

21 THEOREM. *Let e_0 be a (ϕ, Z)-extremal for the special case where $\mathscr{Z} = R^\nu$ and $Z = \bar{R}^\nu_- + z_0$, for some fixed $z_0 \in R^\nu$, and suppose that either Hypothesis* (2.56) *or Hypotheses* (2.58 *and* 59) *hold. Further, suppose that Hypothesis* (1)*—with "finitely" and "finite" both replaced by "Gâteaux"—holds. Then, for every set of Gâteaux semidifferentials $D\phi^i(e_0;\cdot)$, $i = 1, \ldots, \nu$, of the ϕ^i at e_0 of the type described in Hypothesis* (1), *there exist a nonzero vector $\beta \in \bar{R}^\nu_-$ and a linear function $\Lambda = (\Lambda^1, \ldots, \Lambda^\nu): \mathscr{Y} \to R^\nu$ such that*

22 $$\beta \cdot D\phi(e_0; y) \leq 0 \quad \textit{for all} \quad y \in \operatorname{co}\operatorname{co}(\mathscr{E} - e_0),$$

23 $$\beta \cdot \Lambda(e) \leq \beta \cdot \Lambda(e_0) \quad \textit{for all} \quad e \in \mathscr{E},$$

and such that (5) *and* (6) *hold, where $D\phi = (D\phi^1, \ldots, D\phi^\nu)$.*

If we replace Hypothesis (1) *by the assumption that $\phi^1, \ldots, \phi^\nu$ are all defined and convex on $\mathscr{Y}$, then the preceding conclusions remain in effect, with $D\phi^i$ (for each i) denoting the Gâteaux directional differential of ϕ^i at e_0, and, in addition, relations* (5) *and* (22) *and, respectively, equivalent to* (8) *and the relation*

24 $$\beta \cdot \phi(e) \leq \beta \cdot \phi(e_0) \quad \textit{for all} \quad e \in e_0 + \operatorname{co}\operatorname{co}(\mathscr{E} - e_0).$$

We shall leave it to the reader to apply Theorem (21) to obtain multiplier rules for Problems (1.3, 9, and 12) with the equality constraints omitted which are analogous to Theorems (10) and (12) and Corollaries (16) and (17).

We now turn to the proof of Theorem (2). Since Theorem (21) can be proved in almost exactly the same way, we shall leave its proof to the interested reader.

Proof of Theorem 2. If Hypothesis (1) holds, then it follows at once from (I.7.18) and (2.5) that Hypothesis (2.3) [or (2.13)] holds. Hence, by Theorem (2.18), Corollary (2.44) (slightly modified in an evident manner), (I.5.10), and (I.2.12), there exist vectors $\alpha \in R^m$ and $\beta \in \bar{R}^\nu_-$, not both zero, such that (3) and (6) hold.

Alternately, if $\phi^1, \ldots, \phi^\nu$ are all defined and convex on $\mathscr{Y}$, then, by Corollary (I.7.22), Hypothesis (1) is satisfied and (for each i) the function $y \to \phi^i(e_0 + y) - \phi^i(e_0)$ is a finite semidifferential of ϕ^i at e_0 which is defined and convex on $\mathscr{Y}$ [see (I.2.24)]. Consequently, we can conclude as before that there exist vectors $\alpha \in R^m$ and $\beta \in \bar{R}^\nu_-$, not both zero, such that (7) and (6) hold. Further, it follows from Lemma (2.51) that (7) holds here if and only if (3) does, where $D\phi = (D\phi^1, \ldots, D\phi^\nu)$, and $D\phi^i(e_0;\cdot)$ denotes (for each i) the Gâteaux directional differential of ϕ^i at e_0 [which exists and is defined and convex on all of $\mathscr{Y}$ by Lemma (I.7.7)].

Thus, it only remains to show that there exists a linear function $\Lambda = (\Lambda^1, \ldots, \Lambda^\nu): \mathscr{Y} \to R^\nu$ such that (4) and (5) hold, and that, if $\phi^1, \ldots, \phi^\nu$ are defined and convex on $\mathscr{Y}$, then (5) is equivalent to (8). The second of these assertions may be proved in the same way as Lemma (2.51), and the first is a direct consequence of the following lemma.

25 LEMMA. *Let $\mathscr{Y}$ be a linear vector space and $\mathscr{E}_1$ a convex set in $\mathscr{Y}$, let $\tilde{\phi}^1, \ldots, \tilde{\phi}^\nu$ be convex functionals defined on $\mathscr{Y}$, let $\tilde{\phi}$ be a convex functional which is defined on $\mathscr{E}_1$, and let $\beta^1, \ldots, \beta^\nu$ be nonpositive numbers. Suppose that $0 \in \mathscr{E}_1$ and that*

$$\tilde{\phi}^1(0) = \cdots = \tilde{\phi}^\nu(0) = \tilde{\phi}(0) = 0,$$

26 $$\left(-\tilde{\phi} + \sum_{i=1}^{\nu} \beta^i \tilde{\phi}^i\right)(y) \le 0 \quad \text{for all} \quad y \in \mathscr{E}_1.$$

Then there is a linear function $\Lambda = (\Lambda^1, \ldots, \Lambda^\nu): \mathscr{Y} \to R^\nu$ such that

27 $$\left(-\tilde{\phi} + \sum_{i=1}^{\nu} \beta^i \Lambda^i\right)(y) \le 0 \quad \text{for all} \quad y \in \mathscr{E}_1,$$

28 $$\Lambda^i(y) \le \tilde{\phi}^i(y) \quad \text{for all} \quad y \in \mathscr{Y} \quad \text{and each} \quad i.$$

Proof. If $\beta^1 \neq 0$, define the function $f_2: \mathscr{E}_1 \to R$ as follows:

$$f_2 = (-1/\beta^1)\tilde{\phi} + \sum_{i=2}^{\nu} (\beta^i/\beta^1)\tilde{\phi}^i.$$

It follows from our hypotheses and from (I.2.23) that $f_2(0) = 0$ and that f_2 is a convex functional on $\mathscr{E}_1$. If we set $f_1 = \tilde{\phi}^1$ and $A = \mathscr{E}_1$, we conclude, on the basis of Theorem (I.2.25), that there is a linear functional $\Lambda' \in \mathscr{Y}^+$ such that (I.2.26 and 27) hold. If we set $\Lambda^1 = -\Lambda'$, we conclude that $\Lambda^1(y) \leq \tilde{\phi}^1(y)$ for all $y \in \mathscr{Y}$ and that

29 $$\left(-\tilde{\phi} + \beta^1\Lambda^1 + \sum_{i=2}^{v} \beta^i\tilde{\phi}^i\right)(y) \leq 0 \quad \text{for all} \quad y \in \mathscr{E}_1.$$

If $\beta^1 = 0$, we set $A = \{0\}$, $f_1 = \tilde{\phi}^1$, and $f_2(0) = 0$. Again appealing to Theorem (I.2.25), we conclude that there is a linear functional $\Lambda' \in \mathscr{Y}^+$ such that (I.2.26 and 27) hold. If we set $\Lambda^1 = -\Lambda'$ as before, then $\Lambda^1(y) \leq \tilde{\phi}^1(y)$ for all $y \in \mathscr{Y}$, and (29) follows trivially from (26) (since $\beta^1 = 0$).

If $\beta^2 \neq 0$, we again appeal to Theorem (I.2.25), with $A = \mathscr{E}_1$, $f_1 = \tilde{\phi}^2$, and with $f_2 : \mathscr{E}_1 \to R$ defined by

$$f_2 = (-1/\beta^2)\tilde{\phi} + (\beta^1/\beta^2)\Lambda^1 + \sum_{i=3}^{v} (\beta^i/\beta^2)\tilde{\phi}^i,$$

and conclude that there is a functional $\Lambda' \in \mathscr{Y}^+$ satisfying (I.2.26 and 27). If we set $\Lambda^2 = -\Lambda'$, it follows that $\Lambda^2(y) \leq \tilde{\phi}^2(y)$ for all $y \in \mathscr{Y}$, and that

30 $$\left(-\tilde{\phi} + \sum_{i=1}^{2} \beta^i\Lambda^i + \sum_{i=3}^{v} \beta^i\tilde{\phi}^i\right)(y) \leq 0 \quad \text{for all} \quad y \in \mathscr{E}_1.$$

If $\beta^2 = 0$, we set $A = \{0\}$, $f_1 = \tilde{\phi}^2$, and $f_2(0) = 0$, and then conclude, as before, that there is a functional $\Lambda^2 \in \mathscr{Y}^+$ such that $\Lambda^2(y) \leq \tilde{\phi}^2(y)$ for all $y \in \mathscr{Y}$ and such that (30) holds.

Continuing in this manner, we can construct linear functionals $\Lambda^1, \ldots, \Lambda^v$ in $\mathscr{Y}^+$, and if we set $\Lambda = (\Lambda^1, \ldots, \Lambda^v)$, then (27) and (28) hold by construction. |||

31 Note that if we weaken Hypothesis (1) by eliminating the requirement that (for each i) some finite semidifferential of ϕ^i at e_0 can be extended (as a convex functional) onto $\mathscr{Y}$, then all of the conclusions of Theorem (2)—as well as of Theorems (10), (12), and (21), and Corollaries (16) and (17)—except for those that pertain to Λ in case Hypothesis (1) is satisfied, remain in effect, and Theorem (10) then essentially includes our first obtained multiplier rule—Theorem (II.1.4).

32 So far in this section, we have confined ourselves to the case where [for a (φ,ϕ,Z)-extremal or a (ϕ,Z)-extremal] $\mathscr{X} = R^v$ and

$Z = \bar{R}^{\nu}_{-} + z_0$. However, similar results can be obtained if, for some linear topological space $\mathscr{X}'$ and some convex body $Z' \subset \mathscr{X}'$, $\mathscr{X} = R^{\nu} \times \mathscr{X}'$ and $Z = (\bar{R}^{\nu}_{-} + z_0) \times Z'$ [see Lemma (I.4.22)]. In this case, let us write $\phi = (\phi^1, \ldots, \phi^{\nu}, \phi')$, where $\phi^1, \ldots, \phi^{\nu}$ are real-valued, and ϕ' takes on its values in $\mathscr{X}'$. Then, corresponding to the strengthened multiplier rule, Theorem (2), we have the following result [which may be proved in essentially the same way as Theorem (2)].

33 THEOREM. *Let e_0 be a (φ,ϕ,Z)-extremal of the special type indicated in* (32), *and suppose that either Hypotheses* (2.1 *and* 2) *or Hypotheses* (2.11, 12, *and* 14) *hold. Further, suppose that Hypothesis* (2.3) *holds—but with W, $\mathscr{X}$, Z, and ϕ replaced by W', $\mathscr{X}'$, Z', and ϕ', respectively—and that Hypothesis* (1) *holds. Then, for every set of finite semidifferentials $D\phi^i(e_0;\cdot)$, $i = 1, \ldots, \nu$, of the ϕ^i at e_0 of the type described in Hypothesis* (1), *and every finite W'-semidifferential $D\phi'(e_0;\cdot)$ of ϕ' at e_0, there exist vectors $\alpha \in R^m$ and $\beta \in \bar{R}^{\nu}_{-}$ together with a linear functional $\ell' \in (\mathscr{X}')^*$, not all zero, and a linear function $\Lambda = (\Lambda^1, \ldots, \Lambda^{\nu})\colon \mathscr{Y} \to R^{\nu}$ such that*

34 $$\alpha \cdot D\varphi(e_0; y) + \beta \cdot D\phi''(e_0; y) + \ell' \circ D\phi'(e_0; y) \leq 0 \quad \text{for all} \quad y \in \text{co co}(\mathscr{E} - e_0),$$

35 $$\alpha \cdot D\varphi(e_0; y) + \beta \cdot \Lambda(y) + \ell' \circ D\phi'(e_0; y) \leq 0 \quad \text{for all} \quad y \in \text{co co}(\mathscr{E} - e_0),$$

36 $$\Lambda^i(y) \leq D\phi^i(e_0; y) \quad \text{for all} \quad y \in \mathscr{Y} \quad \text{and each} \quad i,$$

37 $$\beta \cdot \phi''(e_0) = \beta \cdot z_0,$$

38 $$\ell'(z') \leq \ell' \circ \phi'(e_0) \quad \text{for all} \quad z' \in \bar{Z}',$$

where $D\varphi(e_0;\cdot)$ denotes the finite differential of φ at e_0, $\phi'' = (\phi^1, \ldots, \phi^{\nu})$, and $D\phi'' = (D\phi^1, \ldots, D\phi^{\nu})$.

39 *If we replace the hypotheses in the second sentence of the theorem by the assumptions that* (i) *$\phi^1, \ldots, \phi^{\nu}$ are all defined and convex on $\mathscr{Y}$*, (ii) *ϕ' is defined and W'-convex on $\mathscr{Y}$—where W' is a closed, convex cone in $\mathscr{X}'$ such that* (int Z') + W' = int Z'*—and* (iii) *ϕ' is Gâteaux directionally differentiable at e_0, then the preceding conclusions remain in effect, with $D\phi^i(e_0;\cdot)$ (for each i) and $D\phi'(e_0;\cdot)$ denoting the Gâteaux directional differentials at e_0 of ϕ^i*

and ϕ', *respectively, and, in addition, inequalities* (34) *and* (36) *are, respectively, equivalent to the relations*

40 $$(\alpha \cdot D\varphi + \beta \cdot \phi'' + \ell' \circ \phi')(e) \leq (\alpha \cdot D\varphi + \beta \cdot \phi'' + \ell' \circ \phi')(e_0) \quad \text{for all} \quad e \in e_0 + \text{co co}\,(\mathscr{E} - e_0),$$

41 $$\Lambda^i(y) \leq \phi^i(e_0 + y) - \phi^i(e_0) \quad \text{for all} \quad y \in \mathscr{Y} \quad \text{and each} \quad i.$$

Analogous results hold for (ϕ, Z)-extremals.

If, in Theorem (33), $D\phi'(e_0;\cdot)$ is linear, then the inequality (35) involves only linear functions [see (9)].

Let us specialize Theorem (33) to Problem (1.2). [We shall leave it to the reader to obtain analogous results for Problems (1.8 and 10)].

42 We shall suppose that, in Problem (1.2), there are a linear topological space $\mathscr{X}'$, and a convex body $Z' \subset \mathscr{X}'$, such that $\mathscr{X}_1 = R^\mu \times \mathscr{X}'$ and $Z_1 = \bar{R}^\mu_- \times Z'$, and we shall write $\phi_1 = (\phi^1, \ldots, \phi^\mu, \phi')$, where $\phi^1, \ldots, \phi^\mu$ are real-valued and ϕ' takes on its values in $\mathscr{X}'$. Then, corresponding to Theorem (10), we have the following result.

43 THEOREM. *Let* e_0 *be a solution of Problem* (1.2), *where* $\mathscr{X}_1$, Z_1, *and* ϕ_1 *are as described in* (32)—*but with* $\mathscr{X}$, Z, *and* ϕ *replaced by* $\mathscr{X}_1$, Z_1, *and* ϕ_1, *respectively. Further, suppose that either Hypotheses* (2.1 *and* 2) *or Hypotheses* (2.11, 12, *and* 14) *hold, where* $\phi = (\phi^0, \phi_1)$. *Further, suppose that Hypothesis* (2.3) *holds—but with* W, $\mathscr{X}$, Z, *and* ϕ *replaced by* W', $\mathscr{X}'$, Z', *and* ϕ', *respectively—and that Hypothesis* (1) *holds. Then, for every set of finite semidifferentials* $D\phi^i(e_0;\cdot)$, $i = 0, \ldots, \nu$, *of the* ϕ^i *at* e_0 *of the type described in Hypothesis* (1) *and every finite* W'*-semidifferential* $D\phi'(e_0;\cdot)$ *of* ϕ' *at* e_0, *there exist vectors* $\alpha \in R^m$ *and* $\beta = (\beta^0, \beta^1, \ldots, \beta^\nu) \in \bar{R}^{\nu+1}_-$ *together with a linear functional* $\ell' \in (\mathscr{X}')^*$, *not all zero, and a linear function* $\Lambda = (\Lambda^0, \Lambda^1, \ldots, \Lambda^\nu)\colon \mathscr{Y} \to R^{\nu+1}$ *such that* (34)–(36) *and* (38) *hold and such that*

44 $$\sum_{i=1}^{\nu} \beta^i \phi^i(e_0) = \sum_{i=1}^{\nu} \beta^i z_0^i,$$

where $D\varphi(e_0;\cdot)$ *denotes the finite differential of* φ *at* e_0, *and* $D\phi'' = (D\phi^0, D\phi^1, \ldots, D\phi^\nu)$.

Theorem (43) follows from Theorem (33) in much the same way that Theorem (10) followed from Theorem (2).

If the assumptions in Theorem (43) are modified as indicated in (39), then the conclusions of the theorem still hold, and inequalities corresponding to (40) and (41) are also valid.

Corollaries (16) and (17) and the first observation in (15) have evident analogs here, and the remarks in (20) and (31) carry over essentially without change.

5. Convex Problems and Sufficient Conditions

In this section, we shall consider Problems (1.2), (1.8) (minimax), and (1.10) (vector-valued criterion) under convexity hypotheses similar to those that we made in Section 3 of Chapter II. We shall show that, for such problems, the multiplier rules developed in Sections 2–4 take on a particularly simple form, much as they did in Chapter II. We shall also see that, when the convex problems are "well-posed," then the necessary conditions of our multiplier rules are also sufficient for optimality.

We begin with Problem (1.2).

1 Problem (1.2) will be said to be *convex* if (i) $\mathscr{E}'$ is a linear vector space $\mathscr{Y}$, (ii) $\mathscr{E}$ is a convex set in $\mathscr{Y}$, (iii) ϕ^0 is a convex functional, (iv) there is a closed, convex cone $W_1 \subset \mathscr{Z}_1$ such that (int Z_1) + W_1 = int Z_1 and such that ϕ_1 is W_1-convex, and (v) φ is affine.

2 Note that if ϕ_1 is affine, then condition (iv) in (1) is automatically satisfied, because we can then choose $W_1 = \{0\}$. Alternately, if Z_1 is a (convex) cone in $\mathscr{Y}$, then this condition will be satisfied if ϕ_1 is $\bar{Z}_1$-convex, because we can then choose $W_1 = \bar{Z}_1$ [see (I.2.34) and Lemmas (I.4.19, 23, and 28)].

3 Problem (1.2), if it is convex, will be said to be *well-posed* if (a) there is an element $e^* \in \mathscr{E}$ such that $\varphi(e^*) = 0$ and $\phi_1(e^*) \in$ (int Z_1), and (b) $0 \in \text{int}\, \varphi(\mathscr{E})$. [As we pointed out in Section 3 of Chapter II—see the footnote on page 82—we may assume, without loss of generality, that $\varphi(\mathscr{E})$ has dimension m.]

The reader should compare (1) and (3) with (II.3.1 and 2), noting (I.2.33). As in Chapter II, we may say that, loosely speaking, a convex problem is well-posed if the equality and "inequality" constraints are consistent on the constraint set $\mathscr{E}$, and if none of the equality constraints is redundant.

Note that if Problem (1.2) is convex, then, by (I.2.18), $\varphi(\mathscr{E})$ is convex.

For convex problems, the multiplier rule, Theorem (3.1), takes the following form.

4 THEOREM. *Let* $e_0 \in \mathscr{E}$ *be a solution of Problem* (1.2), *and suppose that this problem is convex. Then there exist a vector* $\alpha \in R^m$, *a number* $\beta^0 \leq 0$, *and a linear functional* $\ell_1 \in \mathscr{Z}_1^*$, *not all zero, such that*

5 $$(\alpha \cdot \varphi + \beta^0 \phi^0 + \ell_1 \circ \phi_1)(e) \leq (\alpha \cdot \varphi + \beta^0 \phi^0 + \ell_1 \circ \phi_1)(e_0) \quad \textit{for all} \quad e \in \mathscr{E},$$

6 $$\ell_1(z_1) \geq \ell_1 \circ \phi_1(e_0) \quad \textit{for all} \quad z_1 \in \bar{Z}_1.$$

Theorem (4) follows from Theorem (3.1) much as Theorem (II.3.3) followed from Theorem (II.1.4), and we shall leave the details to the reader.

The main feature of Theorem (4) is that the principal multiplier rule (5) is not in terms of differentials of φ, ϕ^0, and ϕ_1, but rather is in terms of these functions themselves. Of course, only the strong convexity hypotheses on the problem data in (1) make it possible to obtain such powerful global (rather than local) necessary conditions.

The necessary conditions of Theorem (4) are also sufficient, if $\beta^0 \neq 0$, whether or not the problem is convex, so long as e_0 satisfies the problem constraints. [Of course, if Problem (1.2) is not convex, then these sufficient conditions are not necessary, and there is no reason to expect them to hold.] Indeed, the following theorem holds.

7 THEOREM. *If, for Problem* (1.2) (*whether or not it is convex*), $e_0 \in \mathscr{E}$, $\alpha \in R^m$, $\beta^0 < 0$, *and* $\ell_1 \in \mathscr{Z}_1^*$ *satisfy* (5) *and* (6) *as well as the relations*

8 $$\varphi(e_0) = 0, \quad \phi_1(e_0) \in Z_1,$$

then e_0 *is a solution of the problem.*

Theorem (7) follows almost at once from the hypotheses, much as Theorem (II.3.6) did.

It turns out that if Problem (1.2) is convex and well-posed, then β^0 cannot vanish in Theorem (4), so that we have conditions which are both necessary and sufficient for optimality. This is expressed by the following theorem.

9 THEOREM. *Suppose that Problem* (1.2) *is convex and well-posed. Then* $e_0 \in \mathscr{E}$ *is a solution of this problem if and only if there exist*

a vector $\alpha \in R^m$ *and a linear functional* $\ell_1 \in \mathscr{Z}_1^*$ *such that* (8) *holds and such that*

10 $$(\phi^0 + \alpha \cdot \varphi + \ell_1 \circ \phi_1)(e) \geq (\phi^0 + \alpha \cdot \varphi + \ell_1 \circ \phi_1)(e_0) \quad \textit{for all} \quad e \in \mathscr{E},$$

11 $$\ell_1(z_1) \leq \ell_1 \circ \phi_1(e_0) \quad \textit{for all} \quad z_1 \in \bar{Z}_1.$$

Theorem (9) follows in almost the exact same way as Theorem (II.3.9), once we take into account Lemma (I.5.14).

Arguing essentially as in the proof of Lemma (2.51), we can prove that Theorem (9) (necessary and sufficient conditions) can be rewritten as follows in differential form.

12 THEOREM. *Suppose that Problem* (1.2) *is convex and well-posed, and that* ϕ_1 *is Gâteaux directionally differentiable at each point of* $\mathscr{E}$. *Then* $e_0 \in \mathscr{E}$ *is a solution of this problem if and only if there exist a vector* $\alpha \in R^m$ *and a linear functional* $\ell_1 \in \mathscr{Z}_1^*$ *such that* (8) *and* (11) *hold and such that*

13 $$D\phi^0(e_0; e - e_0) + \alpha \cdot \varphi_L(e - e_0) + \ell_1 \circ D\phi_1(e_0; e - e_0) \geq 0 \quad \textit{for all} \quad e \in \mathscr{E},$$

where $D\phi^0(e_0;\cdot)$ *and* $D\phi_1(e_0;\cdot)$ *denote the Gâteaux directional differentials of* ϕ^0 *and of* ϕ_1, *respectively, at* e_0, *and* φ_L *denote the linear part of* φ.

Arguing as in the second part of the proof of Lemma (2.51), we can derive [on the basis of Theorem (7)] the following sufficiency theorem in differential form.

14 THEOREM. *Suppose that, for Problem* (1.2), (i) $\mathscr{E}'$ *is a linear vector space,* (ii) φ *is affine and* ϕ^0 *is convex,* (iii) *there is a closed, convex cone* $W_1 \subset \mathscr{Z}_1$ *such that* (int Z_1) + W_1 = int Z_1 *and* ϕ_1 *is* W_1*-convex, and* (iv) ϕ_1 *is Gâteaux directionally differentiable at each point of* $\mathscr{E}$. *Further, suppose that* $e_0 \in \mathscr{E}$, $\alpha \in R^m$, $\beta^0 < 0$, *and* $\ell_1 \in \mathscr{Z}_1^*$ *satisfy* (6) *and* (8) *as well as the inequality*

15 $$\alpha \cdot \varphi_L(e - e_0) + \beta^0 D\phi^0(e_0; e - e_0) + \ell_1 \circ D\phi_1(e_0; e - e_0) \leq 0 \quad \textit{for all} \quad e \in \mathscr{E},$$

where $D\phi^0$, $D\phi_1$ *and* φ_L *are as in Theorem* (12). *Then* e_0 *is a solution of the problem.*

Note that, in Theorem (14), no convexity requirement is imposed on the set $\mathscr{E}$, although, in contradistinction to the "global" sufficiency Theorem (7), we do impose convexity requirements on the functions ϕ^0 and ϕ_1, as well as affineness on φ.

For the special case where Problem (1.2) takes the form of Problem (1.3), the results we have so far obtained essentially coincide with those of Section 3 of Chapter II. Based on the results of the preceding section, we can now strengthen these results as follows.

16 THEOREM. *Suppose that Problem* (1.3) *is convex and well-posed in the sense of* (II.3.1 and 2), *where* $\mathscr{Y} = \mathscr{E}'$ *is a linear vector space. Then* $e_0 \in \mathscr{E}$ *is a solution of this problem if and only if there exist vectors* $\alpha \in R^m$ *and* $\beta = (\beta^1, \ldots, \beta^\mu) \in \bar{R}^\mu_+$, *a linear function* $\Lambda = (\Lambda^1, \ldots, \Lambda^\mu)\colon \mathscr{Y} \to R^\mu$, *and a linear functional* $\Lambda^0 \in \mathscr{Y}^+$ *such that*

17 $$\varphi^i(e_0) = 0 \quad \textit{for} \quad i = 1, \ldots, m$$

$$\textit{and} \quad \phi^i(e_0) \le 0 \quad \textit{for} \quad i = 1, \ldots, \mu,$$

18 $$(\Lambda^0 + \alpha \cdot \varphi + \beta \cdot \Lambda)(e) \ge (\Lambda^0 + \alpha \cdot \varphi + \beta \cdot \Lambda)(e_0)$$

$$\textit{for all} \quad e \in \mathscr{E},$$

19 $$\Lambda^i(y) \le \phi^i(e_0 + y) - \phi^i(e_0)$$

$$\textit{for all} \quad y \in \mathscr{Y} \quad \textit{and each} \quad i = 0, 1, \ldots, \mu,$$

20 $$\sum_{i=1}^{\mu} \beta^i \phi^i(e_0) = 0.$$

The "if" part of the conclusion holds whether or not the problem is convex or well-posed.

Proof. The necessity follows directly from Theorems (4.10) and (I.7.10) and Lemma (I.7.9), once we show [as in the proof of Theorem (II.3.9)] that β^0 in Theorem (4.10) does not vanish. Conversely, if $e_0 \in \mathscr{E}$, $\alpha \in R^m$, $\beta \in \bar{R}^\mu_+$, $\Lambda_0 \in \mathscr{Y}^+$, and the linear function $\Lambda = (\Lambda^1, \ldots, \Lambda^\mu)\colon \mathscr{Y} \to R^\mu$ satisfy (17)–(20), then it is easily seen that (8), (10), and (11) hold—where ℓ_1, ϕ_1, and Z_1 have the evident meanings—and the sufficiency now follows from Theorem (9). The last assertion follows at once from Theorem (7). |||

Let us now turn to Problems (1.8) (minimax) and (1.10) (vector-valued criterion function).

21 Problem (1.8) or Problem (1.10) will be said to be *convex* if (i) $\mathscr{E}'$ is a linear vector space $\mathscr{Y}$, (ii) $\mathscr{E}$ is a convex set in $\mathscr{Y}$, (iii) there

is a closed, convex cone $W_1 \subset \mathscr{Z}_1$ such that (int Z_1) + W_1 = int Z_1 and such that ϕ_1 is W_1-convex, (iv) ϕ_2 is W_2-convex, where $W_2 = Z_2^0$ [see (1.17)], and (v) φ is affine.

The reader at this point should review Remark (2).

22 Problem (1.8), if it is convex, will be said to be *well-posed* if the conditions of (3) hold.

23 There does not appear to be any useful characterization for Problem (1.10) (if it is convex) to be well-posed, in the sense that such a characterization gives rise to the possibility of obtaining conditions which are both necessary and sufficient for optimality.

For convex problems, the multiplier rule, Theorem (3.12), takes the following form.

24 THEOREM. *Let $e_0 \in \mathscr{E}$ be a solution of Problem* (1.8) [*respectively,* (1.10)], *and suppose that this problem is convex and that Hypotheses* (1.21 *and* 22) [*respectively,* (1.25)] *hold. Then there exist a vector $\alpha \in R^m$ and linear functionals $\ell_1 \in \mathscr{Z}_1^*$ and $\ell_2 \in \mathscr{Z}_2^*$, not all zero, such that*

25 $$(\alpha \cdot \varphi + \ell_1 \circ \phi_1 + \ell_2 \circ \phi_2)(e) \le (\alpha \cdot \varphi + \ell_1 \circ \phi_1 + \ell_2 \circ \phi_2)(e_0) \quad \textit{for all} \quad e \in \mathscr{E},$$

26 $$\ell_1(z_1) \ge \ell_1 \circ \phi_1(e_0) \quad \textit{for all} \quad z_1 \in \bar{Z}_1,$$

27 $$\ell_2(z_2) \ge \ell_2 \circ \phi_2(e_0) \quad \textit{for all} \quad z_2 \in Z_2^{\kappa o} \quad [\textit{see } (1.17 \textit{ and } 20)]$$

(*respectively,*

28 $$\ell_2(z_2) \ge 0 \textit{ for all } z_2 \in Z_2^0).$$

Note that (28) holds whenever (27) does, since $\phi_2(e_0) \in Z_2^{\kappa o}$ and $Z_2^{\kappa o} + Z_2^0 = Z_2^{\kappa o}$.

Theorem (24) follows from Theorem (3.12) essentially in the same way that Theorem (II.3.3) followed from Theorem (II.1.4).

Note that the principal multiplier rule in Theorem (24), relation (25), is in terms of the functions φ, ϕ_1, ϕ_2 themselves rather than in terms of their derivatives so that, as in the case of Theorem (4), we have obtained a "global" multiplier rule.

The necessary conditions of Theorem (24) are also sufficient, provided that some additional assumptions hold (whether or not the problem is convex).

29 THEOREM. *If, for Problem* (1.8) [*respectively,* (1.10)] (*whether or not it is convex*), *subject to Hypothesis* (1.22) [*respectively, no*

additional hypothesis], $e_0 \in \mathscr{E}$, $\alpha \in R^m$, $\ell_1 \in \mathscr{Z}_1^*$, *and* $\ell_2 \in \mathscr{Z}_2^*$ *satisfy* (8), *and* (25)–(27) [*respectively*, (8), (25), (26), *and* (28)], *where* κ_0 *is given by* (1.20), *as well as* $\ell_2 \neq 0$ (*respectively*,

30 $$\ell_2(z_2) > 0 \quad \textit{for all} \quad z_2 \in Z_2^0, z_2 \neq 0),$$

then e_0 *is a solution of the problem.*

Proof. First consider the case of Problem (1.8). We shall argue by contradiction. Thus, suppose that $e_0 \in \mathscr{E}, \alpha \in R^m$, $\ell_1 \in \mathscr{Z}_1^*$, and $\ell_2 \in \mathscr{Z}_2^*$ satisfy the indicated hypotheses, and that $e_1 \in \mathscr{E}$ satisfies the relations $\varphi(e_1) = 0$, $\phi_1(e_1) \in Z_1$, and $\sup_{\ell \in \mathscr{L}} \ell \circ \phi_2(e_1) < \kappa_0$, where κ_0 is given by (1.20). This implies that $\phi_2(e_1) \in V_2^{\kappa_0}$ [see (1.18)]. Since $\ell_2 \neq 0$, we can conclude, on the basis of Hypothesis (1.22), Lemma (I.5.14), and (27) that $\ell_2 \circ \phi_2(e_1) > \ell_2 \circ \phi_2(e_0)$. But, because (25), (26), and (8) also hold, this implies that

$$\ell_2 \circ \phi_2(e_0) < \ell_2 \circ \phi_2(e_1) \leq -\alpha \cdot \varphi(e_1) - \ell_1 \circ \phi_1(e_1) \\ + \ell_1 \circ \phi_1(e_0) + \ell_2 \circ \phi_2(e_0) \leq \ell_2 \circ \phi_2(e_0),$$

which is absurd.

We also argue by contradiction in the case of Problem (1.10). Thus, suppose that $e_0 \in \mathscr{E}$, $\alpha \in R^m$, $\ell_1 \in \mathscr{Z}_1^*$, and $\ell_2 \in \mathscr{Z}_2^*$ again satisfy the indicated hypotheses, and that $e_1 \in \mathscr{E}$ satisfies the relations $\varphi(e_1) = 0$, $\phi_1(e_1) \in Z_1$, $\ell \circ \phi_2(e_1) \leq \ell \circ \phi_2(e_0)$ for every $\ell \in \mathscr{L}$, and $\tilde{\ell} \circ \phi_2(e_1) < \tilde{\ell} \circ \phi_2(e_0)$ for some $\tilde{\ell} \in \mathscr{L}$. Hence, $(\phi_2(e_1) - \phi_2(e_0)) \in Z_2^0$, and $\phi_2(e_1) - \phi_2(e_0) \neq 0$, which means, by virtue of (30), that $\ell_2 \circ \phi_2(e_1) > \ell_2 \circ \phi_2(e_0)$. But (25), (26), and (8) then give rise to the inequalities

$$\ell_2 \circ \phi_2(e_1) \leq -\alpha \cdot \varphi(e_1) - \ell_1 \circ \phi_1(e_1) + \ell_1 \circ \phi_1(e_0) + \ell_2 \circ \phi_2(e_0) \\ \leq \ell_2 \circ \phi_2(e_0) < \ell_2 \circ \phi_2(e_1),$$

which is again absurd. |||

Unfortunately, there are no convenient conditions which are both necessary and sufficient for optimality in the case of Problem (1.10) (vector-valued criterion function). However, in the case of Problem (1.8), we can conveniently combine Theorems (24) and (29), if the problem is well-posed, to obtain the following result [compare with Theorem (9)].

31 THEOREM. *Suppose that Problem* (1.8) *is convex and well-posed and satisfies Hypotheses* (1.21 *and* 22). *Then* $e_0 \in \mathscr{E}$ *is a solution of*

this problem if and only if there exist a vector $\alpha \in R^m$ *and functionals* $\ell_1 \in \mathscr{Z}_1^*$ *and* $\ell_2 \in \mathscr{Z}_2^*$, *with* $\ell_2 \neq 0$, *such that* (8) *and* (25)–(27) *hold.*

Theorem (31) follows in almost the exact same way as Theorem (II.3.9), if we also take into account Lemma (I.5.14). Theorem (31) may also be written as follows in differential form:

32 THEOREM. *Suppose that Problem* (1.8) *is convex and well-posed and satisfies Hypotheses* (1.21 *and* 22), *and that* ϕ_1 *and* ϕ_2 *are Gâteaux directionally differentiable at each point of* $\mathscr{E}$. *Then* $e_0 \in \mathscr{E}$ *is a solution of this problem if and only if there exist a vector* $\alpha \in R^m$ *and functionals* $\ell_1 \in \mathscr{Z}_1^*$ *and* $\ell_2 \in \mathscr{Z}_2^*$, *with* $\ell_2 \neq 0$, *such that* (8), (26), (27), *and the relation*

33 $$\alpha \cdot \varphi_L(e - e_0) + \ell_1 \circ D\phi_1(e_0; e - e_0) + \ell_2 \circ D\phi_2(e_0; e - e_0) \leq 0 \quad \text{for all} \quad e \in \mathscr{E}$$

hold, where $D\phi_1(e_0;\cdot)$ *and* $D\phi_2(e_0;\cdot)$ *denote the Gâteaux directional differentials of* ϕ_1 *and* ϕ_2, *respectively, at* e_0, *and* φ_L *denotes the linear part of* φ.

Theorem (32), as well as the following Theorem (34), which is a sufficiency theorem in differential form, can be proved using arguments similar to those of Lemma (2.51). These two theorems are, of course, analogous to Theorems (12) and (14).

34 THEOREM. *Suppose that, for Problem* (1.8) [*respectively,* (1.10)] *subject to Hypothesis* (1.22) [*respectively, no additional hypothesis*], (i) $\mathscr{E}'$ *is a linear vector space,* (ii) φ *is affine,* (iii) ϕ_i (*for* $i = 1, 2$) *is* W_i*-convex—for some closed, convex cone* $W_1 \subset \mathscr{Z}_1$ *such that* $(\text{int } Z_1) + W_1 = \text{int } Z_1$, *and with* $W_2 = Z_2^0$ [*see* (1.17)]—*and* (iv) ϕ_1 *and* ϕ_2 *are Gâteaux directionally differentiable at each point of* $\mathscr{E}$. *Further, suppose that* $e_0 \in \mathscr{E}$, $\alpha \in R^m$, $\ell_1 \in \mathscr{Z}_1^*$, *and* $\ell_2 \in \mathscr{Z}_2^*$ *satisfy* (8), (26), (27), *and* (33) [*respectively,* (8), (26), (28), (30), *and* (33)], *as well as* $\ell_2 \neq 0$, *where* $D\phi_1(e_0;\cdot)$, $D\phi_2(e_0;\cdot)$, *and* φ_L *are as in Theorem* (32). *Then* e_0 *is a solution of the problem.*

In Theorem (34), no convexity requirement is imposed on the set $\mathscr{E}$, although, in contradistinction to the "global" sufficiency Theorem (29), we do impose convexity requirements on the functions ϕ_1 and ϕ_2, as well as affineness on φ.

The preceding results can be considerably simplified for the special case of Problems (1.9 and 12), and results similar to those of Section 3 of Chapter II can then be obtained. We shall leave the details to the interested reader. Corresponding to Theorem (16), we have the following result for Problem (1.9).

35 THEOREM. *Suppose that Problem* (1.9) *is convex in the sense of* (II.3.1)—*with* μ *replaced by* ν—*and well-posed in the sense of* (II.3.2), *where* $\mathscr{Y} = \mathscr{E}'$ *is a linear vector space. Then* $e_0 \in \mathscr{E}$ *is a solution of this problem if and only if there exist vectors* $\alpha \in R^m$ *and* $\beta = (\beta^1, \ldots, \beta^\nu) \in \bar{R}^\nu_-$, *with* $\beta^i \neq 0$ *for some* $i = \mu + 1, \ldots, \nu$, *and a linear function* $\Lambda = (\Lambda^1, \ldots, \Lambda^\nu)\colon \mathscr{Y} \to R^\nu$ *such that*

36 $$\varphi^i(e_0) = 0 \quad \textit{for} \quad i = 1, \ldots, m$$

$$\textit{and} \quad \phi^i(e_0) \leq 0 \quad \textit{for} \quad i = 1, \ldots, \mu,$$

37 $$(\alpha \cdot \varphi + \beta \cdot \Lambda)(e) \leq (\alpha \cdot \varphi + \beta \cdot \Lambda)(e_0) \quad \textit{for all} \quad e \in \mathscr{E},$$

38 $$\Lambda^i(y) \leq \phi^i(e_0 + y) - \phi^i(e_0)$$

$$\textit{for all} \quad y \in \mathscr{Y} \quad \textit{and each} \quad i = 1, \ldots, \nu,$$

39 $$\sum_{i=1}^{\mu} \beta^i \phi^i(e_0) = 0,$$

40 $$\sum_{i=\mu+1}^{\nu} \beta^i \phi^i(e_0) = (\max\,\{\phi^j(e_0)\colon j = \mu + 1, \ldots, \nu\}) \cdot \sum_{i=\mu+1}^{\nu} \beta^i.$$

Proof. The necessity follows directly from Theorems (4.12) and (I.7.10) and Lemma (I.7.9) [we can show that $\beta^i \neq 0$ for some $i = \mu + 1, \ldots, \nu$ by arguing as in the proof of Theorem (II.3.9)]. The sufficiency is an almost immediate consequence [see the proof of Theorem (16)] of (1.23), and Theorem (31). |||

41 The sufficiency part of Theorem (35) holds whether or not the problem is convex [see Theorem (29)], and the necessity part holds even if the problem is not well-posed, provided that we weaken the conclusion $\beta^i \neq 0$ for some $i = \mu + 1, \ldots, \nu$ to $|\alpha| + |\beta| > 0$. By the same token, if we replace Problem (1.9) by Problem (1.12) (without necessarily assuming that it is well-posed), then the necessity part of Theorem (35) remains in force, provided that we delete (40) and again weaken the conclusion that $\beta^i \neq 0$ for some $i = \mu + 1, \ldots, \nu$ to $|\alpha| + |\beta| > 0$. The sufficiency part of the theorem also remains in force for Problem (1.12) (even if it is not convex), provided that we strengthen the hypothesis that $\beta^i \neq 0$ for *some* $i > \mu$ to $\beta^i \neq 0$ for *all* $i > \mu$, and, of course, delete (40) [see Theorem (29)].

42 If we modify Problems (1.2, 8, and 10) by deleting the generalized inequality constraint $\phi_1(e) \in Z_1$ [in Problems (1.3, 9, and 12), this means that the inequality constraints $\phi^i(e) \leq 0$ for $i = 1, \ldots, \mu$ are to be deleted], then all of the results so far obtained in this section remain in force with obvious modifications in the form of deletions in the assumptions and in the conclusions [see (3.32) and (4.20)].

On the other hand, if we modify these problems by deleting the equality constraint $\varphi(e) = 0$, then other corresponding modifications must be made. Let us illustrate these results by considering the following problem, which is essentially a "convex" version of Problem (1.2) without equality constraints.

43 *Problem.* Given a convex set $\mathscr{E}$ in a linear vector space $\mathscr{Y}$, a linear topological space $\mathscr{Z}_1$ and a convex body Z_1 in $\mathscr{Z}_1$, a closed, convex cone $W_1 \subset \mathscr{Z}_1$ such that $(\text{int } Z_1) + W_1 = \text{int } Z_1$, a convex functional ϕ^0 on $\mathscr{Y}$, and a W_1-convex function $\phi_1 : \mathscr{Y} \to \mathscr{Z}_1$, find an element $e_0 \in \mathscr{E}$ that achieves a minimum for ϕ^0 (on $\mathscr{E}$) subject to the constraint $\phi_1(e) \in Z_1$.

44 We shall say that Problem (43) is *well-posed* if there is an element $e^* \in \mathscr{E}$ such that $\phi_1(e^*) \in (\text{int } Z_1)$ [compare with (3)].

45 We can obtain necessary conditions and sufficient conditions for Problem (43) which are analogous to those of Theorems (4), (7), (9), (12), and (14), and which generalize those of Theorems (II.3.28, 30, 32, 40, and 41). We shall state one of these theorems and shall leave the details of the remaining ones as an exercise to the interested reader.

46 THEOREM. *Suppose that Problem* (43) *is well-posed. Then* $e_0 \in \mathscr{E}$ *is a solution of this problem if and only if there exists a linear functional* $\ell_1 \in \mathscr{Z}_1^*$ *such that*

47 $$(\phi^0 + \ell_1 \circ \phi_1)(e) \geq (\phi^0 + \ell_1 \circ \phi_1)(e_0) \quad \textit{for all} \quad e \in \mathscr{E},$$

48 $$\ell_1(z_1) \leq \ell_1 \circ \phi_1(e_0) \quad \textit{for all} \quad z_1 \in \bar{Z}_1,$$

49 $$\phi_1(e_0) \in Z_1.$$

The proof of Theorem (46) should now be evident.

We shall also leave it to the interested reader to obtain the corresponding results for convex minimax problems and for convex "vector-valued criterion function" problems without equality constraints.

50 We close this section, and this chapter, with the reminder that the relation $\ell_1(z_1) \geq$ (respectively, $\leq$) $\ell_1 \circ \phi_1(e_0)$ for all $z_1 \in \bar{Z}_1$, which occurs in most of the theorems in this section, takes the form of the two relations $\ell_1 \in -(\bar{Z}_1)^*$ [respectively, $\ell_1 \in (\bar{Z}_1)^*$] and $\ell_1 \circ \phi_1(e_0) = 0$, if Z_1 is a (convex) cone in $\mathscr{Z}_1$ [see Corollary (2.44)]. Similarly, the relation $\ell_2(z_2) \geq \ell_2 \circ \phi_2(e_0)$ for all $z_2 \in Z_2^{\kappa o}$, which occurs in most of the theorems of this section pertaining to Problem (1.8), is equivalent to the two relations $\ell_2 \in -(Z_2^0)^*$ and $\ell_2 \circ \phi_2(e_0) = [\sup_{\ell \in \mathscr{L}} \ell \circ \phi_2(e_0)]\ell_2(\zeta_0)$ if Hypothesis (3.17) holds [see Corollary (3.21)].

CHAPTER IV

Optimization with Operator Equation Restrictions

IN THIS CHAPTER, we shall justify the form of the hypotheses under which we derived the various multiplier rules in the preceding chapter [namely, see Hypotheses (III.2.11–14 and 57–59)].

We shall be concerned with optimization problems (ordinary, minimax, and with vector-valued criterion function) in which we are given a Banach space $\mathscr{X}$, an open subset A of $\mathscr{X}$, and a set $\mathscr{W}$ of continuously differentiable operators $T: A \to \mathscr{X}$, and we wish to find a function $x \in A$ which satisfies the equation $x = Tx$, for some $T \in \mathscr{W}$, as well as some prescribed equality and "inequality" constraints, and which is at the same time optimal in one of the three indicated senses.

We shall reduce these problems to one of the basic problems discussed in Chapter III, Section 1 [i.e., Problem (III.1.2, 8, or 10)], where the underlying set $\mathscr{E}$ is some suitable subset of $\mathscr{W}$, much as we reduced Problem (II.2.1) to Problem (II.1.1) in (II.2.11). We shall make assumptions on the problem data—particularly on the set $\mathscr{W}$—which will ensure that the pertinent hypotheses of Chapter III previously mentioned are satisfied, and, applying the multiplier rules of Section 3 of Chapter III, we shall obtain necessary conditions that solutions of these problems must satisfy. We shall also particularize these conditions to the special case where $\mathscr{X} = \mathscr{C}^n(I)$ (for some compact interval I), and where the equality and inequality constraints have a particular form which is especially important in applications.

The preceding material is the subject matter of Section 1. In Section 2, we shall extend the results of Section 1 to optimization problems which contain parameters, both in the equality and inequality constraints, and in the operator equation (in the sense that the operators from A into $\mathscr{X}$ may also depend on a parameter). In these problems, one wishes to find not only the "best" function $x \in A$, but also the "best" parameter, which is supposed to lie in a given set in some Banach space.

In Sections 3 and 4 (the latter is devoted to problems with parameters), we shall investigate in detail the case where $\mathscr{X} = \mathscr{C}^n(I)$, and where the operators in $\mathscr{W}$ are all of a special kind—so-called Volterra-type operators. The Appendix of this book is devoted to a detailed study of the properties of these operators. Such operators arise, for example, when the operator equation $x = Tx$ represents an ordinary differential equation, a functional differential equation of retarded type, or a Volterra integral equation. By making use of certain special properties of Volterra-type operators, we shall show how the hypotheses of Sections 2 and 3 can—for such operators—be replaced by alternative hypotheses, which are much easier to verify in concrete examples.

In Section 5, we shall introduce the so-called concept of convexity under switching which applies to Volterra-type operators that arise in functional (and, in particular, ordinary) differential equations and in Volterra integral equations. If the underlying set of operators $\mathscr{W}$ is defined in terms of a class of functions which is "convex under switching"—and, as we shall see in subsequent chapters, this is often the case in applications—then, by virtue of the results of Sections 3 and 4, the basic hypotheses of Sections 1 and 2 also hold. This makes it possible to appeal to the multiplier rules of Chapter III in this case to obtain our desired necessary conditions for optimality. In fact, it is the concept of convexity under switching, together with its consequences, which lies behind the more complicated sets of hypotheses in the multiplier rules of Chapter III. In the loose terminology of (III.2.10), if $\mathscr{W}$ has the convex-under-switching property, then $\mathscr{W}$ is "almost" convex, and the corresponding set $\mathscr{E}$ is "almost" finitely open in itself.

Finally, in Section 6, we shall obtain some sufficient conditions for optimality, based on the results of Section 5 of Chapter III, under the assumption that the set $\mathscr{W}$ is of a particular form.

The reader who is not interested in all of the details of the theory can, without great loss, and without sacrifice of continuity, skip or lightly skim Sections 2 and 4 of this chapter dealing with optimization problems with parameters.

1. General Optimization Problems with Operator Equation Constraints

The results of this section extend those in the first part of Section 2 of Chapter II [in particular, see Theorem (II.2.14)], and the reader is advised to review this earlier material before reading what follows.

1 Throughout this chapter, we shall suppose that a Banach space $\mathscr{X}$ and an open set A in $\mathscr{X}$ are given. We shall denote by $\mathscr{T}$ the linear vector space of all operators $T: A \to \mathscr{X}$ which are continuously differentiable [see (I.7.48)]. We shall endow $\mathscr{T}$ with the topology of pointwise convergence, so that $\mathscr{T}$ becomes a locally convex, linear topological space [see (I.6.33)].

2 We shall denote by $\mathscr{T}_0$ the set of all $T \in \mathscr{T}$ that have *at most one* fixed point [see (I.5.37)], and by $\mathscr{T}_1$ the set of all $T \in \mathscr{T}$ that have *exactly one* fixed point. Clearly, $\mathscr{T}_0 \supset \mathscr{T}_1$.

Now consider the following problem [compare with Problem (II.2.1)].

3 *Problem.* Given a subset $\mathscr{W}$ of $\mathscr{T}_0$, a subset A' of A which contains the fixed point of every $T \in \mathscr{T}_1 \cap (\text{co}\, \mathscr{W})$, a linear topological space $\mathscr{Z}_1$, a convex body [see (I.4.18)] Z_1 in $\mathscr{Z}_1$, and functions $\hat{\varphi}: A' \to R^m$, $\hat{\phi}^0: A' \to R$, and $\hat{\phi}_1: A' \to \mathscr{Z}_1$, find a pair $(x_0, T_0) \in A' \times \mathscr{W}$ such that

4 $$x_0 = T_0 x_0,$$

5 $$\hat{\varphi}(x_0) = 0,$$

6 $$\hat{\phi}_1(x_0) \in Z_1,$$

and such that $\hat{\phi}^0(x_0) \leq \hat{\phi}^0(x)$ for all $x \in A'$ that satisfy the relations

7 $$\hat{\varphi}(x) = 0,\ \hat{\phi}_1(x) \in Z_1, \quad \text{and} \quad x = Tx \quad \text{for some} \quad T \in \mathscr{W}.$$

8 Let us identify Problem (3) as a special case of Problem (III.1.2) [compare with (II.2.11)]. Indeed, let us identify $\mathscr{E}'$ with $\mathscr{T}_1 \cap (\text{co}\, \mathscr{W})$, $\mathscr{E}$ with $(\mathscr{T}_1 \cap \mathscr{W})$, and let us denote by φ_0 the map from $\mathscr{E}'$ into A' which assigns to each $T \in \mathscr{E}'$ its fixed point in A'; i.e.,

9 $$\varphi_0(T) = x \quad \text{if and only if} \quad x = Tx \quad (T \in \mathscr{E}').$$

If we identify the maps $\varphi: \mathscr{E}' \to R^m$, $\phi^0: \mathscr{E}' \to R$, and $\phi_1: \mathscr{E}' \to \mathscr{Z}_1$

with the maps $\hat{\varphi} \circ \varphi_0$, $\hat{\phi}^0 \circ \varphi_0$, and $\hat{\phi}_1 \circ \varphi_0$, respectively, then it is trivial to verify that we have achieved the desired identification, with T_0 a solution of Problem (III.1.2) if and only if (x_0, T_0)—where $x_0 = \varphi_0(T_0)$—is a solution of Problem (3).

Most commonly, $\mathscr{X} = \mathscr{C}^n(I)$ for some compact interval $I = [t_1, t_2]$ [see (I.6.5)], and $\mathscr{W}$ is often defined in terms of a set of integral operators that arise in ordinary differential equations [see (II.2.6)], or, more generally, in functional differential equations (of retarded type) and in Volterra integral equations.

In this case, $A' = A$ and the functions $\hat{\varphi}: A \to R^m$ and $\hat{\phi}^0: A \to R$ are typically defined in terms of given continuous functions $\chi': (R^n)^k \to R^m$ and $\chi^0: (R^n)^k \to R$, and a given subset $\{\tau^1, \ldots, \tau^k\}$ of I through the relations

10 $$\hat{\varphi}(x) = \chi'(x(\tau^1), \ldots, x(\tau^k)), \quad x \in A,$$

11 $$\hat{\phi}^0(x) = \chi^0(x(\tau^1), \ldots, x(\tau^k)), \quad x \in A.$$

Further, the linear topological space $\mathscr{Z}_1$, the convex body $Z_1 \subset \mathscr{Z}_1$, and the function $\hat{\phi}_1: A \to \mathscr{Z}_1$ are typically defined as follows: $\mathscr{Z}_1 = R^\mu \times \mathscr{C}^\nu(I)$ (for some positive integers μ and ν) and $Z_1 = \bar{R}^\mu_- \times Z'$, where

12 $$Z' = \{z: z = (z^1, \ldots, z^\nu) \in \mathscr{C}^\nu(I), z^i(t) \leq 0 \text{ for all } t \in I_i \text{ and each } i = 1, \ldots, \nu\},$$

where $I_1, \ldots, I_\nu$ are given closed subsets of I. In this case, it is easy to see that Z' and Z_1 are closed, convex cones in $\mathscr{C}^\nu(I)$ and $\mathscr{Z}_1$, respectively, and that [see (I.3.34, 45, and 46)]

$$\operatorname{int} Z_1 = R^\mu_- \times \{z: z = (z^1, \ldots, z^\nu) \in \mathscr{C}^\nu(I), z^i(t) < 0 \text{ for all } t \in I_i \text{ and each } i = 1, \ldots, \nu\}.$$

The function $\hat{\phi}_1$ is defined in terms of given continuous functions $\chi'': (R^n)^k \to R^\mu$ and $\tilde{\chi}: R^n \times I \to R^\nu$ through the relation

13 $$\hat{\phi}_1(x) = (\chi''(x(\tau^1), \ldots, x(\tau^k)), \hat{\phi}'(x)), \quad x \in A,$$

where the function $\hat{\phi}': A \to \mathscr{C}^\nu(I)$ is defined by

14 $$(\hat{\phi}'(x))(t) = \tilde{\chi}(x(t), t) \quad \text{for all} \quad t \in I \quad \text{and} \quad x \in A.$$

If $\mathscr{X}$, A', $\hat{\varphi}$, $\hat{\phi}^0$, $\mathscr{Z}_1$, Z', Z_1, and $\hat{\phi}_1$ are defined in the just-indicated manner, then, if we set $\chi = (\chi^0, \chi'', \chi')$, Problem (3)

evidently reduces to the following one:

15 *Problem.* Given a subset $\mathscr{W}$ of $\mathscr{T}_0$ [where $\mathscr{X} = \mathscr{C}^n(I)$], continuous functions

$$\chi = (\chi^0, \chi^1, \ldots, \chi^\mu, \ldots, \chi^{\mu+m}): (R^n)^k \to R^{\mu+m+1}$$

and

$$\tilde{\chi} = (\tilde{\chi}^1, \ldots, \tilde{\chi}^\nu): R^n \times I \to R^\nu,$$

a finite subset $\{\tau^1, \ldots, \tau^k\}$ of I, and closed subsets $I_1, \ldots, I_\nu$ of I, find a function $x \in A$ satisfying the constraints

16 $$x = Tx \quad \text{for some} \quad T \in \mathscr{W},$$

17 $$\chi^i(x(\tau^1), \ldots, x(\tau^k)) \leq 0 \quad \text{for} \quad i = 1, \ldots, \mu,$$

18 $$\chi^i(x(\tau^1), \ldots, x(\tau^k)) = 0 \quad \text{for} \quad i = \mu + 1, \ldots, \mu + m,$$

19 $$\tilde{\chi}^i(x(t), t) \leq 0 \quad \text{for all} \quad t \in I_i \quad \text{and each} \quad i = 1, \ldots, \nu,$$

which in so doing, achieves a minimum for $\chi^0(x(\tau^1), \ldots, x(\tau^k))$.

Let us now return to the general Problem (3). We shall try to find hypotheses on the problem data under which the hypotheses of Theorem (III.3.1) hold [with the identification made in (8)], so that we can appeal to this theorem in order to obtain necessary conditions for solutions of Problem (3). Of course, as we have indicated a number of times already, the hypotheses which we shall make must be consistent with the conditions that hold in the optimal control problems which we are interested in solving.

20 We first point out that the map φ_0 can, in an obvious manner, be extended onto $\mathscr{T}_1$, and we shall momentarily suppose that this has been done. We must point out that $\mathscr{T}_1$ is not finitely open in itself and that φ_0 is not continuous on $\mathscr{T}_1$. Thus, we cannot expect that in general the set $\mathscr{T}_1 \cap (\text{co } \mathscr{W})$ will be finitely open in itself, or that φ_0 will be continuous on this set. On the other hand, Hypotheses (III.2.1 and 11), one of which we require to hold in Theorem (III.3.1), demand that $\mathscr{E}'$—for which we have agreed to take $\mathscr{T}_1 \cap (\text{co } \mathscr{W})$—be finitely open in itself. Further, in Theorem (III.3.1), it is required that, if $\mathscr{E}$ (for which have agreed to take $\mathscr{T}_1 \cap \mathscr{W}$) is not finitely open in itself [i.e., if Hypothesis (III.2.1) is not satisfied]—and in general it is unreasonable to suppose that $\mathscr{T}_1 \cap \mathscr{W}$ has this property—then Hypotheses (III.2.11–14), with ϕ, etc., replaced by ϕ_1, etc., must hold. Hypothesis (III.2.14) in particular demands that φ and ϕ_1 must be continuous on certain subsets of $\mathscr{E}'$ (this is also

required for ϕ^0 in the theorem statement), and these subsets must have additional properties relative to finite subsets of $\mathscr{E}$ (as is indicated in the hypothesis). Inasmuch as $\varphi = \hat{\varphi} \circ \varphi_0$, $\phi^0 = \hat{\phi}^0 \circ \varphi_0$, and $\phi_1 = \hat{\phi}_1 \circ \varphi_0$, and we have already noted that φ_0 is *not* continuous on $\mathscr{T}_1$, we shall make some hypotheses that take into consideration the fact that φ_0 *is* continuous on certain suitable subsets of $\mathscr{T}_1$.

Consequently, based on the preceding considerations, and keeping in mind our ultimate applications, we shall make the following three hypotheses on T_0 and $\mathscr{W}$. The justification for these assumptions will become more apparent in Sections 3 and 5 when we discuss Volterra-type operators and functions which are "convex under switching."

21 HYPOTHESIS. *The set* $(\text{co}\,\mathscr{W}) \subset \mathscr{T}_0$.

22 HYPOTHESIS. *For each* $T \in \text{co}\,\mathscr{W}$ *and each* $x \in A$, $[E - DT(x;\cdot)]$*—where* E *denotes the identity operator on* $\mathscr{X}$ *and* $DT(x;\cdot)$ *denotes the Fréchet differential of* T *at* x*—is a (linear) homeomorphism* [*see* (I.3.28)] *of* $\mathscr{X}$ *onto itself.*

23 HYPOTHESIS. *For every finite subset* $\{T_1, \ldots, T_\rho\}$ *of* $\mathscr{W}$, *there exist a subset* $\mathscr{W}_0$ *of* $\mathscr{W}$ *and a number* $\varepsilon_0 > 0$ *with the following properties*:

A. *The function* φ_0 *is continuous on the set* $(\mathscr{R} \cup \mathscr{W}_0) \cap \mathscr{T}_1$, *where*

24 $$\mathscr{R} = \left\{T : T = T_0 + \sum_{i=1}^{\rho} \lambda^i(T_i - T_0), 0 \leq \lambda^i \leq \varepsilon_0 \textit{ for } i = 1, \ldots, \rho\right\}.$$

B. *For every neighborhood* N *of* 0 *in* $\mathscr{T}$, *there exists a continuous function* γ: $\{\lambda : \lambda = (\lambda^1, \ldots, \lambda^\rho) \in R^\rho,\ 0 \leq \lambda^i \leq \varepsilon_0$ *for* $i = 1, \ldots, \rho\} \to N$ *such that*

25 $$T_0 + \sum_{i=1}^{\rho} \lambda^i(T_i - T_0) + \gamma(\lambda^1, \ldots, \lambda^\rho) \in \mathscr{T}_1 \cap \mathscr{W}_0$$
$$\textit{whenever} \quad 0 \leq \lambda^i \leq \varepsilon_0 \quad \textit{for each} \quad i.$$

Finally let us suppose that $\hat{\varphi}$, $\hat{\phi}^0$, and $\hat{\phi}_1$ satisfy the following continuity and differentiability conditions.

26 HYPOTHESIS. *The functions* $\hat{\varphi}$, $\hat{\phi}^0$, *and* $\hat{\varphi}_1$ *are continuous and, in addition, respectively, dually differentiable, dually semidifferentiable, and dually* W_1*-semidifferentiable at* x_0 [*see* (I.7.23, 25, 26)]

for some closed, convex cone $W_1 \subset \mathscr{Z}_1$ *such that* $(\text{int } Z_1) + W_1 = \text{int } Z_1$.

27 If Problem (3) is such that $\mathscr{Z}_1 = R^\mu \times \mathscr{Z}'$ and $Z_1 = \bar{R}^\mu_- \times Z'$, for some linear topological space $\mathscr{Z}'$ and some convex body $Z' \subset \mathscr{Z}'$—so that we can write $\hat{\phi}_1 = (\hat{\phi}^1, \ldots, \hat{\phi}^\mu, \hat{\phi}')$, where $\hat{\phi}^1, \ldots, \hat{\phi}^\mu$ are real-valued and $\hat{\phi}'$ takes on its values in $\mathscr{Z}'$ [note that, in Problem (15), $\mathscr{Z}_1$ and Z_1 are precisely of this form, with $\mathscr{Z}' = \mathscr{C}^\nu(I)$]—then Hypothesis (26) is equivalent to the following one [see (I.7.50, 26, 27), (I.1.16), and (I.3.38)]:

28 HYPOTHESIS. *The functions* $\hat{\varphi}$, $\hat{\phi}^i$ $(i = 0, \ldots, \mu)$, *and* $\hat{\phi}'$ *are continuous and, in addition, respectively, dually differentiable, dually semidifferentiable and dually* W'*-semidifferentiable at* x_0, *for some closed convex cone* $W' \subset \mathscr{Z}'$ *such that* $(\text{int } Z') + W' = \text{int } Z'$.

29 With regard to Hypotheses (27) and (28), the reader is advised to review Theorem (I.7.44) and the remarks in (I.7.45) regarding the relation between dual differentiability, dual W-semidifferentiability, and Fréchet differentiability, and Lemma (I.7.38) on continuity.

On the basis of the multiplier rule Theorem (III.3.1), we may obtain the following necessary conditions for solutions of Problem (3).

30 THEOREM. *Let* (x_0, T_0) *be a solution of Problem* (3), *and suppose that Hypotheses* (21), (22), *and* (26) *hold. Further, suppose that either* $\mathscr{W}$ *is finitely open in itself, or that Hypothesis* (23) *holds. Then, for every dual semidifferential* $D\hat{\phi}^0(x_0;\cdot)$ *of* $\hat{\phi}^0$ *at* x_0 *and every dual* W_1*-semidifferential* $D\hat{\phi}_1(x_0;\cdot)$ *of* $\hat{\phi}_1$ *at* x_0, *there exist a vector* $\alpha \in R^m$, *a number* $\beta^0 \leq 0$, *and a linear functional* $\ell_1 \in \mathscr{Z}_1^*$, *not all zero, such that*

31 $$(\alpha \cdot D\hat{\varphi} + \beta^0 D\hat{\phi}^0 + \ell_1 \circ D\hat{\phi}_1) \circ [E - DT_0]^{-1}(Tx_0 - x_0) \leq 0 \quad \text{for all} \quad T \in \mathscr{W},$$

32 $$\ell_1(z_1) \geq \ell_1 \circ \hat{\phi}_1(x_0) \quad \text{for all} \quad z_1 \in \bar{Z}_1,$$

where, in (31), *we have written* $D\hat{\phi}^0(\cdot)$ *and* $D\hat{\phi}_1(\cdot)$ *for* $D\hat{\phi}^0(x_0;\cdot)$ *and* $D\hat{\phi}_1(x_0;\cdot)$, *respectively,* $D\hat{\varphi}$ *and* DT_0 *denote, respectively, the dual differential of* $\hat{\varphi}$ *and the Fréchet differential of* T_0, *both at* x_0, *and* E *denotes the identity operator on* $\mathscr{X}$.

If, in particular, the data of Problem (3) *are as indicated in* (27)—*so that Hypothesis* (26) *may be replaced by Hypothesis* (28)—*then,* *for every collection of dual semidifferentials* $D\hat{\phi}^i(x_0;\cdot)$ *of* $\hat{\phi}^i$ *at* x_0, *for* $i = 0, 1, \ldots, \mu$, *and every dual W′-semidifferential* $D\hat{\phi}'(x_0;\cdot)$ *of* $\hat{\phi}'$ *at* x_0, *there exist vectors* $\alpha \in R^m$ *and* $\beta = (\beta^0, \beta^1, \ldots, \beta^\mu) \in \bar{R}_-^{\mu+1}$ *and a linear functional* $\ell' \in (\mathscr{Z}')^*$, *not all zero, such that*

33 $$(\alpha \cdot D\hat{\varphi} + \beta \cdot D\hat{\phi}'' + \ell' \circ D\hat{\phi}') \circ [E - DT_0]^{-1}(Tx_0 - x_0) \leq 0 \quad \textit{for all} \quad T \in \mathscr{W},$$

34 $$\sum_{i=1}^{\mu} \beta^i \hat{\phi}^i(x_0) = 0,$$

35 $$\ell'(z') \geq \ell' \circ \hat{\phi}'(x_0) \quad \textit{for all} \quad z' \in \bar{Z}',$$

where, in (33), *we have written* $D\hat{\phi}'(\cdot)$ *for* $D\hat{\phi}'(x_0;\cdot)$, *and* $D\hat{\phi}''(\cdot) = (D\hat{\phi}^0(x_0;\cdot), \ldots, D\hat{\phi}^\mu(x_0;\cdot))$. *If the dual semidifferentials* $D\hat{\phi}^i(x_0;\cdot)$ *can all be extended to* $\mathscr{X}$ *in such a way that they remain convex, continuous functionals, then there is also a continuous linear function* $\Lambda = (\Lambda^0, \Lambda^1, \ldots, \Lambda^\mu): \mathscr{X} \to R^{\mu+1}$ *such that*

36 $$(\alpha \cdot D\hat{\varphi} + \beta \cdot \Lambda + \ell' \circ D\hat{\phi}') \circ [E - DT_0]^{-1}(Tx_0 - x_0) \leq 0 \quad \textit{for all} \quad T \in \mathscr{W},$$

37 $$\Lambda^i(x) \leq D\hat{\phi}^i(x_0; x) \quad \textit{for all} \quad x \in \mathscr{X} \quad \textit{and each} \quad i = 0, \ldots, \mu.$$

Proof. Arguing as in (II.2.12), we can show, on the basis of Hypotheses (21) and (22), that, for any finite subset $\{T_1, \ldots, T_p\}$ of $\mathscr{W}$, there is a number $\varepsilon_0 > 0$ such that the set $\mathscr{R}$ defined by (24) is contained in $\mathscr{T}_1$. For this reason, we may always assume, without loss of generality, that the number $\varepsilon_0 > 0$ in Hypothesis (23) is such that the set $\mathscr{R}$ [defined by (24)] is contained in $\mathscr{T}_1 \cap (\text{co } \mathscr{W})$.

Let us denote the set $\mathscr{T}_1 \cap (\text{co } \mathscr{W})$ by $\mathscr{E}'$. Then, again arguing as in (II.2.12) [on the basis of Hypotheses (21) and (22)], we can show that (i) $\mathscr{E}'$ is finitely open in itself, (ii) the map $\varphi_0 : \mathscr{E}' \to A'$ defined by (9) is finitely differentiable [see (I.7.12), Theorems (I.7.29 and 44), and (I.2.7)] and hence, by Theorem (I.7.10), is also finitely continuous, and (iii) the function

38 $$T \to [E - DT_0]^{-1}(Tx_0), \; T \in \text{cone}(\mathscr{E}' - T_0),$$

is the finite differential of φ_0 at T_0, where DT_0 denotes the Fréchet differential of T_0 at x_0. Note that the linear function (38), which may be considered to be defined on all of $\mathscr{T}$, is continuous. Also, $\varphi_0(T_0) = x_0$.

If we set $\mathscr{T} = \mathscr{Y}$, $\mathscr{T}_1 \cap \mathscr{W} = \mathscr{E}$, $\varphi = \hat{\varphi} \circ \varphi_0$, $\phi^0 = \hat{\phi}^0 \circ \varphi_0$, and $\phi_1 = \hat{\phi}_1 \circ \varphi_0$, then, as we pointed out in (8), Problem (3) becomes a special case of Problem (III.1.2), with T_0 a solution of the latter. By hypothesis (26), $\hat{\varphi}$ is continuous on $\varphi_0(\mathscr{E}')$, so that, by the last sentence of (I.5.26), φ is finitely continuous. Further, it follows from Hypothesis (26) and Theorem (I.7.34) (a chain rule of differentiation) that φ, ϕ^0, and ϕ_1 are, respectively, finitely differentiable, finitely semidifferentiable, and finitely W_1-semidifferentiable at T_0. Further, the finite differential $D\varphi$ of φ at T_0 is, according to Theorem (I.7.34), given by the function

39 $$T \to D\hat{\varphi} \circ [E - DT_0]^{-1}(Tx_0), \quad T \in \text{cone}\,(\mathscr{E}' - T_0),$$

where $D\hat{\varphi}$ denotes the dual differential of $\hat{\varphi}$ at x_0. Since $D\hat{\varphi}$ can be extended so as to become a member of $\mathbf{B}(\mathscr{X},R^m)$ [see (I.7.23) and (I.5.44)], $D\varphi$ can be extended correspondingly [through (39)] so as to be a member of $\mathbf{B}(\mathscr{T},R^m)$, i.e., $D\varphi$ is a continuous linear function. Also, by Theorem (I.7.34), if $D\hat{\phi}^0$ and $D\hat{\phi}_1$ are, respectively, any dual semidifferential of $\hat{\phi}^0$ at x_0 and any dual W_1-semidifferential of $\hat{\phi}_1$ at x_0, then the functions

40 $$T \to D\hat{\phi}^0 \circ [E - DT_0]^{-1}(Tx_0), \quad T \in \text{cone}\,(\mathscr{E}' - T_0),$$

41 $$T \to D\hat{\phi}_1 \circ [E - DT_0]^{-1}(Tx_0), \quad T \in \text{cone}\,(\mathscr{E}' - T_0),$$

are, respectively, a finite semidifferential of ϕ^0 and a finite W_1-semidifferential of ϕ_1, both at T_0. Since $D\hat{\phi}^0$ and $D\hat{\phi}_1$ are defined and continuous on $\overline{\text{co}}$ tancone $(A'; x_0)$ [see (I.7.25) and (I.4.25)], we may extend the functions given by (40) and (41), by still using these formulas, from cone $(\mathscr{E}' - T_0)$ to $\overline{\text{co}}\,\overline{\text{co}}\,(\mathscr{E} - T_0) \subset \overline{\text{cone}}\,(\mathscr{E}' - T_0)$ [see Lemma (I.1.41) and recall that $\mathscr{E} \subset \mathscr{E}'$], and the resulting extensions are continuous.

We can now conclude that Hypotheses (III.2.2, 11, and 12) hold (with $e_0 = T_0$), that Hypothesis (III.2.13) [or (III.2.3)] holds with ϕ, $\mathscr{Z}$, Z, and W replaced by ϕ_1, $\mathscr{Z}_1$, Z_1, and W_1, respectively, and that Hypothesis (III.2.53) holds with $\mathscr{Z} = R \times \mathscr{Z}_1$, $\phi = (\phi^0, \phi_1)$, and $D\phi = (D\phi^0, D\phi_1)$, where $D\phi^0$ and $D\phi_1$ are the functions given by (40) and (41), respectively. Further, if $\mathscr{W}$ is finitely open in itself, then, since $\mathscr{E} = \mathscr{E}' \cap \mathscr{W}$, we can conclude by virtue of (I.1.40) that Hypothesis (III.2.1) holds.

Our desired conclusion now follows immediately from Theorem (III.3.1) and Corollary (III.3.10), by virtue of (4), if we can show that (i) under Hypothesis (23), Hypothesis (III.2.14) also holds with $\phi = (\phi^0, \phi_1)$ and (ii) $(\mathscr{W} - T_0) \subset \overline{\text{co}}\,\overline{\text{co}}\,(\mathscr{E} - T_0)$.

Thus, suppose that Hypothesis (23) holds. Let $\{T_1, \ldots, T_\rho\}$ be an arbitrary finite subset of $\mathscr{W}$, and let $\mathscr{W}_0$ and ε_0 be the corresponding subsets of $\mathscr{W}$ and positive number whose existence is asserted in Hypothesis (23). As we indicated previously, we may suppose (without loss of generality) that the set $\mathscr{R}$ defined by (24) is contained in $\mathscr{T}_1 \cap (\text{co}\,\mathscr{W}) = \mathscr{E}'$. Let $\mathscr{E}_0 = \mathscr{T}_1 \cap \mathscr{W}_0$, so that $\mathscr{E}_0 \subset \mathscr{E}$ and the function φ_0, as well as the functions $\varphi = \hat{\varphi} \circ \varphi_0$, $\phi^0 = \hat{\phi}^0 \circ \varphi_0$, and $\phi_1 = \hat{\phi}_1 \circ \varphi_0$ [by Hypothesis (26) and Theorem (I.3.27)], are continuous on $\mathscr{R} \cup \mathscr{E}_0$. Taking into account Condition B of Hypothesis (23), we now at once conclude that Hypothesis (III.2.14) is satisfied.

By the same token, it now follows that, for every element $T_1 \in \mathscr{W}$ and every neighborhood N_1 of 0 in $\mathscr{T}$, there exist a number $\lambda^1 \in (0,1)$ and a $T_1' \in \mathscr{E}$ such that $T_1' \in T_0 + \lambda^1(T_1 - T_0) + \lambda^1 N_1$, i.e., such that $(1/\lambda_1)(T_1' - T_0) \in (T_1 - T_0 + N_1) \cap \text{co co}\,(\mathscr{E} - T_0)$, which implies that $(\mathscr{W} - T_0) \subset \overline{\text{co}}\,\overline{\text{co}}\,(\mathscr{E} - T_0)$.

If the data of Problem (3) are as indicated in (27), and if Hypothesis (28) holds—which means that Hypothesis (26) also holds—then, by what we have just shown, taking into account (I.5.10 and 25) and (I.7.50), we conclude that, for every collection of dual semidifferentials $D\hat{\phi}^i(x_0;\cdot)$ of $\hat{\phi}^i$ at x_0, for $i = 0, \ldots, \mu$, and every dual W'-semidifferential $D\hat{\phi}'(x_0;\cdot)$ of $\hat{\phi}'$ at x_0, there exist vectors $\alpha \in R^m$ and $\beta = (\beta^0, \beta^1, \ldots, \beta^\mu) \in \bar{R}_- \times R^\mu$ and a linear functional $\ell' \in (\mathscr{Z}')^*$, not all zero, such that (33) holds and such that

42 $$\sum_{i=1}^{\mu} \beta^i \xi^i + \ell'(z') \geq \sum_{i=1}^{\mu} \beta^i \hat{\phi}^i(x_0) + \ell' \circ \hat{\phi}'(x_0)$$
$$\text{for all} \quad (\xi^1, \ldots, \xi^\mu, z') \in \bar{R}_-^\mu \times \bar{Z}'.$$

Now (35) follows at once from (42). Further, since $\hat{\phi}'(x_0) \in Z'$, (42) implies that $\sum_1^\mu \beta^i(\xi^i - \hat{\phi}^i(x_0)) \geq 0$ for all $(\xi^1, \ldots, \xi^\mu) \in \bar{R}_-^\mu$, which implies, because $\hat{\phi}^i(x_0) \leq 0$ for $i = 1, \ldots, \mu$, that $(\beta^1, \ldots, \beta^\mu) \in \bar{R}_-^\mu$ and that (34) holds.

Because $\hat{\phi}'(x_0) \in Z'$ and $Z' + W' \subset \bar{Z}'$ [see Lemma (I.4.19)], (35) implies that $\ell' \in -(W')^*$ [see (I.5.23)]. Since $D\hat{\phi}'$ is W'-convex [see (I.7.25)], this means that the functional $-(\ell' \circ D\hat{\phi}')$, which is defined on $\overline{\text{co}}$ tancone $(A'; x_0)$, is convex.

If the $D\hat{\phi}^i$ can all be extended to $\mathscr{X}$ in such a way that they remain convex, continuous functionals, then we can appeal to Lemma (III.4.25) and conclude that there is a linear function $\Lambda = (\Lambda^0, \Lambda^1, \ldots, \Lambda^\mu): \mathscr{X} \to R^{\mu+1}$ such that (36) and (37) hold. The continuity of Λ follows from (37), the continuity of the $D\hat{\phi}^i$, Lemma (I.5.13), Theorem (I.3.25), and (I.3.38). |||

43 If the data of Problem (3), instead of being as in (27), are such that $\mathscr{Z}_1 = R^\mu$ and $Z_1 = \bar{R}^\mu_-$, so that we can write $\hat{\phi}_1 = (\hat{\phi}^1, \ldots, \hat{\phi}^\mu)$, where $\hat{\phi}^1, \ldots, \hat{\phi}^\mu$ are real-valued, and if $\hat{\varphi}$ and the $\hat{\phi}^i$ satisfy Hypothesis (28), then it is evident that the second set of conclusions in Theorem (30) remain in force if we delete every reference to $\hat{\phi}'$ and ℓ' [including the term $\ell' \circ D\hat{\phi}'$ in (33) and (36)] as well as relation (35).

44 We point out that relations (31), (33), and (36) in Theorem (30) hold, not only for all $T \in \mathscr{W}$, but also for all $T \in T_0 + \text{co co}$ $(\mathscr{W} - T_0)$ [inasmuch as $\overline{\text{co co}}\,(\mathscr{W} - T_0) \subset \overline{\text{co co}}\,(\mathscr{E} - T_0)$].

45 The following corollaries of Theorem (30) follow directly from Corollaries (III.3.4 and 5).

46 COROLLARY. *If, in Theorem* (30), Z_1 *is a cone, then* (32) *holds if and only if* $\ell_1 \in -(\bar{Z}_1)^*$ *and* $\ell_1 \circ \hat{\phi}_1(x_0) = 0$. *If* Z' *is a cone, then* (35) *holds if and only if* $\ell' \in -(\bar{Z}')^*$ *and* $\ell' \circ \hat{\phi}'(x_0) = 0$.

47 COROLLARY. *If, in Theorem* (30), $D\hat{\phi}^0$ *and* $D\hat{\phi}_1$ *are linear, then* (31) *is equivalent to the maximum principle*

48
$$(\alpha \cdot D\hat{\varphi} + \beta^0 D\hat{\phi}^0 + \ell_1 \circ D\hat{\phi}_1) \circ [E - DT_0]^{-1}(Tx_0) \leq (\alpha \cdot D\hat{\varphi} + \beta^0 D\hat{\phi}^0 + \ell_1 \circ D\hat{\phi}_1) \circ [E - DT_0]^{-1}(x_0) \quad \textit{for all} \quad T \in \mathscr{W}.$$

If the data of Problem (3) *are as indicated in* (27), *and* $D\hat{\phi}^0, \ldots, D\hat{\phi}^\mu$ *and* $D\hat{\phi}'$ *are linear, then* (33) *is equivalent to the maximum principle*

$$(\alpha \cdot D\hat{\varphi} + \beta \cdot D\hat{\phi}'' + \ell' \circ D\hat{\phi}') \circ [E - DT_0]^{-1}(Tx_0) \leq (\alpha \cdot D\hat{\varphi} + \beta \cdot D\hat{\phi}'' + \ell' \circ D\hat{\phi}') \circ [E - DT_0]^{-1}(x_0) \quad \textit{for all} \quad T \in \mathscr{W}.$$

It should now be evident that Theorem (II.2.14) is a special case of Theorem (30) and its Corollaries (46) and (47).

The reader is reminded that the remarks in (III.3.31 and 32) remain in effect here.

49 Let us now turn back to the special case of Problem (15). As was already pointed out, we may consider this problem to be a special case of Problem (3), with the data as indicated in (27), if we set $\mathscr{X} = \mathscr{C}^n(I)$, $A' = A$, and $\mathscr{Z}' = \mathscr{C}^\nu(I)$, and define Z' by (12), $\hat{\varphi}$ by (10) [where $\chi' = (\chi^{\mu+1}, \ldots, \chi^{\mu+m})$], $\hat{\phi}^0$ by (11) (and similarly for $\hat{\phi}^1, \ldots, \hat{\phi}^\mu$), and $\hat{\phi}'$ by (14). It follows from the continuity of $\chi = (\chi^0, \chi^1, \ldots, \chi^\mu, \chi')$ and of $\tilde{\chi}$ and from Corollary (I.5.32) that $\hat{\varphi}, \hat{\phi}^0, \ldots, \hat{\phi}^\mu$ and $\hat{\phi}'$ thus defined are continuous. Further, if χ is continuously differentiable, then it is not hard to show, making use of Lemma (I.7.40) and Theorem (I.7.41), that $\hat{\varphi}$, $\hat{\phi}^0, \ldots, \hat{\phi}^\mu$ are Fréchet differentiable, with Fréchet differentials at any $x_0 \in A$ given by the formulas

50 $$D\hat{\varphi}(x_0; \delta x) = \sum_{j=1}^{k} \mathscr{D}_j^0 \chi' \delta x(\tau^j), \quad \delta x \in \mathscr{C}^n(I),$$

51 $$D\hat{\phi}^i(x_0; \delta x) = \sum_{j=1}^{k} \mathscr{D}_j^0 \chi^i \delta x(\tau^j), \quad \delta x \in \mathscr{C}^n(I), \quad i = 0, \ldots, \mu,$$

where $\mathscr{D}_j^0 \chi'$ denotes $\mathscr{D}_j \chi'(x_0(\tau^1), \ldots, x_0(\tau^k))$ for each j [see (I.7.53 and 58–61)] and similarly for $\mathscr{D}_j^0 \chi^i$. Using Theorem (I.7.43) and Corollary (I.5.32), we can show that if $\tilde{\chi}$ is continuously differentiable with respect to its first argument [see (I.7.51)], then $\hat{\phi}'$ is Fréchet differentiable, with Fréchet differential at any $x_0 \in A$ given by the formula

52 $$(D\hat{\phi}'(x_0; \delta x))(t) = \mathscr{D}_1 \tilde{\chi}(x_0(t), t) \delta x(t), \quad t \in I, \quad \delta x \in \mathscr{C}^n(I).$$

Hence [see (29)], Hypothesis (28) is satisfied. [Of course, Hypothesis (28) will hold even under considerably weaker differentiability conditions on $\chi^0, \ldots, \chi^\mu$, and $\tilde{\chi}$.]

If we now appeal to Theorem (30) and its Corollaries (46) and (47), and take into account (I.6.8), (I.5.25), Theorems (I.6.14 and 22), and (I.6.19) in order to obtain a representation for $(Z')^*$, we arrive at the following necessary conditions for Problem (15).

53 THEOREM. *Let x_0 be a solution of Problem* (15), *and let $T_0 \in \mathscr{W}$ be such that $x_0 = T_0 x_0$. Also suppose that Hypotheses* (21) *and* (22) *hold, that χ is continuously differentiable, and that $\tilde{\chi}$ is continuously differentiable with respect to its first argument. Finally, suppose that either $\mathscr{W}$ is finitely open in itself or that Hypothesis*

(23) *holds. Then there exist a row vector* $\alpha = (\alpha^0, \alpha^1, \ldots, \alpha^{\mu+m}) \in \bar{R}_-^{\mu+1} \times R^m$ *and a row-vector-valued function* $\tilde{\lambda} = (\tilde{\lambda}^1, \ldots, \tilde{\lambda}^\nu) \in (NBV(I))^\nu$ [see (I.6.10)] *such that*

(i) $\|\tilde{\lambda}\| + |\alpha| > 0,$

54 $$\text{(ii)} \sum_{j=1}^{k} \alpha \mathscr{D}_j^0 \chi \delta x_T(\tau^j) + \int_{t_1}^{t_2} d\tilde{\lambda}(t) \mathscr{D}_1 \tilde{\chi}(x_0(t),t) \delta x_T(t)$$
$$\leq \sum_{j=1}^{k} \alpha \mathscr{D}_j^0 \chi \delta x_{T_0}(\tau^j) + \int_{t_1}^{t_2} d\tilde{\lambda}(t) \mathscr{D}_1 \tilde{\chi}(x_0(t),t) \delta x_{T_0}(t)$$
$$\textit{for all} \quad T \in \mathscr{W},$$

where, for each $T \in \mathscr{T}$, δx_T *denotes the column-vector-valued function in* $\mathscr{C}^n(I)$ *defined by*

55 $$\delta x_T = [E - DT_0]^{-1}(Tx_0)$$

[E *denotes the identity operator on* $\mathscr{C}^n(I)$ *and* DT_0 *the Fréchet differential of* T_0 *at* x_0], *and* $\mathscr{D}_j^0 \chi$ (*for each* j) *denotes the Jacobian matrix* $\mathscr{D}_j \chi$ *of* χ *at* $(x_0(\tau^1), \ldots, x_0(\tau^k))$,

56 $$\text{(iii)} \sum_{i=1}^{\mu} \alpha^i \chi^i(x_0(\tau^1), \ldots, x_0(\tau^k)) = \int_{t_1}^{t_2} d\tilde{\lambda}(t) \tilde{\chi}(x_0(t),t) = 0,$$

and (iv) *for each* $i = 1, \ldots, \nu$, $\tilde{\lambda}^i$ *is nonincreasing on* I *and is constant on every subinterval of* I *which does not meet* I_i.

57 Inasmuch as relation (54) is the principal necessary condition in the preceding theorem, we shall refer to this result as a *maximum principle.*

58 Note that, in the statement of Problem (15), the inequality constraints (17) could have written in the form $\chi''(x(\tau^1), \ldots, x(\tau^k)) \in \bar{R}_-^\mu$ [where $\chi'' = (\chi^1, \ldots, \chi^\mu)$], and, if $I_i = \hat{I}$ for each $i = 1, \ldots, \nu$ and some closed subset $\hat{I}$ of I, then (19) could have been written in the form $\tilde{\chi}(x(t),t) \in \bar{R}_-^\nu$ for all $t \in \hat{I}$. If the convex cones $\bar{R}_-^\mu$ and $\bar{R}_-^\nu$ are replaced by arbitrary convex bodies in R^μ and R^ν, respectively, we may still make use of Theorem (30) [once we modify the definitions of Z' and Z_1 in an obvious way] to obtain modified necessary conditions for this modified problem. In particular, if $\bar{R}_-^\mu$ is replaced by R^μ and/or $\bar{R}_-^\nu$ is replaced by R^ν—so that we have effectively removed the inequality constraints (17) and/or (19)—then we should, in Theorem (53), set $(\alpha^1, \ldots, \alpha^\mu) = 0$ and/or $\tilde{\lambda} \equiv 0$. In other words, if one or both, of

the constraints (17), (19) are eliminated from Problem (15), then the necessary conditions of Theorem (53) remain in force, except that the terms corresponding to the eliminated constraints should be omitted from relations (54) and (56).

Remark (44) is also, of course, applicable to relation (54).

It is possible to obtain necessary conditions for Problem (15) even if some components χ^i of χ, for $0 \leq i \leq \mu$, are not differentiable, so long as these components are convex, continuous functionals on $(R^n)^k$. Indeed, let us suppose that the set of integers $\mathscr{I} = \{0, 1, \ldots, \mu\}$ can be partitioned into two mutually exclusive and mutually exhaustive subsets $\mathscr{I}_1$ and $\mathscr{I}_2$ (i.e., $\mathscr{I}_1 \cup \mathscr{I}_2 = \mathscr{I}$, $\mathscr{I}_1 \cap \mathscr{I}_2$ is empty) such that χ^i is convex [and hence, by Theorem (I.5.45), is also continuous, but not necessarily differentiable] for $i \in \mathscr{I}_1$, and χ^i is continuously differentiable for $i \in \mathscr{I}_2$ (as well as for $i = \mu + 1, \ldots, \mu + m$). Then, as we indicated in (49), the functionals $\hat{\phi}^i$ defined on $\mathscr{C}^n(I)$ by the formula

59 $$\hat{\phi}^i(x) = \chi^i(x(\tau^1), \ldots, x(\tau^k)), \quad i = 0, 1, \ldots, \mu, \quad x \in \mathscr{C}^n(I),$$

are Fréchet differentiable for $i \in \mathscr{I}_2$ [as is $\hat{\varphi}$, defined by (10) with $\chi' = (\chi^{\mu+1}, \ldots, \chi^{\mu+m})$], and are evidently continuous and convex for $i \in \mathscr{I}_1$. Hence, by Corollary (I.7.33), for each $i \in \mathscr{I}_1$, $\hat{\phi}^i$ is dually semidifferentiable, and the function $x \to \hat{\phi}^i(x_0 + x) - \hat{\phi}^i(x_0)$ is a dual semidifferential of $\hat{\phi}^i$ at x_0. Therefore, if we also assume, as before, that $\tilde{\chi}$ is continuous and continuously differentiable with respect to its first argument, then Hypothesis (28) holds.

Appealing to Theorem (30), we can conclude that, under the indicated alternate hypotheses for $\chi^0, \ldots, \chi^\mu$, the conclusions of Theorem (53) remain in effect, provided that, in (54), we replace the terms $[\alpha \mathscr{D}_j^0 \chi \delta x_T(\tau^j)]$ $(j = 1, \ldots, k)$ by

$$\sum_{i \in \mathscr{I}_3} \alpha^i \mathscr{D}_j^0 \chi^i \delta x_T(\tau^j) + \sum_{i \in \mathscr{I}_1} \alpha^i \Lambda^i(\delta x_T)$$

[and similarly for the terms $\alpha \mathscr{D}_j^0 \chi \delta x_{T_0}(\tau^j)$], where $\mathscr{I}_3 = \mathscr{I}_2 \cup \{\mu + 1, \ldots, \mu + m\}$ and Λ^i, for each $i \in \mathscr{I}_1$, is some suitable linear functional defined on $\mathscr{C}^n(I)$ which satisfies the inequality

60 $$\Lambda^i(x) \leq \chi^i(x_0(\tau^1) + x(\tau^1), \ldots, x_0(\tau^k) + x(\tau^k)) - \chi^i(x_0(\tau^1), \ldots, x_0(\tau^k))$$
$$\text{for all} \quad x \in \mathscr{C}^n(I).$$

Let us show that (60) implies that Λ^i is of a special form [see (61)]. First, we show that if $x_1, x_2 \in \mathscr{C}^n(I)$ are such that $x_1(\tau^j) = x_2(\tau^j)$ for each $j = 1, \ldots, k$, then $\Lambda^i(x_1) = \Lambda^i(x_2)$. Indeed, if x_1 and x_2 are as just indicated, then (60) and the linearity of Λ^i imply that

$$\gamma[\Lambda^i(x_1) - \Lambda^i(x_2)] = \Lambda^i(\gamma(x_1 - x_2)) \leq 0 \quad \text{for all} \quad \gamma \in R,$$

which is clearly only possible if $\Lambda^i(x_1) = \Lambda^i(x_2)$. Now (for each $i \in \mathscr{I}_1$) define the function $\mathscr{D}^0\chi^i : (R^n)^k \to R$ by means of the formula

61 $$\mathscr{D}^0\chi^i(x(\tau^1), \ldots, x(\tau^k)) = \Lambda^i(x) \quad \text{for all} \quad x \in \mathscr{C}^n(I).$$

By what we have just shown, $\mathscr{D}^0\chi^i$ is well defined by (61), and, since $\mathscr{D}^0\chi^i$ is evidently linear, it follows from (I.2.12) that there are (row) vectors $\mathscr{D}_j^0\chi^i$, $j = 1, \ldots, k$, such that

62 $$\mathscr{D}^0\chi^i(\xi_1, \ldots, \xi_k) = \sum_{j=1}^{k} (\mathscr{D}_j^0\chi^i)\xi_j \quad \text{for all} \quad (\xi_1, \ldots, \xi_k) \in (R^n)^k.$$

Combining (60)–(62), we conclude that (for each $i \in \mathscr{I}_1$)

63 $$\sum_{j=1}^{k} (\mathscr{D}_j^0\chi^i)\xi_j \leq \chi^i(x_0(\tau^1) + \xi_1, \ldots, x_0(\tau^k) + \xi_k) - \chi^i(x_0(\tau^1), \ldots, x_0(\tau^k)) \quad \text{for all} \quad (\xi_1, \ldots, \xi_k) \in (R^n)^k,$$

i.e., that $(\mathscr{D}_1^0\chi^i, \ldots, \mathscr{D}_k^0\chi^i)$ is a subgradient [see (I.2.29)] of χ^i at $(x_0(\tau^1), \ldots, x_0(\tau^k))$. Of course, if χ^i is differentiable at $(x_0(\tau^1), \ldots, x_0(\tau^k))$, then (63) can hold only if $\mathscr{D}_j^0\chi^i = \mathscr{D}_j\chi^i(x_0(\tau^1), \ldots, x_0(\tau^k))$ for each j. In summary, we have proved the following:

64 THEOREM. *Let x_0 be a solution of Problem* (15), *and suppose that all of the hypotheses of Theorem* (53) *are satisfied, except that, for i in some subset $\mathscr{I}_1$ of $\{0,1, \ldots, \mu\}$, instead of requiring χ^i to be continuously differentiable, only assume that χ^i is convex. Then all of the conclusions of Theorem* (53) *remain in force, provided that, for each $i \in \mathscr{I}_1$, $(\mathscr{D}_1^0\chi^i, \ldots, \mathscr{D}_k^0\chi^i)$ is to be interpreted as some (suitably chosen) subgradient of χ^i at $(x_0(\tau^1), \ldots, x_0(\tau^k))$, i.e., as some vector in $(R^n)^k$ which satisfies* (63).

In Chapters V and VII we shall "expand" the necessary conditions of Theorems (53) and (64) for special classes $\mathscr{W}$—much as was done in obtaining the necessary conditions of

Theorem (II.2.36)—and shall obtain a maximum principle in a much more convenient form.

Let us now turn to the minimax variant of Problem (3), which may be stated as follows. (We shall keep the notation so far used in this chapter.)

65 *Problem.* Given a subset $\mathscr{W}$ of $\mathscr{T}_0$, a subset A' of A which contains the fixed point of every $T \in \mathscr{T}_1 \cap (\text{co } \mathscr{W})$, linear topological spaces $\mathscr{Z}_1$ and $\mathscr{Z}_2$, a convex body Z_1 in $\mathscr{Z}_1$, a nonempty set $\mathscr{L} \subset \mathscr{Z}_2^*$, and functions $\hat{\varphi}: A' \to R^m$, $\hat{\phi}_1: A' \to \mathscr{Z}_1$, and $\hat{\phi}_2: A' \to \mathscr{Z}_2$, find a pair $(x_0, T_0) \in A' \times \mathscr{W}$ such that (4)–(6) hold and such that

66 $$\sup_{\ell \in \mathscr{L}} \ell \circ \hat{\phi}_2(x_0) \leq \sup_{\ell \in \mathscr{L}} \ell \circ \hat{\phi}_2(x)$$

for all $x \in A'$ that satisfy (7).

67 Let us identify Problem (65) as a special case of Problem (III.1.8) [compare with (8)]. Indeed, let us identify $\mathscr{E}'$ with $\mathscr{T}_1 \cap (\text{co } \mathscr{W})$, $\mathscr{E}$ with $(\mathscr{T}_1 \cap \mathscr{W})$, and let us define the function $\varphi_0: \mathscr{E}' \to A'$ as in (8), i.e., by (9). If we then identify the maps $\varphi: \mathscr{E}' \to R^m$, $\phi_1: \mathscr{E}' \to \mathscr{Z}_1$, and $\phi_2: \mathscr{E}' \to \mathscr{Z}_2$ with the maps $\hat{\varphi} \circ \varphi_0$, $\hat{\phi}_1 \circ \varphi_0$, and $\hat{\phi}_2 \circ \varphi_0$, respectively, then it is trivial to verify that we have achieved the desired identification, with T_0 a solution of Problem (III.1.8) if and only if (x_0, T_0)—where $x_0 = \varphi_0(T_0)$—is a solution of Problem (65).

Most commonly, $\mathscr{X} = \mathscr{C}^n(I)$ for some compact interval $I = [t_1, t_2]$, and $\mathscr{W}$ is, as in the case of Problem (3), often defined in terms of special types of integral operators. In this case, $A' = A$ and the function $\hat{\varphi}: A \to R^m$ is typically defined by (10), where χ' is some given continuous function from $(R^n)^k$ into R^m and $\{\tau^1, \ldots, \tau^k\}$ is some given subset of I. Further, the linear topological spaces $\mathscr{Z}_1$ and $\mathscr{Z}_2$ are typically $R^\mu \times \mathscr{C}^\nu(I)$ and $\mathscr{C}(I)$, respectively, with $Z_1 = \bar{R}^\mu_- \times Z'$, where Z' is defined by (12) (for some given closed subsets $I_1, \ldots, I_\nu$ of I), in which case $\hat{\phi}_1$ is usually defined by (13) and (14), where χ'' and $\tilde{\chi}$ are given continuous functions from $(R^n)^k$ into R^μ and from $R^n \times I$ into R^ν, respectively, $\hat{\phi}_2$ is usually defined through the relation

68 $$(\hat{\phi}_2(x))(t) = \tilde{\chi}^0(x(t),t) \quad \text{for all} \quad t \in I \quad \text{and} \quad x \in A,$$

where $\tilde{\chi}^0$ is some given continuous function from $R^n \times I$ into

R, and $\mathscr{L}$ is typically given as

69 $$\mathscr{L} = \{\ell : \ell = \ell_\tau, \tau \in I_0\},$$

where I_0 is some given closed subset of I and $\ell_\tau \in (\mathscr{C}(I))^*$ is defined (for each $\tau \in I$) by the relation

70 $$\ell_\tau(z) = z(\tau) \quad \text{for all} \quad z \in \mathscr{C}(I).$$

If $\mathscr{X}$, A', $\hat{\varphi}$, $\mathscr{Z}_1$, Z_1, $\mathscr{Z}_2$, $\mathscr{L}$, $\hat{\phi}_1$, and $\hat{\phi}_2$ are defined in the just-indicated manner, then, if we set $\chi = (\chi'', \chi')$ and $\tilde{\chi} = (\tilde{\chi}^0, \tilde{\chi})$, Problem (65) evidently reduces to the following minimax problem:

71 *Problem.* Given a subset $\mathscr{W}$ of $\mathscr{T}_0$ [where $\mathscr{X} = \mathscr{C}^n(I)$], continuous functions $\chi = (\chi^1, \ldots, \chi^\mu, \ldots, \chi^{\mu+m})\colon (R^n)^k \to R^{\mu+m}$ and $\tilde{\chi}' = (\tilde{\chi}^0, \tilde{\chi}^1, \ldots, \tilde{\chi}^\nu)\colon R^n \times I \to R^{\nu+1}$, a finite subset $\{\tau^1, \ldots, \tau^k\}$ of I, and closed subsets $I_0, I_1, \ldots, I_\nu$ of I, find a function $x \in A$ satisfying the constraints (16)–(19) which, in so doing, achieves a minimum for $[\max_{t \in I_0} \tilde{\chi}^0(x(t),t)]$.

Let us now return to the general case of Problem (65), and let us consider the possibility of appealing to Theorem (III.3.12) in order to obtain necessary conditions for solutions of this problem. It turns out that, if Hypotheses (III.1.21 and 22) hold, then the hypotheses which we invoked for Problem (3) are also adequate here, provided that we make an obvious change in Hypothesis (26)—namely, we replace this hypothesis by the following one:

72 HYPOTHESIS. *The functions $\hat{\varphi}$ and $\hat{\phi}_i$ $(i = 1, 2)$ are continuous and, in addition, respectively, dually differentiable and dually W_i-semidifferentiable at x_0, for some closed, convex cone $W_1 \subset \mathscr{Z}_1$ such that* $(\operatorname{int} Z_1) + W_1 = \operatorname{int} Z_1$, *and with $W_2 = Z_2^0$, as defined by* (III.1.17).

73 The reader should here review Theorems (I.7.44) and the remarks in (I.7.45) regarding the relations between the different types of differentiability, as well as Lemma (I.7.38) regarding continuity, and see how they apply to Hypothesis (72).

We may now appeal to Theorem (III.3.12), and, arguing essentially as in the proof of Theorem (30), arrive at the following necessary conditions.

74 THEOREM. *Let (x_0, T_0) be a solution of Problem* (65), *and suppose that Hypotheses* (III.1.21 *and* 22)—*where $\kappa_0 = \sup_{\ell \in \mathscr{L}} \ell \circ \hat{\phi}_2(x_0)$—as well as Hypotheses* (21), (22), *and* (72) *hold. Further,*

suppose that either $\mathscr{W}$ is finitely open in itself or that Hypothesis (23) *holds. Then, for every dual W_i-semidifferentials $D\hat{\phi}_i(x_0;\cdot)$ of $\hat{\phi}_i$ at x_0 $(i = 1, 2)$, there exist a vector $\alpha \in R^m$ and linear functionals $\ell_i \in \mathscr{X}_i^*$ $(i = 1, 2)$, not all zero, such that*

75 $$(\alpha \cdot D\hat{\varphi} + \ell_1 \circ D\hat{\phi}_1 + \ell_2 \circ D\hat{\phi}_2) \circ [E - DT_0]^{-1}(Tx_0 - x_0) \leq 0 \quad \text{for all} \quad T \in \mathscr{W},$$

76 $$\ell_1(z_1) \geq \ell_1 \circ \hat{\phi}_1(x_0) \quad \text{for all} \quad z_1 \in \bar{Z}_1,$$

77 $$\ell_2(z_2) \geq \ell_2 \circ \hat{\phi}_2(x_0) \quad \text{for all} \quad z_2 \in Z_2^{\kappa_0},$$

where, in (75), *we have written $D\hat{\phi}_i(\cdot)$ for $D\hat{\phi}_i(x_0;\cdot)$ $(i = 1, 2)$, $D\hat{\varphi}$ and DT_0 denote, respectively, the dual differential of $\hat{\varphi}$ and the Fréchet differential of T_0, both at x_0, and E denotes the identity operator on $\mathscr{X}$, and, in* (77), *$Z_2^{\kappa_0}$ is defined by* (III.1.17).

The following corollaries of Theorem (74) follow directly from Corollaries (III.3.21, 24, and 25). [Remark (44) is also applicable here to relation (75).]

78 Corollary. *If, in Theorem* (74), *Z_1 is a cone, then* (76) *holds if and only if $\ell_1 \in -(\bar{Z}_1)^*$ and $\ell_1 \circ \hat{\phi}_1(x_0) = 0$. Alternately, if Hypothesis* (III.3.17) *holds, then* (77) *holds if and only if $\ell_2 \in -(Z_2^0)^*$ and*

$$\ell_2 \circ \hat{\phi}_2(x_0) = \kappa_0 \ell_2(\zeta_0).$$

79 Corollary. *If, in Theorem* (74), *$D\hat{\phi}_1$ and $D\hat{\phi}_2$ are linear, then* (75) *is equivalent to the maximum principle*

80 $$\begin{aligned}(\alpha \cdot D\hat{\varphi} + \ell_1 \circ D\hat{\phi}_1 + \ell_2 \circ D\hat{\phi}_2) \circ [E - DT_0]^{-1}(Tx_0) \\ \leq (\alpha \cdot D\hat{\varphi} + \ell_1 \circ D\hat{\phi}_1 + \ell_2 \circ D\hat{\phi}_2) \circ [E - DT_0]^{-1}(x_0) \\ \text{for all} \quad T \in \mathscr{W}.\end{aligned}$$

We again refer the reader to the remarks in (III.3.31 and 32), which are pertinent here with evident minor notational changes.

For the sake of brevity, we have not considered, in Theorem (74) and Corollaries (78) and (79), the special case where ϕ_1, $\mathscr{X}_1$, and Z_1 are as indicated in (27). However, it should be evident how the necessary conditions can be first specialized and then extended for this case, as was done for Problem (3).

81 Let us now return to the special case of Problem (71). Here, if $\chi = (\chi'', \chi')$ is continuously differentiable, and if $\tilde{\chi}' = (\tilde{\chi}^0, \tilde{\chi})$ is continuously differentiable with respect to its first argument,

then, as in (49), we can show that the functions $\hat{\varphi}$, $\hat{\phi}^i$ (for $i = 1, \ldots, \mu$), $\hat{\phi}'$, and $\hat{\phi}_2$ defined on $\mathscr{X} = \mathscr{C}^n(I)$ by (10), (59), (14), and (68), respectively, are continuous and Fréchet differentiable, with Fréchet differentials at any $x_0 \in A$ given by the formulas (50)–(52) and

82 $$(D\hat{\phi}_2(x_0; \delta x))(t) = \mathscr{D}_1\tilde{\chi}^0(x_0(t),t)\delta x(t), \quad t \in I, \quad \delta x \in \mathscr{C}^n(I),$$

where $\mathscr{D}_j^0\chi'$ denotes $\mathscr{D}_j\chi'(x_0(\tau^1), \ldots, x_0(\tau^k))$, and similarly for $\mathscr{D}_j^0\chi^i$. Hence [see (73)], Hypothesis (72) holds with $\hat{\phi}_1 = (\hat{\phi}^1, \ldots, \hat{\phi}^\mu, \hat{\phi}')$. Also, since $\mathscr{L}$ is given by (69) and (70), it is evident that Hypothesis (III.3.17) holds—in fact, we may choose for ζ_0 the function which is identically equal to 1 on I—and that the sets V_2^κ and Z_2^κ (for each $\kappa \in R$), defined by (III.1.17 and 18), are given equivalently by the formulas

83 $$V_2^\kappa = \{z\colon z \in \mathscr{C}(I), \max_{t \in I_0} z(t) < \kappa\},$$
$$Z_2^\kappa = \{z\colon z \in \mathscr{C}(I), z(t) \leq \kappa \text{ for all } t \in I_0\}.$$

From this it is immediately seen that Hypotheses (III.1.21 and 22) hold.

If we now appeal to Theorem (74) and its Corollaries (78) and (79) and take into account (I.5.25), (I.6.8), Theorems (I.6.14 and 22), and (I.6.19) in order to obtain representations for $(Z')^*$ and $(Z_2^0)^*$ [see (12) and (83)], we arrive at the following necessary conditions for Problem (71) [compare with Theorem (53)].

84 THEOREM. *Let x_0 be a solution of Problem* (71), *and let $T_0 \in \mathscr{W}$ be such that $x_0 = T_0 x_0$. Also suppose that Hypotheses* (21) *and* (22) *hold, that χ is continuously differentiable, and that $\tilde{\chi}'$ is continuously differentiable with respect to its first argument. Finally, suppose that either $\mathscr{W}$ is finitely open in itself, or that Hypothesis* (23) *holds. Then there exist a row vector $\alpha = (\alpha^1, \ldots, \alpha^{\mu+m}) \in \bar{R}_-^\mu \times R^m$ and a row-vector-valued function $\tilde{\lambda} = (\tilde{\lambda}^0, \tilde{\lambda}^1, \ldots, \tilde{\lambda}^\nu) \in (NBV(I))^{\nu+1}$* [*see* (I.6.10)] *such that*

(i) $\|\tilde{\lambda}\| + |\alpha| > 0$,

85 (ii) $$\sum_{j=1}^{k} \alpha\mathscr{D}_j^0\chi\delta x_T(\tau^j) + \int_{t_1}^{t_2} d\tilde{\lambda}(t)\mathscr{D}_1\tilde{\chi}'(x_0(t),t)\delta x_T(t)$$
$$\leq \sum_{j=1}^{k} \alpha\mathscr{D}_j^0\chi\delta x_{T_0}(\tau^j) + \int_{t_1}^{t_2} d\tilde{\lambda}(t)\mathscr{D}_1\tilde{\chi}'(x_0(t),t)\delta x_{T_0}(t)$$
$$\textit{for all} \quad T \in \mathscr{W},$$

where, for each $T \in \mathcal{T}$, δx_T *denotes the column-vector-valued function in* $\mathscr{C}^n(I)$ *defined by*

86 $$\delta x_T = [E - DT_0]^{-1}(Tx_0)$$

[*E denotes the identity operator on* $\mathscr{C}^n(I)$ *and* DT_0 *the Fréchet differential of* T_0 *at* x_0], *and* $\mathscr{D}_j^0\chi$ (*for each j*) *denotes the Jacobian matrix* $\mathscr{D}_j\chi$ *of* χ *at* $(x_0(\tau^1), \ldots, x_0(\tau^k))$,

87 $$\text{(iii)} \quad \sum_{i=1}^{\mu} \alpha^i \chi^i(x_0(\tau^1), \ldots, x_0(\tau^k)) = \sum_{i=1}^{\nu} \int_{t_1}^{t_2} d\tilde{\lambda}^i(t)\tilde{\chi}^i(x_0(t),t) = 0,$$

88 $$\text{(iv)} \quad \int_{t_1}^{t_2} d\tilde{\lambda}^0(t)\tilde{\chi}^0(x_0(t),t) = [\tilde{\lambda}^0(t_2) - \tilde{\lambda}^0(t_1)]\left[\max_{t \in I_0} \tilde{\chi}^0(x_0(t),t)\right],$$

and (v) *for each* $i = 0, 1, \ldots, \nu$, $\tilde{\lambda}^i$ *is nonincreasing on I and is constant on every subinterval of I which does not meet* I_i.

89 Inasmuch as relation (85) is the principal necessary condition in the preceding theorem, we shall refer to this result also as a *maximum principle.*

90 If some of the χ^i, for $i = 1, \ldots, \mu$, are not continuously differentiable, but are convex, then the necessary conditions of Theorem (84) remain in force, provided that the $\mathscr{D}_j^0\chi^i$ are suitably reinterpreted. [See Theorem (64).]

91 If we modify, or even eliminate, one, or both, of the inequality constraints (17) and (19) in Problem (71) as indicated in (58), then we can obtain modified necessary conditions, essentially as described in (58). [If the inequality constraints (19) are deleted, then $\tilde{\lambda}^i$, for $i = 1, \ldots, \nu$, but not necessarily for $i = 0$, should be set identically equal to zero.]

As usual, we point out that remark (44) is applicable to relation (85).

In Section 4 of Chapter V, we shall "expand" the necessary conditions of Theorem (69) for certain special classes $\mathscr{W}$ to obtain a maximum principle in a more convenient form [much as in the case of Theorem (II.2.36)].

92 If we modify Problems (3) and (65) by omitting the equality constraint $\hat{\varphi}(x) = 0$, then we can appeal to Theorems (III.3.34 and 37) and obtain necessary conditions analogous to those of Theorems (30) and (74), with the term $\alpha \cdot D\hat{\varphi}$ in (31), (33), (36), and (75) deleted. Of course, in this case, the reference to $\hat{\varphi}$ in Hypotheses (26), (28), and (72) must also be deleted. Further, we may weaken Hypothesis (23) in a manner analogous to that

in which Hypothesis (III.2.14) was weakened to Hypothesis (III.2.59). Correspondingly, if we modify Problems (15) and (71) by eliminating the equality constraints (18)—so that $\chi = (\chi^0, \chi^1, \ldots, \chi^\mu)$ in the former problem, and $\chi = (\chi^1, \ldots, \chi^\mu)$ in the latter—then all of the necessary conditions of Theorems (53), (64), and (84) remain in effect with the coordinates $\alpha^{\mu+1}, \ldots, \alpha^{\mu+m}$ of α deleted. Of course, here also Hypothesis (23) may be weakened in the just-indicated manner.

93 We point out that Problem (71) may be generalized as follows. Instead of supposing that $\tilde{\chi}^0$ is scalar-valued, let $\tilde{\chi}^0$ take on its values in $R^{\nu'}$, for some integer $\nu' \geq 1$, and suppose that we wish to minimize

$$\max_{t \in I_0} \sup_{\xi \in \mathbf{K}} \xi \cdot \tilde{\chi}^0(x(t),t)$$

where $\mathbf{K}$ is some given bounded subset of $R^{\nu'}$. For example, if $\mathbf{K} = \{\xi \colon \xi \in R^{\nu'}, |\xi| = 1\}$, then our problem will consist in minimizing $\{\max_{t \in I_0} |\tilde{\chi}^0(x(t),t)|\}$. We shall leave it to the interested reader to extend Theorem (84) to this problem.

Finally, we mention that, in the same way as has just been done, we can investigate a corresponding vector-valued criterion function problem. We shall only state the problem; the investigation of the corresponding necessary conditions will be left as an exercise to the interested reader.

94 *Problem.* Given a subset $\mathscr{W}$ of $\mathscr{T}_0$, a subset A' of A which contains the fixed point of every $T \in \mathscr{T}_1 \cap (\text{co}\ \mathscr{W})$, linear topological spaces $\mathscr{Z}_1$ and $\mathscr{Z}_2$, a convex body Z_1 in $\mathscr{Z}_1$, a nonempty set $\mathscr{L} \subset \mathscr{Z}_2^*$, and functions $\hat{\varphi}\colon A' \to R^m$, $\hat{\phi}_1\colon A' \to \mathscr{Z}_1$, and $\hat{\phi}_2\colon A' \to \mathscr{Z}_2$, find a pair $(x_0, T_0) \in A' \times \mathscr{W}$ such that (4)–(6) hold and such that there is no element $x \in A'$ satisfying (7) and the relations $\ell \circ \hat{\phi}_2(x) \leq \ell \circ \hat{\phi}_2(x_0)$ for every $\ell \in \mathscr{L}$ and $\tilde{\ell} \circ \hat{\phi}_2(x) < \tilde{\ell} \circ \hat{\phi}_2(x_0)$ for some $\tilde{\ell} \in \mathscr{L}$.

2. Optimization Problems with Operator Equation Constraints and Parameters

In this section, we shall extend the results of Section 1 to problems in which the optimization is carried out not only with respect to a collection of operators, but also with respect to a set of parameters.

The reader who is only interested in the highlights of the theory can omit this section, as well as Section 4, without loss of continuity. We shall use the results of this section in subsequent chapters in order to investigate so-called "free-time" optimal control problems, optimal control problems with parameters, and optimal control problems with so-called "mixed phase-coordinate control" inequality constraints.

1 As in Section 1, we shall suppose that we are given a Banach space $\mathscr{X}$ and an open set $A \subset \mathscr{X}$. We shall correspondingly define the linear topological space $\mathscr{T}$, the subsets $\mathscr{T}_0$ and $\mathscr{T}_1$ of $\mathscr{T}$, and the map $\varphi_0:\mathscr{T}_1 \to A$ as before. In addition, we shall suppose that we are given a Banach space $\mathscr{P}$ (the so-called parameter space) and an open set $\Pi \subset \mathscr{P}$. Let us denote the open set $A \times \Pi$ in $\mathscr{X} \times \mathscr{P}$ by $\tilde{A}$.

2 Let $\hat{\mathscr{T}}$ denote the linear vector space of all operators $\hat{T}:\tilde{A} \to \mathscr{X}$ which are continuously differentiable. We shall endow $\hat{\mathscr{T}}$ with the topology of pointwise convergence, so that $\hat{\mathscr{T}}$ becomes a locally convex linear topological space [see (I.6.33)].

3 For every $\hat{T} \in \hat{\mathscr{T}}$ and $\pi \in \Pi$, we shall denote by $\hat{T}^\pi$ the operator $x \to \hat{T}(x,\pi)$: $A \to \mathscr{X}$ of $\mathscr{T}$. Further, for $j = 0$ or 1, let

$$\mathscr{S}_j = \{(\hat{T},\pi) \colon \hat{T} \in \hat{\mathscr{T}}, \pi \in \Pi, \hat{T}^\pi \in \mathscr{T}_i\},$$

so that $\mathscr{S}_j \subset \hat{\mathscr{T}} \times \mathscr{P}$ for $j = 0$ and 1. Since $\mathscr{T}_0 \supset \mathscr{T}_1, \mathscr{S}_0 \supset \mathscr{S}_1$.

Now consider the following problem. (A minimax problem will be considered at the end of this section.)

4 *Problem.* Given a subset $\tilde{\mathscr{W}}$ of $\mathscr{S}_0$, a subset $\tilde{A}'$ of $\tilde{A}$ which contains all (x,π) such that x is the fixed point of $\hat{T}^\pi$ with $(\hat{T},\pi) \in \mathscr{S}_1 \cap (\text{co } \tilde{\mathscr{W}})$, a linear topological space $\mathscr{Z}_1$, a convex body $Z_1 \subset \mathscr{Z}_1$, and functions $\hat{\varphi}:\tilde{A}' \to R^m$, $\hat{\phi}^0:\tilde{A}' \to R$, and $\hat{\phi}_1:\tilde{A}' \to \mathscr{Z}_1$, find a triple $(x_0,\hat{T}_0,\pi_0) \in A \times \tilde{\mathscr{W}}$ such that

5 $$x_0 = \hat{T}_0^{\pi_0} x_0,$$

6 $$\hat{\varphi}(x_0,\pi_0) = 0,$$

7 $$\hat{\phi}_1(x_0,\pi_0) \in Z_1,$$

and such that $\hat{\phi}^0(x_0,\pi_0) \leq \hat{\phi}^0(x,\pi)$ for all $(x,\hat{T},\pi) \in A \times \tilde{\mathscr{W}}$ that satisfy the relations

8 $$\hat{\varphi}(x,\pi) = 0, \quad \hat{\phi}_1(x,\pi) \in Z_1, \quad \text{and} \quad x = \hat{T}^\pi(x).$$

9 Later in this section, we shall show that Problem (4) can be identified as a special case of Problem (1.3) if, in the latter, we suitably redefine $\mathscr{X}$, A, $\mathscr{T}$, $\hat{\varphi}$, etc. Consequently, we shall be able to conclude that Problem (4) can also be identified as a special case of Problem (III.1.2) [see (1.8)].

As is evident, we can appeal to the multiplier rule, Theorem (III.3.1), in order to obtain necessary conditions for solutions of Problem (4), only if the data of this problem and the solution point satisfy suitable hypotheses. Based on considerations essentially the same as those of Section 1 [see (1.20)], we shall therefore make the following hypotheses. Not surprisingly, these hypotheses bear a striking similarity to Hypotheses (1.21–23 and 26).

10 HYPOTHESIS. *The set* $(\mathrm{co}\, \tilde{\mathscr{W}}) \cap (\hat{\mathscr{T}} \times \Pi) \subset \mathscr{S}_0$.

11 HYPOTHESIS. *For each* $(\hat{T},\pi_1) \in \mathrm{co}\, \tilde{\mathscr{W}}$ *and each* $(x,\pi) \in \tilde{A}$, $[E - D_1\hat{T}(x,\pi;\cdot)]$*—where* E *denotes the identity operator on* $\mathscr{X}$ *and* $D_1\hat{T}(x,\pi;\cdot)$ *denotes the Fréchet partial differential of* $\hat{T}$ *with respect to its first argument at* (x,π) [*see* (I.7.51)]*—is a (linear homeomorphism of* $\mathscr{X}$ *onto itself*.

12 HYPOTHESIS. *For every finite subset* $\{(\hat{T}_1,\pi_1), \ldots, (\hat{T}_\rho,\pi_\rho)\}$ *of* $\tilde{\mathscr{W}}$*, there exist a subset* $\tilde{\mathscr{W}}_0$ *of* $\tilde{\mathscr{W}}$ *and a number* $\varepsilon_0 > 0$ *with the following properties*:

A. *The map* $(\hat{T},\pi) \to \varphi_0(\hat{T}^\pi)\colon (\tilde{\mathscr{R}} \cup \tilde{\mathscr{W}}_0) \cap \mathscr{S}_1 \to A$, *where*

13 $$\tilde{\mathscr{R}} = \Bigg\{(\hat{T},\pi)\colon (\hat{T},\pi) = \left(\hat{T}_0 + \sum_{i=1}^{\rho} \lambda^i(\hat{T}_i - \hat{T}_0), \pi_0 + \sum_{i=1}^{\rho} \lambda^i(\pi_i - \pi_0)\right), \quad 0 \le \lambda^i \le \varepsilon_0 \textit{ for } i = 1, \ldots, \rho\Bigg\},$$

is continuous.

B. *For every neighborhood* $\hat{N}$ *of* 0 *in* $\hat{\mathscr{T}}$, *there exists a continuous function* $\hat{\gamma}\colon \{\lambda\colon \lambda = (\lambda^1, \ldots, \lambda^\rho) \in R^\rho, 0 \le \lambda^i \le \varepsilon_0$ *for* $i = 1, \ldots, \rho\} \to \hat{N}$ *such that*

14 $$\left(\hat{T}_0 + \sum_{i=1}^{\rho} \lambda^i(\hat{T}_i - \hat{T}_0) + \hat{\gamma}(\lambda^1, \ldots, \lambda^\rho), \pi_0 + \sum_{i=1}^{\rho} \lambda^i(\pi_i - \pi_0)\right) \in \mathscr{S}_1 \cap \tilde{\mathscr{W}}_0 \quad \textit{wherever} \quad 0 \le \lambda^i \le \varepsilon_0 \quad \textit{for each} \quad i.$$

15 HYPOTHESIS. *The functions* $\hat{\varphi}$, $\hat{\phi}^0$, *and* $\hat{\phi}_1$ *are continuous and, in addition, respectively, dually differentiable, dually semidifferentiable, and dually* W_1*-simidifferentiable at* (x_0, π_0) [*see* (I.7.23, 25, 26)] *for some closed, convex cone* $W_1 \subset \mathscr{Z}_1$ *such that* $(\text{int}\, Z_1) + W_1 = \text{int}\, Z_1$.

16 If Problem (4) is such that $\mathscr{Z}_1 = R^\mu \times \mathscr{Z}'$ and $Z_1 = \bar{R}^\mu_- \times Z'$, for some linear topological space $\mathscr{Z}'$ and some convex body $Z' \subset \mathscr{Z}'$—so that we can write $\hat{\phi}_1 = (\hat{\phi}^1, \ldots, \hat{\phi}^\mu, \hat{\phi}')$, where $\hat{\phi}^1, \ldots, \hat{\phi}^\mu$ are real-valued and $\hat{\phi}'$ takes on its values in $\mathscr{Z}'$—then Hypothesis (15) is equivalent to the following one [see (I.7.50, 26, 27), (I.1.16), and (I.3.38)].

17 HYPOTHESIS. *The functions* $\hat{\varphi}$, $\hat{\phi}^i$ $(i = 0, \ldots, \mu)$, *and* $\hat{\phi}'$ *are continuous and, in addition, respectively, dually differentiable, dually semidifferentiable, and dually* W'*-semidifferentiable at* (x_0, π_0), *for some closed, convex cone* $W' \subset \mathscr{Z}'$ *such that* $(\text{int}\, Z') + W' = \text{int}\, Z'$.

18 With regard to Hypotheses (15) and (17), the reader should review Theorem (I.7.44), (I.7.45), and Lemma (I.7.38).

Before passing to the promised necessary conditions, let us briefly dwell on two particular cases of Problem (4) for which Hypotheses (10)–(12) hold.

19 The first particular case is the one in which $\hat{\mathscr{W}} = \mathscr{W} \times \Pi_a$, where $\mathscr{W}$ is some subset of $\hat{\mathscr{T}}$ each of whose members is an operator which is independent of $\pi \in \Pi$ (so that, effectively, $\mathscr{W} \subset \mathscr{T}$), and Π_a is some subset of Π which is finitely open in itself. If the set $\mathscr{W}$ and the operator T_0 in $\mathscr{W}$ satisfy Hypothesis (1.21) [respectively, (1.22), (1.23)], then it is easily seen that $\hat{\mathscr{W}} = \mathscr{W} \times \Pi_a$ and $\hat{T}_0 = T_0$ satisfy Hypothesis (10) [respectively, (11), (12)] for every $\pi_0 \in \Pi_a$. [If, in addition, Π_a consists of a single element, then Problem (4) reduces to Problem (1.3), so that we may consider Problem (1.3) to be a special case of Problem (4).]

As an illustration of this particular case, let us consider the following problem, which is closely related to the previously considered Problem (1.15).

20 *Problem.* Given a subset $\mathscr{W}$ of $\mathscr{T}_0$ (where $\mathscr{X} = \mathscr{C}^n(I)$ for some compact interval $I = [t_1, t_2]$), an open interval $\tilde{I} \supset I$, a continuous function $\chi = (\chi^0, \chi^1, \ldots, \chi^\mu, \ldots, \chi^{\mu+m}) : (R^n)^k \times (\tilde{I})^k \to R^{\mu+m+1}$, and a subset I_a of I^k which is finitely open in itself, find

a function $x \in A$ and an element $(\tau^1, \ldots, \tau^k) \in I_a$ satisfying the constraints

21 $$x = Tx, \quad \text{for some} \quad T \in \mathscr{W},$$

22 $$\chi^i(x(\tau^1), \ldots, x(\tau^k), \tau^1, \ldots, \tau^k) \leq 0 \quad \text{for} \quad i = 1, \ldots, \mu,$$

23 $$\chi^i(x(\tau^1), \ldots, x(\tau^k), \tau^1, \ldots, \tau^k) = 0 \quad \text{for} \quad i = \mu + 1, \ldots, \mu + m,$$

which, in so doing, achieve a minimum for

$$\chi^0(x(\tau^1), \ldots, x(\tau^k), \tau^1, \ldots, \tau^k).$$

24 This problem differs from Problem (1.15) in the following two respects. First, whereas in Problem (1.15) the values $\tau^1, \ldots, \tau^k$ were fixed, here these values are free, subject only to the constraint $(\tau^1, \ldots, \tau^k) \in I_a$, and even appear as arguments in the constraint functionals, as well as in the functional being minimized. When the variable t is identified with the time, as is the case in many applications, Problem (1.15) is often termed a fixed-time problem, and Problem (20) a free-time problem. The second difference lies in that Problem (1.15) contains an inequality constraint of the form (1.19) which is absent in Problem (20). Such a constraint can also be adjoined to Problem (20), but we shall not do this here for the sake of avoiding complications.

25 Let us show that Problem (20) can be identified as a special case of Problem (4). To do this, we identify $\mathscr{P}$ with R^k, Π with $(\tilde{I})^k$, Π_a with I_a, $\tilde{\mathscr{W}}$ with $\mathscr{W} \times I_a$ (where $\mathscr{W}$ is to be considered a subset of $\hat{\mathscr{T}}$ in the obvious way), $\mathscr{Z}_1$ with R^μ, Z_1 with $\bar{R}^\mu_-$, $\tilde{A}'$ with $A \times I^k$, set $\chi' = (\chi^{\mu+1}, \ldots, \chi^{\mu+m})$ and $\chi'' = (\chi^1, \ldots, \chi^\mu)$, and define the functions $\hat{\varphi}: \tilde{A}' \to R^m$, $\hat{\phi}^0: \tilde{A}' \to R$, and $\hat{\phi}_1: \tilde{A}' \to R^\mu$ through the relations

26 $$\hat{\varphi}(x, \tau^1, \ldots, \tau^k) = \chi'(x(\tau^1), \ldots, x(\tau^k), \tau^1, \ldots, \tau^k),$$

27 $$\hat{\phi}^0(x, \tau^1, \ldots, \tau^k) = \chi^0(x(\tau^1), \ldots, x(\tau^k), \tau^1, \ldots, \tau^k),$$

28 $$\hat{\phi}_1(x, \tau^1, \ldots, \tau^k) = \chi''(x(\tau^1), \ldots, x(\tau^k), \tau^1, \ldots, \tau^k).$$

It is now evident that, with these identifications, Problem (20) is a particular case of Problem (4) of the type described in (19) and in (16).

29 The second particular case of Problem (4) we shall consider is the one in which $\tilde{\mathscr{W}} = \hat{\mathscr{W}} \times \Pi_a$, where $\hat{\mathscr{W}}$ consists of a *single* operator $\hat{T}_0 \in \hat{\mathscr{T}}$ and Π_a is some subset of Π which is finitely

open in itself. In this case, Problem (4) consists in finding a "best" parameter $\pi_0 \in \Pi_a$. Here it is evident that the set $\tilde{\mathscr{W}}$ is finitely open in itself.

Let us now return to the general case of Problem (4). As usual, we wish to find necessary conditions that a solution $(x_0,\hat{T}_0,\pi_0)$ of this problem—subject to Hypotheses (10)–(12) and (15)—satisfies. We shall do this by showing that Problem (4) can be identified as a special case of Problem (1.3)—provided that we suitably redefine $\mathscr{X}$, A, etc., in the latter—and that, after we make this identification, Hypotheses (1.21, 22, 23, and 26), respectively, hold whenever, respectively, Hypotheses (10), (11), (12), and (15) hold for Problem (4). We shall then obtain our desired necessary conditions from Theorem (1.30).

30 We proceed as follows. Let $\tilde{\mathscr{X}} = \mathscr{X} \times \mathscr{P}$, and let $\tilde{\mathscr{T}}$ denote the linear topological space of all operators $\tilde{T}:\tilde{A} \to \tilde{\mathscr{X}}$ which are continuously differentiable, with the topology of pointwise convergence. We shall denote by $\tilde{\mathscr{T}}_0$ (respectively, $\tilde{\mathscr{T}}_1$) the set of all $\tilde{T} \in \tilde{\mathscr{T}}$ that have at most one (respectively, exactly one) fixed point.

31 For each $\pi \in \mathscr{P}$, let c_π denote the function from $\tilde{A}$ into $\mathscr{P}$ which is identically equal to π on $\tilde{A}$, i.e., $c_\pi(\tilde{x}) = \pi$ for all $\tilde{x} \in \tilde{A}$. It is evident that $\mathscr{P}$ is topologically isomorphic [see (I.4.12)] to the subspace $\{c_\pi\colon \pi \in \mathscr{P}\}$ of the space of all continuous functions from $\tilde{A}$ into $\mathscr{P}$ (with the topology of pointwise convergence).

32 For each $(\hat{T},\pi) \in \hat{\mathscr{T}} \times \mathscr{P}$, let us consider the operator $\tilde{T} = (\hat{T},c_\pi)\colon \tilde{A} \to \tilde{\mathscr{X}}$. Obviously, $(\hat{T},c_\pi) \in \tilde{\mathscr{T}}$ for every $(\hat{T},\pi) \in \hat{\mathscr{T}} \times \mathscr{P}$. Note that, if $\pi \in \Pi$ and $\hat{T} \in \hat{\mathscr{T}}$, then (for $i = 0$ or 1) $(\hat{T},c_\pi) \in \tilde{\mathscr{T}}_i$ if and only if $(\hat{T},\pi) \in \mathscr{S}_i$ [see (3)], and, if $\pi \notin \Pi$, then $(\hat{T},c_\pi)$ has *no* fixed point and consequently belongs to $\tilde{\mathscr{T}}_0$. Further, if $(\hat{T},\pi) \in \hat{\mathscr{T}} \times \Pi$ and $(\hat{T},c_\pi) \in \tilde{\mathscr{T}}_1$, then $(\varphi_0(\hat{T}^\pi),\pi)$ is the fixed point of $(\hat{T},c_\pi)$.

33 In what follows, we shall ourselves the abuse of notation of denoting the subset $\{\tilde{T}\colon \tilde{T} = (\hat{T},c_\pi), (\hat{T},\pi) \in \tilde{\mathscr{W}}\}$ of $\tilde{\mathscr{T}}$ by $\tilde{\mathscr{W}}$, and [in the context of Hypothesis (12)] use the same abuse of notation for $\tilde{\mathscr{W}}_0$ and $\tilde{\mathscr{R}}$. It is now straightforward to verify that $(x_0,\hat{T}_0,\pi_0)$ is a solution of Problem (4) if and only if $(\tilde{x}_0,\tilde{T}_0)$—where $\tilde{x}_0 = (x_0,\pi_0)$ and $\tilde{T}_0 = (\hat{T}_0,c_{\pi_0})$—is a solution of Problem (1.3), if, in the latter problem, we replace $\mathscr{X}$, A, A', $\mathscr{T}$, $\mathscr{T}_0$, $\mathscr{T}_1$, $\mathscr{W}$, x, x_0, T, and T_0 by these terms with tildes. Further, it is easy to see that now Hypotheses (1.21, 22, 23, and 26), respectively, hold (with

evident changes of notation) whenever Hypotheses (10), (11), (12), and (15), respectively, are satisfied.

Thus, we may appeal to Theorem (1.30) and arrive at the following necessary conditions.

34 THEOREM. *Let* $(x_0, \hat{T}_0, \pi_0)$ *be a solution of Problem* (4), *and suppose that Hypotheses* (10), (11), *and* (15) *hold. Further, suppose that either* $\hat{\mathscr{W}}$ *is finitely open in itself, or that Hypothesis* (12) *holds. Then, for every dual semidifferential* $D\hat{\phi}^0(\tilde{x}_0;\cdot)$ *of* $\hat{\phi}^0$ *at* $\tilde{x}_0 = (x_0, \pi_0)$ *and every dual* W_1*-semidifferential* $D\hat{\phi}_1(\tilde{x}_0;\cdot)$ *of* $\hat{\phi}_1$ *at* $\tilde{x}_0$, *there exist a vector* $\alpha \in R^m$, *a number* $\beta^0 \leq 0$, *and a linear functional* $\ell_1 \in \mathscr{Z}_1^*$, *not all zero, such that*

35
$$(\alpha \cdot D\hat{\varphi} + \beta^0 D\hat{\phi}^0 + \ell_1 \circ D\hat{\phi}_1) \circ ([E - D_1\hat{T}_0]^{-1}(\hat{T}(\tilde{x}_0) - x_0 + D_2\hat{T}_0(\pi - \pi_0)), \pi - \pi_0) \leq 0 \quad \textit{for all} \quad (\hat{T}, \pi) \in \hat{\mathscr{W}},$$

36
$$\ell_1(z_1) \geq \ell_1 \circ \hat{\phi}_1(\tilde{x}_0) \quad \textit{for all} \quad z_1 \in \bar{Z}_1,$$

where, in (35), *we have written* $D\hat{\phi}^0$ *and* $D\hat{\phi}_1$ *for* $D\hat{\phi}^0(\tilde{x}_0;\cdot)$ *and* $D\hat{\phi}_1(\tilde{x}_0;\cdot)$, *respectively,* $D\hat{\varphi}$, $D_1\hat{T}_0$, *and* $D_2\hat{T}_0$ *denote, respectively, the dual differential of* $\hat{\varphi}$ *and the Fréchet partial differentials of* $\hat{T}_0$ *with respect to its first and second arguments, all at* $\tilde{x}_0$, *and E denotes the identity operator on* $\mathscr{X}$.

If, in particular, the data of Problem (4) *are as indicated in* (16)—*so that Hypothesis* (15) *may be replaced by Hypothesis* (17)—*then, for every collection of dual semidifferentials* $D\hat{\phi}^i(\tilde{x}_0;\cdot)$ *of* $\hat{\phi}^i$ *at* $\tilde{x}_0$, *for* $i = 0, \ldots, \mu$, *and every dual* W'*-semidifferential* $D\hat{\phi}'(\tilde{x}_0;\cdot)$ *of* $\hat{\phi}'$ *at* $\tilde{x}_0$, *there exist vectors* $\alpha \in R^m$ *and* $\beta = (\beta^0, \beta^1, \ldots, \beta^\mu) \in \bar{R}_-^{\mu+1}$ *and a linear functional* $\ell' \in (\mathscr{Z}')^*$, *not all zero, such that*

37
$$(\alpha \cdot D\hat{\varphi} + \beta \cdot D\hat{\phi}'' + \ell' \circ D\hat{\phi}') \circ ([E - D_1\hat{T}_0]^{-1}(\hat{T}(\tilde{x}_0) - x_0 + D_2\hat{T}_0(\pi - \pi_0)), \pi - \pi_0) \leq 0 \quad \textit{for all} \quad (\hat{T}, \pi) \in \hat{\mathscr{W}},$$

38
$$\sum_{i=1}^{\mu} \beta^i \hat{\phi}^i(\tilde{x}_0) = 0,$$

39
$$\ell'(z') \geq \ell' \circ \hat{\phi}'(\tilde{x}_0) \quad \textit{for all} \quad z' \in \bar{Z}',$$

where, in (37), we have written $D\hat{\phi}'(\cdot)$ *for* $D\hat{\phi}'(\tilde{x}_0;\cdot)$, *and* $D\hat{\phi}'' = (D\hat{\phi}^0(\tilde{x}_0;\cdot), \ldots, D\hat{\phi}^\mu(\tilde{x}_0;\cdot))$. *If the dual semidifferentials* $D\hat{\phi}^i(\tilde{x}_0;\cdot)$ *can be extended to* $\mathscr{X} \times \mathscr{P}$ *in a way that they remain convex, continuous functionals, then there is also a continuous linear function* $\Lambda = (\Lambda^0, \Lambda^1, \ldots, \Lambda^\mu)\colon \mathscr{X} \times \mathscr{P} \to R^{\mu+1}$ *such that*

40 $$(\alpha \cdot D\hat{\varphi} + \beta \cdot \Lambda + \ell' \circ D\hat{\phi}') \circ ([E - D_1\hat{T}_0]^{-1}(\hat{T}(\tilde{x}_0) - x_0 + D_2\hat{T}_0(\pi - \pi_0)), \pi - \pi_0) \leq 0 \quad \text{for all} \quad (\hat{T},\pi) \in \tilde{\mathscr{W}},$$

41 $$\Lambda^i(x,\pi) \leq D\hat{\phi}^i(\tilde{x}_0; x,\pi) \quad \text{for all} \quad (x,\pi) \in \mathscr{X} \times \mathscr{P} \quad \text{and each} \quad i = 0, \ldots, \mu.$$

Proof. By Theorem (1.30), once we have made the identifications indicated in (33), there exist a vector $\alpha \in R^m$, a number $\beta^0 \leq 0$, and a linear functional $\ell_1 \in \mathscr{Z}_1^*$, not all zero, such that (36) holds and such that

42 $$(\alpha \cdot D\hat{\varphi} + \beta^0 D\hat{\phi}^0 + \ell_1 \circ D\hat{\phi}_1) \circ [\tilde{E} - D\tilde{T}_0]^{-1}(\tilde{T}\tilde{x}_0 - \tilde{x}_0) \leq 0 \quad \text{for all} \quad \tilde{T} \in \tilde{\mathscr{W}},$$

where $D\hat{\varphi}, D\hat{\phi}^0$, and $D\hat{\phi}_1$ are as in the theorem statement, $D\tilde{T}_0 = (D\hat{T}_0, 0)$, where $D\hat{T}_0$ denotes the Fréchet differential of $\hat{T}_0$ at $\tilde{x}_0 = (x_0, \pi_0)$, and $\tilde{E}$ denotes the identity operator on $\tilde{\mathscr{X}}$. Note that

43 $$\tilde{T}\tilde{x}_0 - \tilde{x}_0 = (\hat{T}(\tilde{x}_0) - x_0, \pi - \pi_0) \quad \text{for all} \quad \tilde{T} = (\hat{T}, c_\pi) \in \tilde{\mathscr{W}}.$$

Let us consider $[\tilde{E} - D\tilde{T}_0]^{-1}\tilde{x}$ for any $\tilde{x} \in \tilde{\mathscr{X}}$. It is evident that, for any $\tilde{x}_1 = (x_1, \pi_1) \in \tilde{\mathscr{X}}$, $[\tilde{E} - D\tilde{T}_0]^{-1}\tilde{x}_1 = \tilde{x}_2 = (x_2, \pi_2)$ if and only if

$$\begin{aligned}\tilde{x}_1 = (x_1, \pi_1) &= (\tilde{E} - D\tilde{T}_0)(x_2,\pi_2)\\ &= (x_2,\pi_2) - (D\hat{T}_0(x_2,\pi_2),0)\\ &= (x_2 - D_1\hat{T}_0(x_2) - D_2\hat{T}_0(\pi_2),\pi_2),\end{aligned}$$

where $D_1\hat{T}_0$ and $D_2\hat{T}_0$ are the Frechet partial differentials of $\hat{T}_0$ at $\tilde{x}_0$ with respect to its first and second arguments, respectively, i.e., if and only if

44 $$x_2 = [E - D_1\hat{T}_0]^{-1}(x_1 + D_2\hat{T}_0(\pi_2)) \quad \text{and} \quad \pi_1 = \pi_2,$$

where E is the identity operator on $\mathscr{X}$. But this, together with (42) and (43), implies that (35) holds. The remaining conclusions now follow at once from the corresponding ones in Theorem (1.30). |||

45 We point out [see (1.44)] that relations (35), (37), and (40) in Theorem (34) hold, not only for all $(\hat{T},\pi) \in \tilde{\mathscr{W}}$, but also for all $(\hat{T},\pi) \in (\hat{T}_0,\pi_0) + \overline{\text{co}}\,\overline{\text{co}}\,(\tilde{\mathscr{W}} - (\hat{T}_0,\pi_0))$.

The following two corollaries are now standard.

46 COROLLARY. *If, in Theorem* (34), Z_1 *is a cone, then* (36) *holds if and only if* $\ell_1 \in -(\bar{Z}_1)^*$ *and* $\ell_1 \circ \hat{\phi}_1(\tilde{x}_0) = 0$. *If* Z' *is a cone, then* (39) *holds if and only if* $\ell' \in -(\bar{Z}')^*$ *and* $\ell' \circ \hat{\phi}'(\tilde{x}_0) = 0$.

47 COROLLARY. *If, in Theorem* (34), $D\hat{\phi}^0$ *and* $D\hat{\phi}_1$ *are linear, then* (35) *is equivalent to the maximum principle*

48
$$\begin{aligned}
&(\alpha \cdot D_1\hat{\varphi} + \beta^0 D_1\hat{\phi}^0 + \ell_1 \circ D_1\hat{\phi}_1) \\
&\quad \circ [E - D_1\hat{T}_0]^{-1}(\hat{T}(\tilde{x}_0) + D_2\hat{T}_0(\pi)) \\
&\quad + (\alpha \cdot D_2\hat{\varphi} + \beta^0 D_2\hat{\phi}^0 + \ell_1 \circ D_2\hat{\phi}_1)(\pi) \\
&\quad \leq (\alpha \cdot D_1\hat{\varphi} + \beta^0 D_1\hat{\phi}^0 + \ell_1 \circ D_1\hat{\phi}_1) \\
&\quad \circ [E - D_1\hat{T}_0]^{-1}(x_0 + D_2\hat{T}_0(\pi_0)) \\
&\quad + (\alpha \cdot D_2\hat{\varphi} + \beta^0 D_2\hat{\phi}^0 + \ell_1 \circ D_2\hat{\phi}_1)(\pi_0) \quad \textit{for all} \quad (\hat{T},\pi) \in \tilde{\mathscr{W}},
\end{aligned}$$

where $D_1\hat{\varphi}$ *and* $D_2\hat{\varphi}$ *denote the formal dual partial differentials of* $\hat{\varphi}$ *with respect to its first and second arguments at* $\tilde{x}_0$, *and similarly for* $D_1\hat{\phi}^0$, $D_1\hat{\phi}_1$, $D_2\hat{\phi}^0$, *and* $D_2\hat{\phi}_1$ [*see* (I.7.56 *and* 57)]. *If also* $\tilde{\mathscr{W}} = \hat{\mathscr{W}} \times \Pi_a$, *where* $\hat{\mathscr{W}} \subset \mathscr{T}$ *and* $\Pi_a \subset \Pi$, *then* (48) *is equivalent to the pair of inequalities*

49
$$\begin{aligned}
&(\alpha \cdot D_1\hat{\varphi} + \beta^0 D_1\hat{\phi}^0 + \ell_1 \circ D_1\hat{\phi}_1) \circ [E - D_1\hat{T}_0]^{-1}(\hat{T}(\tilde{x}_0)) \\
&\qquad \leq (\alpha \cdot D_1\hat{\varphi} + \beta^0 D_1\hat{\phi}^0 + \ell_1 \circ D_1\hat{\phi}_1) \circ [E - D_1\hat{T}_0]^{-1}(x_0) \\
&\qquad\qquad \text{for all} \quad \hat{T} \in \hat{\mathscr{W}},
\end{aligned}$$

50
$$\begin{aligned}
&((\alpha \cdot D_1\hat{\varphi} + \beta^0 D_1\hat{\phi}^0 + \ell_1 \circ D_1\hat{\phi}_1) \circ [E - D_1\hat{T}_0]^{-1}(D_2\hat{T}_0) \\
&\qquad + (\alpha \cdot D_2\hat{\varphi} + \beta^0 D_2\hat{\phi}^0 + \ell_1 \circ D_2\hat{\phi}_1))(\pi - \pi_0) \leq 0 \\
&\qquad\qquad \textit{for all} \quad \pi \in \Pi_a.
\end{aligned}$$

If the data of Problem (4) *are as indicated in* (16), *and* $D\hat{\phi}^0$, $D\hat{\phi}^1, \ldots, D\hat{\phi}^\mu$, *and* $D\hat{\phi}'$ *are linear, then relations* (37) *and* (40) *can be simplified in an analogous manner.*

We point out that if $\hat{\mathscr{W}}$ consists of a single operator $\hat{T}_0$, then Inequality (49) contains no information. If π_0 is an internal point of Π_a [see (I.1.42)], then, by Lemma (I.2.39), Inequality (50) holds if and only if

$$(\alpha \cdot D_1\hat{\varphi} + \beta^0 D_1\hat{\phi}^0 + \ell_1 \circ D_1\hat{\phi}_1) \circ [E - D_1\hat{T}_0]^{-1}(D_2\hat{T}_0) + \alpha \cdot D_2\hat{\varphi} + \beta^0 D_2\hat{\phi}^0 + \ell_1 \circ D_2\hat{\phi}_1 = 0.$$

51 Under the hypotheses of Corollary (47), relation (48) holds [see (45)] also for all $(\hat{T},\pi) \in (\hat{T}_0,\pi_0) + \overline{\text{co co}}\,(\tilde{\mathscr{W}} - (\hat{T}_0,\pi_0))$, and, if also $\tilde{\mathscr{W}} = \hat{\mathscr{W}} \times \Pi_a$, then (49) holds for all $\hat{T} \in \hat{T}_0 + \overline{\text{co co}}\,(\hat{\mathscr{W}} - \hat{T}_0)$, and (42) holds for all $\pi \in \pi_0 + \overline{\text{co co}}\,(\Pi_a - \pi_0)$ [see Lemma (I.1.30) and (I.3.34)].

Remarks (III.3.31 and 32)—with evident modifications—also carry over here.

52 Let us now turn back to the special case of Problem (20). It follows directly from the continuity of $\chi = (\chi^0, \chi'', \chi')$ that the functions $\hat{\varphi}$, $\hat{\phi}^0$, and $\hat{\phi}_1$ defined on $\mathscr{C}^n(I) \times I^k$ [with $\mathscr{Z}_1 = R^\mu$] by (26)–(28) are continuous. Let us define functions $\hat{\phi}^1, \ldots, \hat{\phi}^\mu$ analogously to $\hat{\phi}^0$ [see (27)], so that $\hat{\phi}_1 = (\hat{\phi}^1, \ldots, \hat{\phi}^\mu)$. Let us now also suppose that the function χ is continuously differentiable. From this it does *not* necessarily follow that the functions $\hat{\varphi}, \hat{\phi}^0, \ldots, \hat{\phi}^\mu$ are dually differentiable at each point of $\tilde{A}' = A \times I^k$. However, $\hat{\varphi}, \hat{\phi}^0, \ldots, \hat{\phi}^\mu$ are dually differentiable at each $(x_0,\tau_0^1, \ldots, \tau_0^k) \in A \times I^k$ such that x_0 is differentiable at τ_0^j for each $j = 1, \ldots, k$. [If $\tau_0^j = t_1$ (respectively, t_2) for some j, then it is sufficient to suppose that x_0 is Gâteaux differentiable [see (I.7.4)], i.e., is differentiable from the right (respectively, left) at τ_0^j.] Indeed, in this case, the map

53 $$(x,\tau^1, \ldots, \tau^k) \to (x(\tau^1), \ldots, x(\tau^k), \tau^1, \ldots, \tau^k) \colon A \times I^k \to (R^n)^k \times R^k$$

is dually differentiable at $(x_0,\tau_0^1, \ldots, \tau_0^k)$, with the map

$$(\delta x,\delta\tau^1, \ldots, \delta\tau^k) \to (\dot{x}_0(\tau_0^1)\delta\tau^1 + \delta x(\tau_0^1), \ldots, \dot{x}_0(\tau_0^k)\delta\tau^k + \delta x(\tau_0^k), \delta\tau^1, \ldots, \delta\tau^k)$$

as dual differential. [The verification of this, which is quite straightforward—see (I.7.49), Lemma (I.7.40), and Theorem (I.7.44)—is left to the reader. It is interesting to note that the map (53) is *not Fréchet* differentiable at any point in $A \times I^k$.] If we denote the map (53) by χ_0, then it is obvious that $\hat{\varphi} = \chi' \circ \chi_0$, $\hat{\phi}^0 = \chi^0 \circ \chi_0$, and $\hat{\phi}^i = \chi^i \circ \chi_0$ for each $i = 0, \ldots, \mu$, so that, by the chain rule of differentiation in Theorem (I.7.35) [also see Theorem (I.7.44)], $\hat{\varphi}, \hat{\phi}^0, \ldots, \hat{\phi}^\mu$ are also dually differentiable at $(x_0,\tau_0^1,\ldots,\tau_0^k)$ with dual differentials at this point given by the formulas

54
$$D\hat{\varphi}(\delta x,\delta\tau^1,\ldots,\delta\tau^k) = \sum_{j=1}^{k} \mathscr{D}_j^0\chi'[\delta x(\tau_0^j) + \dot{x}_0(\tau_0^j)\delta\tau^j] + \sum_{j=1}^{k} \mathscr{D}_{k+j}^0\chi'\delta\tau^j,$$

55
$$D\hat{\phi}^i(\delta x,\delta\tau^1,\ldots,\delta\tau^k) = \sum_{j=1}^{k} \mathscr{D}_j^0\chi^i[\delta x(\tau_0^j) + \dot{x}_0(\tau_0^j)\delta\tau^j] + \sum_{j=1}^{k} \mathscr{D}_{k+j}^0\chi^i\delta\tau^j, \quad i = 0,\ldots,\mu,$$

$$\delta x \in \mathscr{C}^n(I), \quad (\delta\tau^1,\ldots,\delta\tau^k) \in R^k,$$

where $\mathscr{D}_j^0\chi'$ denotes $\mathscr{D}_j\chi(x_0(\tau_0^1),\ldots,x_0(\tau_0^k),\tau_0^1,\ldots,\tau_0^k)$ [see (I.7.58–61)] for each $j = 1,\ldots,2k$, and similarly for $\mathscr{D}_j^0\chi^i$ for each i.

Hence [see (I.7.45)], Hypothesis (17) is satisfied if x_0 and $\pi_0 = (\tau_0^1,\ldots,\tau_0^k)$ satisfy the previously described hypothesis. [Of course, Hypothesis (17) will be satisfied even under considerably weaker differentiability hypotheses on $\chi^0, \chi^1,\ldots,\chi^\mu$.]

If we now appeal to Theorem (34) and its Corollaries (46) and (47), and take into account the remarks in (19) and (51), we arrive at the following necessary conditions for Problem (20).

56 THEOREM. *Let* $(x_0,\tau_0^1,\ldots,\tau_0^k) \in A \times I_a$ *be a solution of Problem* (20), *and let* $T_0 \in \mathscr{W}$ *be such that* $x_0 = T_0x_0$. *Also suppose that Hypotheses* (1.21 *and* 22) *hold, that* χ *is continuously differentiable, and that* x_0 *is differentiable at* τ_0^j *for each* $j = 1,\ldots,k$. *Finally, suppose that either* $\mathscr{W}$ *is finitely open in itself or that Hypothesis* (1.23) *holds. Then there exists a nonzero row vector* $\alpha = (\alpha^0,\alpha^1,\ldots,\alpha^{\mu+m}) \in \bar{R}_-^{\mu+1} \times R^m$ *such that*

57
$$\text{(i)} \sum_{j=1}^{k} \alpha\mathscr{D}_j^0\chi\delta x_T(\tau_0^j) \le \sum_{j=1}^{k} \alpha\mathscr{D}_j^0\chi\delta x_{T_0}(\tau_0^j) \quad \text{for all} \quad T \in \mathscr{W},$$

where, for each $T \in \mathscr{T}$, δx_T *denotes the column-vector-valued function in* $\mathscr{C}^n(I)$ *defined by*

58
$$\delta x_T = [E - DT_0]^{-1}(Tx_0)$$

[E *denotes the identity operator on* $\mathscr{C}^n(I)$ *and* DT_0 *the Fréchet differential of* T_0 *at* x_0] *and* $\mathscr{D}_j^0\chi$ (*for each* j) *denotes the Jacobian matrix* $\mathscr{D}_j\chi$ *of* χ *at* $(x_0(\tau_0^1), \ldots, x_0(\tau_0^k), \tau_0^1, \ldots, \tau_0^k)$,

59 $$\text{(ii)} \sum_{i=1}^{\mu} \alpha^i \chi^i(x_0(\tau_0^1), \ldots, x_0(\tau_0^k), \tau_0^1, \ldots, \tau_0^k) = 0,$$

and

60 $$\text{(iii)} \sum_{j=1}^{k} \alpha[\mathscr{D}_j^0 \chi \dot{x}_0(\tau_0^j) + \mathscr{D}_{k+j}^0 \chi]\delta\tau^j \leq 0$$

$$\textit{for all} \quad (\delta\tau^1, \ldots, \delta\tau^k) \in \overline{I_a - (\tau_0^1, \ldots, \tau_0^k)},$$

where $\mathscr{D}_{k+j}^0\chi$ *(for each* j*) denotes the previously indicated Jacobian matrix.*

61 For evident reasons, the preceding theorem will also be called a maximum principle.

62 The hypothesis in Theorem (56) that x_0 is differentiable at τ_0^j for each $j = 1, \ldots, k$ can be relaxed at the expense of weakening the necessary conditions of the theorem. Namely, if x_0 is differentiable at τ_0^j only for certain j (say, for $j = 1, \ldots, k_1 \leq k$), then the necessary conditions of Theorem (56) remain in force, except that the sum in (60) must be taken over $j = 1, \ldots, k_1$ rather than $1, \ldots, k$. If x_0 is not differentiable at τ_0^j for any $j = 1, \ldots, k$, then the necessary conditions of the theorem remain in force with the exception that (iii) must be deleted. These conclusions follow immediately, once we take into account the fact that $(x_0, \tau_0^1, \ldots, \tau_0^k)$, as a solution of Problem (20), is also a solution of the problem which differs from Problem (20) in that the values of the parameters τ^j, for those j where x_0 is not differentiable at τ^j, are fixed at τ_0^j.

63 If, in the statement of Problem (20), we replace the inequality constraints (22) by $\chi''(x(\tau^1), \ldots, x(\tau^k), \tau^1, \ldots, \tau^k) \in \mathbf{Z}$ $[\chi'' = (\chi^1, \ldots, \chi^\mu)]$, where $\mathbf{Z}$ is some convex body in R^μ, then our results remain in force with certain minor modifications. If $\mathbf{Z} = R^\mu$, so that, effectively, the inequality constraints (22) are removed, then our necessary conditions remain in force with $(\alpha^1, \ldots, \alpha^\mu) = 0$ [so that (59) is automatically satisfied].

By virtue of (51), we can assert that, in Theorem (56), (57) holds not only for all $T \in \mathscr{W}$, but also for all $T \in T_0 + \overline{\text{co}}\,\overline{\text{co}}\,(\mathscr{W} - T_0)$.

In Chapter V, we shall "expand" the necessary conditions of Theorem (56) for certain special classes $\mathscr{W}$, to obtain a maximum principle in a more convenient form.

We can, of course, apply Theorem (34) and its Corollaries (46) and (47) to obtain necessary conditions for problems considerably more general than Problem (20). One such problem is as follows:

64 *Problem.* Let there be given subsets $\hat{\mathscr{W}}$ of $\hat{\mathscr{T}}$ (where $\mathscr{X} = \mathscr{C}^n(I)$ for some compact interval $I = [t_1, t_2]$) and Π_a of Π such that $\hat{\mathscr{W}} \times \Pi_a \subset \mathscr{S}_0$ and such that Π_a is finitely open in itself, and continuous functions $\chi = (\chi^0, \chi^1, \ldots, \chi^{\mu+m})\colon (R^n)^k \times \Pi \to R^{\mu+m+1}$ and $\tau = (\tau^1, \ldots, \tau^k)\colon \Pi \to R^k$, where $\tau(\Pi \cap \operatorname{co} \Pi_a) \subset I^k$. Then find a pair $(x,\pi) \in A \times \Pi_a$ satisfying the constraints

65 $$x = \hat{T}(x,\pi) \quad \text{for some} \quad \hat{T} \in \hat{\mathscr{W}},$$

66 $$\chi^i(x(\tau^1(\pi)), \ldots, x(\tau^k(\pi)),\pi) \le 0 \quad \text{for} \quad i = 1, \ldots, \mu,$$

67 $$\chi^i(x(\tau^1(\pi)), \ldots, x(\tau^k(\pi)),\pi) = 0 \quad \text{for} \quad i = \mu+1, \ldots, \mu+m,$$

which, in so doing, achieves a minimum for

$$\chi^0(x(\tau^1(\pi)), \ldots, x(\tau^k(\pi)),\pi).$$

If we identify Problem (64) with Problem (4)—much as was done for Problem (20)—and appeal to Theorem (34) and its Corollaries (46) and (47), we arrive at the following necessary conditions for Problem (64).

68 THEOREM. *Let $(x_0,\pi_0) \in A \times \Pi_a$ be a solution of Problem (64), and let $\hat{T}_0 \in \hat{\mathscr{W}}$ be such that $x_0 = \hat{T}_0(x_0,\pi_0)$. Also suppose that Hypotheses (10) and (11) hold for $\mathscr{W} = \hat{\mathscr{W}} \times \Pi_a$, that χ and τ are continuously differentiable, and that x_0 is differentiable at $\tau_0^j = \tau^j(\pi_0)$ for each $j = 1, \ldots, k$. Finally, suppose that either $\hat{\mathscr{W}}$ is finitely open in itself or that Hypothesis (12) holds. Then there exists a nonzero row vector $\alpha = (\alpha^0, \alpha^1, \ldots, \alpha^{\mu+m}) \in \bar{R}_-^{\mu+1} \times R^m$ such that*

69 $$\text{(i)} \sum_{j=1}^{k} \alpha \mathscr{D}_j^0 \chi \, \delta x_{\hat{T}}(\tau_0^j) \le \sum_{j=1}^{k} \alpha \mathscr{D}_j^0 \chi \, \delta x_{\hat{T}_0}(\tau_0^j) \quad \textit{for all} \quad \hat{T} \in \hat{\mathscr{W}},$$

where, for each $\hat{T} \in \hat{\mathscr{T}}$, $\delta x_{\hat{T}}$ denotes the column-vector-valued function in $\mathscr{C}^n(I)$ defined by

70 $$\delta x_{\hat{T}} = [E - D_1 \hat{T}_0]^{-1}(\hat{T}(x_0,\pi_0))$$

[E denotes the identity operator on $\mathscr{C}^n(I)$ and $D_1\hat{T}_0$ the Fréchet partial differential of $\hat{T}_0$ with respect to its first argument at (x_0,π_0)—and similarly for $D_2\hat{T}_0$ in (72)] and $\mathscr{D}_j^0\chi$ (for each

$j = 1, \ldots, k$) denotes the Jacobian matrix $\mathscr{D}_j\chi$ of χ at the point $(x_0(\tau_0^1), \ldots, x_0(\tau_0^k), \pi_0)$,

71 $$\text{(ii)} \sum_{j=1}^{k} \alpha\mathscr{D}_j^0\chi[\delta x_{\delta\pi}(\tau_0^j) + \dot{x}_0(\tau_0^j)D^0\tau^j(\delta\pi)] + \alpha D_{k+1}^0\chi(\delta\pi) \leq 0$$
$$\textit{for all} \quad \delta\pi \in \overline{\Pi_a - \pi_0},$$

where $\delta x_{\delta\pi}$, for each $\delta\pi \in \mathscr{P}$, denotes the column-vector-valued function in $\mathscr{C}^n(I)$ defined by

72 $$\delta x_{\delta\pi} = [E - D_1\hat{T}_0]^{-1} \circ D_2\hat{T}_0(\delta\pi),$$

$D_{k+1}^0\chi(\cdot)$ denotes the Fréchet partial differential of χ with respect to its last argument at $(x_0(\tau_0^1), \ldots, x_0(\tau_0^k), \pi_0)$, and $D^0\tau^j(\cdot)$ (for each j) denotes the Fréchet differential of τ^j at π_0, and

73 $$\text{(iii)} \sum_{i=1}^{\mu} \alpha^i\chi^i(x_0(\tau_0^1), \ldots, x_0(\tau_0^k), \pi_0) = 0.$$

If also π_0 is an internal point of Π_a, then (71) may be replaced by the equation

74 $$\sum_{j=1}^{k} \alpha\mathscr{D}_j^0\chi[(E - D_1\hat{T}_0)^{-1} \circ (D_2\hat{T}_0) + \dot{x}_0(\tau_0^j)D^0\tau^j] + \alpha D_{k+1}^0\chi = 0.$$

The remarks in (63) carry over here with evident minor modifications.

75 If we modify Problem (4) by omitting the equality constraint $\hat{\varphi}(x,\pi) = 0$, then [see (1.92)] we can obtain necessary conditions analogous to those of Theorem (34), except that the term $\alpha \cdot D\hat{\varphi}$ in (35), (37), and (40)—as well as the references to $\hat{\varphi}$ in Hypotheses (15) and (17)—should be deleted, and, in the manner indicated in (1.92), Hypothesis (12) can here be weakened.

Correspondingly, if we modify Problem (20) [respectively, (64)] by eliminating the equality constraints (23) [respectively, (67)]—so that $\chi = (\chi^0, \chi^1, \ldots, \chi^\mu)$—then all of the necessary conditions of Theorem (56) [respectively, (68)] remain in effect—with $\alpha = (\alpha^0, \alpha^1, \ldots, \alpha^\mu)$ a nonzero vector in $\bar{R}_-^{\mu+1}$.

76 Let us make some remarks concerning a problem which may be viewed either in the context of this section, or in the context of the preceding section, and let us compare the two sets of necessary conditions that we may thereby obtain. Namely, let

us suppose that, for Problem (4), (i) $\tilde{\mathscr{W}} = \hat{\mathscr{W}} \times \Pi_a$, where $\hat{\mathscr{W}}$ consists of a single element $\hat{T}_0 \in \hat{\mathscr{T}}$ and Π_a is a subset of Π which is finitely open in itself, (ii) $\tilde{A}' = \tilde{A}$, (iii) the functions $\hat{\varphi}$, $\hat{\phi}^0$, and $\hat{\phi}_1$ are independent of π and satisfy Hypothesis (1.26) for any $x_0 \in A$, (iv) Hypothesis (1.22) is satisfied by the set $\mathscr{W} = \{\hat{T}_0^\pi : \pi \in \Pi_a\}$, (v) co $\{\hat{T}_0^\pi : \pi \in (\text{co}\,\Pi_a) \cap \Pi\} \subset \mathscr{T}_0$, and (vi) Hypothesis (11) is satisfied. We can then conclude, on the basis of Theorem (34), that if $(x_0, \hat{T}_0, \pi_0)$ is a solution of this problem, then, for every dual semidifferential $D\hat{\phi}^0$ of $\hat{\phi}^0$ at x_0 and every dual W_1-semidifferential $D\hat{\phi}_1$ of $\hat{\phi}_1$ at x_0, there exist a vector $\alpha \in R^m$, a number $\beta^0 \leq 0$, and a linear functional $\ell_1 \in \mathscr{L}_1^*$, not all zero, such that

77 $$(\alpha \cdot D\hat{\varphi} + \beta^0 D\hat{\phi}_0 + \ell_1 \circ D\hat{\phi}_1) \circ [E - D_1 \hat{T}_0]^{-1} \circ D_2 \hat{T}_0 (\pi - \pi_0) \leq 0 \quad \text{for all} \quad \pi \in \Pi_a,$$

[where $D\hat{\varphi}$, $D_1\hat{T}_0$, $D_2\hat{T}_0$, and E are as in Theorem (34)] and such that (1.32) holds. It is evident that, in this case, $(x_0, \hat{T}_0^{\pi_0})$ is a solution of Problem (1.3), with $\mathscr{W}$ as previously indicated. If Hypothesis (1.23) is satisfied by $\mathscr{W}$ and x_0 [as is true in many applications], then Theorem (1.30) allows us to conclude that there exist a vector $\alpha \in R^m$, a number $\beta^0 \leq 0$, and a linear functional $\ell_1 \in \mathscr{L}_1^*$, not all zero, such that

78 $$(\alpha \cdot D\hat{\varphi} + \beta^0 D\hat{\phi}^0 + \ell_1 \circ D\hat{\phi}_1) \circ [E - D_1 \hat{T}_0]^{-1} (\hat{T}_0(x_0, \pi) - x_0) \leq 0 \quad \text{for all} \quad \pi \in \Pi_a,$$

and such that (1.32) holds. This raises the following question: Are the inequalities (77) and (78) equivalent, and, if not, is one of them stronger than the other? It turns out, as the reader may easily verify, that these inequalities are generally not equivalent, and that (under our hypotheses) (77) holds whenever (78) does; i.e., the necessary conditions of Theorem (1.30) are in this case generally stronger than those of Theorem (34). We emphasize, however, that if the set $\mathscr{W} = \{\hat{T}_0^\pi : \pi \in \Pi_a\}$ does not satisfy Hypothesis (1.23), then we cannot necessarily conclude that the necessary conditions of Theorem (1.30) hold, and we must content ourselves with the weaker necessary conditions of Theorem (34).

We close this section with some brief comments regarding how the results so far obtained in this section carry over to minimax problems. (As before, we leave the details of the vector-valued criterion problem to the reader.)

Consider the following minimax problem, which is analogous to Problem (4), and at the same time generalizes Problem (1.65).

79 *Problem.* Given a subset $\tilde{\mathscr{W}}$ of $\mathscr{S}_0$, a subset $\tilde{A}'$ of $\tilde{A}$ which contains all (x,π) such that x is a fixed point of $\hat{T}^\pi$ with $(\hat{T},\pi) \in \mathscr{S}_1 \cap (\text{co } \tilde{\mathscr{W}})$, linear topological spaces $\mathscr{Z}_1$ and $\mathscr{Z}_2$, a convex body $Z_1 \subset \mathscr{Z}_1$, a nonempty set $\mathscr{L} \subset \mathscr{Z}_2^*$, and functions $\hat{\varphi}: \tilde{A}' \to R^m$, $\hat{\phi}_1 : \tilde{A}' \to \mathscr{Z}_1$, and $\hat{\phi}_2 : \tilde{A}' \to \mathscr{Z}_2$, find a triple $(x_0,\hat{T}_0,\pi_0) \in A \times \tilde{\mathscr{W}}$ such that (5)–(7) hold and such that

$$\sup_{\ell \in \mathscr{L}} \ell \circ \hat{\phi}_2(x_0,\pi_0) \leq \sup_{\ell \in \mathscr{L}} \ell \circ \hat{\phi}_2(x,\pi)$$

for all $(x,\hat{T},\pi) \in A \times \tilde{\mathscr{W}}$ that satisfy (8).

Arguing as in the case of Problem (4), we can show that Problem (79) may be identified as a special case of Problem (1.65), if, in the latter, we suitably redefine $\mathscr{X}$, A, A', $\mathscr{T}$, etc. Thus, much as was done in deriving Theorem (34), we can obtain on the basis of Theorem (1.74), the following theorem which yields necessary conditions for Problem (79).

80 THEOREM. *Let $(x_0,\hat{T}_0,\pi_0)$ be a solution of Problem* (79), *and suppose that Hypotheses* (III.1.21 *and* 22)*—where $\kappa_0 = \sup_{\ell \in \mathscr{L}} \ell \circ \hat{\phi}_2(x_0,\pi_0)$—as well as Hypotheses* (10), (11), *and* (1.72) *hold, with x_0 replaced by (x_0,π_0) in the latter. Further, suppose that either $\tilde{\mathscr{W}}$ is finitely open in itself, or that Hypothesis* (12) *holds. Then, for every dual W_i-semidifferentials $D\hat{\phi}_i(\tilde{x}_0;\cdot)$ of $\hat{\phi}_i$ at $\tilde{x}_0 = (x_0,\pi_0)(i = 1, 2)$, there exist a vector $\alpha \in R^m$ and linear functionals $\ell_i \in \mathscr{Z}_i^*$ $(i = 1, 2)$, not all zero, such that*

81 $$(\alpha \cdot D\hat{\varphi} + \ell_1 \circ D\hat{\phi}_1 + \ell_2 \circ D\hat{\phi}_2) \circ ([E - D_1\hat{T}_0]^{-1}(\hat{T}(\tilde{x}_0) - x_0 + D_2\hat{T}_0(\pi - \pi_0)), \pi - \pi_0) \leq 0 \quad \textit{for all} \quad (\hat{T},\pi) \in \tilde{\mathscr{W}},$$

82 $$\ell_1(z_1) \leq \ell_1 \circ \hat{\phi}_1(\tilde{x}_0) \quad \textit{for all} \quad z_1 \in \bar{Z}_1,$$

83 $$\ell_2(z_2) \geq \ell_2 \circ \hat{\phi}_2(\tilde{x}_0) \quad \textit{for all} \quad z_2 \in Z_2^{\kappa_0},$$

where, in (81), *we have written* $D\hat{\phi}_i(\cdot)$ *for* $D\hat{\phi}_i(\bar{x}_0;\cdot)(i = 1, 2)$, $D\hat{\varphi}$, $D_1\hat{T}_0$, *and* $D_2\hat{T}_0$ *denote, respectively, the dual differential of* $\hat{\varphi}$ *and the Fréchet partial differentials of* $\hat{T}_0$ *w.r. to its first and second arguments, all at* $\bar{x}_0$, *and E denotes the identity operator on* $\mathscr{X}$, *and, in* (83), $Z_2^{\kappa_0}$ *is defined by* (III.1.17).

84 Evidently, corollaries analogous to Corollaries (1.78) and (47) hold here. Also, the remarks in (45), (51), and (75) carry over here, as do those in (III.3.31 and 32), with evident minor modifications.

For the sake of brevity, we did not consider in Theorem (80) the special case where $\hat{\phi}_1$, $\mathscr{Z}_1$, and Z_1 are as indicated in (16). However, it should be clear how the necessary conditions can be specialized and extended in this case, as was done in the case of Problem (4) in the statement of Theorem (34).

3. Volterra-Type Operators

In this section, we return to the "nonparametric" optimization problems considered in Section 1, but we shall confine ourselves to the case where $\mathscr{X} = \mathscr{C}^n(I)$ (for some compact interval $I = [t_1, t_2]$), and where the operators in $\mathscr{W}$ are all of a special kind, which we shall refer to as *Volterra type*. Such operators arise, as we shall see, when the operator equation $x = Tx$ represents an ordinary differential equation, a functional differential equation of retarded type, or a Volterra integral equation. Making use of particular properties of Volterra-type operators, we shall show that Hypotheses (1.21–23) can be replaced by alternate hypotheses which can be verified with much greater ease.

In Section 4, we shall extend the results of this section to the "parametric" problems considered in Section 2.

1 With $\mathscr{X} = \mathscr{C}^n(I)$ and $I = [t_1, t_2]$, let us continue to use the notation introduced in Section 1. Let us denote by $\mathscr{K}$ the class of all conditionally compact subsets of $\mathscr{C}^n(I)$ [see (I.3.14)]. As is well known [see Reference 3 of Chapter I, Theorem 7.57, p. 137], $Q \in \mathscr{K}$ if and only if (i) Q is bounded [see (I.4.37)] and (ii) the set Q (as a set of functions on I) is equicontinuous [see (I.5.28)].

2 An operator $T \in \mathscr{T}$ will be said to be

(a) *causal* if the relations $x_1 \in A$, $x_2 \in A$, $t \in I$, and $x_1(s) = x_2(s)$ for all $s \in [t_1, t]$ imply that $(Tx_1)(t) = (Tx_2)(t)$;

(b) *locally compact* if, for each $x_1 \in A$, there is a number $\zeta > 0$ (possibly depending on x_1, as well as on T) such that $\{Tx : x \in A, \|x - x_1\| \leq \zeta\} \in \mathscr{K}$;

(c) *of fixed initial value* if $(Tx_1)(t_1) = (Tx_2)(t_1)$ for all $x_1, x_2 \in A$.

3 An operator $T \in \mathscr{T}$ which is causal, locally compact, and of fixed initial value will be said to be of *Volterra type.* We shall denote by $\mathscr{V}$ the set of all $T \in \mathscr{T}$ which are of Volterra type and which have the property that, for each $x \in A$, $DT(x;\cdot)$, the Fréchet differential of T at x, is also of Volterra type [with A replaced by $\mathscr{C}^n(I)$]. It is easily seen that $\mathscr{V}$ is a linear subspace of $\mathscr{T}$.

4 Since not every $T \in \mathscr{V}$ has a fixed point, it is useful to introduce the concept of a local solution of the equation

5 $$x = Tx.$$

Namely, we shall say that a function $x_1 \in A$ is a *local solution* of Equation (5), if, for some $t' \in (t_1,t_2]$, $x_1(t) = (Tx_1)(t)$ for all $t \in [t_1,t']$. [Of course, if $t' = t_2$, then x_1 is a solution of Equation (5), i.e., is a fixed point of T.] In this case, we shall also say that x_1 *satisfies Equation* (5) *on* the interval $[t_1,t']$. Clearly, every solution of Equation (5) is also a local solution of this equation.

6 If $x_1 \in A$ is a local solution, but not a solution, of Equation (5), then evidently there is a number $t' \in (t_1,t_2)$ such that x_1 satisfies Equation (5) on $[t_1,t']$ but does not satisfy Equation (5) on $[t_1,\tilde{t}]$ for any $\tilde{t} > t'$. In this case, if $x_2 \in A$ and $t'' \in (t',t_2]$ are such that x_2 satisfies Equation (5) on $(t_1,t'']$, and such that $x_1(t) = x_2(t)$ for all $t \in [t_1,t']$, then we shall say that x_2 is a *continuation* of x_1, and that x_1 can be *continued.*

7 We shall say that an operator $T \in \mathscr{T}$ has the *uniqueness property* if the relations $x_1 \in A$, $x_2 \in A$, $t \in I$, and $x_j(s) = (Tx_j)(s)$ for all $s \in [t_1,t]$ and $j = 1$ and 2 imply that $x_1(s) = x_2(s)$ for all $s \in [t_1,t]$. In more colloquial terms, $T \in \mathscr{T}$ has the uniqueness property if every two local solutions of Equation (5) agree "so long as" they satisfy this equation. We shall denote by $\mathscr{T}^u$ the set of all $T \in \mathscr{T}$ which have the uniqueness property. It is obvious that $\mathscr{T}^u \subset \mathscr{T}_0$.

8 In Appendix A [see Theorems (A.13) and (A.15)], we shall show that, whenever $T \in \mathscr{V}$ is such that $x_1(t_1) = (Tx_1)(t_1)$ for some

$x_1 \in A$, then Equation (5) has at least one local solution, and every local solution which is not also a solution can be continued.

We shall be particularly concerned in the subsequent chapters with two particular subclasses of $\mathscr{V}$.

9 The operators in the first class are defined in terms of an arbitrary function $x^* \in \mathscr{C}^n(I)$ and an arbitrary function $G: A \times I \to R^n$ belonging to a certain family $\mathscr{G}$. Indeed, let $\mathscr{G}$ denote the family of all functions $G\colon A \times I \to R^n$ with the following properties:

10 (i) For each $s \in I$, the function $x \to G(x,s)\colon A \to R^n$ is continuously differentiable.

11 (ii) For each $x \in A$, the function $s \to G(x,s)\colon I \to R^n$ is integrable.

12 (iii) G is *causal* in the sense that the relations $x_1, x_2 \in A$, $t \in I$, and $x_1(s) = x_2(s)$ for all $s \in [t_1,t]$ imply that $G(x_1,t) = G(x_2,t)$.

13 (iv) For each $x_1 \in A$, there exist an integrable function $\tilde{m}\colon I \to R_+$ and a number $\zeta > 0$ (both possibly depending on x_1 as well as on G) such that

14 $$|G(x,s)| + \|D_1 G(x,s;\cdot)\| \leq \tilde{m}(s) \quad \text{for all} \quad s \in I$$
$$\text{and all} \quad x \in A \quad \text{such that} \quad \|x - x_1\| \leq \zeta,$$

where $D_1 G(x,s;\cdot)$ denotes the Fréchet partial differential of G with respect to its first argument at (x,s).

It is easily seen that $\mathscr{G}$ is a linear vector space.

15 For each $x^* \in \mathscr{C}^n(I)$ and $G \in \mathscr{G}$, we define the operator $T_{x^*,G}\colon A \to \mathscr{C}^N(I)$ as follows:

16 $$(T_{x^*,G}x)(t) = x^*(t) + \int_{t_1}^{t} G(x,s)\, ds, \quad t \in I, \quad x \in A.$$

We show in Appendix A [see Theorems (A.1) and (A.29)] that, for all $x^* \in \mathscr{C}^n(I)$ and $G \in \mathscr{G}$, $T_{x^*,G} \in \mathscr{V} \cap \mathscr{T}^u$ and the Fréchet differential of $T_{x^*,G}$ at any $x \in A$ is given by the formula

17 $$(DT_{x^*,G}(x;\delta x))(t) = \int_{t_1}^{t} D_1 G(x,s;\delta x)\, ds, \quad t \in I, \quad \delta x \in \mathscr{C}^n(I).$$

18 If $x^* \in \mathscr{C}^n(I)$, $G \in \mathscr{G}$, and $t_1' \in I$ are such that (i) $x^*(t) = x^*(t_1')$ for all $t \in [t_1',t_2]$ (i.e., x^* is constant on $[t_1',t_2]$) and (ii) $G(x,s) = 0$

for all $(x,s) \in A \times [t_1,t_1')$, then the operator equation

19 $$x = T_{x^*,G}x$$

is easily seen to be equivalent to the functional differential equation (of retarded type)

20 $$\dot{x}(t) = G(x,t), \quad t \geq t_1',$$

with initial condition

21 $$x(t) = x^*(t) \quad \text{for} \quad t_1 \leq t \leq t_1',$$

in the sense that (i) $x \in A$ is a solution of Equation (19) if and only if (a) x satisfies (21), (b) x is absolutely continuous on $[t_1',t_2]$, and (c) x satisfies (20) for almost all $t \in [t_1',t_2]$; and (ii) $x \in A$ is a local solution of Eq. (19), satisfying this equation on $[t_1,t']$ for some $t' \in (t_1',t_2]$, if and only if (a) x satisfies (21), (b) x is absolutely continuous on $[t_1',t']$, and (c) x satisfies (20) for almost all $t \in [t_1',t']$.

22 The operators in the second subclass of $\mathscr{V}$ in which we shall be interested are defined in terms of an arbitrary function $x^* \in \mathscr{C}^n(I)$ and an arbitrary function $F:\mathbf{G} \times I \times I \to R^n$ belonging to a certain family $\mathscr{F}$ (where $\mathbf{G}$ is some given open set in R^n). More precisely, if $\mathbf{G}$ is as just indicated, and A is the open set in $\mathscr{C}^n(I)$ defined by

23 $$A = \{x : x \in \mathscr{C}^n(I),\ x(t) \in \mathbf{G} \text{ for all } t \in I\},$$

then we denote by $\mathscr{F}$ the family of all functions $F:\mathbf{G} \times I \times I \to R^n$ with the following properties:

24 (i) For each $s \in I$, the function $(\xi,t) \to F(\xi,s,t)$: $\mathbf{G} \times I \to R^n$ is continuous and continuously differentiable with respect to its first argument.

25 (ii) For each $(\xi,t) \in \mathbf{G} \times I$, the function $s \to F(\xi,s,t)$: $I \to R^n$ is integrable.

26 (iii) For each compact subset $\mathbf{G}_c$ of $\mathbf{G}$, there exists an integrable function $\tilde{m}: I \to R_+$ (possibly depending on $\mathbf{G}_c$ as well as on F) such that [see (I.7.60) and (I.6.4)]

27 $$|F(\xi,s,t)| + |\mathscr{D}_1F(\xi,s,t)| \leq \tilde{m}(s) \quad \text{for all} \quad \xi \in \mathbf{G}_c \quad \text{and} \quad (s,t) \in I \times I.$$

It is easily seen that $\mathscr{F}$ is a linear vector space.

28 For each $x^* \in \mathscr{C}^n(I)$ and $F \in \mathscr{F}$, we define the operator $T_{x^*,F}$: $A \to \mathscr{C}^N(I)$ as follows:

29 $$(T_{x^*,F}x)(t) = x^*(t) + \int_{t_1}^{t} F(x(s),s,t)\, ds, \quad t \in I, \quad x \in A.$$

We show in the Appendix [see Theorems (A.5) and (A.31)] that, for all $x^* \in \mathscr{C}^n(I)$ and $F \in \mathscr{F}$, $T_{x^*,F} \in \mathscr{V} \cap \mathscr{T}^u$ and the Fréchet differential of $T_{x^*,F}$ at any $x \in A$ is given by the formula

30 $$(DT_{x^*,F}(x; \delta x))(t) = \int_{t_1}^{t} \mathscr{D}_1 F(x(s),s,t)\delta x(s)\, ds, \quad t \in I, \quad \delta x \in \mathscr{C}^n(I).$$

Obviously, for every $x^* \in \mathscr{C}^n(I)$ and $F \in \mathscr{F}$, the operator equation

31 $$x = T_{x^*,F}x$$

is equivalent to the Volterra integral equation

32 $$x(t) = x^*(t) + \int_{t_1}^{t} F(x(s),s,t)\, ds, \quad t \geq t_1,$$

in very much the same way as was discussed in (18).

33 Let us denote by $\mathscr{F}_1$ the set of all $F \in \mathscr{F}$ which are independent of their last argument; i.e., $\mathscr{F}_1$ consists of all functions $F: \mathbf{G} \times I \to R^n$ satisfying the following hypotheses (often referred to as Carathéodory-type hypotheses):

34 (i) For each $s \in I$, the function $\xi \to F(\xi,s)$: $\mathbf{G} \to R^n$ is continuously differentiable.

35 (ii) For each $\xi \in \mathbf{G}$, the function $s \to F(\xi,s)$: $I \to R^n$ is integrable.

36 (iii) For each compact subset $\mathbf{G}_c$ of $\mathbf{G}$, there exists an integrable function $\tilde{m}: I \to R_+$ (possibly depending on $\mathbf{G}_c$ as well as on F) such that

37 $$|F(\xi,s)| + |\mathscr{D}_1 F(\xi,s)| \leq \tilde{m}(s) \quad \text{for all} \quad (\xi,s) \in \mathbf{G}_c \times I.$$

It is easily seen that $\mathscr{F}_1$ is a linear subspace of $\mathscr{F}$.

If $x^* \in \mathscr{C}^n(I)$ is constant on I—say $x^*(t) \equiv \xi_0$—and $F \in \mathscr{F}_1$, then Equation (31) is evidently equivalent [in essentially the same way as in (18)] to the ordinary differential equation

$$\dot{x}(t) = F(x(t),t), \quad t \geq t_1,$$

with initial condition

$$x(t_1) = \xi_0.$$

It is also possible to define yet another subclass of $\mathscr{V}$, which includes the two just-described subclasses of $\mathscr{V}$ as subclasses of itself. The operators in this class are defined in terms of an arbitrary function $x^* \in \mathscr{C}^n(I)$ and an arbitrary function $\tilde{G}: A \times I \times I \to R^n$ belonging to a family $\tilde{\mathscr{G}}$ which has many of the attributes of $\mathscr{G}$ and of $\mathscr{F}$. Indeed, for each $x^* \in \mathscr{C}^n(I)$ and $\tilde{G} \in \tilde{\mathscr{G}}$, we define the operator $T_{x^*,G}: A \to \mathscr{C}^n(I)$ by the formula:

$$(T_{x^*,\tilde{G}}x)(t) = x^*(t) + \int_{t_1}^{t} \tilde{G}(x,s,t)\,ds, \quad t \in I, \quad x \in A.$$

The equation $x = T_{x^*,\tilde{G}}x$ may be viewed as a functional Volterra integral equation. If $\tilde{\mathscr{G}}$ is suitably defined (and we shall omit all of the details), then the operators $T_{x^*,\tilde{G}}$, with $x^* \in \mathscr{C}^n(I)$ and $\tilde{G} \in \tilde{\mathscr{G}}$, belong to $\mathscr{V} \cap \mathscr{T}^u$.

We now define the subsets $\mathscr{V}_1$ and $\mathscr{V}_2$ of $\mathscr{V}$ with the aid of $\mathscr{G}$ and $\mathscr{F}$, respectively, as follows:

38 $$\mathscr{V}_1 = \{T_{x^*,G} : x^* \in \mathscr{C}^n(I),\ G \in \mathscr{G}\},$$

39 $$\mathscr{V}_2 = \{T_{x^*,F} : x^* \in \mathscr{C}^n(I),\ F \in \mathscr{F}\}.$$

Further, we denote by $\mathscr{V}_2'$ the following subset of $\mathscr{V}_2$:

40 $$\mathscr{V}_2' = \{T_{x^*,F}: x^* \in \mathscr{C}^n(I),\ x^* \text{ is constant on } I,\ F \in \mathscr{F}_1\}.$$

Inasmuch as $\alpha T_{x_1^*,G_1} + \beta T_{x_2^*,G_2} = T_{\alpha x_1^* + \beta x_2^*, \alpha G_1 + \beta G_2}$ for all $\alpha, \beta \in R$, $x_1^*, x_2^* \in \mathscr{C}^n(I)$, and $G_1, G_2 \in \mathscr{G}$, and $\mathscr{G}$ is a linear vector space, it follows that $\mathscr{V}_1$ is a linear subspace of $\mathscr{V}$. Similarly, $\mathscr{V}_2$ is also a linear subspace of $\mathscr{V}$, and $\mathscr{V}_2'$ is a linear subspace of $\mathscr{V}_2$. We remind the reader that $(\mathscr{V}_1 \cup \mathscr{V}_2) \subset \mathscr{T}^u$, i.e., all the operators in $\mathscr{V}_1$ and $\mathscr{V}_2$ have the uniqueness property [see (7)].

41 Hence, if $\mathscr{W} \subset \mathscr{V}_j$ for $j = 1$ or 2, then Hypothesis (1.21) is satisfied, and, in addition, by virtue of Corollary (A.28) in the Appendix, Hypothesis (1.22)—with $\mathscr{X} = \mathscr{C}^n(I)$—holds.

We now turn to the question of Hypothesis (1.23), and begin with the following definitions.

42 We shall say that a subset $\mathscr{W}_1$ of $\mathscr{T}$ is *uniformally locally compact* [compare with (2b)] if, for each $x_1 \in A$, there is a number $\zeta > 0$ (possibly depending on x_1) such that

$$\{Tx: x \in A,\ \|x - x_1\| \le \zeta,\ T \in \mathscr{W}_1\} \in \mathscr{K}.$$

43 A subset of $\mathscr{T}$ which is uniformally locally compact and at the same time equicontinuous [see (I.5.28)] will be said to be *regular*.

44 Note that (as is easily verified) any finite subset of $\mathscr{V}$ is regular, and that if $\mathscr{W}_1$ and $\mathscr{W}_2$ are regular subsets of $\mathscr{T}$ and $\mathscr{W}_3 \subset \mathscr{W}_1$, then $\mathscr{W}_3$, co $\mathscr{W}_1$, and $(\mathscr{W}_1 \cup \mathscr{W}_2)$ are also regular.

45 By Theorem (A.32) in the Appendix, we conclude that, if $\mathscr{W}_1$ is a regular subset of $\mathscr{T}^u \cap \mathscr{V}$, then $\mathscr{W}_1 \cap \mathscr{T}_1$ is open in the induced topology on $\mathscr{W}_1$ [see (I.3.9)], and that φ_0 is continuous on $\mathscr{W}_1 \cap \mathscr{T}_1$ [see (I.3.24)].

If $\mathscr{W} \subset \mathscr{V}_1$ or $\mathscr{W} \subset \mathscr{V}_2$, then, as we shall show in Lemma (48), Hypothesis (1.23) may be replaced by the following one:

46 HYPOTHESIS. *For every finite subset* $\{T_1, \ldots, T_\rho\}$ *of* $\mathscr{W}$, *there exists a regular subset* $\mathscr{W}_0$ *of* $\mathscr{W}$ *with the following property: For every neighborhood* N_0 *of* 0 *in* $\mathscr{T}$, *there is a continuous function* $\gamma_0\colon \{\lambda\colon \lambda = (\lambda^0, \lambda^1, \ldots, \lambda^\rho) \in \bar{R}_+^{\rho+1}, \sum_0^\rho \lambda^i = 1\} \to N_0$ *such that*

47 $$\sum_{i=0}^{\rho} \lambda^i T_i + \gamma_0(\lambda^0, \lambda^1, \ldots, \lambda^\rho) \in \mathscr{W}_0 \quad \textit{for all} \quad \lambda = (\lambda^0, \lambda^1, \ldots, \lambda^\rho).$$

Clearly, if Hypothesis (46) holds, then co $\mathscr{W} \subset \bar{\mathscr{W}}$, so that $\mathscr{W}$ is then, loosely speaking, "almost" convex. Of course, Hypothesis (46) asserts even more; for example, that any simplex, all of whose vertices belong to $\mathscr{W}$, can be continuously deformed—with the deformation "arbitrarily small"—into a regular subset of $\mathscr{W}$.

48 LEMMA. *Suppose that* $\mathscr{W}$ *is a subset of* $\mathscr{V}$ *such that* co $\mathscr{W} \subset (\mathscr{T}^u \cap \mathscr{V})$ *(as must be the case if* $\mathscr{W} \subset \mathscr{V}_1$ *or if* $\mathscr{W} \subset \mathscr{V}_2$*), and that, for some* $T_0 \in (\mathscr{W} \cap \mathscr{T}_1)$, *Hypothesis* (46) *is satisfied. Then* T_0 *and* $\mathscr{W}$ *also satisfy Hypotheses* (1.21–23) *in such a way that the sets* $\mathscr{W}_0$ *in Hypotheses* (46) *and* (1.23) *coincide.*

Proof. Since $\mathscr{T}^u \subset \mathscr{T}_0$, Hypothesis (1.21) holds. Further, Hypothesis (1.22) follows from Corollary (A.28) and the definition of $\mathscr{V}$.

To show that Hypothesis (1.23) holds, let $\{T_1, \ldots, T_\rho\}$ be an arbitrary finite subset of $\mathscr{W} \subset (\mathscr{T}^u \cap \mathscr{V})$, and let $\mathscr{W}_0$ be the corresponding regular subset of $\mathscr{W}$ with the properties indicated in Hypothesis (46). Since $\mathscr{W}_0 \cup \text{co}\,\{T_0, T_1, \ldots, T_\rho\}$ is a regular subset of $\mathscr{T}^u \cap \mathscr{V}$ [see (44)], there is a neighborhood N_1 of 0 in $\mathscr{T}$ such that [see (45)] $(T_0 + N_1) \cap (\mathscr{W}_0 \cup \text{co}\,\{T_0, \ldots, T_\rho\}) \subset \mathscr{T}_1$. Let N_2 be a neighborhood of 0 in $\mathscr{T}$ such that $N_2 + N_2 \subset N_1$, let $\varepsilon_0 \in (0, 1/\rho)$ be such that $\sum_1^\rho \lambda^i (T_i - T_0) \in N_2$ whenever $0 \le \lambda^i \le \varepsilon_0$ for $i = 1, \ldots, \rho$ [see Corollaries (I.4.3 and 4)], let N be an arbitrary neighborhood of 0 in $\mathscr{T}$, let $N_0 = N_2 \cap N$, and let γ_0 be the corresponding continuous function with the properties indicated in Hypothesis (46). Then, if we define the function $\gamma: \{\lambda: \lambda = (\lambda^1, \ldots, \lambda^\rho) \in R^\rho, 0 \leq \lambda^i \leq \varepsilon_0 \text{ for } i = 1, \ldots, \rho\} \to \mathscr{T}$ by the formula

$$\gamma(\lambda) = \gamma(\lambda^1, \ldots, \lambda^\rho) = \gamma_0\left(1 - \sum_{i=1}^{\rho} \lambda^i, \lambda^1, \ldots, \lambda^\rho\right) \quad \text{for all} \quad \lambda,$$

we at once see that γ is continuous and takes on its values in $N_0 \subset N$, and that (1.25) holds. Further, the set $\mathscr{R}$ defined by (1.24) is contained in $(T_0 + N_1) \cap \text{co}\{T_0, \ldots, T_\rho\}$. Hence, $\mathscr{R} \subset \mathscr{T}_1$, and, since $(\mathscr{R} \cup \mathscr{W}_0)$ is a regular subset of $\mathscr{T}^u \cap \mathscr{V}$, [see (44) and (45)] φ_0 is continuous on $(\mathscr{R} \cup \mathscr{W}_0) \cap \mathscr{T}_1$. Hence, Hypothesis (1.23) is satisfied. |||

49 It is easily seen that Lemma (48) remains in force if we replace Hypothesis (46) by the following weaker one.

50 HYPOTHESIS. *For every finite subset* $\{T_1, \ldots, T_\rho\}$ *of* $\mathscr{W}$, *there exist a regular subset* $\mathscr{W}_0$ *of* $\mathscr{W}$ *and a number* $\varepsilon_1 > 0$ *with the following property*: *For every neighborhood* N_0 *of* 0 *in* $\mathscr{T}$, *there is a continuous function* $\gamma_0: \{\lambda: \lambda = (\lambda^0, \lambda^1, \ldots, \lambda^\rho) \in \bar{R}_+^{\rho+1}, \sum_0^\rho \lambda^i = 1, \lambda^i \leq \varepsilon_1$ *for each* $i = 1, \ldots, \rho\} \to N_0$ *such that* (47) *holds for all such* λ.

51 Loosely speaking, Hypothesis (50)—if it holds for all $T_0 \in \mathscr{W}$—asserts that $\mathscr{W}$ is "almost" finitely open in itself. In fact, if $\mathscr{W}$ is finitely open in itself and $\mathscr{W} \subset \mathscr{V}$, then Hypothesis (50) is satisfied for every $T_0 \in \mathscr{W}$ inasmuch as we can choose $\mathscr{W}_0 = (\text{co}\{T_0, \ldots, T_\rho\}) \cap \mathscr{W}$ and set $\gamma_0 \equiv 0$ (with $\varepsilon_1 > 0$ sufficiently small).

In order to apply Lemma (48), it is evidently of interest to be able to characterize regular subsets of $\mathscr{V}_1$ or of $\mathscr{V}_2$. With this aim in mind, we introduce the following definitions.

52 A subset $\mathscr{G}'$ of $\mathscr{G}$ [see (9)] will be said to be *dominated* if, for each $x_1 \in A$, there exist an integrable function $\tilde{m}: I \to R_+$ and a number $\zeta > 0$ (both possibly depending on x_1) such that (14) holds for all $G \in \mathscr{G}'$.

53 A subset $\mathscr{F}'$ of $\mathscr{F}$ [see (22)] will be said to be *dominated* if (i) for each $s \in I$, the set of functions

54
$$\{(\xi,t) \to F(\xi,s,t): F \in \mathscr{F}'\}$$

from $\mathbf{G} \times I$ into R^n is equicontinuous, and (ii) for each compact subset $\mathbf{G}_c$ of $\mathbf{G}$, there exists an integrable function $\tilde{m}: I \to R_+$ (possibly depending on $\mathbf{G}_c$) such that (27) holds for all $F \in \mathscr{F}'$.

55 Note that if $\mathscr{F}'_1 \subset \mathscr{F}_1$, then $\mathscr{F}'_1$ is dominated if and only if, for each compact subset $\mathbf{G}_c$ of $\mathbf{G}$, there is an integrable function $\tilde{m}: I \to R_+$ such that (37) holds for all $F \in \mathscr{F}'_1$.

The following two lemmas show how dominated subsets of $\mathscr{G}$ and $\mathscr{F}$ lead to regular subsets of $\mathscr{V}_1$ and $\mathscr{V}_2$.

56 LEMMA. *Suppose that, for some compact subset A' of $\mathscr{C}^n(I)$ and some dominated subset $\mathscr{G}'$ of $\mathscr{G}$, $\mathscr{W}_0 \subset \{T_{x^*,G}: x^* \in A',\ G \in \mathscr{G}'\}$. Then $\mathscr{W}_0$ is a regular subset of $\mathscr{V}_1$.*

57 LEMMA. *Suppose that, for some compact subset A' of $\mathscr{C}^n(I)$ and some dominated subset $\mathscr{F}'$ of $\mathscr{F}$, $\mathscr{W}_0 \subset \{T_{x^*,F}: x^* \in A', F \in \mathscr{F}'\}$. Then $\mathscr{W}_0$ is a regular subset of $\mathscr{V}_2$.*

Proof of Lemma (56). By hypothesis, for each $x_1 \in A$, there exist an integrable function $\tilde{m}: I \to R_+$ and a number $\zeta > 0$ such that (14) holds for all $G \in \mathscr{G}'$. Without loss of generality, we shall suppose that $\{x: \|x - x_1\| \le \zeta\} \subset A$ (Recall that A is open.). Therefore, if $T_{x^*,G} \in \mathscr{W}_0$, and if $x \in A$ satisfies $\|x - x_1\| \le \zeta$, then, by virtue of the mean-value theorem (I.7.42),

$$\begin{aligned}|(T_{x^*,G}x)(t) - (T_{x^*,G}x_1)(t)| &= \left|\int_{t_1}^{t} [G(x,s) - G(x_1,s)]\, ds\right| \\ &\le \int_{t_1}^{t} |G(x,s) - G(x_1,s)|\, ds \\ &\le \|x - x_1\| \int_{t_1}^{t_2} \tilde{m}(s)\, ds\end{aligned}$$

for all $t \in I$, which immediately implies that $\mathscr{W}_0$ is equicontinuous. If, in addition, $t_1 \le t' < t'' \le t_2$, then

58 $$\begin{aligned}|(T_{x^*,G}x)(t'') - (T_{x^*,G}x)(t')| &= \left|x^*(t'') - x^*(t') + \int_{t'}^{t''} G(x,s)\, ds\right| \\ &\le |x^*(t'') - x^*(t')| + \int_{t'}^{t''} \tilde{m}(s)\, ds.\end{aligned}$$

Since $x^* \in A'$, where A' is compact, we can now conclude, on the basis of (58) and (1), that the set

59 $$\{Tx: x \in A, \|x - x_1\| \le \zeta, T \in \mathscr{W}_0\}$$

is an equicontinuous set of functions. By the same token, setting $t' = t_1$ in (58), we see that the set (59) is a bounded subset of $\mathscr{C}^n(I)$, so that this set is in $\mathscr{K}$, which means that [see (42)–(44)] $\mathscr{W}_0$ is regular. |||

Proof of Lemma (57). Let $x_1 \in A$ [see (23)]. Since $\{x_1(t): t \in I\}$ is [see (I.3.45) and Theorem (I.3.26)] a compact subset of $\mathbf{G} \subset R^n$, there is a number $\zeta > 0$ such that the compact set $\{\xi: \xi \in R^n, |\xi - x_1(t)| \leq \zeta \text{ for some } t \in I\}$ in R^n, which we shall denote by $\mathbf{G}_c$, is contained in $\mathbf{G}$. (This easily follows from the definition of compactness.) By hypothesis, there is an integrable function $\tilde{m}: I \to R_+$ such that (27) holds for all $F \in \mathscr{F}'$. It follows from this and the mean value Theorem (I.7.42) that, if $\xi \in R^n$ satisfies $|\xi - x_1(s)| \leq \zeta$ for some $s \in I$, then

$$|F(\xi,s,t) - F(x_1(s),s,t)| \leq \tilde{m}(s)|\xi - x_1(s)|$$
$$\text{for all} \quad t \in I \quad \text{and} \quad F \in \mathscr{F}'.$$

Consequently, if $T_{x^*,F} \in \mathscr{W}_0$ and if $x \in A$ is such that $\|x - x_1\| \leq \zeta$, then

$$\begin{aligned} |(T_{x^*,F}x)(t) - (T_{x^*,F}x_1)(t)| &= \left|\int_{t_1}^{t} [F(x(s),s,t) - F(x_1(s),s,t)]\, ds\right| \\ &\leq \int_{t_1}^{t} |F(x(s),s,t) - F(x_1(s),s,t)|\, ds \\ &\leq \int_{t_1}^{t_2} \tilde{m}(s)|x(s) - x_1(s)|\, ds \\ &\leq \|x - x_1\| \int_{t_1}^{t_2} \tilde{m}(s)\, ds \end{aligned}$$

for all $t \in I$, from which it follows at once that $\mathscr{W}_0$ is equicontinuous. If, in addition, $t_1 \leq t' < t'' \leq t_2$, then

$$\begin{aligned} &|(T_{x^*,F}x)(t'') - (T_{x^*,F}x)(t')| \\ &\quad = \left|x^*(t'') - x^*(t') + \int_{t_1}^{t''} F(x(s),s,t'')\, ds - \int_{t_1}^{t'} F(x(s),s,t')\, ds\right| \\ &\quad \leq |x^*(t'') - x^*(t')| + \int_{t_1}^{t'} |F(x(s),s,t'') - F(x(s),s,t')|\, ds \\ &\quad + \int_{t'}^{t''} |F(x(s),s,t'')|\, ds. \end{aligned}$$

Inasmuch as $|F(x(s),s,t'')| \leq \tilde{m}(s)$ for all $(s,t'') \in I^2$ and since $x^* \in A' \in \mathscr{K}$, we shall be able to conclude that the set defined by (59) is here an equicontinuous set of functions once we demonstrate that

$$\int_{t_1}^{t'} |F(x(s),s,t'') - F(x(s),s,t')|\, ds \xrightarrow[t'' \to t'+]{} 0$$

uniformly with respect to $F \in \mathscr{F}'$ and $x \in \{x: x \in A, \|x - x_1\| \leq \zeta\}$.

But, for each $(s,t',t'') \in I^3$, $F_0 \in \mathscr{F}'$, and $x \in A$ satisfying $\|x - x_1\| \leq \zeta$, we have that

60 $$|F_0(x(s),s,t'') - F_0(x(s),s,t')| \leq \sup_{F \in \mathscr{F}'} \sup_{\xi \in \mathbf{G}_c} |F(\xi,s,t'') - F(\xi,s,t')|.$$

Denoting the term in the right-hand side of (60) by $\Delta(s,t',t'')$, it follows at once that

61 $$\sup_{F_0 \in \mathscr{F}'} \sup_{\substack{x \in A \\ \|x - x_1\| \leq \zeta}} \int_{t_1}^{t'} |F_0(x(s),s,t'') - F_0(x(s),s,t')| \, ds \leq \int_{t_1}^{t'} \Delta(s,t',t'') \, ds$$

for all $(t',t'') \in I^2$. Now, by hypothesis, for each $s \in I$, the set of functions (54) is equicontinuous and hence, since $\mathbf{G}_c \times I$ is compact [see (I.3.35)], is uniformly equicontinuous on $\mathbf{G}_c \times I$ [see (I.5.28 and 30) and Theorem (I.5.31)]. Hence, $\Delta(s,t',t'') \to 0$ as $t'' \to t'^+$ for each $s \in I$. Also, $\Delta(s,t',t'') \leq 2\tilde{m}(s)$ for all $(s,t',t'') \in I^3$ because (27) holds for all $F \in \mathscr{F}'$. Consequently, by the Lebesgue dominated convergence theorem, the right-hand and left-hand terms in (61) tend to zero as $t'' \to t'^+$. Hence, the set (59) is equicontinuous.

Since $(T_{x^*,F}x)(t_1) = x^*(t_1)$ for all $x \in A$ whenever $T_{x^*,F} \in \mathscr{V}_2$, and since $A' \in \mathscr{K}$, so that $\{x^*(t): x^* \in A', t \in I\}$ is a bounded set in R^n, we can conclude that the set (59) is bounded in $\mathscr{C}^n(I)$, i.e., belongs to $\mathscr{K}$, which means that $\mathscr{W}_0$ is regular [see (42)–(44)]. |||

The following corollary to Lemma (57) follows at once from the remark in (55).

62 COROLLARY. *Suppose that A' is a compact subset of $\mathscr{C}^n(I)$, and that $\mathscr{F}'_1$ is a subset of $\mathscr{F}_1$ with the following property: For each compact subset $\mathbf{G}_c$ of $\mathbf{G}$, there exists an integrable function $\tilde{m}: I \to R_+$ such that* (37) *holds for all $F \in \mathscr{F}'_1$. Then if $\mathscr{W}_0 \subset \{T_{x^*,F}: x^* \in A', F \in \mathscr{F}'_1\}$, $\mathscr{W}_0$ is a regular subset of $\mathscr{V}_2$.*

63 We remind the reader that the purpose of Lemma (48) was to find conditions for the set $\mathscr{W}$ which are sufficient for Hypotheses (1.21–23) to hold, in order that we may appeal to the theorems of Section 1 of this chapter, and from them obtain necessary conditions that solutions of Problems (1.3, 15, 65, 71, and 94) must satisfy. In order to make use of this lemma, we must still exhibit regular subsets $\mathscr{W}_0$ of $\mathscr{W}$ for each finite subset $\{T_1, \ldots, T_\rho\}$ of $\mathscr{W}$, as is called for in Hypotheses (46) or (50). In Section 5, we shall show that when $\mathscr{W}$ has the so-called property of "convexity under switching", a property that $\mathscr{W}$ does indeed have in many impor-

tant optimal control applications, then, by virtue of Lemmas (56) or (57), the sets $\mathscr{W}_0$ with the required properties do exist.

4. Volterra-Type Operators and Problems with Parameters

In this section, we shall consider the optimization problems with parameters discussed in Section 2. As in the last section, we shall confine ourselves to the case where $\mathscr{X} = \mathscr{C}^n(I)$, and where the operators $\hat{T}^\pi$, for each $(\hat{T},\pi) \in \hat{\mathscr{W}}$, are of Volterra type. As in the preceding section, our aim here is to replace Hypotheses (2.10–12) by alternate hypotheses which are easier to verify.

We shall continue to use the notation introduced in Sections 1–3, with $\mathscr{X} = \mathscr{C}^n(I)$.

1 An operator $\hat{T} \in \hat{\mathscr{T}}$ will be said to be *locally compact* if, for every $\pi_1 \in \Pi$ and each finite subset $\{\pi'_1, \ldots, \pi'_\rho\}$ of $\mathscr{P}$, there is a number $\varepsilon_1 > 0$ (possibly depending on $\pi_1, \pi'_1, \ldots, \pi'_\rho$, as well as on $\hat{T}$) such that

2 $$\left\{\hat{T}^\pi : \pi \in \Pi, \pi = \pi_1 + \sum_{i=1}^{\rho} \lambda^i \pi'_i, 0 \leq \lambda^i \leq \varepsilon_1 \text{ for each } i = 1, \ldots, \rho\right\}$$

is a uniformly locally compact [see (3.42)] subset of $\mathscr{T}$.

3 Since Π is open by hypothesis, we may suppose that $\varepsilon_1 > 0$ in (1) is such that $\pi_1 + \sum_1^\rho \lambda^i \pi'_i \in \Pi$ whenever $0 \leq \lambda^i \leq \varepsilon_1$ for each $i = 1, \ldots, \rho$ [see Corollaries (I.4.3 and 7)]. It then follows at once from Corollary (I.5.32) that the set of operators (2) is equicontinuous, and hence also regular [see (3.43)] (if it is uniformly locally compact).

4 Also note that if $\hat{T} \in \hat{\mathscr{T}}$ is locally compact, then $\hat{T}^\pi \in \mathscr{T}$ (for every $\pi \in \Pi$) is also locally compact [see (3.2)]. Further, a locally compact operator $T \in \mathscr{T}$ is also locally compact when viewed (in the obvious way) as an element of $\hat{\mathscr{T}}$.

5 We shall denote by $\hat{\mathscr{V}}$ the set $\{\hat{T}: \hat{T} \in \hat{\mathscr{T}}, \hat{T}$ locally compact, $\hat{T}^\pi \in \mathscr{V}$ for all $\pi \in \Pi\}$, and by $\hat{\mathscr{C}}^n$ the linear vector space of all continuous functions $\hat{x}^*: I \times \Pi \to R^n$ which are continuously differentiable [see (I.7.52)] with respect to their second argument.

6 In this section, we shall concern ourselves with the case where $\hat{\mathscr{W}} \subset \hat{\mathscr{V}} \times \Pi$. In particular, we shall consider two special subclasses of $\hat{\mathscr{V}}$ which bear a close resemblance to the classes $\mathscr{V}_1$ and $\mathscr{V}_2$ introduced in the preceding section.

7 The operators in the first class are defined in terms of an arbitrary function $\hat{x}^* \in \hat{\mathscr{C}}^n$ and an arbitrary function $\hat{G}: A \times I \times \Pi \to R^n$ belonging to a certain family $\hat{\mathscr{G}}$. Indeed, let $\hat{\mathscr{G}}$ denote the family of all functions $\hat{G}: A \times I \times \Pi \to R^n$ with the following properties:

8 (i) For each $s \in I$, the function $(x,\pi) \to \hat{G}(x,s,\pi)$: $\tilde{A} \to R^n$ is continuously differentiable.

9 (ii) For each $(x,\pi) \in \tilde{A}$, the function $s \to \hat{G}(x,s,\pi)$: $I \to R^n$ is integrable.

10 (iii) $\hat{G}$ is *causal* in the sense that the relations $x_1,x_2 \in A$, $t \in I$, and $x_1(s) = x_2(s)$ for all $s \in [t_1,t]$ imply that $\hat{G}(x_1,t,\pi) = \hat{G}(x_2,t,\pi)$ for every $\pi \in \Pi$.

11 (iv) For each $(x_1,\pi_1) \in \tilde{A}$, there exist an integrable function $\tilde{m}: I \to R_+$ and positive numbers ζ_1, ζ_2—where $\tilde{m}$ and ζ_1 may depend on (x_1,π_1) and $\hat{G}$, but ζ_2 may depend only on π_1 and $\hat{G}$—such that

12 $$|\hat{G}(x,s,\pi)| + \|D_1\hat{G}(x,s,\pi;\cdot)\| + \|D_3\hat{G}(x,s,\pi;\cdot)\| \leq \tilde{m}(s)$$

$$\text{for all} \quad s \in I \quad \text{and all} \quad (x,\pi) \in \tilde{A} \quad \text{such that} \quad \|x - x_1\| \leq \zeta_1 \quad \text{and} \quad \|\pi - \pi_1\| \leq \zeta_2,$$

where $D_j\hat{G}(x,s,\pi;\cdot)$ (for $j = 1$ or 3) denotes the Fréchet partial differential of $\hat{G}$ with respect to its j-th argument at (x,s,π).

13 It is easily seen that $\hat{\mathscr{G}}$ is a linear vector space and that, for each $\hat{G} \in \hat{\mathscr{G}}$ and every $\pi \in \Pi$, the function $(x,s) \to \hat{G}(x,s,\pi): A \times I \to R^n$ is in $\mathscr{G}$ [see (3.9)].

14 For each $\hat{x}^* \in \hat{\mathscr{C}}^n$ and $\hat{G} \in \hat{\mathscr{G}}$, define the operator $\hat{T}_{\hat{x}^*,\hat{G}}: \tilde{A} \to \mathscr{C}^n(I)$ as follows:

15 $$(\hat{T}_{\hat{x}^*,\hat{G}}(x,\pi))(t) = \hat{x}^*(t,\pi) + \int_{t_1}^{t} \hat{G}(x,s,\pi)\, ds, \quad t \in I, \quad (x,\pi) \in \tilde{A}.$$

We show in the Appendix [see Theorem (A.7)] that—for all $\hat{x}^* \in \hat{\mathscr{C}}^n$ and $\hat{G} \in \hat{\mathscr{G}}$—$\hat{T}_{\hat{x}^*,\hat{G}} \in \hat{\mathscr{V}}$ and the Fréchet partial differentials of $\hat{T}_{\hat{x}^*,\hat{G}}$ with respect to its first and second arguments at any $(x,\pi) \in \tilde{A}$ are given by the formulas:

16 $$(D_1\hat{T}_{\hat{x}^*,\hat{G}}(x,\pi;\delta x))(t) = \int_{t_1}^{t} D_1\hat{G}(x,s,\pi;\delta x)\, ds, \quad t \in I, \quad \delta x \in \mathscr{C}^n(I),$$

17 $$(D_2\hat{T}_{\hat{x}^*,\hat{G}}(x,\pi;\delta\pi))(t) = D_2\hat{x}^*(t,\pi;\delta\pi) + \int_{t_1}^{t} D_3\hat{G}(x,s,\pi;\delta\pi)\, ds, \quad t \in I,\ \delta\pi \in \mathscr{P}.$$

18 Inasmuch as $\mathscr{V}_1 \subset (\mathscr{V} \cap \mathscr{T}^u)$ [see (3.15 and 38)], and the function $t \to \hat{x}^*(t,\pi): I \to R^n$ is in $\mathscr{C}^n(I)$ for each $\hat{x}^* \in \hat{\mathscr{C}}^n$ and $\pi \in \Pi$, it follows at once from (13) that $\hat{T}^{\pi}_{\hat{x}^*,\hat{G}} \in \mathscr{V}_1 \subset (\mathscr{V} \cap \mathscr{T}^u)$ and that $(\hat{T}_{\hat{x}^*,\hat{G}},\pi) \in \mathscr{S}_0$ [see (2.3)] for all $\hat{x}^* \in \hat{\mathscr{C}}^n$, $\hat{G} \in \hat{\mathscr{G}}$, and $\pi \in \Pi$.

19 If $\hat{x}^* \in \hat{\mathscr{C}}^n$, $\hat{G} \in \hat{\mathscr{G}}$, and $t_1' \in I$ are such that (i) $\hat{x}^*(t,\pi) = \hat{x}^*(t_1',\pi)$ for all $t \in [t_1',t_2]$ and $\pi \in \Pi$, and (ii) $\hat{G}(x,s,\pi) = 0$ for all $(x,s,\pi) \in A \times [t_1,t_1') \times \Pi$, then the operator equation

20 $$x = \hat{T}_{\hat{x}^*,\hat{G}}(x,\pi)$$

is easily seen to be equivalent to the functional differential equation (of retarded type) with parameter

21 $$\dot{x}(t) = G(x,t,\pi), \quad t \geq t_1',$$

with initial condition

22 $$x(t) = \hat{x}^*(t,\pi) \quad \text{for} \quad t_1 \leq t \leq t_1',$$

in the same sense that Eq. (3.19) was equivalent to Eqs. (3.20 and 21).

23 The operators in the second subclass of $\hat{\mathscr{V}}$ in which we shall be interested are defined in terms of an arbitrary function $\hat{x}^* \in \hat{\mathscr{C}}^n$ and an arbitrary function $\hat{F}:\mathbf{G} \times I \times \Pi \to R^n$ belonging to a certain family $\hat{\mathscr{F}}_1$ (where $\mathbf{G}$ is some given open set in R^n). More precisely, if $\mathbf{G}$ is as just indicated, and A is the open set in $\mathscr{C}^n(I)$ defined by (3.23), then we denote by $\hat{\mathscr{F}}_1$ the family of all functions $\hat{F}:\mathbf{G} \times I \times \Pi \to R^n$ with the following properties:

24 (i) For each $s \in I$, the function $(\xi,\pi) \to \hat{F}(\xi,s,\pi): \mathbf{G} \times \Pi \to R^n$ is continuously differentiable.

25 (ii) For each $(\xi,\pi) \in \mathbf{G} \times \Pi$, the function $s \to \hat{F}(\xi,s,\pi): I \to R^n$ is integrable.

26 (iii) For each compact subset $\mathbf{G}_c$ of $\mathbf{G}$ and each $\pi_1 \in \Pi$, there exist an integrable function $\tilde{m}:I \to R_+$ and a number $\zeta > 0$—where $\tilde{m}$ may depend on $\mathbf{G}_c,\pi_1$, and $\hat{F}$, but ζ may depend only on π_1 and $\hat{F}$—such that

27 $$|\hat{F}(\xi,s,\pi)| + |\mathscr{D}_1\hat{F}(\xi,s,\pi)| + \|D_3\hat{F}(\xi,s,\pi;\cdot)\| \leq \tilde{m}(s)$$
$$\text{for all} \quad (\xi,s) \in \mathbf{G}_c \times I \quad \text{and all} \quad \pi \in \Pi$$
$$\text{such that} \quad \|\pi - \pi_1\| \leq \zeta,$$

where $D_3\hat{F}(\xi,s,\pi;\cdot)$ denotes the Fréchet partial differential of $\hat{F}$ with respect to its third argument at (ξ,s,π) [also see (I.7.60) and (I.6.4)].

28 It is easily seen that $\hat{\mathscr{F}}_1$ is a linear vector space and that, for each $\hat{F} \in \hat{\mathscr{F}}_1$ and every $\pi \in \Pi$, the function $(\xi,s) \to \hat{F}(\xi,s,\pi)$: $\mathbf{G} \times I \to R^n$ is in $\mathscr{F}_1 \subset \mathscr{F}$ [see (3.33)]. (For the sake of brevity, we shall not consider the set which corresponds to $\mathscr{F}$ in the same way that $\hat{\mathscr{F}}_1$ corresponds to $\mathscr{F}_1$. However, the reader who is interested in this case should have little difficulty extending our definitions and results.)

29 For each $\hat{x}^* \in \hat{\mathscr{C}}^n$ and $\hat{F} \in \hat{\mathscr{F}}_1$, define the operators $\hat{T}_{\hat{x}^*,\hat{F}}: \tilde{A} \to \mathscr{C}^n(I)$ as follows:

30 $$(\hat{T}_{\hat{x}^*,\hat{F}}(x,\pi))(t) = \hat{x}^*(t,\pi) + \int_{t_1}^{t} \hat{F}(x(s),s,\pi)\,ds, \quad t \in I, \quad (x,\pi) \in \tilde{A}.$$

We show in the Appendix [see Theorem (A.11)] that—for all $\hat{x}^* \in \hat{\mathscr{C}}^n$ and $\hat{F} \in \hat{\mathscr{F}}_1$—$\hat{T}_{\hat{x}^*,\hat{F}} \in \hat{\mathscr{V}}$ and the Fréchet partial differentials of $\hat{T}_{\hat{x}^*,\hat{F}}$ with respect to its first and second arguments at any $(x,\pi) \in \tilde{A}$ are given by the formulas

31 $$(D_1\hat{T}_{\hat{x}^*,\hat{F}}(x,\pi;\,\delta x))(t) = \int_{t_1}^{t} \mathscr{D}_1\hat{F}(x(s),s,\pi)\delta x(s)\,ds,$$
$$t \in I, \quad \delta x \in \mathscr{C}^n(I),$$

32 $$(D_2\hat{T}_{\hat{x}^*,\hat{F}}(x,\pi;\,\delta\pi))(t) = D_2\hat{x}^*(t,\pi;\,\delta\pi) + \int_{t_1}^{t} D_3\hat{F}(x(s),s,\pi;\,\delta\pi)\,ds,$$
$$t \in I, \quad \delta\pi \in \mathscr{P}.$$

33 Since $\mathscr{V}_2 \subset \mathscr{T}^u$ [see (3.28 and 39)], it follows directly from (28) that $\hat{T}^{\pi}_{\hat{x}^*,\hat{F}} \in \mathscr{V}_2 \subset (\mathscr{V} \cap \mathscr{T}^u)$ and that $(\hat{T}_{\hat{x}^*,\hat{F}},\pi) \in \mathscr{S}_0$ [see (2.3)] for all $\hat{x}^* \in \hat{\mathscr{C}}^n$, $\hat{F} \in \hat{\mathscr{F}}_1$, and $\pi \in \Pi$.

34 If $\hat{x}^* \in \hat{\mathscr{C}}^n$ is independent of t—say $\hat{x}^*(t,\pi) = \xi_0(\pi)$ for all $(t,\pi) \in I \times \Pi$ and for some continuously differentiable function $\xi_0: \Pi \to R^n$—and if $\hat{F} \in \hat{\mathscr{F}}_1$, then the operator equation

$$x = \hat{T}_{\hat{x}^*,\hat{F}}(x,\pi)$$

is evidently equivalent to the ordinary differential equation (with parameter)

$$\dot{x}(t) = \hat{F}(x(t),t,\pi), \quad t \geq t_1,$$

with initial condition (depending on a parameter)

$$x(t_1) = \xi_0(\pi).$$

We now define the subsets $\hat{\mathscr{V}}_1$ and $\hat{\mathscr{V}}_2$ of $\hat{\mathscr{V}}$ with the aid of $\hat{\mathscr{G}}$ and $\hat{\mathscr{F}}_1$, respectively, as follows:

35 $$\hat{\mathscr{V}}_1 = \{\hat{T}_{\hat{x}^*,\hat{G}} : \hat{x}^* \in \hat{\mathscr{C}}^n, \hat{G} \in \hat{\mathscr{G}}\},$$

36 $$\hat{\mathscr{V}}_2 = \{\hat{T}_{\hat{x}^*,\hat{F}} : \hat{x}^* \in \hat{\mathscr{C}}^n, \hat{F} \in \hat{\mathscr{F}}_1\}.$$

37 Because $\hat{\mathscr{G}}$, $\hat{\mathscr{C}}^n$, and $\hat{\mathscr{F}}_1$ are linear vector spaces, it easily follows that $\hat{\mathscr{V}}_1$ and $\hat{\mathscr{V}}_2$ are linear subspaces of $\hat{\mathscr{V}}$. Also, we recall that $\hat{T}^\pi \in (\mathscr{V}_j \cap \mathscr{T}^u)$ whenever $\hat{T} \in \hat{\mathscr{V}}_j$ (for $j = 1$ or 2) and $\pi \in \Pi$, and that $(\hat{\mathscr{V}}_1 \cup \hat{\mathscr{V}}_2) \times \Pi \subset \mathscr{S}_0$ [see (18) and (33)].

38 Hence, if $\mathscr{W} \subset \hat{\mathscr{V}}_j \times \Pi$ for $j = 1$ or 2, then Hypothesis (2.10) holds, and, in addition, by Corollary (A.28), Hypothesis (2.11)—with $\mathscr{X} = \mathscr{C}^n(I)$—holds.

We now turn to Hypothesis (2.12), and begin with the following definition [compare with (3.43) and (1)].

39 A subset $\hat{\mathscr{W}}_1$ of $\hat{\mathscr{T}}$ will be said to be *regular* if it is equicontinuous and if, for every $\pi_1 \in \Pi$ and each finite subset $(\pi'_1, \ldots, \pi'_\rho)$ of $\mathscr{P}$, there is a number $\varepsilon_1 > 0$ (possibly depending on $\pi_1, \pi'_1, \ldots, \pi'_\rho$) such that

40 $$\left\{\hat{T}^\pi : \hat{T} \in \hat{\mathscr{W}}_1, \pi \in \Pi, \pi = \pi_1 + \sum_{i=1}^{\rho} \lambda^i \pi'_i, \; 0 \le \lambda^i \le \varepsilon_1 \text{ for each } i = 1, \ldots, \rho\right\}$$

is a uniformly locally compact [see (3.42)] subset of $\mathscr{T}$.

41 Arguing as in (3), we see that $\varepsilon_1 > 0$ in (39) may be chosen such that $\pi_1 + \sum_1^\rho \lambda^i \pi'_i \in \Pi$ whenever $0 \le \lambda^i \le \varepsilon_1$ for each $i = 1, \ldots, \rho$, and that then the set (40) is equicontinuous and hence also regular (as a subset of $\mathscr{T}$).

42 Note that [compare with (3.44)] any finite subset of $\hat{\mathscr{V}}$ is regular, and that, if $\hat{\mathscr{W}}_1$ and $\hat{\mathscr{W}}_2$ are regular subsets of $\hat{\mathscr{T}}$, and $\hat{\mathscr{W}}_3 \subset \hat{\mathscr{W}}_1$, then $\hat{\mathscr{W}}_3$, (co $\hat{\mathscr{W}}_1$), and $(\hat{\mathscr{W}}_1 \cup \hat{\mathscr{W}}_2)$ are also regular.

If $\mathscr{W} \subset (\hat{\mathscr{V}}_j \times \Pi)$ (for $j = 1$ or 2), then, as we shall show in Lemma (45), Hypothesis (2.12) may be replaced by the following one [compare with Hypothesis (3.50)].

43 HYPOTHESIS. *For every finite subset* $\{(\hat{T}_1,\pi_1), \ldots, (\hat{T}_\rho,\pi_\rho)\}$ *of* $\mathscr{W}$, *there exist a regular subset* $\hat{\mathscr{W}}_0$ *of* $\hat{\mathscr{V}}_j$ *and a number* $\varepsilon' > 0$ *with the following property: For every neighborhood* $\hat{N}_0$ *of* 0 *in* $\hat{\mathscr{T}}$, *there*

is a continuous function $\hat{\gamma}_0: \{\lambda: \lambda = (\lambda^0, \lambda^1, \ldots, \lambda^\rho) \in \bar{R}_+^{\rho+1}, \sum_0^\rho \lambda^i = 1, \lambda^i \leq \varepsilon'$ *for each* $i = 1, \ldots, \rho\} \to \hat{N}_0$ *such that*

44
$$\left(\sum_{i=0}^{\rho} \lambda^i \hat{T}_i + \hat{\gamma}_0(\lambda^0, \lambda^1, \ldots, \lambda^\rho), \sum_{i=0}^{\rho} \lambda^i \pi_i\right) \in \tilde{\mathscr{W}} \cap (\hat{\mathscr{W}}_0 \times \Pi)$$

$$\textit{for all} \quad \lambda = (\lambda^0, \lambda^1, \ldots, \lambda^\rho).$$

The reader should here review the remarks in (3.51) and observe that analogous comments can be made here.

45 LEMMA. *Suppose that* $\tilde{\mathscr{W}}$ *is a subset of* $(\hat{\mathscr{V}}_j \times \Pi)$ $(j = 1$ *or* $2)$ *and that, for some* $(\hat{T}_0, \pi_0) \in (\tilde{\mathscr{W}} \cap \mathscr{S}_1)$, *Hypothesis* (43) *is satisfied. Then* $(\hat{T}_0, \pi_0)$ *and* $\tilde{\mathscr{W}}$ *also satisfy Hypotheses* (2.10–12) *in such a way that the set* $\tilde{\mathscr{W}}_0$ *of Hypothesis* (2.12) *and the set* $\hat{\mathscr{W}}_0$ *of Hypothesis* (43) *satisfy* $\tilde{\mathscr{W}}_0 \subset (\hat{\mathscr{W}}_0 \times \Pi)$.

Proof. As we have already pointed out in (38), Hypotheses (2.10 and 11) are satisfied under the hypotheses of the lemma. To show that Hypothesis (2.12) holds, let $\{(\hat{T}_1, \pi_1), \ldots, (\hat{T}_\rho, \pi_\rho)\}$ be an arbitrary finite subset of $\tilde{\mathscr{W}} \subset (\hat{\mathscr{V}}_j \times \Pi)$, and let $\hat{\mathscr{W}}_0$ and $\varepsilon' > 0$ be the corresponding regular subset of $\hat{\mathscr{V}}_j$ and a number with the properties indicated in Hypothesis (43). Let $\hat{\mathscr{W}}_1 = \hat{\mathscr{W}}_0 \cup (\text{co}\,\{\hat{T}_0, \hat{T}_1, \ldots, \hat{T}_\rho\})$. Since $\hat{\mathscr{W}}_1$ is a regular subset of $\hat{\mathscr{V}}_j$ [see (42)], there is a number $\varepsilon_1 \in (0, \varepsilon')$ such that [see (41)] the set $\Pi_c = \{\pi: \pi = \pi_0 + \sum_1^\rho \lambda^i(\pi_i - \pi_0),\ 0 \leq \lambda^i \leq \varepsilon_1 \text{ for } i = 1, \ldots, \rho\}$ is contained in Π and such that the set $\mathscr{W}_1 = \{\hat{T}^\pi: \hat{T} \in \hat{\mathscr{W}}_1, \pi \in \Pi_c\}$ is a regular subset of $\mathscr{V}_j \subset (\mathscr{T}^u \cap \mathscr{V})$ [see (37)].

Hence [see (3.45)], there is a neighborhood N_1 of 0 in $\mathscr{T}$ such that $(\hat{T}_0^{\pi_0} + N_1) \cap \mathscr{W}_1 \subset \mathscr{T}_1$. Let N_2 be a neighborhood of 0 in $\mathscr{T}$ such that $N_2 + N_2 + N_2 \subset N_1$, and let $\varepsilon_0 \in (0, 1/\rho)$ be such that (i) $\varepsilon_0 < \varepsilon_1$, (ii) $\sum_1^\rho \lambda^i(\hat{T}_i^{\pi_0} - \hat{T}_0^{\pi_0}) \in N_2$ whenever $0 \leq \lambda^i \leq \varepsilon_0$ for $i = 1, \ldots, \rho$, and (iii) $(\hat{T}^\pi - \hat{T}^{\pi_0}) \in N_2$ whenever $\hat{T} \in \hat{\mathscr{W}}_1$ and $\pi = \pi_0 + \sum_1^\rho \lambda^i(\pi_i - \pi_0)$ with $0 \leq \lambda^i \leq \varepsilon_0$ for each i [see Corollaries (I.4.3 and 4) and recall that $\hat{\mathscr{W}}_1$, because it is regular, is an equicontinuous family]. This at once implies that the set $\tilde{\mathscr{R}}$ defined by (2.13) is contained in $\mathscr{S}_1$.

Let $\tilde{\mathscr{W}}_0 = \tilde{\mathscr{W}} \cap (\hat{\mathscr{W}}_0 \times \Pi_c)$. It remains to show that Conditions A and B of Hypothesis (2.12) are satisfied. To verify Condition B, let $\hat{N}$ be an arbitrary neighborhood of 0 in $\hat{\mathscr{T}}$ and let $\hat{N}_2 = \{\hat{T}: \hat{T} \in \hat{\mathscr{T}}, \hat{T}^{\pi_0} \in N_2\}$. Since the map $\hat{T} \to \hat{T}^{\pi_0}: \hat{\mathscr{T}} \to \mathscr{T}$ is evidently continuous, it follows from Theorem (I.3.25B) that $\hat{N}_2$ is a neighborhood of 0 in $\hat{\mathscr{T}}$. Further, let $\hat{N}_0 = \hat{N}_2 \cap \hat{N}$, and let $\hat{\gamma}_0$ be the corresponding function with the properties indicated in Hypothesis (43). Then, if we define the function $\hat{\gamma}$: $\{\lambda: \lambda = (\lambda^1, \ldots, \lambda^\rho) \in R^\rho,\ 0 \leq \lambda^i \leq \varepsilon_0 \text{ for } i = 1, \ldots, \rho\} \to \hat{\mathscr{T}}$ by the formula

$$\hat{\gamma}(\lambda) = \hat{\gamma}(\lambda^1, \ldots, \lambda^\rho) = \hat{\gamma}_0\left(1 - \sum_1^\rho \lambda^i, \lambda^1, \ldots, \lambda^\rho\right) \quad \text{for all} \quad \lambda,$$

we see that $\hat{\gamma}$ is evidently continuous and takes on its values in $\hat{N}_0 \subset \hat{N}$, and that (2.14) holds.

To verify Condition A of Hypothesis (2.12), we use arguments very similar to the preceding ones (we shall omit the details), taking into account [see (3.45)] that φ_0 is continuous on $\mathscr{W}_1 \cap \mathscr{T}_1$. |||

If $\mathscr{W} = \hat{\mathscr{W}} \times \Pi_a$, where $\hat{\mathscr{W}} \subset \hat{\mathscr{V}}_j$ ($j = 1$ or 2) and Π_a is a subset of Π which is finitely open in itself, then Hypothesis (43) can be replaced by the following one [compare with Hypothesis (3.50)].

46 HYPOTHESIS. *For every finite subset* $\{\hat{T}_1, \ldots, \hat{T}_\rho\}$ *of* $\hat{\mathscr{W}}$, *there exists a regular subset* $\hat{\mathscr{W}}_0$ *of* $\hat{\mathscr{W}}$ *and a number* $\varepsilon' > 0$ *with the following property: For every neighborhood* $\hat{N}_0$ *of* 0 *in* $\hat{\mathscr{T}}$, *there is a continuous function* $\hat{\gamma}_0 = \{\lambda: \lambda = (\lambda^0, \lambda^1, \ldots, \lambda^\rho) \in \bar{R}_+^{\rho+1}, \sum_0^\rho \lambda^i = 1, \lambda^i \leq \varepsilon'$ *for each* $i = 1, \ldots, \rho\} \to \hat{N}_0$ *such that*

$$\sum_{i=0}^{\rho} \lambda^i \hat{T}_i + \hat{\gamma}_0(\lambda^0, \lambda^1, \ldots, \lambda^\rho) \in \hat{\mathscr{W}}_0 \quad \textit{for all} \quad \lambda = (\lambda^0, \lambda^1, \ldots, \lambda^\rho).$$

Indeed, we have the following lemma [which is an immediate consequence of Lemma (45)].

47 LEMMA. *Suppose that* $\mathscr{W} = \hat{\mathscr{W}} \times \Pi_a$, *where* $\Pi_a \subset \Pi$ *is finitely open in itself and* $\hat{\mathscr{W}}$ *is a subset of* $\hat{\mathscr{V}}_j$ ($j = 1$ *or* 2), *and also suppose that, for some* $\hat{T}_0 \in \hat{\mathscr{W}}$, *Hypothesis* (46) *is satisfied. Then, for every* $\pi_0 \in \Pi_a$ *such that* $\hat{T}_0^{\pi_0} \in \mathscr{T}_1$, *Hypotheses* (2.10–12) *are also satisfied in such a way that the set* $\mathscr{W}_0$ *of Hypothesis* (2.12) *and the set* $\hat{\mathscr{W}}_0$ *of Hypothesis* (46) *satisfy* $\mathscr{W}_0 \subset (\hat{\mathscr{W}}_0 \times \Pi)$.

In order to apply Lemmas (45) and (47), it is evidently of interest to be able to characterize regular subsets of $\hat{\mathscr{V}}_1$ and $\hat{\mathscr{V}}_2$. With this aim in mind, we introduce the following definitions.

48 A subset $\hat{\mathscr{G}}'$ of $\hat{\mathscr{G}}$ [see (7)] will be said to be *dominated* if, for each $(x_1,\pi_1) \in \tilde{A}$, there exist an integrable function $\tilde{m}: I \to R_+$ and positive numbers ζ_1, ζ_2—where $\tilde{m}$ and ζ_1 may depend on (x_1,π_1) but ζ_2 may depend only on π_1—such that (12) holds for all $\hat{G} \in \hat{\mathscr{G}}'$.

49 A subset $\hat{\mathscr{F}}'_1$ of $\hat{\mathscr{F}}_1$ [see (23)] will be said to be *dominated* if, for each compact subset $\mathbf{G}_c$ of $\mathbf{G}$ and each $\pi_1 \in \Pi$, there exist an integrable function $\tilde{m}: I \to R_+$ and a number $\zeta > 0$—where $\tilde{m}$ may depend on $\mathbf{G}_c$ and π_1 but ζ may depend only on π_1—such that (27) holds for all $\hat{F} \in \hat{\mathscr{F}}'_1$.

50 A subset $\hat{A}'$ of $\mathscr{C}^n$ will be said to be *regular* if it is an equicontinuous family of functions and if, for every $\pi_1 \in \Pi$, there is a number $\zeta > 0$ (possibly depending on π_1) such that $|\hat{x}^*(t_1,\pi_1)| < \zeta$ for all $\hat{x}^* \in \hat{A}'$.

The following two lemmas show how dominated subsets of $\hat{\mathscr{G}}$ and $\hat{\mathscr{F}}_1$ lead to regular subsets of $\hat{\mathscr{V}}_1$ and $\hat{\mathscr{V}}_2$. These lemmas are analogous to Lemmas (3.56 and 57), and may be proved very similarly. We shall omit their proofs for this reason.

51 LEMMA. *Suppose that, for some regular subset $\hat{A}'$ of $\mathscr{C}^n$ and some dominated subset $\hat{\mathscr{G}}'$ of $\hat{\mathscr{G}}$, $\hat{\mathscr{W}}_0 \subset \{\hat{T}_{\hat{x}^*,\hat{G}} : \hat{x}^* \in \hat{A}',\ \hat{G} \in \hat{\mathscr{G}}'\}$. Then $\hat{\mathscr{W}}_0$ is a regular subset of $\hat{\mathscr{V}}_1$.*

52 LEMMA. *Suppose that, for some regular subset $\hat{A}'$ of $\mathscr{C}^n$ and some dominated subset $\hat{\mathscr{F}}'_1$ of $\hat{\mathscr{F}}_1$, $\hat{\mathscr{W}}_0 \subset \{\hat{T}_{\hat{x}^*,\hat{F}} : \hat{x}^* \in \hat{A}',\ \hat{F} \in \hat{\mathscr{F}}'_1\}$. Then $\hat{\mathscr{W}}_0$ is a regular subset of $\hat{\mathscr{V}}_2$.*

To find the regular subsets of $\hat{\mathscr{V}}_1$ or $\hat{\mathscr{V}}_2$ which are called for in Hypotheses (43) and (46), we shall make use of the concept of "convexity under switching," which is the subject of the next section.

5. Convexity under Switching

As we have seen, the theorems of Section 1 of this chapter provide necessary conditions that solutions of Problems (1.3, 15, 65, 71, and 94) must satisfy, so long as the problem data satisfy certain hypotheses. One of these hypotheses is that either $\mathscr{W}$ is finitely open in itself, or that $\mathscr{W}$, together with T_0—where (x_0,T_0) is a solution of the problem—satisfy Hypothesis (1.23). Inasmuch as, in a large number of applications, $\mathscr{W}$ is not finitely open in itself, we must rely on Hypothesis (1.23) being fulfilled. In Section 3, we saw that if $\mathscr{X} = \mathscr{C}^n(I)$, and if the operators in $\mathscr{W}$ are of a special kind, i.e., if $\mathscr{W} \subset \mathscr{V}_1$ or $\mathscr{W} \subset \mathscr{V}_2$, then Hypothesis (1.23) [as well as Hypotheses (1.21 and 22)] are satisfied so long as $\mathscr{W}$ and T_0 satisfy the much simpler Hypothesis (3.50) [see Lemma (3.48) and (3.49)].

In this section, making use of Lemmas (3.56 and 57), we shall show that if $\mathscr{X} = \mathscr{C}^n(I)$, and if $\mathscr{W}$ is contained either in $\mathscr{V}_1$ or in $\mathscr{V}_2$ and, in addition, satisfies a condition which turns out to hold in a large number of important applications, then $\mathscr{W}$ and T_0 satisfy Hypothesis (3.50), and consequently satisfy Hypotheses (1.21–23) as well.

The results which we shall obtain in this section will also be useful for the problems with parameters discussed in Section 2, when we wish to make use of the theorems of that section together with Lemma (4.47).

1 We begin with some mathematical preliminaries. If $\{H_0,H_1,\ldots,H_\rho\}$ is a given finite set of functions from some given compact interval $I = [t_1,t_2]$ into R^{m_0}—where m_0 is some given positive integer—then a function $H: I \to R^{m_0}$ will be said to be a *chattering combination* of $H_0, H_1, \ldots, H_\rho$ if there are mutually disjoint subsets $I_0, I_1, \ldots, I_\rho$ of I, each one a (possible empty) finite union of subintervals of I, with $I = \bigcup_{i=0}^{\rho} I_i$, such that $H(t) = H_i(t)$ for all $t \in I_i$ and each $i = 0, 1, \ldots, \rho$. Loosely speaking, a chattering combination of $H_0, H_1, \ldots, H_\rho$ is a function H which "chatters" among the functions H_i or "switches" from one H_i to another.

2 We shall denote by Ch $\{H_0,H_1,\ldots,H_\rho\}$ the set of all chattering combinations of $H_0, H_1, \ldots, H_\rho$.

The basic result of this section is the following "chattering lemma," which essentially asserts that every element in co $\{H_0,H_1,\ldots,H_\rho\}$ can be approximated arbitrarily closely, in a certain weak sense [see (4)], by an element of Ch $\{H_0,H_1,\ldots,H_\rho\}$.

3 LEMMA. *Let $H_0, H_1, \ldots, H_\rho$ be given integrable functions from I into R^{m_0}, and let $\varepsilon_0 > 0$ be given. Then there exist functions $H_\lambda \in$ Ch $\{H_0,H_1,\ldots,H_\rho\}$, defined for each $\lambda = (\lambda^0, \lambda^1, \ldots, \lambda^\rho) \in \bar{R}_+^{\rho+1}$ with $\sum_0^\rho \lambda^i = 1$, such that*

4
$$\max_{t\in I} \left| \int_{t_1}^{t} \left[H_\lambda(s) - \sum_{i=0}^{\rho} \lambda^i H_i(s) \right] ds \right| < \varepsilon_0$$

for all $\lambda = (\lambda^0, \lambda^1, \ldots, \lambda^\rho)$,

and such that

5
$$\text{meas}\,\{t: H_\lambda(t) \neq H_{\lambda'}(t)\} \xrightarrow[\lambda\to\lambda']{} 0 \quad \textit{for all} \quad \lambda'.$$

Proof. Since H_i is integrable for each i, there exist piecewise-constant functions $\tilde{H}_i : I \to R^{m_0}$ such that

6 $$\int_{t_1}^{t_2} |\tilde{H}_i(s) - H_i(s)| \, ds < \frac{\varepsilon_0}{4(\rho + 1)} \quad \text{for} \quad i = 0, 1, \ldots, \rho.$$

Let

7 $$\hat{H}(s) = \max_{0 \leq i \leq \rho} (|H_i(s)| + |\tilde{H}_i(s)|).$$

Evidently, $\hat{H}$ is integrable and non-negative on I, so that there are points $s_1, \ldots, s_\mu$ in I, with $t_1 = s_1 < s_2 < \cdots < s_\mu = t_2$, such that

8 $$\int_{s_{j-1}}^{s_j} \hat{H}(s) \, ds < \frac{\varepsilon_0}{4} \quad \text{for each} \quad j = 2, \ldots, \mu.$$

Without loss of generality, we shall suppose that all of the points of discontinuity of all of the functions $\tilde{H}_i$ are included among the s_j, so that $\tilde{H}_i$ is constant on (s_{j-1}, s_j) for all i and j.

Let us denote the intervals $[s_{j-1}, s_j)$ by J_j for each $j = 2, \ldots, \mu - 1$, and denote $[s_{\mu-1}, s_\mu]$ by J_μ. For each $\lambda = (\lambda^0, \lambda^1, \ldots, \lambda^\rho) \in \bar{R}_+^{\rho+1}$ with $\sum_0^\rho \lambda^i = 1$ and each $j = 2, \ldots, \mu$, let us partition J_j into $(\rho + 1)$ subintervals $J_{j,0}^\lambda, J_{j,1}^\lambda, \ldots, J_{j,\rho}^\lambda$ such that (i) the $J_{j,i}^\lambda$ are mutually disjoint, (ii) $\bigcup_{i=0}^\rho J_{j,i}^\lambda = J_j$, (iii) $J_{j,i+1}^\lambda$ adjoins $J_{j,i}^\lambda$ on the right for each $i = 0, 1, \ldots, \rho - 1$, and (iv) the measure of $J_{j,i}^\lambda = \lambda^i \cdot$ (measure of J_j) for each i. Let us denote $\bigcup_{j=2}^\mu J_{j,i}^\lambda$ by I_i^λ (for each i and λ), and let us define the function $H_\lambda \in \mathrm{Ch}\,\{H_0, H_1, \ldots, H_\rho\}$ for each λ by the formula

$$H_\lambda(s) = H_i(s) \quad \text{for all} \quad s \in I_i^\lambda \; (i = 0, \ldots, \rho).$$

Evidently,

9 $$|H_\lambda(s)| \leq \hat{H}(s) \quad \text{for all} \quad s \in I \quad \text{and all} \quad \lambda,$$

and (5) is satisfied.

It remains to show that (4) holds. Let us fix $\lambda = (\lambda^0, \ldots, \lambda^\rho)$ and $t \in I$. Thus, $t \in J_{j_0}$ for some $j_0 = 2, \ldots, \mu$, and

10 $$\left| \int_{t_1}^{t} \left[H_\lambda(s) - \sum_{i=0}^{\rho} \lambda^i H_i(s) \right] ds \right| \leq \sum_{j=2}^{j_0} \left| \int_{s_{j-1}}^{s_j} \left(H_\lambda - \sum_{i=0}^{\rho} \lambda^i H_i \right) ds \right| + \int_t^{s_{j_0}} \left| H_\lambda - \sum_{i=0}^{\rho} \lambda^i H_i \right| ds.$$

But [see (7)–(9)]

11
$$\int_t^{s_{j_0}} \left| H_\lambda - \sum_{i=0}^{\rho} \lambda^i H_i \right| ds \leq \int_t^{s_{j_0}} \left[\hat{H} + \sum_{i=0}^{\rho} \lambda^i \hat{H} \right] ds$$
$$\leq 2 \int_{s_{j_0 - 1}}^{s_{j_0}} \hat{H}(s)\, ds < \frac{\varepsilon_0}{2}.$$

Further, for each $j \geq 2$,

$$\left| \int_{s_{j-1}}^{s_j} \left(H_\lambda - \sum_{i=0}^{\rho} \lambda^i H_i \right) ds \right| = \left| \sum_{i=0}^{\rho} \left(\int_{J_{j,i}^{\lambda}} H_i\, ds - \lambda^i \int_{J_j} H_i\, ds \right) \right|$$
$$\leq \sum_{i=0}^{\rho} \int_{J_{j,i}^{\lambda}} |H_i - \tilde{H}_i|\, ds$$
$$+ \left| \sum_{i=0}^{\rho} \left(\int_{J_{j,i}^{\lambda}} \tilde{H}_i\, ds - \lambda^i \int_{J_j} \tilde{H}_i\, ds \right) \right|$$
$$+ \sum_{i=0}^{\rho} \lambda^i \int_{J_j} |\tilde{H}_i - H_i|\, ds.$$

But $\tilde{H}_i$ is constant on (s_{j-1}, s_j) and meas $(J_{j,i}^{\lambda}) = \lambda^i \cdot$ meas (J_j) for each i and j, so that

$$\int_{J_{j,i}^{\lambda}} \tilde{H}_i\, ds = \lambda^i \int_{J_j} \tilde{H}_i\, ds \quad \text{for all } i \text{ and } j.$$

Hence, since $0 \leq \lambda^i \leq 1$ for each i, we have, by virtue of (6), that,

12
$$\sum_{j=2}^{j_0} \left| \int_{s_{j-1}}^{s_j} \left(H_\lambda - \sum_{i=0}^{\rho} \lambda^i H_i \right) ds \right| \leq \sum_{j=2}^{j_0} \sum_{i=0}^{\rho} (1 + \lambda^i) \int_{J_j} |\tilde{H}_i - H_i|\, ds$$
$$\leq \sum_{i=0}^{\rho} 2 \int_{t_1}^{s_{j_0}} |\tilde{H}_i - H_i|\, ds$$
$$< 2(\rho + 1) \frac{\varepsilon_0}{4(\rho + 1)} = \frac{\varepsilon_0}{2}.$$

Consequently, combining (10)–(12), we conclude that

$$\left| \int_{t_1}^{t} \left[H_\lambda(s) - \sum_{i=0}^{\rho} \lambda^i H_i(s) \right] ds \right| < \varepsilon_0.$$

Since $t \in I$ and λ were arbitrarily chosen, (4) holds. |||

In order to make use of the chattering lemma, we must extend the concept of chattering combinations and introduce the concept of convexity under switching.

13 Let **A** be an arbitrary set, and let $\{\mathbf{F}_0, \mathbf{F}_1, \ldots, \mathbf{F}_\rho\}$ be some given collection of functions from $\mathbf{A} \times I$ into R^{m_0} (for some given

positive integer m_0). Then a function $\mathbf{F}:\mathbf{A} \times I \to R^{m_0}$ will be said to be a *chattering combination* of $\mathbf{F}_0, \mathbf{F}_1, \dots, \mathbf{F}_\rho$ if there are mutually disjoint subsets $I_0, I_1, \dots, I_\rho$ of I, each one a (possibly empty) finite union of subintervals of I, with $I = \bigcup_{i=0}^{\rho} I_i$, such that $\mathbf{F}(\mathbf{x},t) = \mathbf{F}_i(\mathbf{x},t)$ for all $(\mathbf{x},t) \in \mathbf{A} \times I_i$ and each $i = 0, 1, \dots, \rho$. We shall again denote by Ch $\{\mathbf{F}_0,\mathbf{F}_1, \dots ,\mathbf{F}_\rho\}$ the set of all chattering combinations of $\mathbf{F}_0,\mathbf{F}_1, \dots ,\mathbf{F}_\rho$.

14 A set $\mathbf{\Gamma}$ of functions from $\mathbf{A} \times I$ into R^{m_0} will be said to be *convex under switching* if Ch $\{\mathbf{F}_0,\mathbf{F}_1, \dots ,\mathbf{F}_\rho\} \subset \mathbf{\Gamma}$ for every finite subset $\{\mathbf{F}_0,\mathbf{F}_1, \dots ,\mathbf{F}_\rho\}$ of $\mathbf{\Gamma}$.

We point out that the concept of convexity under switching obviously also applies to a class of functions from I into R^{m_0}. Further, convexity under switching can be defined also (in an identical manner to that indicated previously) if the basic interval is open (and possibly unbounded) instead of compact.

15 If, in (13), we replace the words "finite union of subintervals of I" by "measurable subset of I," then we shall say that $\mathbf{F}$ is a *strong chattering combination* of $\mathbf{F}_0, \mathbf{F}_1, \dots, \mathbf{F}_\rho$, and shall use the notation **Ch** $\{\mathbf{F}_0,\mathbf{F}_1, \dots ,\mathbf{F}_\rho\}$ to denote the set of all strong chattering combinations of $\mathbf{F}_0, \dots, \mathbf{F}_\rho$. A set $\mathbf{\Gamma}$ of functions from $\mathbf{A} \times I \to R^{m_0}$ will be said to be *strongly convex under switching* if **Ch** $\{\mathbf{F}_0,\mathbf{F}_1, \dots ,\mathbf{F}_\rho\} \subset \mathbf{\Gamma}$ for every finite subset $\{\mathbf{F}_0,\mathbf{F}_1, \dots ,\mathbf{F}_\rho\}$ of $\mathbf{\Gamma}$. Clearly, a set which is strongly convex under switching is also convex under switching.

Let us first consider the problems of Section 1 where $\mathscr{X} = \mathscr{C}^n(I)$ and $\mathscr{W} \subset \mathscr{V}_1$. The following lemma is immediate.

16 LEMMA. *If* $\{G_0,G_1, \dots ,G_\rho\}$ *is any finite subset of* $\mathscr{G}$ [see (3.9)], *then* Ch $\{G_0,G_1, \dots ,G_\rho\}$ *is a dominated subset of* $\mathscr{G}$ [*see* (3.52)].

Our first important result is contained in the following theorem, which asserts, roughly speaking, that if $\mathscr{W}$ is a subset of $\mathscr{V}_1$ which is defined in terms of a subset of $\mathscr{G}$ which is convex under switching and in terms of a subset of $\mathscr{C}^n(I)$ which is finitely open in itself, then $\mathscr{W}$ is "almost" finitely open in itself, in the sense that Hypothesis (3.50) holds [see (3.51)].

17 THEOREM. *Suppose that, for some subset A' of $\mathscr{C}^n(I)$ which is finitely open in itself and some subset $\mathscr{G}'$ of $\mathscr{G}$ which is convex under switching,* $\mathscr{W} = \{T_{x^*,G}: x^* \in A',\ G \in \mathscr{G}'\}$. *Then, for every* $T_0 \in \mathscr{W}$, *Hypothesis* (3.50) *is satisfied.*

Proof. Let $T_0 \in \mathscr{W}$ and $\{T_1, \dots ,T_\rho\} \subset \mathscr{W}$ be arbitrary but fixed. Let $T_i = T_{x_i^*,G_i}$, where $x_i^* \in A'$ and $G_i \in \mathscr{G}'$, for each

$i = 0, \ldots, \rho$. Since A' is finitely open in itself, there is a number $\varepsilon_1 > 0$ such that the set

18 $$A_0 = \left\{x^*: x^* = x_0^* + \sum_{i=1}^{\rho} \lambda^i(x_i^* - x_0^*),\ 0 \le \lambda^i \le \varepsilon_1 \text{ for each } i = 1, \ldots, \rho\right\},$$

which is compact in $\mathscr{C}^n(I)$, is contained in A'. Let

19 $$\mathscr{W}_0 = \{T_{x^*,G}: x^* \in A_0,\ G \in \mathrm{Ch}\,\{G_0, G_1, \ldots, G_\rho\}\}$$

It follows from our hypotheses and from Lemmas (16) and (3.56) that $\mathscr{W}_0$ is a regular subset of $\mathscr{W}$.

Now let N_0 be any neighborhood of 0 in $\mathscr{T}$, and let $x_1, \ldots, x_\nu \in A$ and $\varepsilon_0 > 0$ be such that

20 $$N_0 \supset \left\{T: T \in \mathscr{T},\ \max_{1 \le j \le \nu} \|Tx_j\| < \varepsilon_0\right\}.$$

By definition of $\mathscr{G}$, the functions $t \to G_i(x_j,t): I \to R^n$ are integrable for each i and j. For each $i = 0, \ldots, \rho$, let H_i be the function defined on I by the formula

$$H_i(t) = (G_i(x_1,t), \ldots, G_i(x_\nu,t)), \quad t \in I,$$

so that H_i is an integrable function from I into $R^{\nu n}$ for each i. By Lemma (3), there exist functions $H_\lambda \in \mathrm{Ch}\,\{H_0, H_1, \ldots, H_\rho\}$, defined for each $\lambda = (\lambda^0, \lambda^1, \ldots, \lambda^\rho) \in \bar{R}_+^{\rho+1}$ with $\sum_0^\rho \lambda^i = 1$, such that (4) and (5) hold. If, for each λ, we define G_λ as a chattering combination of the G_i in the same way that H_λ is a chattering combination of the H_i, so that $G_\lambda \in \mathrm{Ch}\,\{G_0, G_1, \ldots, G_\rho\}$, then, by virtue of (4),

21 $$\max_{t \in I} \left| \int_{t_1}^{t} [G_\lambda(x_j,s) - \sum_{i=0}^{\rho} \lambda^i G_i(x_j,s)]\, ds \right| < \varepsilon_0 \quad \text{for } j = 1, \ldots, \nu \text{ and all } \lambda,$$

and, by (5),

22 $$\mathrm{meas}\,\{t: G_\lambda(x,t) \ne G_{\lambda'}(x,t)\} \xrightarrow[\lambda \to \lambda']{} 0 \quad \text{for every } x \in A \text{ and } \lambda'.$$

For each $\lambda = (\lambda^0, \ldots, \lambda^\rho) \in R^{\rho+1}$, let

23 $$x_\lambda^* = x_0^* + \sum_{i=1}^{\rho} \lambda^i(x_i^* - x_0^*).$$

Finally, for each λ belonging to the set

24 $$\Lambda^* = \left\{\lambda\colon \lambda = (\lambda^0, \lambda^1, \ldots, \lambda^\rho) \in \bar{R}_+^{\rho+1}, \sum_0^\rho \lambda^i = 1, \lambda^i \le \varepsilon_1 \text{ for each } i = 1, \ldots, \rho\right\},$$

let $T_\lambda = T_{x_\lambda^*, G_\lambda}$. Evidently, $T_\lambda \in \mathscr{W}_0$ for every $\lambda \in \Lambda^*$ [see (18) and (19)], and it follows at once from (22), (23), and Lemma (16) that the map $\lambda \to T_\lambda\colon \Lambda^* \to \mathscr{T}$ is continuous. Let us denote by γ_0 the map $\lambda \to T_\lambda - \sum_0^\rho \lambda^i T_i = T_{0, G_\lambda - \sum_0^\rho \lambda^i G_i}\colon \Lambda^* \to \mathscr{T}$, so that γ_0 is continuous and (3.47) holds, and [see (20) and (21)] γ_0 takes on its values in N_0. |||

25 Note that, if A' is convex, then we may replace in the preceding proof the set A_0 in (18) by co $\{x_0^*, x_1^*, \ldots, x_\rho^*\}$, and Λ^* by $\{\lambda\colon \lambda = (\lambda^0, \lambda^1, \ldots, \lambda^\rho) \in \bar{R}_+^{\rho+1}, \sum_0^\rho \lambda^i = 1\}$, and thereby conclude that Hypothesis (3.46) is satisfied.

In almost the same way that we proved Theorem (17), we may prove the following analogous theorem for problems with parameters.

26 THEOREM. *Suppose that, for some subset $\hat{A}'$ of $\hat{\mathscr{C}}^n$ which is finitely open in itself and some subset $\hat{\mathscr{G}}'$ of $\hat{\mathscr{G}}$ which is convex under switching, $\hat{\mathscr{W}} = \{\hat{T}_{\hat{x}^*, \hat{G}}\colon \hat{x}^* \in \hat{A}', \hat{G} \in \hat{\mathscr{G}}'\}$. Then, for every $\hat{T}_0 \in \hat{\mathscr{W}}$, Hypothesis (4.46) is satisfied.*

We now turn to the problems of Section 1 where $\mathscr{X} = \mathscr{C}^n(I)$ and $\mathscr{W} \subset \mathscr{V}_2$. [In the sequel, a chattering combination of elements $F_0, F_1, \ldots, F_\rho \in \mathscr{F}$ is to be understood with $\mathbf{A} = \mathbf{G} \times I$ and with $\mathbf{x} = (\xi, t)$, where ξ is the first argument of F and t is the third.] The following lemma is immediate [compare with Lemma (16)].

27 LEMMA. *If $\{F_0, F_1, \ldots, F_\rho\}$ is any finite subset of $\mathscr{F}$ [see (3.22)], then* Ch $\{F_0, F_1, \ldots, F_\rho\}$ *is a dominated subset of $\mathscr{F}$ [see (3.53)].*

Our next result, which is analogous to Theorem (17), asserts, roughly speaking, that, if $\mathscr{W}$ is a subset of $\mathscr{V}_2$ which is defined in terms of a subset of $\mathscr{F}$ which is convex under switching and in terms of a subset of $\mathscr{C}^n(I)$ which is finitely open in itself, then $\mathscr{W}$ is "almost" finitely open in itself, in the sense that Hypothesis (3.50) holds [see (3.51)].

28 THEOREM. *Suppose that, for some subset A' of $\mathscr{C}^n(I)$ which is finitely open in itself and some subset $\mathscr{F}'$ of $\mathscr{F}$ which is convex*

under switching, $\mathscr{W} = \{T_{x^*,F}: x^* \in A', F \in \mathscr{F}'\}$. *Then, for every* $T_0 \in \mathscr{W}$, *Hypothesis* (3.50) *is satisfied.*

Proof. Let $T_0 \in \mathscr{W}$ and $\{T_1, \ldots, T_\rho\} \subset \mathscr{W}$ be arbitrary but fixed. Let $T_i = T_{x_i^*,F_i}$, where $x_i^* \in A'$ and $F_i \in \mathscr{F}'$, for each $i = 0, \ldots, \rho$. Since A' is finitely open in itself, there is a number $\varepsilon_1 > 0$ such that the set A_0 defined by (18), which is compact in $\mathscr{C}^n(I)$, is contained in A'. Let

29 $$\mathscr{W}_0 = \{T_{x^*,F}: x^* \in A_0, F \in \text{Ch}\{F_0, F_1, \ldots, F_\rho\}\}.$$

It follows from our hypotheses and from Lemmas (27) and (3.57) that $\mathscr{W}_0$ is a regular subset of $\mathscr{W}$.

Now let N_0 be any neighborhood of 0 in $\mathscr{T}$, and let $x_1, \ldots, x_\nu \in A$ and $\varepsilon_0 > 0$ be such that (20) holds. Let $\mathscr{W}_1 = \mathscr{W}_0 \cup (\text{co}\{T_0, T_1, \ldots, T_\rho\})$. By (3.44), $\mathscr{W}_1$ is regular and hence uniformly locally compact [see (3.42 and 43)], which means that the sets $\{Tx_j: T \in \mathscr{W}_1\}$, for each $j = 1, \ldots, \nu$, belong to $\mathscr{K}$ [see (3.1)]. Hence [see (I.3.15)], $\{Tx_j: T \in \mathscr{W}_1, j = 1, \ldots, \nu\} \in \mathscr{K}$, so that there exists a $\delta_0 > 0$ such that [see Theorem (I.5.31)]

$$|(Tx_j)(t') - (Tx_j)(t'')| < \frac{\varepsilon_0}{3} \quad \text{for all} \quad T \in \mathscr{W}_1 \quad \text{and} \quad j = 1, \ldots, \nu$$

whenever $t', t'' \in I$ are such that $|t' - t''| < \delta_0$. Let $\{s_1, \ldots, s_\mu\} \subset I$ be such that $t_1 = s_1 < s_2 < \cdots < s_\mu = t_2$ and such that $s_k - s_{k-1} < \delta_0$ for each $k = 2, \ldots, \mu$. Thus, in particular,

30 $$|(Tx_j)(s) - (Tx_j)(s_k)| < \frac{\varepsilon_0}{3} \quad \text{for all} \quad T \in \mathscr{W}_1 \quad \text{and} \quad j = 1, \ldots, \nu$$
$$\text{whenever} \quad s \in [s_{k-1}, s_k] \quad \text{for some} \quad k = 2, \ldots, \mu.$$

It follows at once from the definition of $\mathscr{F}$ [see (3.24 and 25)] that, for each i, j, and k, the function $s \to F_i(x_j(s), s, s_k): I \to R^n$ is integrable. For each $i = 0, \ldots, \rho$, let H_i be the function defined on I by the formula

$$\begin{aligned} H_i(s) = (&F_i(x_1(s), s, s_1), \ldots, F_i(x_1(s), s, s_\mu), \\ &F_i(x_2(s), s, s_1), \ldots, F_i(x_2(s), s, s_\mu), \\ &\ldots, F_i(x_\nu(s), s, s_1), \ldots, F_i(x_\nu(s), s, s_\mu)), \; s \in I, \end{aligned}$$

so that H_i is an integrable function from I into $R^{\mu\nu n}$ for each i. By Lemma (3), there exist functions $H_\lambda \in \text{Ch}\{H_0, H_1, \ldots, H_\rho\}$, defined for each $\lambda = (\lambda^0, \lambda^1, \ldots, \lambda^\rho) \in \bar{R}_+^{\rho+1}$ with $\sum_0^\rho \lambda^i = 1$, such that (4) and (5) hold, but with ε_0 replaced by $(\varepsilon_0/3)$ in (4). If, for each λ, we define F_λ as a chattering combination of the F_i in the same way that H_λ is a chattering combination of the H_i, so that $F_\lambda \in \text{Ch}\,\{F_0, F_1, \ldots, F_\rho\}$, then, by virtue of (4),

31
$$\left| \int_{t_1}^{s_k} \left[F_\lambda(x_j(s),s,s_k) - \sum_{i=0}^{\rho} \lambda^i F_i(x_j(s),s,s_k)\right] ds \right| < \frac{\varepsilon_0}{3}$$

for each $j = 1, \ldots, \nu$ and $k = 1, \ldots, \mu$ and every λ,

and, for every λ', by virtue of (5),

32
$$\text{meas}\,\{s\colon F_\lambda(x(s),s,t) \neq F_{\lambda'}(x(s),s,t)\} \xrightarrow[\lambda \to \lambda']{} 0$$

for all $x \in A$ and $t \in I$

and uniformly with respect to such (x,t).

For each $\lambda \in \Lambda^*$, where Λ^* is defined by (24), let x_λ^* be defined by (23), and let $T_\lambda = T_{x_\lambda^*, F_\lambda}$. Evidently, $T_\lambda \in \mathscr{W}_0$ for every $\lambda \in \Lambda^*$ [see (18) and (29)], and it follows from (23), (32), and Lemma (27) that the map $\lambda \to T_\lambda\colon \Lambda^* \to \mathscr{T}$ is continuous. Let us denote by γ_0 the map $\lambda \to T_\lambda - \sum_0^\rho \lambda^i T_i = T_{0,F_\lambda - \sum_0^\rho \lambda^i F_i}\colon \Lambda^* \to \mathscr{T}$, so that γ_0 is continuous and (3.47) holds and [see (20), (30), and (31)] γ_0 takes on its values in N_0. |||

The remarks in (25) apply here also.

In almost the same way that we proved Theorem (28), we may prove the following analogous theorem for problems with parameters.

33 THEOREM. *Suppose that, for some subset $\hat{A}'$ of $\hat{\mathscr{C}}^n$ which is finitely open in itself and some subset $\hat{\mathscr{F}}'$ of $\hat{\mathscr{F}}_1$ which is convex under switching, $\hat{\mathscr{W}} = \{\hat{T}_{\hat{x}^*,\hat{F}}\colon \hat{x}^* \in \hat{A}', \hat{F} \in \hat{\mathscr{F}}'\}$. Then, for every $\hat{T}_0 \in \hat{\mathscr{W}}$, Hypothesis (4.46) is satisfied.*

Combining Theorems (17) and (28), Lemma (3.48), and (3.49), we arrive at the following result.

34 THEOREM. *Suppose that either*

(A) *$\mathscr{W} = \{T_{x^*,G}\colon x^* \in A', G \in \mathscr{G}'\}$, where A' is a subset of $\mathscr{C}^n(I)$ which is finitely open in itself and $\mathscr{G}'$ is a subset of $\mathscr{G}$ which is convex under switching, or*

(B) *$\mathscr{W} = \{T_{x^*,F}\colon x^* \in A', F \in \mathscr{F}'\}$, where A' is as in* (A) *and $\mathscr{F}'$ is a subset of $\mathscr{F}$ which is convex under switching.*

Then, for every $T_0 \in \mathscr{W} \cap \mathscr{T}_1$, Hypotheses (1.21–23) *are satisfied in such a way that if $\mathscr{W}$ is of the form of* (A) [*respectively,* (B)], *then the set* $\mathscr{W}_0$ *in Hypothesis* (1.23)—*for any finite subset* $\{T_1,\ldots,T_\rho\}$ *of* $\mathscr{W}$—*is given by* (19) [*respectively*, (29)], *with* A_0 *given by* (18) *for some* $\varepsilon_1 > 0$, *where* $T_i = T_{x_i^*,G_i}$ (*respectively*, $T_i = T_{x_i^*,F_i}$) *for each* $i = 0,\ldots,\rho$.

Combining Theorems (26) and (33) and Lemma (4.47), we arrive at the following result.

35 THEOREM. *Suppose that* $\tilde{\mathscr{W}} = \hat{\mathscr{W}} \times \Pi_a$, *where* $\Pi_a \subset \Pi$ *is finitely open in itself, and either*

(A) $\hat{\mathscr{W}} = \{\hat{T}_{\hat{x}^*,\hat{G}}: \hat{x}^* \in \hat{A}', \hat{G} \in \hat{\mathscr{G}}'\}$, *where* $\hat{A}'$ *is some subset of* $\mathscr{C}^n$ *which is finitely open in itself and* $\hat{\mathscr{G}}'$ *is a subset of* $\hat{\mathscr{G}}$ *which is convex under switching, or*

(B) $\hat{\mathscr{W}} = \{\hat{T}_{\hat{x}^*,\hat{F}}: \hat{x}^* \in \hat{A}', \hat{F} \in \hat{\mathscr{F}}'\}$, *where* $\hat{A}'$ *is as in* (A), *and* $\hat{\mathscr{F}}'$ *is a subset of* $\hat{\mathscr{F}}_1$ *which is convex under switching.*

Then, for every $(\hat{T}_0,\pi_0) \in \tilde{\mathscr{W}} \cap \mathscr{S}_1$, *Hypotheses* (2.10–12) *are satisfied in such a way that if* $\hat{\mathscr{W}}$ *is of the form of* (A) [*respectively*, (B)], *then the set* $\tilde{\mathscr{W}}_0$ *in Hypothesis* (2.12)—*for any finite subset* $\{(\hat{T}_1,\pi_1),\ldots,(\hat{T}_\rho,\pi_\rho)\}$ *of* $\tilde{\mathscr{W}}$—*is contained in the set*

36 $$\{\hat{T}_{\hat{x}^*,\hat{G}}: \hat{x}^* \in \operatorname{co}\{\hat{x}_0^*,\ldots,\hat{x}_\rho^*\}, \quad \hat{G} \in \operatorname{Ch}\{\hat{G}_0,\hat{G}_1,\ldots,\hat{G}_\rho\}\} \times \Pi,$$

where $\hat{T}_i = \hat{T}_{\hat{x}_i^*,\hat{G}_i}$ *for each* $i = 0,\ldots,\rho$ (*respectively, in the set*

37 $$\{\hat{T}_{\hat{x}^*,\hat{F}}: \hat{x}^* \in \operatorname{co}\{\hat{x}_0^*,\ldots,\hat{x}_\rho^*\}, \quad \hat{F} \in \operatorname{Ch}\{\hat{F}_0,\hat{F}_1,\ldots,\hat{F}_\rho\} \times \Pi,$$

where $\hat{T}_i = \hat{T}_{\hat{x}_i^*,\hat{F}_i}$ *for each* $i = 0,\ldots,\rho$).

In the succeeding chapters, we shall exploit Theorems (34) and (35) in conjunction with the necessary conditions of Sections 1 and 2.

6. Sufficient Conditions for Optimization Problems with Operator Equation Constraints

In this, the last section of the Chapter, we shall apply some of the sufficiency results developed in Section 5 of Chapter III to Problems (1.3, 15, 65, and 71). Our results will generalize many of those obtained in Section 4 of Chapter II.

We begin with Problem (1.3). Let us suppose that the data of this problem satisfy the following hypotheses [compare with Hypotheses (II.4.1–3)].

1 HYPOTHESIS. *The sets* A *and* A' *both coincide with* $\mathscr{X}$.

2 HYPOTHESIS. *The function* $\hat{\varphi}$ *is affine.*

3 HYPOTHESIS. *The functional* $\hat{\phi}^0$ *is convex.*

4 HYPOTHESIS. *There is a closed, convex cone* $W_1 \subset \mathscr{Z}_1$, *such that* $(\operatorname{int} Z_1) + W_1 = \operatorname{int} Z_1$ *and such that* $\hat{\phi}_1$ *is* W_1*-convex. Further,* $\hat{\phi}_1$ *is Gâteaux directionally differentiable* [*see* (I.7.3)].

5 HYPOTHESIS. *The set* $\mathscr{W}$ *consists of all operators* $T:\mathscr{X} \to \mathscr{X}$ *of the form*

6 $$Tx = Lx + \tilde{x}, \quad x \in \mathscr{X},$$

where L is some fixed operator in $\mathbf{B}(\mathscr{X},\mathscr{X})$ [*see* (I.5.1)], *and* $\tilde{x}$ *belongs to some given subset* $\tilde{\mathscr{X}}$ *of* $\mathscr{X}$. *We shall also suppose that L is such that* $(E - L)$*—where E denotes the identity operator on* $\mathscr{X}$*—is a (linear) homeomorphism of* $\mathscr{X}$ *onto itself.*

7 If we identify Problem (1.3), subject to Hypotheses (1)—(5), as a special case of Problem (III.1.2) in the manner indicated in (1.8), then $\mathscr{E} = \mathscr{W}$ (since, evidently, $\mathscr{W} \subset \mathscr{T}_1$), and

8 $$\varphi_0(T) = (E - L)^{-1}T(0) \quad \text{for all} \quad T \in \mathscr{W}.$$

Let us extend φ_0 to all of $\mathscr{T}$ by formula (8) [so that then φ_0 is evidently a linear map from $\mathscr{T}$ into $\mathscr{X}$], extend the maps φ, ϕ^0, and ϕ_1 to $\mathscr{T}$ by keeping the formulas $\varphi = \hat{\varphi} \circ \varphi_0$, $\phi^0 = \hat{\phi}^0 \circ \varphi_0$, and $\phi_1 = \hat{\phi}_1 \circ \varphi_0$, and redefine $\mathscr{E}'$ as $\mathscr{T}$. Then it follows from (I.2.17, 23, and 34) that φ, ϕ^0, and ϕ_1 are, respectively, affine, convex, and W_1-convex. Further, it follows from Hypotheses (3) and (4) and Lemma (I.7.7) that ϕ^0 and ϕ_1 are Gâteaux directionally differentiable. If we now appeal to Theorem (III.5.14), we immediately arrive at the following result [compare with Theorem (II.4.6)].

9 THEOREM. *Suppose that, for Problem* (1.3), *Hypotheses* (1)–(5) *are satisfied. Further, suppose that* $x_0 \in \mathscr{X}$, $\tilde{x}_0 \in \tilde{\mathscr{X}}$, $\alpha \in R^m$, $\beta^0 \leq 0$, *and* $\ell_1 \in \mathscr{Z}_1^*$ *satisfy the relations*

10 $$x_0 = Lx_0 + \tilde{x}_0,$$

11 $$\hat{\varphi}(x_0) = 0 \quad \text{and} \quad \hat{\phi}_1(x_0) \in Z_1,$$

12 $$\beta^0 \neq 0$$

13 $$\ell_1(z_1) \geq \ell_1 \circ \hat{\phi}_1(x_0) \quad \text{for all} \quad z_1 \in \bar{Z}_1,$$

14 $$\alpha \cdot \hat{\varphi}_L(x - x_0) + \beta^0 D\hat{\phi}^0(x_0; x - x_0) + \ell_1 \circ D\hat{\phi}_1(x_0; x - x_0) \leq 0$$
$$\text{for all} \quad x \in \mathscr{X} \quad \text{such that} \quad x = Lx + \tilde{x} \quad \text{for some} \quad \tilde{x} \in \tilde{\mathscr{X}},$$

where $D\hat{\phi}^0(x_0;\cdot)$ *and* $D\hat{\phi}_1(x_0;\cdot)$ *denote the Gâteaux directional differentials of* $\hat{\phi}^0$ *and* $\hat{\phi}_1$, *respectively, at* x_0, *and* $\hat{\varphi}_L$ *denotes the linear part of* $\hat{\varphi}$. *Then, if* $T_0:\mathscr{X} \to \mathscr{X}$ *is the operator defined by*

15 $$T_0 x = Lx + \tilde{x}_0, \quad x \in \mathscr{X},$$

(x_0, T_0) *is a solution of the problem.*

Let us now turn to the minimax problem (1.65). Here, we replace Hypothesis (3) by the following one:

16 HYPOTHESIS. *The function* $\hat{\phi}_2$ *is* Z_2^0*-convex* [*where* Z_2^0 *is given by* (III.1.17)] *and Gâteaux directionally differentiable.*

In a manner analogous to that used in the derivation of Theorem (9), we can derive the following theorem on the basis of Theorem (III.5.34).

17 THEOREM. *Suppose that, for Problem* (1.65) *subject to Hypothesis* (III.1.22), *Hypotheses* (1), (2), (4), (5), *and* (16) *are satisfied. Further, suppose that* $x_0 \in \mathscr{X}$, $\tilde{x}_0 \in \tilde{\mathscr{X}}$, $\alpha \in R^m$, $\ell_1 \in \mathscr{Z}_1^*$, *and* $\ell_2 \in \mathscr{Z}_2^*$ *satisfy the relations* (10), (11), (13), *as well as*

18 $$\alpha \cdot \hat{\varphi}_L(x - x_0) + \ell_1 \circ D\hat{\phi}_1(x_0; x - x_0) + \ell_2 \circ D\hat{\phi}_2(x_0; x - x_0) \le 0$$

$$\text{for all} \quad x \in \mathscr{X} \quad \text{such that} \quad x = Lx + \tilde{x} \quad \text{for some} \quad \tilde{x} \in \tilde{\mathscr{X}},$$

19 $$\ell_2 \neq 0,$$

20 $$\ell_2(z_2) \ge \ell_2 \circ \hat{\phi}_2(x_0) \quad \text{for all} \quad z_2 \in Z_2^{\kappa_0},$$

where, in (18), $D\hat{\phi}_i(x_0;\cdot)$ $(i = 1$ *or* $2)$ *denotes the Gâteaux directional differential of* $\hat{\phi}_i$ *at* x_0 *and* $\hat{\varphi}_L$ *denotes the linear part of* $\hat{\varphi}$, *and, in* (20), $\kappa_0 = \sup_{\ell \in \mathscr{L}} \ell \circ \hat{\phi}_2(x_0)$. *Then, if* $T_0:\mathscr{X} \to \mathscr{X}$ *is the operator defined by* (15), (x_0, T_0) *is a solution of the problem.*

Let us compare the sufficient conditions of Theorems (9) and (17) with the corresponding necessary conditions of Theorems (1.30 and 74). In this regard, let us make the following additional hypotheses on the problem data and on x_0 and $\tilde{x}_0$.

21 HYPOTHESIS. *For every finite subset* $\tilde{x}_1, \ldots, \tilde{x}_\rho$ *of* $\tilde{\mathscr{X}}$, *there is a number* $\varepsilon_0 \in (0, 1/\rho)$ *with the following property: for every* $\varepsilon > 0$, *there is a continuous function* γ: $\{\lambda : \lambda = (\lambda^1, \ldots, \lambda^\rho) \in R^\rho, 0 \le \lambda^i \le \varepsilon_0$ *for each* $i = 1, \ldots, \rho\} \to \mathscr{X}$ *such that*

$$\tilde{x}_0 + \sum_{i=1}^{\rho} \lambda^i(\tilde{x}_i - \tilde{x}_0) + \gamma(\lambda^1, \ldots, \lambda^\rho) \in \tilde{\mathscr{X}}$$

$$\text{and} \quad \|\gamma(\lambda^1, \ldots, \lambda^\rho)\| < \varepsilon \quad \text{whenever}$$

$$0 \le \lambda^i \le \varepsilon_0 \quad \text{for each} \quad i.$$

Note that Hypothesis (21) is satisfied whenever $\tilde{\mathscr{X}}$ is finitely open in itself (and, in particular, whenever $\tilde{\mathscr{X}}$ is convex), because we can then set $\gamma \equiv 0$. Roughly speaking, Hypothesis (21) asserts that $\tilde{\mathscr{X}}$ is "almost" finitely open in itself.

22 HYPOTHESIS. *The functions $\hat{\varphi}$ and $\hat{\phi}_1$ are continuous, and the Gâteaux directional differential of $\hat{\phi}_1$ at x_0 is at the same time a dual W_1-semidifferential.*

23 HYPOTHESIS. *The function $\hat{\phi}^0$ is continuous, and the Gâteaux directional differential of $\hat{\phi}^0$ at x_0 is at the same time a dual semidifferential.*

24 HYPOTHESIS. *The function $\hat{\phi}_2$ is continuous, and the Gâteaux directional differential of $\hat{\phi}_2$ at x_0 is at the same time a dual W_2-semidifferential where $W_2 = Z_2^0$.*

If we now consider Problem (1.3), subject to Hypotheses (1)–(5) and (21)–(23), then we can appeal to Theorem (1.30) [it is easily seen that Hypotheses (1.21–23 and 26) are consequences of Hypotheses (5) and (21)–(23)—see Lemma (I.7.40)], and conclude that if (x_0,T_0) is a solution of the problem, with T_0 given by (15) for some $\tilde{x}_0 \in \tilde{\mathscr{X}}$, then there exist a vector $\alpha \in R^m$, a number $\beta^0 \leq 0$, and a linear functional $\ell_1 \in \mathscr{Z}_1^*$, not all zero, such that (10), (11), (13), and (14) hold.

By the same token, if we consider Problem (1.65), subject to Hypotheses (III.1.21 and 22), (1), (2), (4), (5), (16), (21), (22), and (24), then we can appeal to Theorem (1.74) and conclude that if (x_0,T_0) is a solution of the problem, with T_0 given by (15) for some $\tilde{x}_0 \in \tilde{\mathscr{X}}$, then there exist a vector $\alpha \in R^m$ and linear functionals $\ell_1 \in \mathscr{Z}_1^*$ and $\ell_2 \in \mathscr{Z}_2^*$, not all zero, such that (10), (11), (13), (18), and (20) hold.

Thus, we have shown that if the data of Problems (1.3 and 65) satisfy the just-indicated hypotheses, then the necessary conditions of Theorems (1.30 and 74)—strengthened by the additional requirement that (12) or (19), respectively, hold—are also sufficient for (x_0,T_0) to be a solution of the problem.

We leave it to the reader to obtain analogous results for Problem (1.94) (vector-valued criterion), and for problems which differ from Problems (1.3, 65, and 94) in that the equality constraint $\hat{\varphi}(x) = 0$ is omitted. Further, we mention that a number of the other abstract results of Section 5 of Chapter III can be applied to Problems (1.3, 65, and 94).

If we wish to specialize Theorem (9) to Problem (1.15), it is necessary to suppose that the data of Problem (1.15) satisfy the following two hypotheses [as well as Hypotheses (1) and (5)]:

25 HYPOTHESIS. *The functions* χ^i, *for* $i = \mu + 1, \ldots, \mu + m$, *are of the form*

26 $$\chi^i(\xi_1, \ldots, \xi_k) = \sum_{j=1}^{k} \eta_{ij} \cdot \xi_j - v_i$$

$$\textit{for all} \quad (\xi_1, \ldots, \xi_k) \in (R^n)^k,\ i = \mu + 1, \ldots, \mu + m,$$

where the η_{ij} $(i = \mu + 1, \ldots, \mu + m;\ j = 1, \ldots, k)$ *are given fixed vectors in* $(R^n)^k$, *and* $v_{\mu+1}, \ldots, v_{\mu+m}$ *are given fixed real numbers.*

27 HYPOTHESIS. *The functions* χ^i, *for* $i = 0, \ldots, \mu$, *as well as the functions* $\xi \to \tilde{\chi}^i(\xi,t)$, *for every* $t \in I$ *and each* $i = 1, \ldots, \nu$, *are convex, and* $\tilde{\chi}$ *is continuously differentiable w.r. to its first argument.*

We shall leave it to the reader to obtain the result analogous to Theorem (9) for Problem (1.15), subject to Hypotheses (25) and (26), and to make a comparison of the resultant sufficient conditions with the necessary conditions of Theorems (1.53 and 64).

Similarly, Theorem (17) may be specialized to Problem (1.71)—subject to suitable assumptions on the data of the problem.

Finally, we point out that analogous results can be obtained for problems with parameters.

CHAPTER V

Optimal Control Problems with Ordinary Differential Equation Constraints

1. Introduction

In this chapter we shall carry out a detailed investigation of some particular cases of the general optimization problems which were discussed in Chapter IV. Namely, we shall study the case where the operator equations $x = Tx$ with $T \in \mathscr{W}$ represent ordinary differential equations with a so-called control variable.

1 More precisely, we shall concern ourselves with problems which are defined in terms of some given n-vector-valued function f, whose domain is $\mathbf{G} \times V \times I$, where $\mathbf{G}$ is a given open set in R^n, V is a given arbitrary set, and I is a given compact interval $[t_1,t_2]$, and in terms of some given class $\mathscr{U}$ of functions $u : I \to V$.

Our optimization problem will consist of finding a "best," or "optimal" (in a sense to be made precise), pair $(x,u) \in \mathscr{C}^n(I) \times \mathscr{U}$, subject to the constraints that (i) x is absolutely continuous and satisfies the ordinary differential equation

2 $$\dot{x}(t) = f(x(t),u(t),t) \quad \text{a.e. in} \quad I,$$

and (ii) x satisfies certain additional restrictions which are typically of the form

3 $$\chi^i(x(\tau^1), \ldots, x(\tau^k)) \leq 0 \quad \text{for} \quad i = 1, \ldots, \mu,$$

4 $$\chi^i(x(\tau^1), \ldots, x(\tau^k)) = 0 \quad \text{for} \quad i = \mu + 1, \ldots, \mu + m,$$

5 $$\tilde{\chi}^i(x(t),t) \leq 0 \quad \text{for all} \quad t \in I_i \quad \text{and each} \quad i = 1, \ldots, \nu,$$

where $\chi^1, \ldots, \chi^{\mu+m}$ are given continuous functions from $(R^n)^k$ into R, $\tilde{\chi}^1, \ldots, \tilde{\chi}^\nu$ are given continuous functions from $R^n \times I$ into R, and $I_1, \ldots, I_\nu$ are given closed subsets of I. Further, $\tau^1, \ldots, \tau^k$ may either be given numbers in I, or else may be free (and thus also subject to an optimum choice), so long as $(\tau^1, \ldots, \tau^k)$ belongs to some given subset I_a of I^k.

6 This type of problem arises very commonly in the theory of optimal control and of the optimization of trajectories. In the typical physical model, t represents the time and the vector $x(t)$ represents the "condition" of the physical system at the time t, in the sense that the n coordinates of $x(t)$ completely characterize the state of the system (at time t). Thus, the differential equation (2) describes the manner in which the system evolves with time. Since the right-hand side of Eq. (2) contains the variable u, the evolution of the system is generally influenced by the choice of u. Said differently, the behavior of x can be controlled by a proper choice of u. Consequently, u is commonly referred to as a *control* variable, whereas x is called a *state* variable, or *phase vector*. Inasmuch as our problem consists largely in finding a best control function, the preceding problem is commonly referred to as an *optimal control problem.*

7 If the times $\tau^1, \ldots, \tau^k$ are preassigned, then the problem will be referred to as *fixed-time*; if one or more of the τ^i is free (even if subject to some constraint), then the problem will be referred to as *free-time*.

Constraints of the form of (3) and (4) have already been encountered in Chapter IV [see (IV.1.17 and 18) and (IV.2.22 and 23)]. Evidently, they represent restrictions on the permitted values of the state variable at the times $\tau^1, \ldots, \tau^k$. For example, it is quite common that the initial and terminal values of x are completely prescribed. Constraints of the form of (5) have also been encountered before—see (IV.1.19). (Problems with such constraints are usually referred to as problems with *restricted phase coordinates.*) They arise when the state variable must lie within some given region in R^n during certain time periods.

The restriction on the control functions u (that they must belong to a preassigned class $\mathscr{U}$, which will be referred to as the class of *admissible controls*) generally is used to model existing physical constraints on the actual "controllers." In most applications, $V \subset R^r$ (for some positive integer r).

In Chapter VI (see Section 3), we shall also consider constraints of the form

$$\tilde{\tilde{\chi}}^i(x(t),u(t),t) \leq 0 \quad \text{for almost all} \quad t \in I_i' \quad \text{and each} \quad i = 1, \ldots, \nu',$$

where $\tilde{\tilde{\chi}}^1, \ldots, \tilde{\tilde{\chi}}^{\nu'}$ are given continuous functions from $\mathbf{G} \times V \times I$

into R, and $I'_1, I'_2, \ldots, I'_{\nu'}$ are given Lebesgue measurable subsets of I. (Such problems are said to have mixed control-phase variable inequality constraints.) In Chapter VII, we shall investigate problems in which the ordinary differential equation (3) is replaced by a functional differential equation, by a Volterra integral equation, or by a difference equation.

In the next section, we shall investigate free-time problems of the type described earlier, in the absence of restrictions on the phase coordinates. In Section 3, we shall concern ourselves with fixed-time problems, both with and without restricted phase coordinates. In both sections, we shall consider the problem wherein we wish to minimize a functional of the form

8 $$\chi^0(x(\tau^1), \ldots, x(\tau^k)),$$

where χ^0 is some given continuous functional on $(R^n)^k$; i.e., "best" is to be understood in the sense of achieving a minimum for the functional (8). In the free-time problem, we shall allow the functions χ^i in (2), (3), and (8) to also depend on $\tau^1, \ldots, \tau^k$.

It turns out that, in many concrete problems, the "cost" functional is not of the form of (8), but rather is of the more general form

9 $$\chi^0(x(\tau^1), \ldots, x(\tau^k)) + \int_{\tau^1}^{\tau^k} \hat{f}^0(x(t), u(t), t)\, dt$$

[where $\tau^1 = \min_i \tau^i$ and $\tau^k = \max_i \tau^i$], and the constraints (3) and (4) have the more general form

10 $$\chi^i(x(\tau^1), \ldots, x(\tau^k)) + \int_{\tau^1}^{\tau^k} \hat{f}^i(x(t), u(t), t)\, dt \leq 0 \quad \text{for} \quad i = 1, \ldots, \mu,$$

11 $$\chi^i(x(\tau^1), \ldots, x(\tau^k)) + \int_{\tau^1}^{\tau^k} \hat{f}^i(x(t), u(t), t)\, dt = 0 \quad \text{for} \quad i = \mu + 1, \ldots, \mu + m,$$

where the $\hat{f}^i$ are given functionals defined on $\mathbf{G} \times V \times I$ and the χ^i are as before (e.g., χ^i may vanish identically for some i).

12 Let us show that a problem with cost given by (9) and with constraints given by (10), (11), and (5) can be reduced to a problem with cost given by (8) and constraints given by (3)–(5) (with suitably redefined x, f, and χ^i). Namely, in case integrals appear in the cost and constraint functionals in the manner indicated,

let us replace x by $\mathbf{x} = (\hat{x}, x)$, where $\hat{x} = (\hat{x}^0, \hat{x}^1, \ldots, \hat{x}^{\mu+m}) \in \mathscr{C}^{\mu+m+1}(I)$ [so that $\mathbf{x} \in \mathscr{C}^{n+\mu+m+1}(I)$], replace f by $\mathbf{f} = (\hat{f}, f)$ where $\hat{f} = (\hat{f}^0, \hat{f}^1, \ldots, \hat{f}^{\mu+m})$ [so that $\hat{f}: \mathbf{G} \times V \times I \to R^{\mu+m+1}$], and the differential equation (2) by the equation

$$\dot{\mathbf{x}}(t) = \mathbf{f}(x(t), u(t), t) \quad \text{a.e. in} \quad I,$$

replace the constraints (10) and (11) by

13 $$\hat{x}^i(\tau^1) = 0 \quad \text{for} \quad i = 0, \ldots, \mu + m,$$
$$\chi^i(x(\tau^1), \ldots, x(\tau^k)) + \hat{x}^i(\tau^k) \leq 0 \quad \text{for} \quad i = 1, \ldots, \mu,$$
$$\chi^i(x(\tau^1), \ldots, x(\tau^k)) + \hat{x}^i(\tau^k) = 0 \quad \text{for} \quad i = \mu + 1, \ldots, \mu + m,$$

and the cost functional (9) by

14 $$\chi^0(x(\tau^1), \ldots, x(\tau^k)) + \hat{x}^0(\tau^k),$$

where x, $\mathbf{x}$, and $\hat{x}$ are related by the equation $\mathbf{x} = (\hat{x}, x)$.

It is immediately seen that, with this reformulation, we arrive at a problem which is equivalent to the one originally posed, but which now has a cost functional of the form of (8), and inequality and equality constraints of the form of (3) and (4). Of course, the dimension of the phase vector has been increased from n to $(n + \mu + m + 1)$, and $(\mu + m + 1)$ additional equality constraints have had to be introduced [see (13)].

In order to obtain necessary conditions for our optimal control problem, we shall have to make certain assumptions on the character of the function f [see Hypotheses (16)–(18)]. Inasmuch as the procedure outlined in (12) required us to "adjoin" the function $\hat{f} = (\hat{f}^0, \ldots, \hat{f}^{\mu+m})$ to the original function f, we can use this procedure to advantage [for problems with constraints defined by (10) and (11) and cost by (9)] only if the functions $\hat{f}^i$ have the properties we require for f. In Section 7, we shall see that if $\hat{f}^0, \ldots, \hat{f}^\mu$ are all independent of $v \in V$, and are, in addition, convex as functions of their first argument, then we can obtain necessary conditions directly from the general theory developed in Chapter IV—without having to resort to the reformulation procedure described in (12)—and can then dispense with the differentiability requirement on $\hat{f}^0, \ldots, \hat{f}^\mu$ which we impose on f in Hypothesis (16).

In Section 4, we shall consider optimal control problems with a different type of cost functional. Namely, we shall consider problems wherein we wish to minimize

15 $$\max_{t \in I_0} \tilde{\chi}^0(x(t),t),$$

where $\tilde{\chi}^0$ is a given continuous functional defined on $R^n \times I$, and I_0 is a given closed subset of I. This is clearly a minimax problem of the type of Problem (IV.1.71). We shall investigate this problem both with and without restricted phase coordinates.

To obtain necessary conditions for the problems we have just described, we shall have to make certain hypotheses on the class $\mathscr{U}$ and on the functions f, χ^i, and $\tilde{\chi}^i$. In Sections 2–4, we shall suppose that $\mathscr{U}$ is convex under switching (See Chapter IV, Section 5). In Section 5, we shall consider the case in which the class of functions $\{F_u: u \in \mathscr{U}\}$—where, for each $u \in \mathscr{U}$, F_u denotes the function $(\xi,t) \to f(\xi,u(t),t)$: $\mathbf{G} \times I \to R^n$—is finitely open in itself. Throughout Sections 2–5, we shall suppose that the functions χ^i and $\tilde{\chi}^i$ possess the continuity and differentiability properties indicated in (IV.1.49 and 81) and (IV.2.52), but in Section 6 we shall also consider the case covered by Theorem (IV.1.64). In addition, we shall in Section 6 indicate some ways in which our results can be extended, for example, by generalizing the inequality constraints (3) and (5) [see (IV.1.58) and (IV.2.63)]. In Section 8, we shall derive some sufficient conditions for fixed-time problems with restricted phase coordinates, subject to certain convexity assumptions.

Throughout this chapter, we shall suppose that f and $\mathscr{U}$ satisfy the following three hypotheses:

16 HYPOTHESIS. *For each* $(v,t) \in V \times I$, *the function* $\xi \to f(\xi,v,t)$: $\mathbf{G} \to R^n$ *is continuously differentiable.*

17 HYPOTHESIS. *For every* $\xi \in \mathbf{G}$ *and each* $u \in \mathscr{U}$, *the function* $t \to f(\xi,u(t),t)$: $I \to R^n$ *is measurable.*

18 HYPOTHESIS. *For each* $u \in \mathscr{U}$ *and each compact subset* $\mathbf{G}_c$ *of* $\mathbf{G}$, *there exists an integrable function* $\tilde{m}: I \to R_+$ *(possibly depending on* u *and on* $\mathbf{G}_c$*) such that*

19 $$|f(\xi,u(t),t)| + |\mathscr{D}_1 f(\xi,u(t),t)| \le \tilde{m}(t) \quad \textit{for all} \quad (\xi,t) \in \mathbf{G}_c \times I.$$

20 If V is a topological space and f is continuous, then Hypothesis (17) holds so long as each $u \in \mathscr{U}$ is measurable [in the sense that $u^{-1}(O)$ is a (Lebesgue) measurable subset of I for each open subset O of V]. If also f is continuously differentiable with respect to its

first argument [see (I.7.51)]—so that Hypothesis (16) certainly holds—and if, for each $u \in \mathscr{U}$, there is a compact subset V_u of V such that $u(t) \in V_u$ for all $t \in I$, then, by (I.3.35 and 46), Hypothesis (18) holds, and we may even choose $\tilde{m}$ to be constant.

21 It is clear that if f and $\mathscr{U}$ satisfy Hypotheses (16)–(18), then, for each $u \in \mathscr{U}$, the function $F_u:\mathbf{G} \times I \to R^n$ defined by the formula

22 $$F_u(\xi,t) = f(\xi,u(t),t), \quad (\xi,t) \in \mathbf{G} \times I,$$

belongs to the set $\mathscr{F}_1$ defined in (IV.3.33).

23 Further, if $\mathscr{U}$ is convex under switching [see (IV.5.14)], then evidently the set $\{F_u\colon u \in \mathscr{U}\}$—where F_u is defined by (22) for each $u \in \mathscr{U}$—is also convex under switching.

We shall make use of (21) and (23) in Sections 2–4 in order to apply Theorem (IV.5.34).

2. A General Free-Time Optimal Control Problem

In this section, we shall use Theorems (IV.5.34) and (IV.2.56) to obtain necessary conditions for the following free-time optimal control problem.

1 *Problem.* Let there be given a function $f:\mathbf{G} \times V \times I \to R^n$ [where $\mathbf{G}$, V, and I are as indicated in (1.1)] satisfying Hypothesis (1.16), an open set $A \subset \mathscr{C}^n(I)$ such that $x(t) \in \mathbf{G}$ for all $x \in A$ and $t \in I$, a class $\mathscr{U}$ of functions $u:I \to V$ which is convex under switching and satisfies Hypotheses (1.17 and 18), an open interval $\tilde{I} \supset I$, a continuously differentiable function $\chi = (\chi^0,\chi^1,\dots,\chi^{\mu+m})$: $(R^n)^k \times (\tilde{I})^k \to R^{\mu+m+1}$, and a subset I_a of I^k which is finitely open in itself. Then find an absolutely continuous function $x_0 \in A$, a function $u_0 \in \mathscr{U}$, and an element $(\tau_0^1,\dots,\tau_0^k) \in I_a$ such that

2 $$\dot{x}_0(t) = f(x_0(t),u_0(t),t) \quad \text{for almost all} \quad t \in I,$$

3 $$\chi^i(x_0(\tau_0^1),\dots,x_0(\tau_0^k),\tau_0^1,\dots,\tau_0^k) \le 0 \quad \text{for} \quad i = 1,\dots,\mu,$$

4 $$\chi^i(x_0(\tau_0^1),\dots,x_0(\tau_0^k),\tau_0^1,\dots,\tau_0^k) = 0 \quad \text{for} \quad i = \mu+1,\dots,\mu+m,$$

and such that

5 $$\chi^0(x_0(\tau_0^1),\dots,x_0(\tau_0^k),\tau_0^1,\dots,\tau_0^k) \le \chi^0(x(\tau^1),\dots,x(\tau^k),\tau^1,\dots,\tau^k)$$

for all $(x,\tau^1,\dots,\tau^k) \in A \times I_a$ with the properties that (i) x is an absolutely continuous function satisfying the differential equation

6 $$\dot{x}(t) = f(x(t),u(t),t) \quad \text{for almost all} \quad t \in I$$

for some $u \in \mathscr{U}$, (ii) $\chi^i(x(\tau^1),\ldots,x(\tau^k),\tau^1,\ldots,\tau^k) \leq 0$ for $i = 1,\ldots,\mu$, and (iii) $\chi^i(x(\tau^1),\ldots,x(\tau^k),\tau^1,\ldots,\tau^k) = 0$ for $i = \mu + 1,\ldots,\mu + m$.

7 If $(x_0,u_0,\tau_0^1,\ldots,\tau_0^k)$ is a solution of Problem (1), then x_0 will be called an *optimal trajectory* and u_0 an *optimal control.*

In this and the succeeding sections of this chapter, we shall use the notation of Chapter IV, with $\mathscr{X} = \mathscr{C}^n(I)$.

8 If we set

9 $$\mathscr{W} = \{T_{x^*,F_u} : u \in \mathscr{U},\ x^* \in \mathscr{C}^n(I) \text{ and } x^* \text{ is constant on } I\},$$

where, for each $u \in \mathscr{U}$, F_u is the function in $\mathscr{F}_1$ defined by (1.22) [see (1.21) and (IV.3.33)], and the operator $T_{x^*,F} : A \to \mathscr{C}^n(I)$ [for each $x^* \in \mathscr{C}^n(I)$ and $F \in \mathscr{F}_1$] is defined by

10 $$(T_{x^*,F}x)(t) = x^*(t) + \int_{t_1}^{t} F(x(s),s)\, ds, \quad t \in I, \quad x \in A,$$

then, inasmuch as Eq. (6) is clearly equivalent to the equation

11 $$x(t) = (T_{x^*,F_u}x)(t) \quad \text{for all} \quad t \in I,$$

where $x^*(t) = x(t_1)$ for all $t \in I$, Problem (1) has been recast in the form of Problem (IV.2.20). [Recall that, from the results of Appendix A, $\mathscr{W} \subset \mathscr{T}^u \subset \mathscr{T}_0$—see (IV.1.2) and (IV.3.7 and 28).]

In order to obtain necessary conditions for Problem (1), we shall appeal to Theorems (IV.2.56) and (IV.5.34).

12 Thus, suppose that $(x_0,u_0,\tau_0^1,\ldots,\tau_0^k) \in A \times \mathscr{U} \times I_a$ is a solution of Problem (1) such that x_0 is differentiable at τ_0^j for each $j = 1,\ldots,k$. Thus, (2)–(4) hold. For ease of notation, we shall suppose that $\tau_0^1 \leq \tau_0^2 \leq \cdots \leq \tau_0^k$. As we have just seen, $(x_0,\tau_0^1,\ldots,\tau_0^k)$ is then a solution of Problem (IV.2.20), where $\mathscr{W}$ is defined by (9), (10), and (1.22). By Theorem (IV.5.34) and (1.23), Hypotheses (IV.1.21–23) are satisfied by the operator $T_0 \in \mathscr{W}$ defined by

13 $$(T_0x)(t) = x_0(t_1) + \int_{t_1}^{t} f(x(s),u_0(s),s)\, ds, \quad t \in I, \quad x \in A.$$

Further, Eq. (2) is clearly equivalent to the operator equation $x_0 = T_0x_0$, so that $T_0 \in \mathscr{T}_1$. Then, by Theorem (IV.2.56), there exists a nonzero row vector $\alpha = (\alpha^0, \alpha^1, \ldots, \alpha^{\mu+m}) \in \bar{R}_-^{\mu+1} \times R^m$

such that

14 $$\text{(i)} \sum_{j=1}^{k} \alpha \mathscr{D}_j^0 \chi \delta x_{\xi,u}(\tau_0^j) \leq \sum_{j=1}^{k} \alpha \mathscr{D}_j^0 \chi \delta x_{x_0(t_1),u_0}(\tau_0^j)$$

$$\text{for all} \quad \xi \in R^n \quad \text{and} \quad u \in \mathscr{U},$$

where, for each $\xi \in R^n$ and $u \in \mathscr{U}$, $\delta x_{\xi,u}$ denotes the column-vector-valued function in $\mathscr{C}^n(I)$ defined by [see (IV.2.58) and (IV.3.30)]

15 $$\delta x_{\xi,u}(t) = \int_{t_1}^{t} \mathscr{D}_1 f(x_0(s),u_0(s),s)\delta x_{\xi,u}(s)\, ds + \xi + \int_{t_1}^{t} f(x_0(s),u(s),s)\, ds \quad \text{for all} \quad t \in I,$$

and $\mathscr{D}_j^0\chi$ (for each j) denotes the Jacobian matrix $\mathscr{D}_j\chi$ of χ at $(x_0(\tau_0^1), \ldots ,x_0(\tau_0^k),\tau_0^1, \ldots ,\tau_0^k)$,

$$\text{(ii)} \sum_{i=1}^{\mu} \alpha^i \chi^i(x_0(\tau_0^1), \ldots ,x_0(\tau_0^k),\tau_0^1, \ldots ,\tau_0^k) = 0,$$

and

16 $$\text{(iii)} \sum_{j=1}^{k} \alpha[\mathscr{D}_j^0 \chi \dot{x}_0(\tau_0^j) + \mathscr{D}_{k+j}^0 \chi]\delta\tau^j \leq 0$$

$$\text{for all} \quad (\delta\tau^1, \ldots ,\delta\tau^k) \in \overline{I_a - (\tau_0^1, \ldots ,\tau_0^k)}.$$

Note that (ii) implies—by virtue of (3), since $\alpha^i \leq 0$ for $i = 1, \ldots , \mu$—that

17 $$\alpha^i \chi^i(x_0(\tau_0^1), \ldots ,x_0(\tau_0^k),\tau_0^1, \ldots ,\tau_0^k) = 0 \quad \text{for} \quad i = 1, \ldots , \mu.$$

It is evident that Eq. (15) is equivalent to the linear, inhomogeneous, ordinary differential equation

18 $$\frac{d}{dt}(\delta x_{\xi,u}(t)) = \mathscr{D}^0 f(t)\delta x_{\xi,u}(t) + f(x_0(t),u(t),t) \quad \text{a.e.} \quad \text{in } I$$

with initial condition

19 $$\delta x_{\xi,u}(t_1) = \xi,$$

where, for ease of notation, we have set

20 $$\mathscr{D}_1 f(x_0(t),u_0(t),t) = \mathscr{D}^0 f(t), \quad t \in I.$$

Using the variations of parameters formula [see Theorem (A.51) in the Appendix], we can write the solution of (18) and (19) in the form

21 $$\delta x_{\xi,u}(t) = \Phi(t)\xi + \Phi(t) \int_{t_1}^{t} \Phi^{-1}(s) f(x_0(s),u(s),s)\, ds, \quad t \in I,$$

where Φ is the absolutely continuous $(n \times n)$-matrix-valued function defined on I that satisfies

22 $$\dot{\Phi}(t) = \mathscr{D}^0 f(t)\Phi(t) \quad \text{a.e. on} \quad I, \quad \Phi(t_1) = \text{the identity matrix.}$$

Consequently, we can rewrite (14) as follows:

23 $$\sum_{j=1}^{k} \alpha\mathscr{D}_j^0\chi\Phi(\tau_0^j)\left\{\xi + \int_{t_1}^{\tau_0^j} \Phi^{-1}(s)[f(x_0(s),u(s),s) - f(x_0(s),u_0(s),s)]\,ds\right\} \le 0 \quad \text{for all} \quad \xi \in R^n \quad \text{and} \quad u \in \mathscr{U}.$$

Since $u_0 \in \mathscr{U}$, (23) implies, in particular, that

$$\sum_{j=1}^{k} \alpha\mathscr{D}_j^0\chi\Phi(\tau_0^j)\xi \le 0 \quad \text{for all} \quad \xi \in R^n,$$

which is possible only if

24 $$\sum_{j=1}^{k} \alpha\mathscr{D}_j^0\chi\Phi(\tau_0^j) = 0.$$

Since (23) must also hold for $\xi = 0$, we further conclude that

25 $$\int_{t_1}^{\tau_0^k} \psi(t)f(x_0(t),u(t),t)\,dt \le \int_{t_1}^{\tau_0^k} \psi(t)f(x_0(t),u_0(t),t)\,dt \quad \text{for all} \quad u \in \mathscr{U},$$

where $\psi(\cdot)$ is the piecewise-continuous n (row)-vector-valued function defined on $[t_1,\tau_0^k]$ which is given by the formulas

26 $$\psi(t) = \sum_{j=1}^{k} \alpha\mathscr{D}_j^0\chi\Phi(\tau_0^j)\Phi^{-1}(t) \quad \text{for} \quad t_1 \le t < \tau_0^1,$$

27 $$\psi(t) = \sum_{j=i}^{k} \alpha\mathscr{D}_j^0\chi\Phi(\tau_0^j)\Phi^{-1}(t) \quad \text{for} \quad \tau_0^{i-1} \le t < \tau_0^i \quad \text{and each} \quad i = 2, \ldots, k,$$

28 $$\psi(\tau_0^k) = \alpha\mathscr{D}_k^0\chi.$$

It follows from (24) and (26) that $\psi(t) = 0$ for $t_1 \le t < \tau_0^1$, so that (25) can be rewritten in the form

29 $$\int_{\tau_0^1}^{\tau_0^k} \psi(t)f(x_0(t),u_0(t),t)\,dt \ge \int_{\tau_0^1}^{\tau_0^k} \psi(t)f(x_0(t),u(t),t)\,dt \quad \text{for all} \quad u \in \mathscr{U}.$$

We also see from (27) and (28) that ψ is absolutely continuous on each of the intervals $[\tau_0^1,\tau_0^2), [\tau_0^2,\tau_0^3), \ldots, [\tau_0^{k-2},\tau_0^{k-1}), [\tau_0^{k-1},\tau_0^k]$, and, since [see Theorem (A.47) in the Appendix]

30 $$\frac{d}{dt}(\Phi^{-1}(t)) = -\Phi^{-1}(t)\mathscr{D}^0 f(t) \quad \text{a.e. on} \quad I,$$

ψ satisfies the following linear ordinary differential equation on each of the aforementioned subintervals of I [see (20)]:

31 $$\dot{\psi}(t) = -\psi(t)\mathscr{D}_1 f(x_0(t),u_0(t),t) \quad \text{a.e.}$$

Further, (27) and (24) imply that

32 $$\psi(\tau_0^1) = -\alpha\mathscr{D}_1^0\chi.$$

If $\tau_0^1 < \tau_0^2 < \cdots < \tau_0^k$, then (27) also yields

33 $$\psi(\tau_0^j) - \psi(\tau_0^{j-}) = -\alpha\mathscr{D}_j^0\chi \quad \text{for} \quad j = 2, \ldots, k-1.$$

[Formula (33) must be modified in an evident manner if $\tau_0^{j-1} = \tau_0^j$ for some $j = 2, \ldots, k-1$.]

Thus, we have proved the following theorem.

34 THEOREM. *Let $(x_0,u_0,\tau_0^1,\ldots,\tau_0^k)$ be a solution of Problem* (1) *such that x_0 is differentiable at τ_0^j for each $j = 1, \ldots, k$, and suppose that $\tau_0^1 < \tau_0^2 < \cdots < \tau_0^k$. Then there exist a nonzero row vector $\alpha = (\alpha^0, \alpha^1, \ldots, \alpha^{\mu+m}) \in \bar{R}_-^{\mu+1} \times R^m$ and an n (row)-vector-valued function ψ defined on $[\tau_0^1,\tau_0^k]$ such that*

(i) *on each of the intervals $[\tau_0^1,\tau_0^2), \ldots, [\tau_0^{k-2},\tau_0^{k-1})$, and $[\tau_0^{k-1},\tau_0^k]$, $\psi(\cdot)$ is absolutely continuous and satisfies the linear ordinary differential equation* (31),

(ii) *u_0 satisfies the maximum condition* (29),

(iii) *ψ satisfies the boundary conditions* (28) *and* (32) *as well as the jump conditions* (33), *where* (*for each* $j = 1, \ldots, k$) *$\mathscr{D}_j^0\chi$ denotes the Jacobian matrix $\mathscr{D}_j\chi$ of χ at $(x_0(\tau_0^1), \ldots, x_0(\tau_0^k), \tau_0^1, \ldots, \tau_0^k)$, and*

(iv) *relations* (16) *and* (17) *hold.*

35 We repeat that our hypothesis that $\tau_0^1 \leq \tau_0^2 \leq \cdots \leq \tau_0^k$ is only one of convenience, and that if $\tau_0^{j-1} = \tau_0^j$ for some $j = 2, \ldots, k$, then (33) must be modified in an obvious manner. Also, if x_0 is not differentiable at τ_0^j for some j, then [see (IV.2.62)], the necessary conditions of Theorem (34) remain in force except that, in (16), the sum is to be taken only over the "nonsingular"

indices. If x_0 is not differentiable at τ_0^j for any $j = 1, \ldots, k$, then, in the necessary conditions, (16) should be deleted. Further, if Problem (1) is modified by eliminating either the inequality constraints (3) or the equality constraints (4), or both, then the necessary conditions of Theorem (34) still hold, but with μ, or m, or both, equal to zero—so that (17) is to be deleted if there are no inequality constraints [see (IV.2.63 and 75)].

36 The function ψ in Theorem (34) is commonly known as an *adjoint variable.* Note that Eqs. (31) and (33), together with *either* Eq. (32) or Eq. (28), completely specify ψ, at least in terms of α, x_0, u_0, and the τ_0^i [see Theorem (A.47)]. The fundamental inequality (29)—which is the essential result in the theorem—is referred to as a *maximum principle in integral form* [compare with (II.2.37)]. As we shall see in Theorem (44), under suitable additional hypotheses, (29) may also be written as a "pointwise" maximum principle. Equations (28), (32), and (16)—the last of which is peculiar to free-time problems—are referred to as *transversality conditions*, and Eqs. (33), which describe the discontinuity of ψ at τ_0^j for each $j = 2, \ldots, k-1$, will for obvious reasons be referred to as *jump conditions.*

37 Note that, if $k = 2$ (so that the problem constraint and "cost" functionals, which appear in (3)–(5), depend only on the initial time τ^1, the final time τ^2, and the values of x at these times), then the adjoint variable ψ is absolutely continuous on the entire interval $[\tau_0^1, \tau_0^2]$, and the jump conditions (33) must be deleted from the necessary conditions.

38 The necessary conditions of Theorem (34) are clearly trivial if $\psi(t) \equiv 0$ on $[\tau_0^1, \tau_0^k]$, and it is therefore pertinent to ask: Under what conditions will $\psi(t) \equiv 0$? By (28), (32), (33), and (16), if $\psi \equiv 0$ on $[\tau_0^1, \tau_0^k]$, then

$$\sum_{i=0}^{\mu+m} \alpha^i(\mathscr{D}_1^0\chi^i, \ldots, \mathscr{D}_k^0\chi^i) = 0 \quad \text{in} \quad (R^n)^k,$$

$$\sum_{i=0}^{\mu+m} \alpha^i \sum_{j=1}^{k} (\mathscr{D}_{k+j}^0\chi^i)(\delta\tau^j) \le 0$$

$$\text{for all} \quad (\delta\tau^1, \ldots, \delta\tau^k) \in \overline{I_a - (\tau_0^1, \ldots, \tau_0^k)}.$$

Consequently, if the $\mathscr{D}_j^0\chi^i$ satisfy some condition under which there can be *no* nonzero vector $\tilde{\alpha} = (\tilde{\alpha}^0, \tilde{\alpha}^1, \ldots, \tilde{\alpha}^{\mu+m}) \in \bar{R}_-^{\mu+1} \times R^m$ which satisfies the relations

39 $$\sum_{i=0}^{\mu+m} \tilde{\alpha}^i(\mathscr{D}_1^0\chi^i, \ldots, \mathscr{D}_k^0\chi^i) = 0 \quad \text{in} \quad (R^n)^k,$$

40 $$\sum_{i=0}^{\mu+m} \tilde{\alpha}^i \sum_{j=1}^{k} (\mathscr{D}_{k+j}^0\chi^i)(\delta\tau^j) \leq 0$$

$$\text{for all} \quad (\delta\tau^1, \ldots, \delta\tau^k) \in \overline{I_a - (\tau_0^1, \ldots, \tau_0^k)},$$

41 $$\tilde{\alpha}^i\chi^i(x_0(\tau_0^1), \ldots, x_0(\tau_0^k), \tau_0^1, \ldots, \tau_0^k) = 0 \quad \text{for} \quad i = 1, \ldots, \mu,$$

then (under this condition) the function ψ in Theorem (34) cannot vanish identically, and, in fact, one of the vectors $\psi(\tau_0^1)$, $\psi(\tau_0^k)$, $\psi(\tau_0^2) - \psi(\tau_0^{2-}), \ldots, \psi(\tau_0^{k-1}) - \psi((\tau_0^{k-1})^-)$ must be different from zero, in which case, by Corollary (A.49), $\psi(t) \neq 0$ for all $t \in [\tau_0^i, \tau_0^{i+1})$ for some $i = 1, \ldots, k-1$.

With this in mind, we shall invoke the following condition [compare with Condition (II.2.40)].

42 CONDITION. *There is at least one element* $(\xi_1, \ldots, \xi_k, \delta\tau^1, \ldots, \delta\tau^k) \in (R^n)^k \times \overline{[I_a - (\tau_0^1, \ldots, \tau_0^k)]}$ *such that*

$$\sum_{j=1}^{k} [(\mathscr{D}_{k+j}^0\chi^i)(\delta\tau^j) + (\mathscr{D}_j^0\chi^i)\cdot\xi_j] \begin{cases} < 0 & \text{for } i = 0 \text{ and each } i = 1, \ldots, \mu \\ & \text{such that } \chi^i(x_0(\tau_0^1), \ldots, x_0(\tau_0^k), \tau_0^1, \ldots, \tau_0^k) \\ = 0 & \text{for } i = \mu+1, \ldots, \mu+m, \end{cases}$$

and the m vectors $(\mathscr{D}_1^0\chi^i, \ldots, \mathscr{D}_k^0\chi^i)$, $i = \mu+1, \ldots, \mu+m$, *are linearly independent in* $(R^n)^k$. [*If* $(\tau_0^1, \ldots, \tau_0^k)$ *is an interior point of* I_a, *in which case* (40) *takes the form* $\sum_{i=0}^{\mu+m} \tilde{\alpha}^i\mathscr{D}_{k+j}^0\chi^i = 0$ *for* $j = 1, \ldots, k$, *linear independence of the vectors* $(\mathscr{D}_1^0\chi^i, \ldots, \mathscr{D}_k^0\chi^i)$ *for* $i = \mu+1, \ldots, \mu+m$ *may be weakened to linear independence of the vectors* $(\mathscr{D}_1^0\chi^i, \ldots, \mathscr{D}_k^0\chi^i, \mathscr{D}_{k+1}^0\chi^i, \ldots, \mathscr{D}_{2k}^0\chi^i)$ (*for* $i = \mu+1, \ldots, \mu+m$) *in* $(R^n)^k \times R^k$].

Condition (42) essentially asserts that Problem (1) is well-posed in the sense that the problem equality and inequality constraints are, to "first order," compatible at $(x_0, \tau_0^1, \ldots, \tau_0^k)$, both with one another and with the functional being minimized.

It is not hard to see that, if Condition (42) holds, then there is no nonzero vector $\tilde{\alpha} = (\tilde{\alpha}^0, \tilde{\alpha}^1, \ldots, \tilde{\alpha}^{\mu+m}) \in \bar{R}_-^{\mu+1} \times R^m$ which satisfies (39)–(41), so that $\psi(t) \neq 0$ for all $t \in [\tau_0^i, \tau_0^{i+1})$ for some $i = 1, \ldots, k-1$.

We point out that if the vectors $(\mathscr{D}_1^0\chi^i, \ldots, \mathscr{D}_k^0\chi^i)$, as i ranges over 0 and over those integers from 1 to $\mu + m$ for which $\chi^i(x_0(\tau_0^1), \ldots, x_0(\tau_0^k), \tau_0^1, \ldots, \tau_0^k) = 0$, are linearly independent in $(R^n)^k$, then Condition (42) is satisfied.

If $\mathscr{U}$ and f in Problem (1) satisfy certain additional conditions (as is the case in many important applications), then the maximum principle in integral form, (29), can be rewritten in an equivalent "pointwise" form. Thus, we introduce the following hypothesis.

43 HYPOTHESIS. *There are a subset U of V and a subset I^* of I whose complement with respect to I has Lebesgue measure zero such that* (i) *$u(t) \in U$ for all $u \in \mathscr{U}$ and every $t \in I$,* (ii) *every constant function $u: I \to U$ belongs to $\mathscr{U}$, and either* (iiia) *for every $v \in U$, the function $(\xi,t) \to f(\xi,v,t): \mathbf{G} \times I^* \to R^n$ is continuous, or* (iiib) *there is a finite or countable subset $\tilde{U}$ of U such that $f(\xi,\tilde{U},t)$ is dense in $f(\xi,U,t)$ for all $(\xi,t) \in \mathbf{G} \times I^*$.*

We point out that part (iiib) of Hypothesis (43) is satisfied if U is a topological space with a countable dense subset and if the function $v \to f(\xi,v,t): U \to R^n$ is continuous for each $(\xi,t) \in \mathbf{G} \times I$.

44 THEOREM. *Let $(x_0, u_0, \tau_0^1, \ldots, \tau_0^k)$ be a solution of Problem* (1), *and suppose that $\tau_0^1 \leq \tau_0^j \leq \tau_0^k$ for each $j = 2, \ldots, k-1$ and that Hypothesis* (43) *holds. Then the maximum condition* (29) *of Theorem* (34) *is equivalent to the "pointwise" maximum principle*

45 $$\psi(t)f(x_0(t),u_0(t),t) = \max_{v \in U} \psi(t)f(x_0(t),v,t)$$

$$\textit{for almost all} \quad t \in [\tau_0^1, \tau_0^k].$$

Proof. It is obvious that (45)—in the presence of Hypothesis (43)—implies (29). To prove the converse, we argue by contradiction, much as was done in the proof of relation (II.2.41). Thus, suppose that the set

$$\hat{I} = \left\{t: t \in [\tau_0^1, \tau_0^k],\ \psi(t)f(x_0(t),u_0(t),t) < \sup_{v \in U} \psi(t)f(x_0(t),v,t)\right\}$$

has positive measure. Let us denote the function

$$t \to \psi(t)f(x_0(t),u_0(t),t): I \to R$$

by H_0.

First suppose that alternative (iiib) in Hypothesis (43) holds. Let $\tilde{U} = \{v_1, v_2, \ldots\}$, and let us denote (for each $j = 1, 2, \ldots$) the function $t \to \psi(t)f(x_0(t), v_j, t): I \to R$ by H_j. It follows from Hypotheses (1.17 and 18), conclusion (i) of Theorem (34), and part (ii) of Hypotheses (43) that $H_0, H_1, H_2, \ldots,$ are in $L_1(I)$. Let us denote (for each j) the function $t \to \int_{t_1}^{t} H_j(s)\, ds: I \to R$ by $\mathbf{H}_j$, so that $\mathbf{H}_j$ is absolutely continuous and $\dot{\mathbf{H}}_j(t) = H_j(t)$ for almost all $t \in I$. Since a countable union of sets of measure zero still has measure zero, and the measure of I^* is the same as that of I, there is a $\hat{t} \in (\hat{I} \cap I^*)$ such that $\dot{\mathbf{H}}_j(\hat{t}) = H_j(\hat{t})$ for each $j = 0, 1, \ldots$.

By definition of $\hat{I}$ and by part (iiib) of Hypothesis (43),

$$\begin{aligned}\dot{\mathbf{H}}_0(\hat{t}) = H_0(\hat{t}) &= \psi(\hat{t})f(x_0(\hat{t}), u_0(\hat{t}), \hat{t}) < \psi(\hat{t})f(x_0(\hat{t}), v_j, \hat{t}) \\ &= H_j(\hat{t}) = \dot{\mathbf{H}}_j(\hat{t})\end{aligned}$$

for some $j = 1, 2, \ldots$. Hence, for some $\delta > 0$ sufficiently small,

$$\mathbf{H}_0(\hat{t} + \delta) - \mathbf{H}_0(\hat{t}) < \mathbf{H}_j(\hat{t} + \delta) - \mathbf{H}_j(\hat{t}),$$

i.e.,

46 $$\int_{\hat{t}}^{\hat{t}+\delta} H_0(t)\, dt < \int_{\hat{t}}^{\hat{t}+\delta} H_j(t)\, dt.$$

Let us define the function $\tilde{u}: I \to U$ by the relation

47 $$\tilde{u}(t) = \begin{cases} u_0(t) & \text{if } t \in I, \quad t \notin [\hat{t}, \hat{t} + \delta], \\ v_j & \text{for } \hat{t} \le t \le \hat{t} + \delta. \end{cases}$$

Since $\mathscr{U}$ is convex under switching, it follows from part (ii) of Hypothesis (43) that $\tilde{u} \in \mathscr{U}$. But (46) and (47) imply that

$$\int_{\tau_0^1}^{\tau_0^k} \psi(t)f(x_0(t), u_0(t), t)\, dt < \int_{\tau_0^1}^{\tau_0^k} \psi(t)f(x_0(t), \tilde{u}(t), t)\, dt,$$

contradicting (29).

On the other hand, if alternative (iiia) in Hypothesis (43) holds, we choose a point $\hat{t} \in (\hat{I} \cap I^*)$ which is a point of continuity of ψ and is also such that $\dot{\mathbf{H}}_0(\hat{t}) = H_0(\hat{t})$, so that

$$\dot{\mathbf{H}}_0(\hat{t}) = H_0(\hat{t}) = \psi(\hat{t})f(x_0(\hat{t}), u_0(\hat{t}), \hat{t}) < \psi(\hat{t})f(x_0(\hat{t}), \hat{v}, \hat{t})$$

for some $\hat{v} \in U$. If we now define $\tilde{u} \in \mathscr{U}$ by (47)—but with v_j replaced by $\hat{v}$—where $\delta > 0$ is sufficiently small, we can obtain a contradiction essentially in the same way as before. |||

48 A type of pointwise maximum principle can be obtained even in the absence of Hypothesis (43). Namely, let $\{u_j : j = 1, 2, \ldots\}$ be an arbitrary (but fixed) denumerable subset of $\mathscr{U}$, and let us define the sets $R_0(t)$ and $V_0(t)$ for each $t \in I$ as follows:

$$R_0(t) = \overline{\{f(x_0(t), u_j(t), t) : j = 0, 1, 2, \ldots\}},$$

$$V_0(t) = \{v : v \in V, f(x_0(t), v, t) \in R_0(t)\}.$$

Arguing as in the proof of Theorem (44), we can show that condition (29) in Theorem (34) implies that the "pointwise" maximum principle (45)—with U replaced by $V_0(t)$—holds. From this, we can conclude that, for every $u \in \mathscr{U}$,

$$\psi(t)f(x_0(t), u_0(t), t) \geq \psi(t)f(x_0(t), u(t), t) \quad \text{for almost all} \quad t \in I.$$

We can now also immediately conclude that if Hypothesis (43) is replaced by the following one:

HYPOTHESIS. *There is a function U from I into the class of all subsets of V such that $u(t) \in U(t)$ for almost all $t \in I$ and every $u \in \mathscr{U}$, and there is a denumerable or finite subset $\{u_j : j = 1, 2, \ldots\}$ of $\mathscr{U}$ such that*

$$\overline{\{f(\xi, u_j(t), t) : j = 1, 2, \ldots\}} \supset f(\xi, U(t), t) \quad \textit{for every} \quad (\xi, t) \in \mathbf{G} \times I,$$

then (29) is equivalent to (45), provided that we replace U by $U(t)$ in the latter.

49 In a large number of applications, $V \subset R^r$ (for some positive integer r), f is continuous on $\mathbf{G} \times V \times I$ and continuously differentiable w.r. to its first argument, and $\mathscr{U}$ consists of all essentially bounded and measurable (or all piecewise-continuous, or all piecewise-constant) functions $u : I \to U$, where U is a subset of V such that $\bar{U} \subset V$. In this situation, [see (1.20) and (I.3.45)] Hypotheses (1.16–18) and (43) hold, and $\mathscr{U}$ is convex under switching, so that $\mathscr{U}$ and f satisfy all of the requirements imposed on them in the statement of Problem (1), and the pointwise maximum principle holds.

Let us now turn to the transversality condition (16), which is associated with the fact that our problem is free-time. If $(\tau_0^1, \ldots, \tau_0^k)$ is an interior point of I_a, then (16) is evidently equivalent to the relations

50 $$\alpha \mathscr{D}_j^0 \chi \dot{x}_0(\tau_0^j) + \alpha \mathscr{D}_{k+j}^0 \chi = 0 \quad \text{for} \quad j = 1, \ldots, k.$$

If $I_a = I^k$, then (16) is equivalent to the relations

51 $$\alpha[\mathscr{D}_j^0\chi\dot{x}_0(\tau_0^j) + \mathscr{D}_{k+j}^0\chi]\delta\tau^j \leq 0$$
$$\text{for all} \quad \delta\tau^j \in I - \tau_0^j \quad \text{and each} \quad j = 1, \ldots, k,$$

which can hold if, and only if, for each $j = 1, \ldots, k$,

52 $$\alpha\mathscr{D}_j^0\chi\dot{x}_0(\tau_0^j) + \alpha\mathscr{D}_{k+j}^0\chi \begin{cases} \leq 0 & \text{if} \quad \tau_0^j = t_1, \\ = 0 & \text{if} \quad t_1 < \tau_0^j < t_2, \\ \geq 0 & \text{if} \quad \tau_0^j = t_2. \end{cases}$$

Note that the terms $\alpha\mathscr{D}_j^0\chi$ in (16), (50), and (52) may be replaced by expressions involving ψ, if we make use of (28), (32), and (33).

53 If $I_a = \{(\tau^1, \ldots, \tau^k)\colon t_1 \leq \tau^1 \leq \tau^j \leq \tau^k \leq t_2$ for each $j = 2, \ldots, k-1\}$, and if $\chi^0(\xi_1, \ldots, \xi_k, \tau^1, \ldots, \tau^k) = \tau^k - \tau^1$, so that we are trying to minimize the effective duration, $(\tau^k - \tau^1)$, of the control process, then the optimal control problem will be referred to as a *time-optimal* problem. For the sake of simplicity, let us suppose that, in this problem, $\tau_0^1 < \tau_0^j < \tau_0^k$ for each $j = 2, \ldots, k-1$. Then (16) implies that (52) here also holds for each $j = 1, \ldots, k$ (in fact, the middle alternative holds if $1 < j < k$), so that, by (32) and (28),

54 $$\psi(\tau_0^1)\dot{x}_0(\tau_0^1) \geq \alpha\mathscr{D}_{k+1}^0\chi = -\alpha^0 + \sum_{j=1}^{\mu+m} \alpha^j\mathscr{D}_{k+1}^0\chi^j \geq \sum_{j=1}^{\mu+m} \alpha^j\mathscr{D}_{k+1}^0\chi^j,$$

55 $$\psi(\tau_0^k)\dot{x}_0(\tau_0^k) \geq -\alpha\mathscr{D}_{2k}^0\chi = -\alpha^0 - \sum_{j=1}^{\mu+m} \alpha^j\mathscr{D}_{2k}^0\chi^j \geq -\sum_{j=1}^{\mu+m} \alpha^j\mathscr{D}_{2k}^0\chi^j,$$

with equality holding in the first inequality of (54) [respectively, (55)] if $t_1 < \tau_0^1$ (respectively, $t_2 > \tau_0^k$). If $\mathscr{D}_{2k}^0\chi^j = 0$ for each $j > 0$ (e.g., if the χ^j, for $j > 0$, are independent of τ^k), then (55) obviously implies that

56 $$\psi(\tau_0^k)\dot{x}_0(\tau_0^k) \geq 0.$$

By the same token,

57 $$\psi(\tau_0^1)\dot{x}_0(\tau_0^1) \geq 0$$

if $\mathscr{D}_{k+1}^0\chi^j = 0$ for each $j > 0$ (e.g., if the χ^j, for $j > 0$, are independent of τ^1).

Some of the preceding conditions may be stated more concisely with the use of the Hamiltonian function which was introduced in (II.2.44). We recall that the Hamiltonian function

$H: R^n \times \mathbf{G} \times V \times I \to R$ is defined by the relation

58 $$H(\eta,\xi,v,t) = \eta \cdot f(\xi,v,t).$$

In terms of H, the maximum principles (29) and (45) may, respectively, be written in the equivalent forms [compare with (II.2.45 and 46)]

59 $$\int_{\tau_0^1}^{\tau_0^k} H(\psi(t),x_0(t),u_0(t),t)\,dt = \max_{u \in \mathscr{U}} \int_{\tau_0^1}^{\tau_0^k} H(\psi(t),x_0(t),u(t),t)\,dt,$$

60 $$H(\psi(t),x_0(t),u_0(t),t) = \max_{v \in U} H(\psi(t),x_0(t),v,t)$$

$$\text{for almost all} \quad t \in [\tau_0^1,\tau_0^k].$$

Let us denote by $H_0(t)$ the Hamiltonian "along" the optimal trajectory, i.e.,

61 $$H_0(t) = H(\psi(t),x_0(t),u_0(t),t) \quad \text{for all} \quad t \in I.$$

If, for some $j = 1, \ldots, k$,

62 $$\dot{x}_0(\tau_0^j) = f(x_0(\tau_0^j),u_0(\tau_0^j),\tau_0^j),$$

then relations (50), (52), and (54)–(57) can be written concisely in terms of the function H_0. Namely, if (62) holds for $j = 1$ and/or k, then, in (54)–(57), we can write $H_0(\tau_0^1)$ instead of $[\psi(\tau_0^1)\dot{x}_0(\tau_0^1)]$ and/or $H_0(\tau_0^k)$ instead of $[\psi(\tau_0^k)\dot{x}_0(\tau_0^k)]$, and, by virtue of (32) and (28), in (50) and (52) we can write $[-H_0(\tau_0^1)]$ instead of $\alpha\mathscr{D}_1^0\chi\dot{x}_0(\tau_0^1)$ and/or $H_0(\tau_0^k)$ instead of $\alpha\mathscr{D}_k^0\chi\dot{x}_0(\tau_0^k)$. Further, if, for some $j = 2, \ldots, k-1$, $\tau_0^{j-1} < \tau_0^j < \tau_0^{j+1}$ [so that (33) holds] and the function $t \to f(x_0(t),u_0(t),t): I \to R^n$ is continuous at τ_0^j [so that (62) holds], then (50) or the middle alternative in (52) can be written as

63 $$H_0(\tau_0^j) - H_0(\tau_0^{j-}) = \alpha\mathscr{D}_{k+j}^0\chi,$$

so that, if $\mathscr{D}_{k+j}^0\chi = 0$, then H_0 is continuous at τ_0^j (even though ψ may not be).

From an applications standpoint, the following problem is sometimes more natural than Problem (1).

64 *Problem.* Let there be given an open (not necessarily bounded) interval $\tilde{I}$, an open set $\mathbf{G}$ in R^n, an arbitrary set V, a function $f: \mathbf{G} \times V \times \tilde{I} \to R^n$ which satisfies Hypothesis (1.16)—with I replaced by $\tilde{I}$—a class $\mathscr{U}$ of functions $u: \tilde{I} \to V$ which is convex

under switching and is such that Hypotheses (1.17 and 18) hold for every compact subinterval I of $\tilde{I}$ [the function $\tilde{m}$ in Hypothesis (1.18) may depend on I as well as on u and on $\mathbf{G}_c$], a continuously differentiable function $\chi = (\chi^0, \chi^1, \ldots, \chi^{\mu+m}): (R^n)^k \times (\tilde{I})^k \to R^{\mu+m+1}$, and a subset I_a of $(\tilde{I})^k$ which is finitely open in itself. Then find an element $(\tau_0^1, \ldots, \tau_0^k) \in I_a$, a function $u_0 \in \mathscr{U}$, and an absolutely continuous function $x_0:[\tau_0^*, \tau_0^{**}] \to \mathbf{G}$—where $\tau_0^* = \min_j \tau_0^j$ and $\tau_0^{**} = \max_j \tau_0^j$—such that

$$\dot{x}_0(t) = f(x_0(t), u_0(t), t) \quad \text{for almost all} \quad t \in [\tau_0^*, \tau_0^{**}],$$

such that (3) and (4) hold, and such that (5) holds for all $(x, \tau^1, \ldots, \tau^k)$ with the properties that (i) $(\tau^1, \ldots, \tau^k) \in I_a$, (ii) x is an absolutely continuous function from $[\min_j \tau^j, \max_j \tau^j]$ into $\mathbf{G}$ satisfying the differential equation

$$\dot{x}(t) = f(x(t), u(t), t) \quad \text{for almost all} \quad t \in [\min_j \tau^j, \max_j \tau^j]$$

for some $u \in \mathscr{U}$, (iii) $\chi^i(x(\tau^1), \ldots, x(\tau^k), \tau^1, \ldots, \tau^k) \leq 0$ for $i = 1, \ldots, \mu$, and (iv) $\chi^i(x(\tau^1), \ldots, x(\tau^k), \tau^1, \ldots, \tau^k) = 0$ for $i = \mu + 1, \ldots, \mu + m$.

Problem (64) differs from Problem (1) in that the "allowed" interval is the open (possibly unbounded) interval $\tilde{I}$, rather than the compact interval I, and in that the differential equation $\dot{x} = f(x,u,t)$ need not be satisfied on the entire "allowed" interval, but must only be satisfied during the course of the control process, i.e., on the interval $[\min \tau^j, \max \tau^j]$.

65 If $(\tau_0^1, \ldots, \tau_0^k)$, u_0, x_0 constitute a solution of Problem (64), then, by (IV.3.28) and Theorem (A.15), there are a number $\varepsilon_0 > 0$ and an absolutely continuous function $\tilde{x}_0:[\tau_0^* - \varepsilon_0, \tau_0^{**} + \varepsilon_0] \to \mathbf{G}$ such that (i) $[\tau_0^* - \varepsilon_0, \tau_0^{**} + \varepsilon_0] \subset \tilde{I}$, (ii) $x_0(t) = \tilde{x}_0(t)$ for all $t \in [\tau_0^*, \tau_0^{**}]$, and (iii) $\dot{\tilde{x}}_0(t) = f(\tilde{x}_0(t), u_0(t), t)$ a.e. on $[\tau_0^* - \varepsilon_0, \tau_0^{**} + \varepsilon_0]$. If we set $I = [\tau_0^* - \varepsilon_0, \tau_0^{**} + \varepsilon_0]$, then it is easily seen that $(\tilde{x}_0, \tilde{u}_0, \tau_0^1, \ldots, \tau_0^k)$—where $\tilde{u}_0$ is the restriction of u_0 to I—constitute a solution of Problem (1) [defined in terms of the data of Problem (64) in an evident manner, and with I as just indicated], so that we can appeal to Theorem (34) to obtain necessary conditions which $\tilde{x}_0$, $\tilde{u}_0$, and $(\tau_0^1, \ldots, \tau_0^k)$ must satisfy. Since these conditions only involve the values of $\tilde{x}_0$ and $\tilde{u}_0$ on $[\tau_0^*, \tau_0^{**}]$, and, on this interval, $x_0 = \tilde{x}_0$ and $u_0 = \tilde{u}_0$, Theorem (34) also applies to solutions of Problem (64).

3. Fixed-Time Optimal Control Problems

This section is devoted to obtaining necessary conditions for the following fixed-time optimal control problem with restricted phase coordinates.

1 *Problem.* Let there be given a function $f:\mathbf{G} \times V \times I \to R^n$ [where $\mathbf{G}$, V, and I are as indicated in (1.1)] satisfying Hypothesis (1.16), an open set $A \subset \mathscr{C}^n(I)$ such that $x(t) \in \mathbf{G}$ for all $x \in A$ and $t \in I$, a class $\mathscr{U}$ of functions $u:I \to V$ which is convex under switching and satisfies Hypotheses (1.17 and 18), an open interval $\tilde{I} \supset I$, continuously differentiable functions $\chi = (\chi^0, \chi^1, \ldots, \chi^{\mu+m})$: $(R^n)^k \to R^{\mu+m+1}$ and $\tilde{\chi} = (\tilde{\chi}^1, \ldots, \tilde{\chi}^\nu)$: $R^n \times \tilde{I} \to R^\nu$ such that $\mathscr{D}_1\tilde{\chi}$ and $\mathscr{D}_2\tilde{\chi}$ are continuously differentiable, a finite subset $I_a = \{\tau^1, \ldots, \tau^k\}$ of I such that $t_1 = \tau^1 < \tau^2 < \cdots < \tau^k = t_2$, and closed subsets $I_1, \ldots, I_\nu$ of I. Then find an absolutely continuous function $x_0 \in A$ and a function $u_0 \in \mathscr{U}$ such that

2 $$\dot{x}_0(t) = f(x_0(t),u_0(t),t) \quad \text{for almost all} \quad t \in I,$$

3 $$\chi^i(x_0(\tau^1), \ldots ,x_0(\tau^k)) \leq 0 \quad \text{for} \quad i = 1, \ldots, \mu,$$

4 $$\chi^i(x_0(\tau^1), \ldots ,x_0(\tau^k)) = 0 \quad \text{for} \quad i = \mu + 1, \ldots, \mu + m,$$

5 $$\tilde{\chi}^i(x_0(t),t) \leq 0 \quad \text{for all} \quad t \in I_i \quad \text{and each} \quad i = 1, \ldots, \nu,$$

and such that

6 $$\chi^0(x_0(\tau^1), \ldots ,x_0(\tau^k)) \leq \chi^0(x(\tau^1), \ldots ,x(\tau^k))$$

for all absolutely continuous functions $x \in A$ satisfying the differential equation

7 $$\dot{x}(t) = f(x(t),u(t),t) \quad \text{for almost all} \quad t \in I$$

for some $u \in \mathscr{U}$, together with the constraints

8 $$\chi^i(x(\tau^1), \ldots ,x(\tau^k)) \leq 0 \quad \text{for} \quad i = 1, \ldots, \mu$$

$$\text{and} \quad \chi^i(x(\tau^1), \ldots ,x(\tau^k)) = 0 \quad \text{for} \quad i = \mu + 1, \ldots, \mu + m,$$

9 $$\tilde{\chi}^i(x(t),t) \leq 0 \quad \text{for all} \quad t \in I_i \quad \text{and each} \quad i = 1, \ldots, \nu.$$

The fixed-time problem without restricted phase coordinates may be considered as a special case of Problem (1)—merely by considering the sets $I_1, \ldots, I_\nu$ to be empty. [It could also have been considered to be a special case of Problem (2.1), but, for convenience, we shall consider it in this section]. Since this problem is perhaps the most common optimal control problem, we shall state it separately in its own right:

10 *Problem.* Suppose that the sets $\mathbf{G}$, V, I, I_a, A, and $\mathscr{U}$ and the functions f and χ are as indicated in Problem (1). Then find an absolutely continuous function $x_0 \in A$ and a function $u_0 \in \mathscr{U}$ such that (2)–(4) hold, and such that (6) is satisfied for every absolutely continuous function $x \in A$ which satisfies (7) for some $u \in \mathscr{U}$, together with the constraints (8).

11 We shall obtain necessary conditions for Problems (1) and (10) by appealing to Theorems (IV.5.34) and (IV.1.53). Let us begin by studying the general Problem (1). We shall then specialize our results to the particular case of Problem (10).

12 If (x_0,u_0) is a solution of Problem (1) or of Problem (10), then, as in Section 2, x_0 will be called an *optimal trajectory*, and u_0 an *optimal control.*

13 As in Section 2, let us define $\mathscr{W}$ by (2.9)—also see (1.22) and (2.10). As we indicated in Section 2, $\mathscr{W} \subset \mathscr{T}^u \subset \mathscr{T}_0$. With this choice of $\mathscr{W}$, Problem (1) has evidently been cast in the form of Problem (IV.1.15).

14 Let (x_0,u_0) be a solution of Problem (1), so that (2)–(5) hold. Let $T_0 \in \mathscr{W}$ be defined by (2.13), so that Eq. (2) is equivalent to the operator equation $x_0 = T_0 x_0$, and $T_0 \in \mathscr{T}_1$. Then, by Theorem (IV.5.34) and (1.23), Hypotheses (IV.1.21–23) hold. Appealing to Theorem (IV.1.53), we conclude that there exist a row vector $\alpha = (\alpha^0, \alpha^1, \ldots, \alpha^{\mu+m}) \in R^{\mu+m+1}$ and a row-vector-valued function $\tilde{\lambda} = (\tilde{\lambda}^1, \ldots, \tilde{\lambda}^\nu) \in (NBV(I))^\nu$ [see (I.6.10)] such that

15 (i) $\|\tilde{\lambda}\| + |\alpha| > 0,$

16 (ii) $\alpha^i \leq 0 \quad \text{for} \quad i = 0, 1, \ldots, \mu,$

17 (iii)
$$\sum_{j=1}^{k} \alpha \mathscr{D}_j^0 \chi \delta x_{\xi,u}(\tau^j) + \int_{t_1}^{t_2} d\tilde{\lambda}(t) \mathscr{D}_1 \tilde{\chi}(x_0(t),t) \delta x_{\xi,u}(t)$$
$$\leq \sum_{j=1}^{k} \alpha \mathscr{D}_j^0 \chi \delta x_{x_0(t_1),u_0}(\tau^j) + \int_{t_1}^{t_2} d\tilde{\lambda}(t) \mathscr{D}_1 \tilde{\chi}(x_0(t),t) \delta x_{x_0(t_1),u_0}(t)$$
$$\text{for all} \quad \xi \in R^n \quad \text{and} \quad u \in \mathscr{U},$$

where, for each $\xi \in R^n$ and $u \in \mathscr{U}$, $\delta x_{\xi,u}$ denotes the column-vector-valued function in $\mathscr{C}^n(I)$ defined by [see (IV.1.55) and (IV.3.30)] (2.15), and $\mathscr{D}_j^0 \chi$ (for each j) denotes the Jacobian matrix

$\mathscr{D}_j\chi$ of χ at $(x_0(\tau^1), \ldots, x_0(\tau^k))$,

18 (iv) $\sum_{i=1}^{\mu} \alpha^i\chi^i(x_0(\tau^1), \ldots, x_0(\tau^k)) = \int_{t_1}^{t_2} d\tilde{\lambda}(t)\tilde{\chi}(x_0(t),t) = 0,$ and

(v) for each $i = 1, \ldots, \nu$, $\tilde{\lambda}^i$ is nonincreasing on I and is constant on any subinterval of I which does not meet I_i. It follows from (18), by virtue of (v), (16), (3), and (5), that

19
$$\alpha^i\chi^i(x_0(\tau^1), \ldots, x_0(\tau^k)) = 0 \quad \text{for} \quad i = 1, \ldots, \mu,$$

20
$$\int_{t_1}^{t_2} d\tilde{\lambda}^i(t)\tilde{\chi}^i(x_0(t),t) = 0 \quad \text{for} \quad i = 1, \ldots, \nu.$$

But (20), (5), and (v) imply that, for each $i = 1, \ldots, \nu$, $\tilde{\lambda}^i$ is constant on any subinterval of I which does not meet the set I_i^0 defined as follows:

21
$$I_i^0 = \{t : t \in I_i, \tilde{\chi}^i(x_0(t),t) = 0\}, \quad i = 1, \ldots, \nu.$$

[Note that, since I_i is closed and $\tilde{\chi}^i$ is continuous, I_i^0 is closed.] Thus, in particular, $\tilde{\lambda}^i$ is constant on any subinterval of I_i on which $\tilde{\chi}^i(x_0(t),t) < 0$.

Recall that Eq. (2.15) is equivalent to (2.18–20), whose solution is given by (2.21 and 22). Consequently, (17) can be rewritten as follows:

22
$$\sum_{j=1}^{k} \alpha\mathscr{D}_j^0\chi\Phi(\tau^j)\left\{\xi + \int_{t_1}^{\tau^j} \Phi^{-1}(s)\delta f_u^0(s)\,ds\right\} + \int_{t_1}^{t_2} d\tilde{\lambda}(t)\mathscr{D}_1\tilde{\chi}(x_0(t),t)\Phi(t)\left\{\xi + \int_{t_1}^{t} \Phi^{-1}(s)\delta f_u^0(s)\,ds\right\} \leq 0$$
$$\text{for all} \quad \xi \in R^n \quad \text{and} \quad u \in \mathscr{U},$$

where, for each $u \in \mathscr{U}$,

23
$$\delta f_u^0(s) = f(x_0(s),u(s),s) - f(x_0(s),u_0(s),s) \quad \text{for all} \quad s \in I.$$

Since $u_0 \in \mathscr{U}$, and $\delta f_{u_0}^0 \equiv 0$, (22) implies, in particular, that

24
$$\left\{\sum_{j=1}^{k} \alpha\mathscr{D}_j^0\chi\Phi(\tau^j) + \int_{t_1}^{t_2} d\tilde{\lambda}(t)\mathscr{D}_1\tilde{\chi}(x_0(t),t)\Phi(t)\right\}\xi \leq 0 \quad \text{for all} \quad \xi \in R^n,$$

which can only hold if

25
$$\sum_{j=1}^{k} \alpha\mathscr{D}_j^0\chi\Phi(\tau^j) + \int_{t_1}^{t_2} d\tilde{\lambda}(t)\mathscr{D}_1\tilde{\chi}(x_0(t),t)\Phi(t) = 0.$$

Since (22) must also hold for $\xi = 0$, we further conclude that (since $t_1 = \tau^1 < \tau^2 < \cdots < \tau^k = t_2$)

26 $$\int_{t_1}^{t_2} \psi(s)\delta f_u^0(s)\,ds + \int_{t_1}^{t_2} d\tilde{\lambda}(t)\left(\psi_1(t)\int_{t_1}^{t} \Phi^{-1}(s)\delta f_u^0(s)\,ds\right) \le 0$$
$$\text{for all} \quad u \in \mathscr{U},$$

where $\psi(\cdot)$ is the n (row)-vector-valued function defined on I by the formulas [compare with (2.27 and 28)]

27 $$\psi(t) = \sum_{j=i}^{k} \alpha\mathscr{D}_j^0\chi\Phi(\tau^j)\Phi^{-1}(t) \quad \text{for} \quad \tau^{i-1} \le t < \tau^i$$
$$\text{and each} \quad i = 2, \ldots, k,$$

28 $$\psi(\tau^k) = \psi(t_2) = \alpha\mathscr{D}_k^0\chi,$$

and $\psi_1(\cdot)$ is the $(v \times n)$-matrix-valued function defined on I by the formula

29 $$\psi_1(t) = \mathscr{D}_1\tilde{\chi}(x_0(t),t)\Phi(t) \quad \text{for all} \quad t \in I.$$

Note that ψ_1 is absolutely continuous because x_0 and Φ are and because of our differentiability hypothesis on $\tilde{\chi}$.

As we saw in Section 2, it follows from (27) and (28) that, on each of the intervals $[t_1,\tau^2)$, $[\tau^2,\tau^3)$, . . . , $[\tau^{k-1},t_2]$, ψ is absolutely continuous and satisfies the linear ordinary differential equation

30 $$\dot{\psi}(t) = -\psi(t)\mathscr{D}_1 f(x_0(t),u_0(t),t) \quad \text{a.e.}$$

Further, (27) clearly implies that [compare with (2.33)]

31 $$\psi(\tau^j) - \psi(\tau^{j-}) = -\alpha\mathscr{D}_j^0\chi \quad \text{for} \quad j = 2, \ldots, k-1.$$

If we interchange the order of integration in the double integral in (26), and define the functions $\hat{\psi}$ and $\bar{\psi}$ on I through the formulas

32 $$\hat{\psi}(s) = \int_{s}^{t_2} d\tilde{\lambda}(t)\psi_1(t)\Phi^{-1}(s), \quad s \in I,$$

33 $$\bar{\psi}(s) = \psi(s) + \hat{\psi}(s), \quad s \in I,$$

then (26) takes the form [see (23)]

34 $$\int_{t_1}^{t_2} \bar{\psi}(s)f(x_0(s),u_0(s),s)\,ds \ge \int_{t_1}^{t_2} \bar{\psi}(s)f(x_0(s),u(s),s)\,ds$$
$$\text{for all} \quad u \in \mathscr{U}.$$

Further, by (27), (2.22), (25), and (28),

35 $$\bar{\psi}(t_1) = \bar{\psi}(\tau^1) = -\alpha\mathscr{D}_1^0\chi, \quad \bar{\psi}(\tau^k) = \bar{\psi}(t_2) = \alpha\mathscr{D}_k^0\chi.$$

Because (for each $i = 1, \ldots, \nu$) $\tilde{\lambda}^i$ may be discontinuous at certain points of I_i at which $\tilde{\chi}^i(x_0(t),t) = 0$, the same holds for the functions $\hat{\psi}$ and $\bar{\psi}$. However, the function $\tilde{\psi}$ defined on I through the formula

36 $$\tilde{\psi}(s) = \bar{\psi}(s) + \tilde{\lambda}(s)\mathscr{D}_1\tilde{\chi}(x_0(s),s) \quad \text{for} \quad s \in I,$$

is (as we shall see) continuous—and even satisfies a linear inhomogeneous ordinary differential equation—on each of the intervals $[t_1,\tau^2)$, $[\tau^2,\tau^3), \ldots, [\tau^{k-1},t_2]$. Indeed, if we integrate by parts in (32) and recall that [see (I.6.10)] $\tilde{\lambda}(t_2) = 0$, we conclude that

37 $$\tilde{\psi}(s) = \psi(s) - \int_s^{t_2} \tilde{\lambda}(t)\dot{\psi}_1(t)\, dt\Phi^{-1}(s) \quad \text{for} \quad s \in I.$$

Hence, on each of the aforementioned intervals, $\tilde{\psi}$ is absolutely continuous. Further, on the basis of (29), (30), (2), (2.20, 22, and 30), we can convince ourselves, after some tedious but straightforward manipulations, that, on each of these intervals, $\tilde{\psi}$ satisfies the differential equation

38 $$\dot{\tilde{\psi}}(t) = -\tilde{\psi}(t)\mathscr{D}_1 f(x_0(t),u_0(t),t) + \tilde{\lambda}(t)\mathscr{D}_1 p(x_0(t),u_0(t),t) \quad \text{a.e.},$$

where p is the function from $\mathbf{G} \times V \times I$ into R^ν which is given by the formula

39 $$p(\xi,v,t) = \mathscr{D}_1\tilde{\chi}(\xi,t)f(\xi,v,t) + \mathscr{D}_2\tilde{\chi}(\xi,t) \quad \text{for all} \quad (\xi,v,t) \in \mathbf{G} \times V \times I.$$

Also, it follows from (35) and (36) that

40 $$\tilde{\psi}(t_1) = -\alpha\mathscr{D}_1^0\chi + \tilde{\lambda}(t_1)\mathscr{D}_1\tilde{\chi}(x_0(t_1),t_1),$$

and from (37), (28), and (31) that

41 $$\tilde{\psi}(t_2) = \alpha\mathscr{D}_k^0\chi,$$

42 $$\tilde{\psi}(\tau^j) - \tilde{\psi}(\tau^{j-}) = -\alpha\mathscr{D}_j^0\chi \quad \text{for} \quad j = 2, \ldots, k-1.$$

Further, (34) can be written in terms of $\tilde{\psi}$ as follows [see (36)]:

43 $$\int_{t_1}^{t_2} [\tilde{\psi}(t) - \tilde{\lambda}(t)\mathscr{D}_1\tilde{\chi}(x_0(t),t)]f(x_0(t),u_0(t),t)\, dt$$
$$\geq \int_{t_1}^{t_2} [\tilde{\psi}(t) - \tilde{\lambda}(t)\mathscr{D}_1\tilde{\chi}(x_0(t),t)]f(x_0(t),u(t),t)\, dt \quad \text{for all} \quad u \in \mathscr{U}.$$

Thus, we have proved the following theorem:

44 THEOREM. *Let* (x_0,u_0) *be a solution of Problem* (1). *Then there exist a row-vector* $\alpha = (\alpha^0, \alpha^1, \ldots, \alpha^{\mu+m}) \in \bar{R}_-^{\mu+1} \times R^m$ *and row-*

vector-valued functions $\tilde{\lambda} = (\tilde{\lambda}^1, \dots, \tilde{\lambda}^\nu) \in [NBV(I)]^\nu$ *and* $\tilde{\psi} \in L^n_\infty(I)$ *such that*:

(i) $\|\tilde{\lambda}\| + |\alpha| > 0$;

(ii) *on each of the intervals* $[t_1,\tau^2)$, $[\tau^2,\tau^3), \dots, [\tau^{k-1},t_2]$, $\tilde{\psi}(\cdot)$ *is absolutely continuous and satisfies the linear inhomogeneous ordinary differential equation* (38), *where p is defined by* (39);

(iii) u_0 *satisfies the maximum condition* (43);

(iv) $\tilde{\psi}$ *satisfies the boundary conditions* (40) *and* (41) *as well as the jump conditions* (42), *where* (*for each* $j = 1, \dots, k$) $\mathscr{D}_j^0\chi$ *denotes the Jacobian matrix* $\mathscr{D}_j\chi$ *of* χ *at* $(x_0(\tau^1), \dots, x_0(\tau^k))$;

(v) *Eqs.* (19) *hold*;

(vi) *for each* $i = 1, \dots, \nu$, $\tilde{\lambda}^i$ *is nonincreasing on I and is constant on every subinterval of I which does not meet the closed subset* I_i^0 *of I defined by* (21).

45 The function $\tilde{\psi}$ in Theorem (44) may be considered an adjoint variable which is analogous to the adjoint variable ψ of Theorem (2.34). However, note that, whereas the function ψ of Theorem (2.34) satisfies a linear *homogeneous* ordinary differential equation on each of the intervals $[\tau_0^1,\tau_0^2)$, etc. [see Eq. (2.31)], the function $\tilde{\psi}$ in Theorem (44) satisfies a linear *inhomogeneous* ordinary differential equation [Eq. (38)] on each of the intervals $[\tau^1,\tau^2)$, etc. Further, the maximum condition (2.29) in Theorem (2.34) differs from the maximum condition (43) of Theorem (44) in that the latter has the additional term $\tilde{\lambda}(t)\mathscr{D}_1\tilde{\chi}(x_0(t),t)$. Finally, the boundary conditions on the adjoint variable at the left-hand endpoint in the two theorems—(40) and (2.32)—differ in a similar respect. Of course, if $\tilde{\lambda} \equiv 0$ on I, as must be the case if $\tilde{\chi}^i(x_0(t),t) < 0$ for all $t \in I_i$ and each $i = 1, \dots, \nu$, then these differences in the properties of the adjoint variables disappear. Note that the condition (2.16) of Theorem (2.34) has no counterpart in Theorem (44), because this condition is peculiar to free-time problems.

46 Because the maximum condition (43) has the equivalent form (34), where $\bar{\psi}$ and $\tilde{\psi}$ are related through (36), the function $\bar{\psi}$ is sometimes regarded as an adjoint variable. The disadvantage of this viewpoint is that $\bar{\psi}$ does not in general satisfy a differential equation (except possibly in some formal sense), and, in fact, generally is not continuous on the intervals $[\tau^1,\tau^2)$, etc. On the other hand, if $\tilde{\lambda}$ is constant on some subinterval I' of $[\tau^j,\tau^{j+1})$, for some j—as must be the case if $I' \cap I_i = \varnothing$ for each i, or if,

for each i, $\tilde{\chi}^i(x_0(t),t) < 0$ for all $t \in I' \cap I_i$—then it follows at once from (33), (32), (30), and (2.30) that $\bar{\psi}$ is absolutely continuous on I' and satisfies the differential equation [compare with (2.31)]

47 $$\dot{\bar{\psi}}(t) = -\bar{\psi}(t)\mathscr{D}_1 f(x_0(t),u_0(t),t) \quad \text{a.e. on} \quad I'.$$

Of course, if $\tilde{\lambda} \equiv 0$ on I, then $\tilde{\psi} \equiv \bar{\psi} \equiv \psi$ [see (36) and (37)].

48 Note that the function $\tilde{\psi}$ is completely determined by Eqs. (38), (39), and (42), together with either Eq. (40) or Eq. (41), at least in terms of α, x_0, u_0, and $\tilde{\lambda}$ [see Theorem (A.51)].

49 The fundamental inequality (43) [or, equivalently, (34)], which is the essential result in Theorem (44), is, as usual, referred to as a *maximum principle in integral form.* [See Theorem (70) for a pointwise maximum principle.] Equations (40) and (41) are referred to (as usual) as *transversality conditions*, and Eqs. (42) as *jump conditions*. Of course, if $k = 2$—so that $\tau^1 = t_1$ and $\tau^2 = t_2$, and the equality and inequality constraints (8), as well as the cost functional appearing in (6), do not involve any "intermediate" values of x—then (42) must be deleted from the necessary conditions, and it follows that $\tilde{\psi}$ is absolutely continuous on I.

50 If Problem (1) is modified by eliminating either the inequality or the equality constraints in (8) (or both), then the necessary conditions of Theorem (44) remain in force, but with μ or m (or both) equal to zero, so that (19) is to be deleted from the necessary conditions if there are no inequality constraints in (8) [see (IV.1.58 and 92)]. If the "phase" inequality constraint (9) is deleted, then we arrive at Problem (10), which we shall deal with later in this section.

51 The necessary conditions of Theorem (44) are clearly trivial if $\tilde{\psi}(t) = \tilde{\lambda}(t)\mathscr{D}_1\tilde{\chi}(x_0(t),t)$ for almost all $t \in I$, or, equivalently, if $\bar{\psi} = 0$ a.e. on I, and it is therefore pertinent to ask under what conditions can this singular situation fail to arise. With this in mind, let us invoke the following two conditions [compare with Conditions (II.2.40) and (2.42)]:

52 CONDITION. *For every* $t \in \bigcap_{j=1}^{\nu} I_j^0$ [*see* (21)], *there is at least one column vector* $\xi_t \in R^n$ *such that*

53 $$\mathscr{D}_1\tilde{\chi}^i(x_0(t),t)\xi_t < 0$$

for each $i = 1, \ldots, \nu$ *such that* $t \in I_i^0$.

54 CONDITION. *There is at least one vector* $(\xi_1, \ldots, \xi_k) \in (R^n)^k$ *such that*

55
$$\sum_{j=1}^{k} (\mathscr{D}_j^0 \chi^i) \cdot \xi_j \begin{cases} < 0 & \text{for } i = 0 \text{ and each } i = 1, \ldots, \mu \\ & \text{such that } \chi^i(x_0(\tau^1), \ldots, x_0(\tau^k)) = 0, \\ = 0 & \text{for } i = \mu + 1, \ldots, \mu + m, \end{cases}$$

56
$$(\mathscr{D}_1 \tilde{\chi}^i(x_0(\tau^j), \tau^j)) \cdot \xi_j < 0 \quad \text{for each } i = 1, \ldots, \nu$$
$$\text{and each } j = 1, \ldots, k \quad \text{such that } \tau^j \in I_i^0 \quad [\text{see } (21)],$$

and the m vectors $(\mathscr{D}_1^0 \chi^i, \ldots, \mathscr{D}_k^0 \chi^i)$, $i = \mu + 1, \ldots, \mu + m$, *are linearly independent in* $(R^n)^k$.

Condition (52) essentially asserts that the constraints (9) are, to "first-order," compatible at x_0. Condition (54) asserts that the aforementioned constraints are also (to "first-order," at x_0) compatible with the remaining inequality constraints and equality constraints (8) as well as with the cost functional.

Let us show that, if Conditions (52) and (54) hold, then $\bar{\psi}(s) \neq 0$ for all s in a subset of I of positive measure. Note that, since $\tilde{\lambda}$ and $\tilde{\psi}$ are continuous from the right in (t_1, t_2) [see (I.6.10)], $\bar{\psi}$ also is [by virtue of (36)]. By the same token, $\lim_{\tau \to t_1^+} \bar{\psi}(\tau)$ exists, as does $\lim_{\tau \to s-} \bar{\psi}(\tau)$ for every $s \in (t_1, t_2]$.

Let us denote the function $\bar{\psi}\Phi$ by $\bar{\psi}_1$.

We first consider the case where $\alpha = 0$; i.e., where $\alpha^i = 0$ for $i = 0, \ldots, \mu + m$. Then [see (15)] $\tilde{\lambda} \not\equiv 0$ on I and [see (27), (28), (32), (33), and (35)]

57
$$\bar{\psi}_1(s) = \int_s^{t_2} d\tilde{\lambda}(t) \psi_1(t) \quad \text{for all} \quad s \in I,$$

58
$$\bar{\psi}_1(t_1) = \bar{\psi}_1(t_2) = 0.$$

If $\tilde{\lambda}$ has a discontinuity at t_2—which implies that $t_2 \in \bigcup_{j=1}^{\nu} I_j^0$—then, by (57) and (29),

59
$$\bar{\psi}_1(t_2^-) = \sum_{i=1}^{\nu} [\tilde{\lambda}^i(t_2) - \tilde{\lambda}^i(t_2^-)] \mathscr{D}_1 \tilde{\chi}^i(x_0(t_2), t_2) \Phi(t_2).$$

If we postmultiply both sides of Eq. (59) by $\Phi^{-1}(t_2)\xi_{t_2}$, where ξ_{t_2} is the (column) vector whose existence is asserted in Condition (52), and take into account conclusion (vi) of Theorem (44), we conclude that $\bar{\psi}_1(t_2^-)\Phi^{-1}(t_2)\xi_{t_2} > 0$, which implies that $\bar{\psi}(t_2^-) \neq 0$. Hence, $\bar{\psi}(s) \neq 0$ for all $s \in (t_2 - \varepsilon, t_2)$ for some $\varepsilon > 0$. Similarly, we can show that, if $\tilde{\lambda}$ is discontinuous at t_1, then $\bar{\psi}(s) \neq 0$ for all $s \in (t_1, t_1 + \varepsilon)$ and some $\varepsilon > 0$. If $\tilde{\lambda}$ is continuous both at t_1 and t_2, then, for some $\bar{t} \in (t_1,t_2)$ and some $j = 1, \ldots, \nu$, $\tilde{\lambda}^j(\bar{t} - \varepsilon) > \tilde{\lambda}^j(\bar{t})$ for all $\varepsilon \in (0,\bar{t} - t_1)$. Consequently, $\bar{t} \in I_j^0$. Let $\xi_{\bar{t}}$ be the corresponding (column) vector whose existence is asserted in Condition (52). We shall show that, for all $\varepsilon > 0$ sufficiently small,

60 $$[\bar{\psi}_1(\bar{t} - \varepsilon) - \bar{\psi}_1(\bar{t})]\Phi^{-1}(\bar{t})\xi_{\bar{t}} > 0,$$

which will evidently imply that $\bar{\psi}_1$, and hence $\bar{\psi}$, cannot vanish a.e. in I. Indeed, we choose an $\varepsilon_1 > 0$ with the property that, whenever $\bar{t} - \varepsilon_1 \leq t \leq \bar{t}$, then (i) $t \notin I_i^0$ for each $i = 1, \ldots, \nu$ such that $\bar{t} \notin I_i^0$, and (ii) $\mathscr{D}_1\tilde{\chi}^i(x_0(t),t)\Phi(t)\Phi^{-1}(\bar{t})\xi_{\bar{t}} < 0$ for each $i = 1, \ldots, \nu$ such that $\bar{t} \in I_i^0$. [Such an ε_1 exists because I_i^0 is closed for each i and the functions $x_0, \mathscr{D}_1\tilde{\chi}$, and Φ are continuous.] By (57) and (29), if $0 < \varepsilon < \varepsilon_1$, then

61 $$[\bar{\psi}_1(\bar{t} - \varepsilon) - \bar{\psi}_1(\bar{t})]\Phi^{-1}(\bar{t})\xi_{\bar{t}} = \int_{\bar{t}-\varepsilon}^{\bar{t}} \sum_{i=1}^{\nu} d\tilde{\lambda}^i(t)\mathscr{D}_1\tilde{\chi}^i(x_0(t),t)\Phi(t)\Phi^{-1}(\bar{t})\xi_{\bar{t}}.$$

By conclusion (vi) of Theorem (44) and the definition of $\bar{t}$, the right-hand side of (61) is positive, so that (60) holds whenever $0 < \varepsilon < \varepsilon_1$, as was to be shown.

Now suppose that $\alpha \neq 0$. We shall show that at least one of the vectors

62 $$\bar{\psi}(t_1^+), \bar{\psi}(t_2^-), \bar{\psi}(\tau^2), \bar{\psi}(\tau^{2-}), \ldots, \bar{\psi}(\tau^{k-1}), \bar{\psi}((\tau^{k-1})^-)$$

is nonzero, which, by virtue of the previously indicated continuity properties of $\bar{\psi}$, will immediately imply our desired conclusion that $\bar{\psi}(t) \neq 0$ for t in some subset of I of positive measure. But, by (35), (36), and (42), and the continuity of $\tilde{\psi}$ at t_1 and t_2,

63 $$\bar{\psi}(t_1^+) = -\alpha\mathscr{D}_1^0\chi + [\tilde{\lambda}(t_1) - \tilde{\lambda}(t_1^+)]\mathscr{D}_1\tilde{\chi}(x_0(t_1),t_1),$$

64 $$\bar{\psi}(\tau^j) - \bar{\psi}(\tau^{j-}) = -\alpha\mathscr{D}_j^0\chi + [\tilde{\lambda}(\tau^{j-}) - \tilde{\lambda}(\tau^j)]\mathscr{D}_1\tilde{\chi}(x_0(\tau^j),\tau^j) \quad \text{for} \quad j = 2, \ldots, k-1,$$

65 $$\bar{\psi}(t_2^-) = \alpha\mathscr{D}_k^0\chi - [\tilde{\lambda}(t_2^-) - \tilde{\lambda}(t_2)]\mathscr{D}_1\tilde{\chi}(x_0(t_2),t_2).$$

Let $(\xi_1, \ldots, \xi_k) \in (R^n)^k$ be the vector whose existence is asserted in Condition (54). It then follows from (63)–(65) that

66 $$-\tilde{\psi}(t_1^+)\cdot\xi_1 - \sum_{j=2}^{k-1}[\tilde{\psi}(\tau^j) - \tilde{\psi}(\tau^{j-})]\cdot\xi_j + \tilde{\psi}(t_2^-)\cdot\xi_k$$
$$= \sum_{i=0}^{\mu+m}\alpha^i\sum_{j=1}^{k}(\mathscr{D}_j^0\chi^i)\cdot\xi_j$$
$$- \sum_{j=2}^{k}\sum_{i=1}^{\nu}[\tilde{\lambda}^i(\tau^{j-}) - \tilde{\lambda}^i(\tau^j)]\mathscr{D}_1\tilde{\chi}^i(x_0(\tau^j),\tau^j)\cdot\xi_j$$
$$- \sum_{i=1}^{\nu}[\tilde{\lambda}^i(t_1) - \tilde{\lambda}^i(t_1^+)]\mathscr{D}_1\tilde{\chi}^i(x_0(t_1),t_1)\cdot\xi_1.$$

We now argue by contradiction, and thus suppose that all the vectors (62) vanish, so that the expression in (66) also does. Since (16) and (19) hold, this implies, by virtue of Conclusion (vi) of Theorem (44) together with (55) and (56), that (i) $\alpha^i = 0$ for $i = 0,\ldots,\mu$, (ii) $\tilde{\lambda}^i(\tau^{j-}) - \tilde{\lambda}^i(\tau^j) = 0$ for $i = 1,\ldots,\nu$ and $j = 2,\ldots,k$, and (iii) $\tilde{\lambda}^i(t_1) - \tilde{\lambda}^i(t_1^+) = 0$ for $i = 1,\ldots,\nu$. Since (by hypothesis) all the vectors (62) vanish, it now follows from (63)–(65) that

67 $$\alpha\mathscr{D}_j^0\chi = \sum_{i=\mu+1}^{\mu+m}\alpha^i\mathscr{D}_j^0\chi^i = 0 \quad \text{for} \quad j = 1,\ldots,k.$$

Further, because $\alpha = 0$, at least one of the numbers $\alpha^{\mu+1},\ldots,\alpha^{\mu+m}$ is different from zero. But (67) then contradicts the last assertion of Condition (54), so that we have proved the following lemma.

68 LEMMA. *Let* (x_0,u_0) *be a solution of Problem* (1) *such that Conditions* (52) *and* (54) *are satisfied. Then the functions* $\tilde{\psi}$ *and* $\tilde{\lambda}$ *in Theorem* (44) *have the property that*

69 $$\tilde{\psi}(s) - \tilde{\lambda}(s)\mathscr{D}_1\tilde{\chi}(x_0(s),s) = 0$$

for all s *in a subset of* I *of positive measure.*

We point out that Condition (52) will be satisfied if, for every $t \in \bigcup_{j=1}^{\nu} I_j^0$, the vectors $\{\mathscr{D}_1\tilde{\chi}^i(x_0(t),t)\colon i = 1,\ldots,\nu,\ t \in I_i^0\}$ are linearly independent in R^n. Further, Condition (54) will be satisfied if the vectors $(\mathscr{D}_1^0\chi^i,\ldots,\mathscr{D}_k^0\chi^i)$, as i ranges over 0 and over those integers from 1 to $\mu + m$ for which $\chi^i(x_0(\tau^1),\ldots,x_0(\tau^k)) = 0$, together with the vectors $(\mathscr{D}_1\tilde{\chi}^i(x_0(\tau^1),\tau^1),0,\ldots,0)$ as i ranges over those integers from 1 to ν for which $\tau^1 = t_1 \in I_i^0$, the vectors

$(0,\mathscr{D}_1\tilde{\chi}^i(x_0(\tau^2),\tau^2),0,\ldots,0)$ as i ranges over those integers from 1 to v for which $\tau^2 \in I_i^0$, etc., are linearly independent in $(R^n)^k$.

Corresponding to Theorem (2.44), we have the following theorem, which can be proved in essentially the same way as Theorem (2.44), once we take into account that $\bar{\psi}$ is continuous from the right and bounded in (t_1,t_2). [The remarks in (2.48–49) also carry over.]

70 THEOREM. *Let (x_0,u_0) be a solution of Problem* (1) *such that Hypothesis* (2.43) *holds. Then the maximum condition* (43) *of Theorem* (44) *is equivalent to the pointwise maximum principle*

71 $$[\tilde{\psi}(t) - \tilde{\lambda}(t)\mathscr{D}_1\tilde{\chi}(x_0(t),t)]f(x_0(t),u_0(t),t) = \max_{v\in U}\,[\tilde{\psi}(t) - \tilde{\lambda}(t)\mathscr{D}_1\tilde{\chi}(x_0(t),t)]f(x_0(t),v,t) \quad \textit{for almost all} \quad t \in I,$$

which may also be written as

72 $$\bar{\psi}(t)f(x_0(t),u_0(t),t) = \max_{v\in U} \bar{\psi}(t)f(x_0(t),v,t) \quad \textit{for almost all} \quad t \in I.$$

Relations (34), (43), (71), and (72) may be written compactly in terms of the Hamiltonian function H defined by (2.58). For example, (72) may be rewritten as [compare with (2.60)]

73 $$H(\bar{\psi}(t),x_0(t),u_0(t),t) = \max_{v\in U} H(\bar{\psi}(t),x_0(t),v,t) \quad \textit{for almost all} \quad t \in I.$$

74 Let us now turn to Problem (10). By what was said in (IV.1.58), the necessary conditions for Problem (10) differ from those for Problem (1) only in that the function $\tilde{\lambda}$ should be set identically equal to zero. Thus, by (37), $\tilde{\psi} \equiv \psi$ in this case, and we have the following theorem.

75 THEOREM. *Let (x_0,u_0) be a solution of Problem* (10). *Then there exist a nonzero row-vector $\alpha = (\alpha^0, \alpha^1, \ldots, \alpha^{\mu+m}) \in \bar{R}_-^{\mu+1} \times R^m$ and an n (row)-vector-valued function ψ defined on I such that*

(i) *on each of the intervals $[t_1,\tau^2), [\tau^2,\tau^3), \ldots, [\tau^{k-1},t_2]$, $\psi(\cdot)$ is absolutely continuous and satisfies the linear ordinary differential equation*

76 $$\dot{\psi}(t) = -\psi(t)\mathscr{D}_1 f(x_0(t),u_0(t),t) \quad \text{a.e.},$$

(ii) *u_0 satisfies the maximum condition*

77 $$\int_{t_1}^{t_2} \psi(t)f(x_0(t),u_0(t),t)\,dt \geq \int_{t_1}^{t_2} \psi(t)f(x_0(t),u(t),t)\,dt \quad \textit{for all} \quad u \in \mathscr{U},$$

(iii) ψ *satisfies the boundary conditions*

78 $$\psi(t_1) = -\alpha \mathscr{D}_1^0 \chi \quad \text{and} \quad \psi(t_2) = \alpha \mathscr{D}_k^0 \chi$$

as well as the jump conditions

79 $$\psi(\tau^j) - \psi(\tau^{j-}) = -\alpha \mathscr{D}_j^0 \chi \quad \text{for} \quad j = 2, \ldots, k-1,$$

where (for each $j = 1, \ldots, k$) $\mathscr{D}_j^0 \chi$ denotes the Jacobian matrix $\mathscr{D}_j \chi$ of χ at $(x_0(\tau^1), \ldots, x_0(\tau^k))$, and

(iv) *Eqs.* (19) *hold.*

80 As usual, the function ψ in Theorem (75) is considered an adjoint variable, the basic inequality (77) is referred to as a maximum principle in integral form [also see Theorem (84)], Eqs. (78) are referred to as transversality conditions, and Eqs. (79) as jump conditions. If $k = 2$—so that $\tau^1 = t_1$ and $\tau^2 = t_2$, and the problem constraints (8) and the cost functional in (6) do not involve any "intermediate" values of x—then (79) must be deleted from the necessary conditions, and it follows that ψ is absolutely continuous on I.

The necessary conditions of Theorem (75) essentially, and very naturally, are very similar to those of Theorem (2.34), the main difference being the absence here of the transversality conditions (2.16), which are peculiar to free-time problems.

Note that ψ is completely determined by Eqs. (76) and (79) and either of relations (78), at least in terms of α, x_0, and u_0.

81 If Problem (10) is modified by eliminating either the inequality or the equality constraints in (8) (or both), then the necessary conditions of Theorem (75) remain in force, but with μ or m (or both) equal to zero, so that (19) is to be deleted if there are no inequality constraints [see (IV.1.58 and 92)].

To assure that $\psi \not\equiv 0$ on I, we shall impose the following condition [compare with Conditions (II.2.40), (2.42), and (54)].

82 CONDITION. *There is at least one vector $(\xi_1, \ldots, \xi_k) \in (R^n)^k$ such that*

83 $$\sum_{j=1}^{k} (\mathscr{D}_j^0 \chi^i) \cdot \xi_j \begin{cases} < 0 & \text{for} \quad i = 0 \quad \text{and each} \quad i = 1, \ldots, \mu \\ & \quad \text{such that } \chi^i(x_0(\tau^1), \ldots, x_0(\tau^k)) = 0, \\ = 0 & \text{for} \quad i = \mu + 1, \ldots, \mu + m, \end{cases}$$

and the m vectors $(\mathscr{D}_1^0 \chi^i, \ldots, \mathscr{D}_k^0 \chi^i)$, $i = \mu + 1, \ldots, \mu + m$, are linearly independent in $(R^n)^k$.

Condition (82) essentially asserts that the constraints (8) are, to "first-order," at x_0, compatible with each other and with the cost functional. It is then not hard to show [see the proof of Lemma (68)] that if (x_0,u_0) is a solution of Problem (10) such that Condition (82) is satisfied, then at least one of the vectors (62)—with $\bar{\psi}$ replaced by ψ—is nonzero, so that [by Corollary (A.49)] $\psi(t) \neq 0$ for all t in one of the intervals $[t_1,\tau^2)$, $[\tau^2,\tau^3)$, . . . , $[\tau^{k-1},t_2]$. Note that condition (82) will be satisfied if the vectors $(\mathscr{D}_1^0\chi^i, \ldots, \mathscr{D}_k^0\chi^i)$, as i ranges over 0 and over those integers from 1 to $\mu + m$ for which $\chi^i(x_0(\tau^1), \ldots, x_0(\tau^k)) = 0$, are linearly independent in $(R^n)^k$.

If Hypothesis (2.43) is satisfied [also see (2.48–49)], then, as usual, the pointwise maximum principle holds:

84 THEOREM. *Let* (x_0,u_0) *be a solution of Problem* (10) *such that Hypothesis* (2.43) *holds. Then the maximum condition* (77) *of Theorem* (75) *is equivalent to the pointwise maximum principle*

85 $$\psi(t)f(x_0(t),u_0(t),t) = \max_{v\in U} \psi(t)f(x_0(t),v,t) \quad \textit{for almost all} \quad t \in I.$$

Finally, relations (77) and (85) may be written in terms of the Hamiltonian function H defined by (2.58) as [compare with (2.59 and 60) and (73)]

86 $$\int_{t_1}^{t_2} H(\psi(t),x_0(t),u_0(t),t)\,dt = \max_{u\in\mathscr{U}} \int_{t_1}^{t_2} H(\psi(t),x_0(t),u(t),t)\,dt,$$

87 $$H(\psi(t),x_0(t),u_0(t),t) = \max_{v\in U} H(\psi(t),x_0(t),v,t) \quad \text{for almost all} \quad t \in I.$$

In the next chapter [see Theorem (VI.1.59)] we shall see that if we replace the requirement in Problem (10) that $\mathscr{U}$ is convex under switching by a requirement which is, in essence, a weaker one, then a weakened form of the maximum principle [Theorem (75)] holds.

4. Minimax Optimal Control Problems

In this rather brief section, we shall obtain necessary conditions for minimax optimal control problems. Both our methods and our results will be very similar to those of the preceding section. Specifically, we shall investigate the following problem:

1 *Problem.* Let there be given a function $f\colon \mathbf{G} \times V \times I \to R^n$ [where $\mathbf{G}$, V, and I are as indicated in (1.1)] satisfying Hypothesis

(1.16), an open set $A \subset \mathscr{C}^n(I)$ such that $x(t) \in \mathbf{G}$ for all $x \in A$ and $t \in I$, a class $\mathscr{U}$ of functions $u: I \to V$ which is convex under switching and satisfies Hypotheses (1.17 and 18), an open interval $\tilde{I} \supset I$, continuously differentiable functions $\chi = (\chi^1, \ldots, \chi^{\mu+m})$: $(R^n)^k \to R^{\mu+m}$ and $\tilde{\chi} = (\tilde{\chi}^0, \tilde{\chi}^1, \ldots, \tilde{\chi}^\nu)$: $R^n \times \tilde{I} \to R^{\nu+1}$ such that $\mathscr{D}_1\tilde{\chi}$ and $\mathscr{D}_2\tilde{\chi}$ are continuously differentiable, a finite subset $I_a = \{\tau^1, \ldots, \tau^k\}$ of I such that $t_1 = \tau^1 < \tau^2 < \cdots < \tau^k = t_2$, and closed subsets $I_0, I_1, \ldots, I_\nu$ of I. Then find an absolutely continuous function $x_0 \in A$ and a function $u_0 \in \mathscr{U}$ such that

2 $$\dot{x}_0(t) = f(x_0(t),u_0(t),t) \quad \text{for almost all} \quad t \in I,$$

3 $$\chi^i(x_0(\tau^1), \ldots, x_0(\tau^k)) \leq 0 \quad \text{for} \quad i = 1, \ldots, \mu,$$

4 $$\chi^i(x_0(\tau^1), \ldots, x_0(\tau^k)) = 0 \quad \text{for} \quad i = \mu + 1, \ldots, \mu + m,$$

5 $$\tilde{\chi}^i(x_0(t),t) \leq 0 \quad \text{for all} \quad t \in I_i \quad \text{and each } i = 1, \ldots, \nu,$$

and such that

6 $$\max_{t \in I_0} \tilde{\chi}^0(x_0(t),t) \leq \max_{t \in I_0} \tilde{\chi}^0(x(t),t)$$

for all absolutely continuous functions $x \in A$ satisfying the differential equation

7 $$\dot{x}(t) = f(x(t),u(t),t) \quad \text{for almost all} \quad t \in I$$

for some $u \in \mathscr{U}$, together with the constraints

8 $$\chi^i(x(\tau^1), \ldots, x(\tau^k)) \leq 0 \quad \text{for} \quad i = 1, \ldots, \mu,$$

$$\chi^i(x(\tau^1), \ldots, x(\tau^k)) = 0 \quad \text{for} \quad i = \mu + 1, \ldots, \mu + m,$$

9 $$\tilde{\chi}^i(x(t),t) \leq 0 \quad \text{for all} \quad t \in I_i \quad \text{and each} \quad i = 1, \ldots, \nu.$$

We shall obtain necessary conditions for Problem (1) by appealing to Theorems (IV.5.34) and (IV.1.84), and then arguing as in Section 3.

As in Sections 2 and 3, let us define $\mathscr{W}$ by (2.9) [Also see (1.22) and (2.10).], so that $\mathscr{W} \subset \mathscr{T}^u \subset \mathscr{T}_0$. With this choice of $\mathscr{W}$, Problem (1) has evidently been cast in the form of Problem (IV.1.71), if we identify $\tilde{\chi}'$ with $\tilde{\chi}$.

10 Let (x_0,u_0) be a solution of Problem (1) and let $T_0 \in \mathscr{W}$ be defined by (2.13), so that Eq. (2) is equivalent to the equation $x_0 = T_0 x_0$, and $T_0 \in \mathscr{T}_1$. By Theorem (IV.5.34) and (1.23), Hypotheses (IV.1.21–23) hold. Appealing to Theorem (IV.1.84),

and arguing essentially as in the proof of Theorem (3.44), we can prove the following theorem.

11 THEOREM. *Let* (x_0, u_0) *be a solution of Problem* (1). *Then there exist a row-vector* $\alpha = (\alpha^1, \ldots, \alpha^{\mu+m}) \in \bar{R}^{\mu}_{-} \times R^m$, *and row-vector-valued functions* $\tilde{\lambda} = (\tilde{\lambda}^0, \tilde{\lambda}, \ldots, \tilde{\lambda}^{\nu}) \in [NBV(I)]^{\nu+1}$ *and* $\bar{\psi} \in L^n_{\infty}(I)$ *such that conclusions* (i)–(vi) *of Theorem* (3.44) *hold, with* (vi) *holding for* $i = 0$ *as well as for* $i = 1, \ldots, \nu$, *where the closed subset* I^0_0 *of* I *is defined by*

$$I^0_0 = \left\{ t \colon t \in I_0,\, \tilde{\chi}^0(x_0(t),t) = \max_{\tau \in I_0} \tilde{\chi}^0(x_0(\tau),\tau) \right\}.$$

The remarks in (3.45, 46, and 48–50)—except for the next-to-last sentence in (3.45) and the last sentence in (3.50)—carry over to this section with evident minor modifications. If Problem (1) is modified by deleting the "phase" inequality constraint (9), then the necessary conditions of Theorem (11) remain in force with $\nu = 0$.

Lemma (3.68), which provides sufficient conditions for $\bar{\psi}$ to be different from zero on a subset of I of positive measure, also carries over, provided that Conditions (3.52 and 54) are modified by letting an index run from 0 to ν wherever it ran from 1 to ν, and by deleting $i = 0$ in relation (3.55).

Theorem (3.70), pertaining to the pointwise maximum principle, also carries over here without change, as do the subsequent remarks regarding the Hamiltonian function H.

12 Finally, we remind the reader that similar results can be obtained for Problem (1) generalized as indicated in (IV.1.93).

5. Optimal Control Problems with Convexity in Place of Convexity under Switching

The necessary conditions developed in the preceding three sections have been based on Theorems (IV.2.56) and (IV.1.53 and 84). In each case, we have used Theorem (IV.5.34) to show that the set $\mathscr{W}$ of "admissible operators" defined by (2.9) satisfied Hypotheses (IV.1.21–23) because of the convexity under switching of the set $\mathscr{U}$.

Now since, in these problems, $\mathscr{W} \subset \mathscr{V}_2$, Hypotheses (IV.1.21 and 22) are satisfied, whether or not $\mathscr{U}$ is convex under switching [see (IV.3.41)]. Further, Theorems (IV.2.56) and (IV.1.53 and 84) admit an alternative hypothesis to Hypothesis (IV.1.23), namely that the set $\mathscr{W}$ be finitely open in itself. Consequently, if, in Problems (2.1), (3.1), (3.10), or (4.1), we replace the requirement that $\mathscr{U}$ is convex under switching by the requirement that the set $\mathscr{W}$ defined by (2.9) is finitely open in itself [but keep the requirement that $\mathscr{U}$ satisfies Hypotheses (1.17 and 18)], then the necessary conditions of Theorems (2.34), (3.44 and 75), and (4.11) remain in force.

Note that if $\mathscr{U}$ is such that the set [see (1.22)]

1 $$\{F_u : u \in \mathscr{U}\}$$

is finitely open in itself (or, in particular, convex), then $\mathscr{W}$, defined by (2.9 and 10), is also finitely open in itself. In fact, recall that in our investigation of a similar problem in Chapter II [see Hypothesis (II.2.19)], we did hypothesize that the set (1) is convex.

Of course, $\mathscr{U}$ and f may be such that the set (1) is convex, while $\mathscr{U}$ is not convex under switching. As an example, in which the maximum principle in integral form has a rather interesting consequence, suppose that $V = R$, that

$$f(\xi,v,t) = g(\xi,t) + b(t)v \quad \text{for all} \quad (\xi,v,t),$$

where g is some given continuous function from $\mathbf{G} \times I$ into R^n which is continuously differentiable with respect to its first argument and b is a given continuous function from I into R^n, and that, for some integer $p \geq 2$,

2 $$\mathscr{U} = \left\{u : u \in L_p(I), \int_I |u(t)|^p \, dt \leq 1\right\}$$

[see (I.6.23)]. Since $L_p(I) \subset L_1(I)$, it easily follows that Hypotheses (1.16–18) are satisfied, and, because of the convexity of $\mathscr{U}$ and of the linearity of f in v, the set (1) is convex. (However, $\mathscr{U}$ is evidently not convex under switching.)

The maximum principle in integral form—say for Problem (3.10), i.e., relation (3.77)—then takes the form

3 $$\int_{t_1}^{t_2} \psi(t)b(t)u_0(t) \, dt = \max_{u \in \mathscr{U}} \int_{t_1}^{t_2} \psi(t)b(t)u(t) \, dt,$$

which, by virtue of the Hölder inequality, implies that (provided that $\psi b \not\equiv 0$)

4 $$u_0(t) = \kappa|\psi(t)b(t)|^{1/(p-1)}(\operatorname{sgn}[\psi(t)b(t)]) \quad \text{for almost all} \quad t \in I,$$

where

5 $$\kappa = \left(\int_{t_1}^{t_2} |\psi(t)b(t)|^{p/(p-1)}\right)^{1/p} \quad \text{and} \quad \operatorname{sgn}\gamma = \begin{cases} 1 & \text{if} \quad \gamma > 0, \\ 0 & \text{if} \quad \gamma = 0, \\ -1 & \text{if} \quad \gamma < 0. \end{cases}$$

Analogous results can be obtained for Problems (3.1), (2.1), and (4.1). Further, if $\mathscr{U}$, instead of being given by (2), is an arbitrary convex (or even finitely open in itself) set in $L_1(I)$, then (3)—although not (4)—still follows from the maximum principle in integral form.

6. Some Generalizations

Let us discuss in this section how some of the assumptions which we made on the problem data in Sections 2–5 can be relaxed.

1 The problems which we considered in Sections 2–4 all included a differential equation constraint of the form of Eq. (2.6), in which the function u could be chosen at will from some given class $\mathscr{U}$ (of functions from I into V), where $\mathscr{U}$ was convex under switching and, together with the given function f, satisfied Hypotheses (1.16–18). We found it expedient to reformulate this differential equation constraint, which was given in terms of the single function f and the class $\mathscr{U}$, into the equivalent differential equation constraint

$$\dot{x}(t) = F(x(t),t)$$

where F could be chosen at will from the class $\mathscr{F}' = \{F_u : u \in \mathscr{U}\}$—where F_u is given by (1.22). It followed from Hypotheses (1.16–18) [see (1.21 and 23)] that the class $\mathscr{F}'$ "of admissible 'right-hand sides'" was a subset of $\mathscr{F}_1$ [see (IV.3.33)] and was convex under switching. In the arguments of Sections 2–4, no specific properties of $\mathscr{F}'$, other than those in the preceding sentence, were ever used. [See (2.12), (3.14), and (4.10)]. Hence, all of the necessary conditions developed in Sections 2–4 remain in force if, in the problem statements, we replace the constraint that (2.6) must hold, for some $u \in \mathscr{U}$—for a given function f and class $\mathscr{U}$ with the properties indicated earlier—by the constraint that

$$\dot{x}(t) = F(x(t),t) \quad \text{a.e. in} \quad I$$

for some $F \in \mathscr{F}'$, where $\mathscr{F}'$ is a given subset of $\mathscr{F}_1$ which is convex under switching. (Even the requirement that $\mathscr{F}'$ is convex under switching can be weakened to the requirement that the set $\mathscr{W} = \{T_{x^*,F}\colon F \in \mathscr{F}',\ x^* \in \mathscr{C}^n(I) \text{ and } x^* \text{ is constant on } I\}$ must satisfy Hypothesis (IV.3.46) for every $T_0 \in \mathscr{W} \cap \mathscr{T}_1$ [see Lemma (IV.3.48) and Corollary (IV.3.62)].) Of course, in this case, some evident notational changes must be made in the necessary conditions. For example, if $(x_0,F_0) \in A \times \mathscr{F}'$ is a solution of Problem (3.10)—modified in the above-indicated manner—then, in Theorem (3.75), Eq. (3.76) must be rewritten in the form

2 $$\dot{\psi}(t) = -\psi(t)\mathscr{D}_1 F_0(x_0(t),t) \quad \text{a.e.,}$$

and the maximum condition (3.77) takes the form

3 $$\int_{t_1}^{t_2} \psi(t)F_0(x_0(t),t)\,dt \geq \int_{t_1}^{t_2} \psi(t)F(x_0(t),t)\,dt \quad \text{for all} \quad F \in \mathscr{F}',$$

while there is generally no analog of (3.85), the maximum principle in pointwise form.

4 We may also generalize Problems (2.1), (3.1), (3.10), or (4.1) by replacing the inequality constraints (2.3) or (3.3) by constraints of the form

5 $$\chi''(x_0(\tau_0^1), \ldots, x_0(\tau_0^k), \tau_0^1, \ldots, \tau_0^k) \in \mathbf{Z}$$

or

6 $$\chi''(x_0(\tau^1), \ldots, x_0(\tau^k)) \in \mathbf{Z},$$

respectively, where $\chi'' = (\chi^1, \ldots, \chi^\mu)$ and $\mathbf{Z}$ is a given convex body in R^μ. In this case [see (IV.2.63) and (IV.1.58 and 91)], the necessary conditions of Theorems (2.34), (3.44), (3.75), and (4.11) remain in force, with the exception that the transversality and jump conditions (2.17, 18, 32, and 33) and (3.19, 40–42, 78, and 79) must be modified. A similar assertion holds for the phase inequality constraints (3.9) in Problems (3.1) and (4.1). We leave the details to the interested reader.

7 In Problems (3.1), (3.10), and (4.1), we may replace the requirement that all of the functions χ^i for $i \leq \mu$ are continuously differentiable by the requirement that, for each $i \leq \mu$, χ^i is either continuously differentiable or convex. In this case, the necessary conditions of Theorems (3.44), (3.75), and (4.11) remain

in force [see Theorem (IV.1.64) and (IV.1.90)], provided that $(\mathscr{D}_1^0\chi^i, \ldots, \mathscr{D}_k^0\chi^i)$ is interpreted as a (suitably chosen) subgradient of χ^i [see (I.2.29)] at $(x_0(\tau^1), \ldots, x_0(\tau^k))$—in case χ^i is not differentiable at $(x_0(\tau^1), \ldots, x_0(\tau^k))$.

8 In all of the problems discussed in this chapter, we have assumed that the function f [or the functions F in (1)] satisfies the differentiability requirement of Hypotheses (1.16 and 18). This requirement was actually used only twice in the preceding arguments—first to ensure that the operators in $\mathscr{W}$ [see (2.9 and 10) and (1.22)] are continuously differentiable, and second, with the aid of Theorem (A.31) and Corollary (A.28) in the Appendix, to ensure that Hypotheses (IV.1.21 and 22) are satisfied [see (IV.3.28 and 41)]. It is clear that $\mathscr{W}$ will have these desired properties even if we considerably weaken the differentiability requirement on f. Of course, in order that the form of the necessary conditions be essentially the same, the Fréchet differential of the operator $T_{x_0(t_1),F_{u_0}} \in \mathscr{W}$ at x_0 should be of the form [compare with (IV.3.30)]

9 $$(DT_{x_0(t_1),F_{u_0}}(x_0;\delta x))(t) = \int_{t_1}^{t} \mathscr{D}^0 f(s)\delta x(s)\,ds, \quad t \in I, \quad \delta x \in \mathscr{C}^n(I),$$

for some integrable function $\mathscr{D}^0 f$ from I into the set of all $(n \times n)$ matrices. In this case, $\mathscr{D}^0 f(s)$ may be looked upon as a "generalized" derivative of f at $(x_0(s),u_0(s),s)$. (If f satisfies Hypothesis (1.16) [as well as Hypotheses (1.17 and 18)], then $\mathscr{D}^0 f$ must be given by (2.20) for almost all $t \in I$, if (9) is to hold.) Then, of course, evident notational changes must be made in the adjoint differential equations (2.31) and (3.38 and 76) which appear in the necessary conditions.

10 Finally, we point out that we can also obtain necessary conditions, very analogous to those of Sections 2–5, for optimal control problems with a vector-valued criterion function [see (IV.1.94)].

7. An Unorthodox Fixed-Time Optimal Control Problem

In this section, we shall consider a variant of Problem (3.1) in which the functionals which define the inequality constraints, as well as the functional being minimized, have an additional integral term. Specifically, we shall consider the following problem.

1 *Problem.* Suppose that the sets **G**, V, I, $\tilde{I}$, I_a, $I_1, \ldots, I_\nu$, A, and $\mathscr{U}$ and the functions f, χ, and $\tilde{\chi}$ are as in Problem (3.1). Further, let $\hat{f}^0, \ldots, \hat{f}^\mu$ be given continuous functionals defined on $R^n \times I$ such that (for each $i = 0, \ldots, \mu$ and every $t \in I$) the function $\xi \to \hat{f}^i(\xi,t)\colon R^n \to R$ is convex. Then find an absolutely continuous function $x_0 \in A$ and a function $u_0 \in \mathscr{U}$ such that (3.2, 4, and 5), and the relations

2 $$\chi^i(x_0(\tau^1), \ldots, x_0(\tau^k)) + \int_{t_1}^{t_2} \hat{f}^i(x_0(t),t)\, dt \leq 0 \quad \text{for} \quad i = 1, \ldots, \mu,$$

hold and such that

3 $$\chi^0(x_0(\tau^1), \ldots, x_0(\tau^k)) + \int_{t_1}^{t_2} \hat{f}^0(x_0(t),t)\, dt \leq \chi^0(x(\tau^1), \ldots, x(\tau^k)) + \int_{t_1}^{t_2} \hat{f}^0(x(t),t)\, dt$$

for all absolutely continuous functions $x \in A$ satisfying the differential equation (3.7) for some $u \in \mathscr{U}$, as well as the constraints (3.9) together with

4 $$\chi^i(x(\tau^1), \ldots, x(\tau^k)) + \int_{t_1}^{t_2} \hat{f}^i(x(t),t)\, dt \leq 0 \quad \text{for} \quad i = 1, \ldots, \mu,$$

5 $$\chi^i(x(\tau^1), \ldots, x(\tau^k)) = 0 \quad \text{for} \quad i = \mu + 1, \ldots, \mu + m.$$

6 If each functional $\hat{f}^i$ ($i = 0, \ldots, \mu$) is continuously differentiable with respect to its first argument, then, using the procedure described in (1.12), we may reformulate Problem (1) so that it becomes a special case of Problem (3.1). Since we are not imposing this differentiability requirement on the $\hat{f}^i$, we cannot use this procedure [because the "augmented" function **f** described in (1.12) will not satisfy Hypothesis (1.16)] and, for this reason, our problem is unorthodox. If the $\hat{f}^i$ all vanish identically, then Problem (1) clearly is equivalent to Problem (3.1).

To obtain necessary conditions for Problem (1), we shall appeal to Theorem (IV.1.30). Note that Problem (1), as it stands, is not a special case of Problem (IV.1.15), so that we cannot, as we could in the case of Problem (3.1), appeal to Theorem (IV.1.53).

As usual, we define $\mathscr{W}$ by (2.9), so that $\mathscr{W} \subset \mathscr{T}^u \subset \mathscr{T}_0$ and $\mathscr{W} \subset \mathscr{V}_2' \subset \mathscr{V}_2$. Let us set $\mathscr{X} = \mathscr{C}^n(I)$, $A' = A$, $\mathscr{Z}' = \mathscr{C}^\nu(I)$, and define Z' by (IV.1.12). Further, let us define the functions

$\hat{\varphi}:A \to R^m$ and $\hat{\phi}':A \to \mathscr{X}'$ by (IV.1.10 and 14), where $\chi' = (\chi^{\mu+1}, \ldots, \chi^{\mu+m})$, and the functionals $\hat{\phi}^i_j$ and $\hat{\phi}^i$, $i = 0, \ldots, \mu$ and $j = 1, 2$, on A by the relations

7 $$\hat{\phi}^i_1(x) = \chi^i(x(\tau^1), \ldots, x(\tau^k)), \quad x \in A, \quad i = 0, \ldots, \mu,$$

8 $$\hat{\phi}^i_2(x) = \int_{t_1}^{t_2} \hat{f}^i(x(t),t)\, dt, \quad x \in A, \quad i = 0, \ldots, \mu,$$

9 $$\hat{\phi}^i = \hat{\phi}^i_1 + \hat{\phi}^i_2, \quad i = 0, \ldots, \mu.$$

As was pointed out in (IV.1.49), the functions $\hat{\varphi}$, $\hat{\phi}'$, and $\hat{\phi}^i_1$ thus defined, are, by virtue of the assumed properties of χ and $\tilde{\chi}$, continuous and Fréchet differentiable, with Fréchet differentials at any $x_0 \in A$ given by (IV.1.50–52), with $D\hat{\phi}^i$ replaced by $D\hat{\phi}^i_1$, where $\mathscr{D}^0_j\chi'$ and $\mathscr{D}^0_j\chi^i$ have their usual meanings. Further, by the convexity and continuity of the $\hat{f}^i$, $\hat{\phi}^i_2$ (for each i) is continuous and convex, and hence [see Corollary (I.7.33)] dually semidifferentiable, with dual semidifferential at any $x_0 \in A$ given by the functional

$$\delta x \to \int_{t_1}^{t_2} [\hat{f}^i(x_0(t) + \delta x(t),t) - \hat{f}^i(x_0(t),t)]\, dt : \mathscr{C}^n(I) \to R,$$

so that [see (9), (I.7.47), Theorem (I.7.44), and (I.7.45)] $\hat{\phi}^i$ (for each i) is continuous and dually semidifferentiable with dual semidifferential at any $x_0 \in A$ given by the functional

$$D\hat{\phi}^i(x_0;\delta x) = \sum_{j=1}^{k} \mathscr{D}^0_j\chi^i \delta x(\tau^j) + \int_{t_1}^{t_2} [\hat{f}^i(x_0(t) + \delta x(t),t) - \hat{f}^i(x_0(t),t)]\, dt, \quad \delta x \in \mathscr{C}^n(I), \quad i = 0, \ldots, \mu.$$

It is now clear that Problem (1) is of the form of Problem (IV.1.3) with the data as described in (IV.1.27). Let (x_0,u_0) be a solution of Problem (1), and let $T_0 \in \mathscr{W}$ be defined by (2.13), so that $x_0 = T_0x_0$ and $T_0 \in \mathscr{T}_1$, and (x_0,T_0) is a solution of the corresponding Problem (IV.1.3). By Theorem (IV.5.34) and (1.23), Hypotheses (IV.1.21–23) hold, and, as we have just seen, Hypothesis (IV.1.28) also holds.

Appealing to Theorem (IV.1.30)—see (IV.1.36 and 37)—and its Corollary (IV.1.46), taking into account Theorems (I.6.14 and 22) and (I.6.19), we can conclude that there exists a row vector $\alpha = (\alpha^0, \alpha^1, \ldots, \alpha^{\mu+m}) \in R^{\mu+m+1}$ and row-vector-valued functions $\tilde{\lambda} = (\tilde{\lambda}^1, \ldots, \tilde{\lambda}^\nu) \in [NBV(I)]^\nu$ and $\hat{\lambda}_i$ (for $i = 0, \ldots, \mu$) in

$[NBV(I)]^n$ such that (3.15 and 16) hold and such that

10 $$\sum_{i=\mu+1}^{\mu+m} \sum_{j=1}^{k} \alpha^i \mathscr{D}_j^0 \chi^i \delta x_{\xi,u}(\tau^j) + \int_{t_1}^{t_2} \left[d\tilde{\lambda}(t) \mathscr{D}_1 \tilde{\chi}(x_0(t),t) + \sum_{i=0}^{\mu} \alpha^i d\hat{\lambda}_i(t) \right] \delta x_{\xi,u}(t) \leq \sum_{i=\mu+1}^{\mu+m} \sum_{j=1}^{k} \alpha^i \mathscr{D}_j^0 \chi^i \delta x_{x_0(t_1),u_0}(\tau^j) + \int_{t_1}^{t_2} \left[d\tilde{\lambda}(t) \mathscr{D}_1 \tilde{\chi}(x_0(t),t) + \sum_{i=0}^{\mu} \alpha^i \, d\hat{\lambda}_i(t) \right] \delta x_{x_0(t_1),u_0}(t)$$

$$\text{for all} \quad \xi \in R^n \quad \text{and} \quad u \in \mathscr{U},$$

where $\delta x_{\xi,u}$ (for every $\xi \in R^n$ and $u \in \mathscr{U}$) denotes the column-vector-valued function in $\mathscr{C}^n(I)$ defined by (2.15) [see (IV.3.30)],

11 $$\int_{t_1}^{t_2} d\hat{\lambda}_i(t) \delta x(t) \leq \sum_{j=1}^{k} \mathscr{D}_j^0 \chi^i \delta x(\tau^j) + \int_{t_1}^{t_2} [\hat{f}^i(x_0(t) + \delta x(t),t) - \hat{f}^i(x_0(t),t)] \, dt$$

$$\text{for all} \quad \delta x \in \mathscr{C}^n(I) \quad \text{and each} \quad i = 0, \ldots, \mu,$$

12 $$\sum_{i=1}^{\mu} \alpha^i \left[\chi^i(x_0(\tau^1), \ldots, x_0(\tau^k)) + \int_{t_1}^{t_2} \hat{f}^i(x_0(t),t) \, dt \right] = \int_{t_1}^{t_2} d\tilde{\lambda}(t) \tilde{\chi}(x_0(t),t) = 0,$$

and, for each $i = 1, \ldots, v$, $\tilde{\lambda}^i$ is nonincreasing on I and is constant on any subinterval of I which does not meet I_i.

As in Section 3, we can show that (12), (2), and (3.5 and 16) imply that

13 $$\alpha^i \left[\chi^i(x_0(\tau^1), \ldots, x_0(\tau^k)) + \int_{t_1}^{t_2} \hat{f}^i(x_0(t),t) \, dt \right] = 0$$

$$\text{for} \quad i = 1, \ldots, \mu,$$

and that, for each $i = 1, \ldots, v$, $\tilde{\lambda}^i$ is constant on every subinterval of I which does not meet the set I_i^0 defined by (3.21).

Now, for each $i = 0, \ldots, \mu$, let $\hat{\lambda}_i''$ denote the function in $[NBV(I)]^n$ which is constant on each of the intervals (t_1,τ^2), $[\tau^2,\tau^3), \ldots, [\tau^{k-1},t_2)$ and which satisfies the relations

14 $$\hat{\lambda}_i''(t_1^+) - \hat{\lambda}_i''(t_1) = \mathscr{D}_1^0 \chi^i,$$

15 $$\hat{\lambda}_i''(\tau^j) - \hat{\lambda}_i''(\tau^{j-}) = \mathscr{D}_j^0 \chi^i \quad \text{for} \quad j = 2, \ldots, k,$$

and let $\hat{\lambda}_i' = \hat{\lambda}_i - \hat{\lambda}_i''$. Clearly, $\hat{\lambda}_i' \in [NBV(I)]^n$ for each i. We shall show that (for each i) $\hat{\lambda}_i'$ is Lipschitz continuous, and hence, absolutely continuous, and that, for almost all $t \in I$, $d\hat{\lambda}_i'(t)/dt$ is a subgradient of the function $\xi \to \hat{f}^i(\xi,t)$ at $x_0(t)$.

Indeed, it follows from (11), (14), and (15) that

16 $$\int_{t_1}^{t_2} d\hat{\lambda}_i'(t)\delta x(t) \leq \int_{t_1}^{t_2} [\hat{f}^i(x_0(t) + \delta x(t),t) - \hat{f}^i(x_0(t),t)]\, dt$$
$$\text{for all} \quad \delta x \in \mathscr{C}^n(I), \quad i = 0, \ldots, \mu.$$

Let us arbitrarily fix numbers τ', τ'' in I with $\tau'' > \tau'$. For every $\delta \in (0, \tau'' - \tau')$ and every $\xi \in R^n$, let us define the function $\delta x(\cdot\,; \delta,\xi) \in \mathscr{C}^n(I)$ as follows:

17 $$\delta x(t;\delta,\xi) = \begin{cases} 0 & \text{if } t \in [t_1,\tau'] \text{ or if } \tau'' + \delta \leq t \leq t_2 \\ \xi & \text{if } t \in [\tau' + \delta, \tau''] \\ \delta^{-1}\xi(t - \tau') & \text{if } t \in [\tau', \tau' + \delta] \\ \delta^{-1}\xi(\tau'' + \delta - t) & \text{if } t \in [\tau'', \tau'' + \delta]. \end{cases}$$

(If $\tau'' = t_2$, we of course omit the last alternative in (17). If $\tau' = t_1$, then we replace the first two alternatives in (17) by: ξ if $t \in [\tau',\tau'']$.) Applying (16) to functions of this form and, for fixed ξ, passing to the limit as $\delta \to 0^+$ [keeping in mind that $\hat{\lambda}_i'$ is continuous from the right in (t_1,t_2)], we conclude that

18 $$[\hat{\lambda}_i'(\tau'') - \hat{\lambda}_i'(\tau')]\xi \leq \int_{\tau'}^{\tau''} [\hat{f}^i(x_0(t) + \xi, t) - \hat{f}^i(x_0(t),t)]\, dt$$
$$\text{for all} \quad \xi \in R^n.$$

It evidently follows from (18) that $|\hat{\lambda}_i'(\tau'') - \hat{\lambda}_i'(\tau')| \leq \kappa_i(\tau'' - \tau')$, where [see (I.3.46)]

$$\kappa_i = \max_{\substack{t \in I \\ |\xi| = 1}} |\hat{f}^i(x_0(t) + \xi, t) - \hat{f}^i(x_0(t),t)|.$$

Since τ', τ'' in I were arbitrary (so long as $\tau'' > \tau'$), this means that $\hat{\lambda}_i'$ is Lipschitz continuous on I. Consequently, $\hat{\lambda}_i'$ is absolutely continuous on I, $\dot{\hat{\lambda}}_i'$ (which is defined for almost all $t \in I$) is in $L_\infty^n(I)$, and (18) can be written in the form

19 $$\frac{1}{\tau'' - \tau'} \int_{\tau'}^{\tau''} \dot{\hat{\lambda}}_i'(s)\, ds\xi \leq \frac{1}{\tau'' - \tau'} \int_{\tau'}^{\tau''} [\hat{f}^i(x_0(s) + \xi, s) - \hat{f}^i(x_0(s),s)]\, ds$$
$$\text{for all} \quad \xi \in R^n.$$

If $t \in (t_1, t_2)$ is a regular point for $\dot{\hat{\lambda}}_i'$ [see (II.2.42)]—and almost all points of (t_1,t_2) are such—then, if we let $\tau' < t < \tau''$ and pass to the limit in (19) as $(\tau'' - \tau') \to 0$, we conclude that $\dot{\hat{\lambda}}_i'(t)\xi \leq \hat{f}^i(x_0(t) + \xi, t) - \hat{f}^i(x_0(t),t)$ for all $\xi \in R^n$, i.e., that $\dot{\hat{\lambda}}_i'(t)$ is a subgradient of the function $\xi \to \hat{f}^i(\xi,t)$ at $x_0(t)$ [see (I.2.29)].

For each $i = 0, \ldots, \mu$, let us denote the function $\dot{\hat{\lambda}}_i' \in L_\infty^n(I)$ by $\mathscr{D}^0\hat{f}^i$. Then (10) can be rewritten in the form

$$\sum_{j=1}^{k} \alpha\mathscr{D}_j^0\chi[\delta x_{\xi,u}(\tau^j) - \delta x_{x_0(t_1),u_0}(\tau^j)]$$

$$+ \int_{t_1}^{t_2}\left[d\tilde{\lambda}(t)\mathscr{D}_1\tilde{\chi}(x_0(t),t) + \sum_{i=0}^{\mu} \alpha^i\mathscr{D}^0\hat{f}^i(t)\,dt\right][\delta x_{\xi,u}(t) - \delta x_{x_0(t_1),u_0}(t)]$$

$$\text{for all} \quad \xi \in R^n \quad \text{and} \quad u \in \mathscr{U},$$

where $\mathscr{D}_j^0\chi$ has the usual meaning.

Performing essentially the same computations as in the derivation of Theorem (3.44), we can prove the following theorem:

20 THEOREM. *Let* (x_0,u_0) *be a solution of Problem* (1). *Then there exist a row vector* $\alpha = (\alpha^0, \alpha^1, \ldots, \alpha^{\mu+m}) \in \bar{R}_-^{\mu+1} \times R^m$ *and row-vector-valued functions* $\tilde{\lambda} = (\tilde{\lambda}^1, \ldots, \tilde{\lambda}^\nu) \in [NBV(I)]^\nu$, $\mathscr{D}^0\hat{f}^i$ $(i = 0, \ldots, \mu)$ *in* $L_\infty^n(I)$, *and* $\tilde{\psi} \in L_\infty^n(I)$ *such that conclusions* (i)–(vi) *of Theorem* (3.44) *hold except that, in* (v), *Eq.* (3.19) *should be replaced by* (13), *and in* (ii), *Eq.* (3.38) *should be replaced by the differential equation*

21 $$\dot{\tilde{\psi}}(t) = -\tilde{\psi}(t)\mathscr{D}_1 f(x_0(t),u_0(t),t) - \sum_{i=0}^{\mu} \alpha^i\mathscr{D}^0\hat{f}^i(t) + \tilde{\lambda}(t)\mathscr{D}_1 p(x_0(t),u_0(t),t) \quad \text{a.e.}$$

In addition, for each $i = 0, \ldots, \mu$ *and almost every* $t \in I$, $\mathscr{D}^0\hat{f}^i(t)$ *is a subgradient of the function* $\xi \to \hat{f}^i(\xi,t)$ *at* $x_0(t)$.

Note that the necessary conditions of Theorem (20) differ from those of Theorem (3.44) only in that (a) the adjoint equation (21)—in comparison with Eq. (3.38)—contains the extra term $[-\sum\alpha^i\mathscr{D}^0\hat{f}^i]$, where the $\mathscr{D}^0\hat{f}^i(t)$ are certain subgradients, and (b) relations (3.19) must be replaced by relations (13). Of course, this result should not be too surprising, since Problem (3.1) may be looked upon as the special case of Problem (1) with $\hat{f}^i \equiv 0$ for $i = 0, \ldots, \mu$.

If (for some $i = 0, \ldots, \mu$ and some $t \in I$) the function $\xi \to \hat{f}^i(\xi,t)$ is Fréchet differentiable at $x_0(t)$, then the subgradient $\mathscr{D}^0\hat{f}^i(t)$ must coincide with the Jacobian matrix $\mathscr{D}_1\hat{f}^i(x_0(t),t)$ [see Lemma (I.7.62)]. If each one of the functionals $\hat{f}^0, \ldots, \hat{f}^\mu$ is continuously differentiable with respect to its first argument, then as was indicated in (6), we could reduce Problem (1) to Problem (3.1). In this case, the necessary conditions of Theorem (20) coincide with those obtainable from Theorem (3.44).

22 If we complicate Problem (1) by adding a term of the form

23
$$\int_{t_1}^{t_2} \tilde{f}^i(x(t),u(t),t)\,dt$$

to the left-hand side of each of the constraint relations (4) and (5), and similarly modify (3) [i.e., if we add the integral (23) with $i = 0$ to the cost functional as well], (2), and (3.4), where $\tilde{f}^0, \ldots, \tilde{f}^{\mu+m}$ are given functionals defined on $\mathbf{G} \times V \times I$ which satisfy the requirements of Hypotheses (1.16–18), then we can reduce this complicated problem to one of the form of Problem (1) by using essentially the same procedure as that described in (1.12). Thus, we can obtain necessary conditions for this complicated problem as well.

24 We may also complicate Problem (1) in another fashion, by replacing the term $\int_{t_1}^{t_2} \hat{f}^i(x(t),t)\,dt$ in (4) by a term of the form $\int_{t_1}^{t_2} \hat{f}^i(x(t),u(t),t)\,dt$, and similarly modify (2) and (3), where $\hat{f}^0, \ldots, \hat{f}^\mu$ are now given functionals defined on $R^n \times V \times I$. However, instead of requiring the functions $\hat{f}^i$ to satisfy the differentiability Hypothesis (1.16) [as was done in (23)], we may instead suppose that V is a linear vector space, and that (for each $i = 0, \ldots, \mu$ and every $t \in I$) the function $(\xi,v) \to \hat{f}^i(\xi,v,t)\colon R^n \times V \to R$ is convex. This problem will be treated in Section 2 of Chapter VI under the assumptions that $V = R^r$, that $\mathscr{U}$ is a subset of $L^r_\infty(I)$ which is finitely open in itself, and that the $\hat{f}^i$ are continuous. Of course, if the functions $\hat{f}^i$ are independent of $v \in V$, then we are back at Problem (1).

Returning to the original problem (1), we point out that the remarks in (3.46 and 48) regarding the adjoint variable carry over here, provided that, in (3.47), we add the term $-\sum_0^\mu \alpha^i \mathscr{D}^0\hat{f}^i(t)$ to the right-hand side. Further, Lemma (3.68), which yields conditions under which the maximum condition (3.43) is not trivial, also carries over, provided that, in relation (55) of Condition (54), we replace $\chi^i(x_0(\tau^1), \ldots, x_0(\tau^k)) = 0$ by $\chi^i(x_0(\tau^1), \ldots, x_0(\tau^k)) + \int_{t_1}^{t_2} \hat{f}^i(x_0(t),t)\, dt = 0$. If Hypothesis (2.43) holds, then the maximum principle in integral form is equivalent to the maximum principle in pointwise form here also [see Theorem (3.70) as well as (2.48–49)]. If we modify Problem (1) by eliminating either the equality constraints (5) or the phase inequality constraints (3.9), or both, then the evident modifications in the necessary conditions should be made [see (3.50) and Theorem (3.75)]. Finally, most of the remarks of Section 6 carry over here also, particularly those that pertain to the possibility of some of the functions $\chi^0, \ldots, \chi^\mu$ being convex, without necessarily being continuously differentiable.

25 We close this section with the observation that analogous results, with evident modifications, can be obtained if we generalize Problems (2.1) and (4.1) in the same way that we generalized Problem (3.1) in this section.

8. Sufficient Conditions for a "Convex" Fixed-Time Optimal Control Problem

In this section, we shall use the results of Section 6 of Chapter IV in order to obtain sufficient conditions for solutions of the fixed-time optimal control problem which was considered in the preceding section. [A similar result can be obtained for the minimax problem mentioned in (7.20).] The sufficient conditions will be seen to "almost" coincide with the necessary conditions of Theorem (7.20). Of course, to obtain these sufficient conditions, we shall have to make some rather severe hypotheses on the problem data. Our results will generalize those that were obtained for Problem (II.4.14) in Theorem (II.4.32).

Thus, let us consider the following problem.

1 *Problem.* Let there be given a compact interval $I = [t_1, t_2]$, an integrable $(n \times n)$-matrix-valued function $\mathbf{A}$ defined on I, an arbitrary set V, a function $h: V \times I \to R^n$, a class $\mathscr{U}$ of functions $u: I \to V$ such that the function $t \to h(u(t),t): I \to R^n$ is in $L_1^n(I)$ for each $u \in \mathscr{U}$, convex and continuously differentiable functions $\chi^i: (R^n)^k \to R$ for $i = 0, 1, \ldots, \mu$, row vectors $\eta_{ij} \in R^n$ and real numbers v_i for $i = 1, \ldots, m$ and $j = 1, \ldots, k$, an open interval

$\tilde{I} \supset I$, a continuously differentiable function $\tilde{\chi} = (\tilde{\chi}^1, \ldots, \tilde{\chi}^\nu)$: $R^n \times \tilde{I} \to R^\nu$ such that $\mathscr{D}_1\tilde{\chi}$ and $\mathscr{D}_2\tilde{\chi}$ are continuously differentiable and such that the functions $\xi \to \tilde{\chi}^i(\xi,t)$: $R^n \to R$ (for each $i = 1, \ldots, \nu$ and every $t \in I$) are convex, a finite subset $I_a = \{\tau^1, \ldots, \tau^k\}$ of I such that $t_1 = \tau^1 < \tau^2 < \cdots < \tau^k = t_2$, closed subsets $I_1, \ldots, I_\nu$ of I, and a continuous function $\hat{f} = (\hat{f}^0, \hat{f}^1, \ldots, \hat{f}^\mu)$: $R^n \times I \to R^{\mu+1}$ which is continuously differentiable with respect to its first argument and is such that the function $\xi \to \hat{f}^i(\xi,t)$: $R^n \to R$ is convex for each $t \in I$ and every $i = 0, \ldots, \mu$. Then find an absolutely continuous function $x_0 \in \mathscr{C}^n(I)$ and a function $u_0 \in \mathscr{U}$ such that

2 $$\dot{x}_0(t) = \mathbf{A}(t)x_0(t) + h(u_0(t),t) \quad \text{a.e. in} \quad I,$$

3 $$\chi^i(x_0(\tau^1), \ldots, x_0(\tau^k)) + \int_{t_1}^{t_2} \hat{f}^i(x_0(t),t)\,dt \le 0 \quad \text{for} \quad i = 1, \ldots, \mu,$$

4 $$\sum_{j=1}^{k} \eta_{ij} x_0(\tau^j) = v_i \quad \text{for} \quad i = 1, \ldots, m,$$

5 $$\tilde{\chi}^i(x_0(t),t) \le 0 \quad \text{for all} \quad t \in I_i \quad \text{and each} \quad i = 1, \ldots, \nu,$$

and such that

6 $$\chi^0(x_0(\tau^1), \ldots, x_0(\tau^k)) + \int_{t_1}^{t_2} \hat{f}^0(x_0(t),t)\,dt \le \chi^0(x(\tau^1), \ldots, x(\tau^k)) + \int_{t_1}^{t_2} \hat{f}^0(x(t),t)\,dt$$

for all absolutely continuous functions $x \in \mathscr{C}^n(I)$ satisfying the differential equation

7 $$\dot{x}(t) = \mathbf{A}(t)x(t) + h(u(t),t) \quad \text{a.e. on} \quad I$$

for some $u \in \mathscr{U}$, together with the constraints (3)–(5) with x_0 replaced by x.

8 If $\mathscr{U}$ is convex under switching, then Problem (1) is evidently a special case of Problem (7.1). Of course, we have put some heavy restrictions on the data of Problem (7.1): that $\mathbf{G} = R^n$ and $A = \mathscr{C}^n(I)$, that f is of the special form $f(\xi,v,t) = \mathbf{A}(t)\xi + h(v,t)$ (i.e., that f is linear in ξ), that the functions χ^i are convex for $i = 0, \ldots, \mu$ and affine for $i = \mu + 1, \ldots, \mu + m$ [so that the equality constraints (3.4) are linear], that the functions $\tilde{\chi}^i$ are convex in their first argument, and that the functions $\hat{f}^i$ are continuously differentiable [as well as convex—as was already assumed in Problem (7.1)] with respect to their first argument.

Note that Problem (II.4.14) is a special case of Problem (1).

We shall obtain sufficient conditions for a pair $(x_0,u_0) \in \mathscr{C}^n(I) \times \mathscr{U}$ to be a solution of Problem (1). These sufficient conditions will be seen to "almost" coincide with the necessary conditions supplied by Theorem (7.20). More precisely, we shall prove the following theorem.

9 THEOREM. *Suppose that, for Problem* (1), $x_0 \in \mathscr{C}^n(I)$, *which is absolutely continuous, and* $u_0 \in \mathscr{U}$ *satisfy the problem constraints* (2)–(5). *Further, suppose that there exist row vectors* $\beta = (\beta^0, \ldots, \beta^\mu) \in \bar{R}_-^{\mu+1}$ *and* $\alpha = (\alpha^1, \ldots, \alpha^m) \in R^m$, *and row-vector-valued functions* $\tilde{\lambda} = (\tilde{\lambda}^1, \ldots, \tilde{\lambda}^\nu) \in [NBV(I)]^\nu$ *and* $\tilde{\psi} \in L_\infty^n(I)$ *such that*

(i) $\beta^0 < 0$,

(ii) *on each of the intervals* $[t_1,\tau^2), [\tau^2,\tau^3), \ldots, [\tau^{k-1},\tau^k]$, $\tilde{\psi}$ *is absolutely continuous and satisfies the equation*

10 $$\dot{\tilde{\psi}}(t) = -\tilde{\psi}(t)\mathbf{A}(t) - \beta\mathscr{D}_1\hat{f}(x_0(t),t) + \tilde{\lambda}(t)\mathscr{D}_1 p(x_0(t),u_0(t),t) \quad \text{a.e.},$$

where

11 $$p(\xi,v,t) = \mathscr{D}_1\tilde{\chi}(\xi,t)[\mathbf{A}(t)\xi + h(v,t)] + \mathscr{D}_2\tilde{\chi}(\xi,t)$$

$$\text{for all} \quad (\xi,v,t) \in R^n \times V \times I,$$

12 $$\text{(iii)} \int_{t_1}^{t_2} [\tilde{\psi}(t) - \tilde{\lambda}(t)\mathscr{D}_1\tilde{\chi}(x_0(t),t)]h(u_0(t),t)\,dt$$

$$\geq \int_{t_1}^{t_2} [\tilde{\psi}(t) - \tilde{\lambda}(t)\mathscr{D}_1\tilde{\chi}(x_0(t),t)]h(u(t),t)\,dt \quad \text{for all} \quad u \in \mathscr{U},$$

13 $$\text{(iv)}\ \tilde{\psi}(t_1) = -\beta\mathscr{D}_1^0\chi - \sum_{\ell=1}^{m} \alpha^\ell \eta_{\ell 1} + \tilde{\lambda}(t_1)\mathscr{D}_1\tilde{\chi}(x_0(t_1),t_1),$$

14 $$\tilde{\psi}(t_2) = \beta\mathscr{D}_k^0\chi + \sum_{\ell=1}^{m} \alpha^\ell \eta_{\ell k},$$

15 $$\tilde{\psi}(\tau^j) - \tilde{\psi}(\tau^{j-}) = -\beta\mathscr{D}_j^0\chi - \sum_{\ell=1}^{m} \alpha^\ell \eta_{\ell j}$$

$$\text{for} \quad j = 2, \ldots, k-1,$$

where $\mathscr{D}_j^0\chi$, *for each* $j = 1, \ldots, k$, *denotes* $\mathscr{D}_j\chi(x_0(\tau^1), \ldots, x_0(\tau^k))$ *and* $\chi = (\chi^0, \chi^1, \ldots, \chi^\mu)$,

16 $$\text{(v)}\ \beta^i\left[\chi^i(x_0(\tau^1), \ldots, x_0(\tau^k)) + \int_{t_1}^{t_2} \hat{f}^i(x_0(t),t)\,dt\right] = 0$$

$$\text{for} \quad i = 1, \ldots, \mu,$$

(vi) *for each* $i = 1, \ldots, \nu$, $\tilde{\lambda}^i$ *is nonincreasing on* I *and is constant on every subinterval of* I *which does not meet the closed subset* I_i^0 *of* I *defined by* (3.21). *Then* (x_0, u_0) *is a solution of Problem* (1).

The sufficient conditions of Theorem (9) differ from the necessary conditions of Theorem (7.20) only in that the necessary condition $\|\tilde{\lambda}\| + |\alpha| + |\beta| > 0$ is replaced by the stronger condition $\beta^0 < 0$.

Theorem (9) includes Theorem (II.4.32) as a special case.

Proof of Theorem (9). Let Φ be the absolutely continuous $(n \times n)$-matrix-valued function defined on I that satisfies

17 $$\dot{\Phi}(t) = \mathbf{A}(t)\Phi(t) \quad \text{a.e. on } I, \quad \Phi(t_1) = \text{the identity matrix.}$$

By Theorems (A.47 and 51) in the Appendix,

18 $$\frac{d}{dt}(\Phi^{-1}(t)) = -\Phi^{-1}(t)\mathbf{A}(t)$$
$$\text{a.e. on } I, \quad \Phi^{-1}(t_1) = \text{the identity matrix,}$$

and the solution of Eq. (7)—for any $u \in \mathscr{U}$—is given by

19 $$x(t) = \Phi(t)\left[x(t_1) + \int_{t_1}^{t} \Phi^{-1}(s)h(u(s),s)\, ds\right], \quad t \in I.$$

Now, by Theorem (A.51) of the Appendix, the function $\tilde{\psi}$ that satisfies conditions (ii) and (iv) of the theorem statement is uniquely specified by Eqs. (10), (11), (14), and (15), and it may then be verified by direct substitution into these equations that

20 $$\tilde{\psi}(t) = \psi(t) + \left\{\int_t^{t_2} \beta\mathscr{D}_1\hat{f}(x_0(s),s)\Phi(s)\, ds - \int_t^{t_2} \tilde{\lambda}(s)\frac{d}{ds}[\mathscr{D}_1\tilde{\chi}(x_0(s),s)\Phi(s)]\, ds\right\}\Phi^{-1}(t)$$
$$\text{for all} \quad t \in I,$$

where

21 $$\psi(t) = \sum_{j=1}^{k}\left[\beta\mathscr{D}_j^0\chi + \sum_{\ell=1}^{m} \alpha^\ell \eta_{\ell j}\right]\Phi(\tau^j)\Phi^{-1}(t)$$
$$\text{for} \quad \tau^{i-1} \le t < \tau^i \quad \text{and each} \quad i = 2, \ldots, k,$$

22 $$\psi(t_2) = \psi(\tau^k) = \beta\mathscr{D}_k^0\chi + \sum_{\ell=1}^{m} \alpha^\ell \eta_{\ell k}.$$

Integrating by parts in the second integral in (20), and taking into account that $\tilde{\lambda}(t_2) = 0$, we obtain

23 $$\tilde{\psi}(t) = \psi(t) + \left\{\int_t^{t_2} \beta\mathscr{D}_1\hat{f}(x_0(s),s)\Phi(s)\,ds + \int_t^{t_2} d\tilde{\lambda}(s)\mathscr{D}_1\tilde{\chi}(x_0(s),s)\Phi(s)\right\}\Phi^{-1}(t) + \tilde{\lambda}(t)\mathscr{D}_1\tilde{\chi}(x_0(t),t) \quad \text{for all} \quad t \in I.$$

Combining (13), (21), and (23), we conclude, on the basis of (18), that

24 $$\sum_{j=1}^{k}\left[\beta\mathscr{D}_j^0\chi + \sum_{\ell=1}^{m}\alpha^\ell\eta_{\ell j}\right]\Phi(\tau^j) + \int_{t_1}^{t_2}\beta\mathscr{D}_1\hat{f}(x_0(s),s)\Phi(s)\,ds + \int_{t_1}^{t_2} d\tilde{\lambda}(s)\mathscr{D}_1\tilde{\chi}(x_0(s),s)\Phi(s) = 0.$$

If we substitute (23) into (12), interchange the order of integration in each of the resulting double integrals, and take into account (19), (21), and (24), we obtain that

25 $$\sum_{j=1}^{k}\left[\beta\mathscr{D}_j^0\chi + \sum_{\ell=1}^{m}\alpha^\ell\eta_{\ell j}\right][x_{\xi,u}(\tau^j) - x_0(\tau^j)] + \int_{t_1}^{t_2}\beta\mathscr{D}_1\hat{f}(x_0(s),s)[x_{\xi,u}(s) - x_0(s)]\,ds + \int_{t_1}^{t_2} d\tilde{\lambda}(s)\mathscr{D}_1\tilde{\chi}(x_0(s),s)[x_{\xi,u}(s) - x_0(s)] \leq 0 \quad \text{for all} \quad \xi \in R^n \quad \text{and} \quad u \in \mathscr{U},$$

where, for each $\xi \in R^n$ and $u \in \mathscr{U}$, $x_{\xi,u}$ is the solution of Eq. (7) with initial value $x(t_1) = \xi$, so that

26 $$x_{\xi,u}(t) = \xi + \int_{t_1}^{t}\mathbf{A}(s)x_{\xi,u}(s)\,ds + \int_{t_1}^{t} h(u(s),s)\,ds \quad \text{for all} \quad t \in I, \xi \in R^n, \quad \text{and} \quad u \in \mathscr{U}.$$

Let us now appeal to Theorem (IV.6.9). We first identify Problem (1) with Problem (IV.1.3) in such a way that Hypotheses (IV.6.1–5) hold.

Thus, we set $\mathscr{X} = \mathscr{C}^n(I) = A = A'$, define the linear operator $L:\mathscr{X} \to \mathscr{X}$ by the relation

$$(Lx)(t) = \int_{t_1}^{t}\mathbf{A}(s)x(s)\,ds, \quad t \in I, \quad x \in \mathscr{X},$$

denote by $\tilde{\mathscr{X}}$ the set of all $\tilde{x} \in \mathscr{X}$ of the form

$$\tilde{x}(t) = \xi + \int_{t_1}^{t} h(u(s),s)\, ds, \quad t \in I,$$

where $\xi \in R^n$ and $u \in \mathscr{U}$, and denote by $\mathscr{W}$ the set of all operators $T:\mathscr{X} \to \mathscr{X}$ of the form

$$Tx = Lx + \tilde{x}, \quad x \in \mathscr{X},$$

where $\tilde{x} \in \tilde{\mathscr{X}}$. Further, let us set $\mathscr{Z}_1 = R^\mu \times \mathscr{C}^\nu(I)$ and $Z_1 = \bar{R}^\mu_- \times Z'$, where the closed convex cone Z' is given by (IV.1.12), let us define the functionals $\hat{\phi}^i_j$ and $\hat{\phi}^i$ $(i = 0, \ldots, \mu; j = 1, 2)$ on $\mathscr{X}$ by (7.7–9) and the function $\hat{\phi}':\mathscr{X} \to \mathscr{C}^\nu(I)$ by (IV.1.14), let us set $\hat{\phi}_1 = (\hat{\phi}^1, \ldots, \hat{\phi}^\mu, \hat{\phi}')$, and let us define the function $\hat{\varphi} = (\hat{\varphi}^1, \ldots, \hat{\varphi}^m)$: $\mathscr{X} \to R^m$ by the formulas

$$\hat{\varphi}^i(x) = \sum_{j=1}^{k} \eta_{ij} x(\tau^j) - v_i, \quad i = 1, \ldots, m.$$

Then it is easily seen that Problem (1) is of the form of Problem (IV.1.3), and that, by Corollary (A.28), Hypothesis (IV.6.5) holds. Hypotheses (IV.6.1 and 2) evidently hold. Further, it easily follows from our hypotheses on $\tilde{\chi}$, $\hat{f}$, and the χ^i that the functionals $\hat{\phi}^i_j$ and $\hat{\phi}^i$ $(i = 0, \ldots, \mu$: $j = 1, 2)$ are convex, and that $\hat{\phi}'$ is Z'-convex, so that $\hat{\phi}_1$ is Z_1-convex [see (I.2.33–35)]. Further, our hypotheses on $\tilde{\chi}$, $\hat{f}$, and the χ^i imply [see the discussion in (IV.1.49)] that $\hat{\phi}'$ and the $\hat{\phi}^i$ are Fréchet differentiable, with Fréchet differentials [which must also be Gâteaux directional differentials—see Theorems (I.7.29 and 44) and (I.7.5)] at any $x_0 \in \mathscr{X}$ given by (IV.1.52) and the formula

27 $$D\hat{\phi}^i(x_0; \delta x) = \sum_{j=1}^{k} \mathscr{D}^0_j \chi^i \delta x(\tau^j) + \int_{t_1}^{t_2} \mathscr{D}_1 \hat{f}^i(x_0(t),t)\delta x(t)\, dt,$$

$$\delta x \in \mathscr{X}, \quad i = 0, \ldots, \mu,$$

so that, by (I.7.49), $\hat{\phi}_1$ is Gâteaux directionally differentiable. Hence, Hypotheses (IV.6.3 and 4) hold [see Lemmas (I.4.19 and 28)].

If we now define $\tilde{x}_0 \in \tilde{\mathscr{X}}$ by

$$\tilde{x}_0(t) = x_0(t_1) + \int_{t_1}^{t} h(u_0(s),s)\, ds, \quad t \in I,$$

and define $\ell_1 \in \mathscr{X}_1^*$ by the relation

$$\ell_1(\xi^1, \ldots, \xi^\mu, z) = \sum_{i=1}^{\mu} \beta^i \xi^i + \int_{t_1}^{t_2} d\tilde{\lambda}(t) z(t),$$

then it follows directly from our hypotheses as well as relations (25)–(27) and (IV.1.52), that relations (IV.6.10–14) hold. Our desired conclusion now follows at once from Theorem (IV.6.9). |||

28 Corollary. *If, in Problem* (1), *there are subsets $U(t)$ of V, defined for each $t \in I$, such that $u(t) \in U(t)$ for all $u \in \mathscr{U}$ and $t \in I$, then Theorem* (9) *remains in force if we replace* (12) *by the condition*

29
$$\begin{aligned}[\tilde{\psi}(t) - \tilde{\lambda}(t)\mathscr{D}_1\tilde{\chi}(x_0(t),t)]h(u_0(t),t) \\ = \max_{v \in U(t)} [\tilde{\psi}(t) - \tilde{\lambda}(t)\mathscr{D}_1\tilde{\chi}(x_0(t),t)]h(v,t)\end{aligned}$$

for almost all $\quad t \in I$.

Proof. The proof is immediate since the "pointwise" maximum condition (29) is clearly stronger than the integral maximum condition (12). |||

30 If we modify Problem (1) by deleting the phase coordinate inequality constraint (5), then the sufficient conditions of Theorem (9) remain in force, provided that we set $\tilde{\lambda} \equiv 0$ and omit (11) and all terms involving $\tilde{\chi}$. Similarly, if we delete the equality constraints (4) and/or the inequality constraints (3), then Theorem (9) still holds, provided that we make some evident modifications in relations (13)–(16).

31 We may, in the statement of Problem (1), remove the differentiability requirements on the functions χ^i and $\hat{f}^i$. Then the sufficient conditions of Theorem (9) remain in force, provided that $(\mathscr{D}_1^0\chi^i, \ldots, \mathscr{D}_k^0\chi^i)$, for each $i = 0, \ldots, \mu$, is interpreted as a subgradient of χ^i at $(x_0(\tau^1), \ldots, x_0(\tau^k))$, and $\mathscr{D}_1\hat{f}(x_0(t),t)$ in Eq. (10) is replaced by some function $\mathscr{D}^0\hat{f} = (\mathscr{D}^0\hat{f}^0, \ldots, \mathscr{D}^0\hat{f}^\mu)$ such that (for each $i = 0, \ldots, \mu$) $\mathscr{D}^0\hat{f}^i \in L_\infty^n(I)$, and $\mathscr{D}^0\hat{f}^i(t)$ is a subgradient of the function $\xi \to \hat{f}^i(\xi,t)$ at $x_0(t)$ for almost every $t \in I$. [Thus, the resultant sufficient conditions "almost" agree with those of Theorem (7.20).] This result may be proved essentially in the same way as Theorem (9), because one may derive an inequality which differs from (25) only in the notational changes described above, and this modified inequality is easily seen to imply that (IV.6.14) holds. [Note that $\hat{\phi}^i$, for each $i = 0, \ldots, \mu$, as a convex

functional on $\mathscr{X}$, is here Gâteaux directionally differentiable by Lemma (I.7.7), even if the χ^i and $\hat{f}^i$ are not differentiable].

32 We may also modify Problem (1) by adding a term of the form $\int_{t_1}^{t_2} \hat{h}^i(u_0(t),t)\,dt$ to the left-hand side of the inequality constraints (3), and similarly modify the cost functional appearing in (6), where $\hat{h} = (\hat{h}^0, \ldots, \hat{h}^\mu)$ is some given function from $V \times I$ into $R^{\mu+1}$ with the property that, for each $u \in \mathscr{U}$, the function $t \to \hat{h}(u(t),t)\colon I \to R^{\mu+1}$ is in $L_1^{\mu+1}(I)$. We may reduce this problem to the form of the original Problem (1) by using the procedure described in (1.12). Once we do this, we obtain essentially the same sufficient conditions as for the original problem, except that now (i) the maximum condition (12) should be replaced by the condition

33 $$\int_{t_1}^{t_2} \{[\tilde{\psi}(t) - \tilde{\lambda}(t)\mathscr{D}_1\tilde{\chi}(x_0(t),t)]h(u_0(t),t) + \beta\hat{h}(u_0(t),t)\}\,dt$$
$$\geq \int_{t_1}^{t_2} \{[\tilde{\psi}(t) - \tilde{\lambda}(t)\mathscr{D}_1\tilde{\chi}(x_0(t),t)]h(u(t),t) + \beta\hat{h}(u(t),t)\}\,dt$$
$$\text{for all} \quad u \in \mathscr{U},$$

and (ii) $\hat{f}^i(x_0(t),t)$ must be replaced by $[\hat{f}^i(x_0(t),t) + \hat{h}^i(u_0(t),t)]$ in (16).

34 Finally, we point out that results analogous to those of this section can be obtained for the minimax problem in which we try to minimize the cost functional

$$\max_{t \in I_0} \tilde{\chi}^0(x_0(t),t)$$

[instead of $\hat{\phi}^0$, as defined by (7.7–9) with $i = 0$], where I_0 is some given closed subset of I, and $\tilde{\chi}^0$ is a given function from $R^n \times \tilde{I}$ into R with the same properties as those we hypothesized for $\tilde{\chi}^1, \ldots, \tilde{\chi}^\nu$ [see Theorem (IV.6.17)]. We leave the details to the reader.

CHAPTER VI

Optimal Control Problems with Parameters and Related Problems

IN THIS CHAPTER, we shall continue our investigation of optimal control problems with ordinary differential equation constraints. But now we shall exclusively be concerned with problems that can be expressed in parametric form, using the viewpoint of Section 2 of Chapter IV. This "parameter" will, in much (though not all) of this chapter turn out to be the control function.

In Section 1, we shall examine a free-time optimal control problem which is in the spirit of Problem (V.2.1), but in which the function f in the right-hand side of the differential equation, together with the functions χ^i (which define the constraints) and the free times τ^i depend also on a parameter π, and our optimization is carried out both with respect to the admissible controls $u \in \mathscr{U}$ and the parameter π (which is constrained to lie in some given set Π_a in a Banach space $\mathscr{P}$). Using the results of Section 2 of Chapter IV, together with Theorem (IV.5.35), we shall obtain necessary conditions that solutions of this problem must satisfy. We shall specialize these conditions to the case of parameter optimization problems only; i.e., to problems in which the controls are absent. We shall show how it is sometimes advantageous to view an optimal control problem as a parameter optimization problem—by considering the control to be a parameter. In fact, this latter viewpoint is used to investigate optimal control problems in Sections 2–4 of this chapter.

In Section 2, we shall consider an optimal control problem which generalizes Problem (V.7.1) by permitting the integrands $\hat{f}^i$—which appear both in the inequality constraints and in the cost functional—to depend on the control variable $u(t)$ as well as on the phase variable $x(t)$ [see (V.7.24)]. This problem will be unorthodox in that we shall replace the usual differentiability requirement on the $\hat{f}^i$ by a convexity condition, and our necessary conditions—which will be weaker than usual—will be in terms of subgradients of the $\hat{f}^i$.

In Section 3, we shall examine a fixed-time optimal control problem with so-called mixed control-phase variable inequality constraints, i.e., with constraints of the form

$$\tilde{\tilde{\chi}}^i(x(t),u(t),t) \leq 0 \quad \text{for almost all} \quad t \in I'_i \quad \text{and each} \quad i = 1, \ldots, \nu',$$

where $\tilde{\tilde{\chi}}^1, \ldots, \tilde{\tilde{\chi}}^{\nu'}$ are given continuous functionals, and $I'_1, \ldots, I'_{\nu'}$ are given Lebesgue measurable subsets of I. In Section 4, we shall consider a minimax problem in which we wish to minimize a functional of the form

$$\operatorname*{ess\,sup}_{t \in I'_0} \max_{\xi \in \mathbf{K}} \left[\xi \cdot \tilde{\tilde{\chi}}(x(t),u(t),t)\right],$$

where $\tilde{\tilde{\chi}}$ is a given continuous ν'-vector-valued function, $\mathbf{K}$ is a given compact subset of $R^{\nu'}$, and I'_0 is a given Lebesgue measurable subset of I. Not surprisingly, the necessary conditions which we shall obtain in Sections 3 and 4 will turn out to be very similar.

1. Optimal Control Problems with Parameters

1 Throughout this section, we shall suppose that we are given an open set $\mathbf{G} \subset R^n$, an arbitrary set V, and a compact interval $I = [t_1, t_2]$. We shall also suppose that we are given a Banach space $\mathscr{P}$ (the so-called parameter space), and an open set $\Pi \subset \mathscr{P}$.

Corresponding to the function f of Chapter V, we shall in this section suppose that we are given a function $f: \mathbf{G} \times V \times I \times \Pi \to R^n$ and a class $\mathscr{U}$ of "admissible" control functions $u: I \to V$. Corresponding to Hypotheses (V.1.16–18), we shall make the following hypotheses:

2 HYPOTHESIS. *For each* $(v,t) \in V \times I$, *the function* $(\xi,\pi) \to f(\xi,v,t,\pi)$: $\mathbf{G} \times \Pi \to R^n$ *is continuously differentiable.*

3 HYPOTHESIS. *For every* $(\xi,\pi) \in \mathbf{G} \times \Pi$ *and each* $u \in \mathscr{U}$, *the function* $t \to f(\xi,u(t),t,\pi)$: $I \to R^n$ *is measurable.*

4 HYPOTHESIS. *For each* $u \in \mathscr{U}$, *each compact subset* $\mathbf{G}_c$ *of* $\mathbf{G}$, *and each* $\pi_1 \in \Pi$, *there exist an integrable function* $\tilde{m}: I \to R_+$ *and a number* $\zeta > 0$—*where* $\tilde{m}$ *may depend on* $\mathbf{G}_c$, π_1, *and* u, *but* ζ *may*

depend only on π_1 *and* u*—such that*

5 $$|f(\xi,u(t),t,\pi)| + |\mathscr{D}_1 f(\xi,u(t),t,\pi)| + \|D_3 f(\xi,u(t),t,\pi;\cdot)\| \leq \tilde{m}(t)$$

for all $(\xi,t) \in \mathbf{G}_c \times I$ *and all* $\pi \in \Pi$ *such that* $\|\pi - \pi_1\| \leq \zeta$.

6 If V is a topological space and f is continuous, then Hypothesis (3) holds so long as each $u \in \mathscr{U}$ is measurable [in the sense that $u^{-1}(O)$ is a (Lebesgue) measurable subset of I for each open subset O of V]. If also $\mathscr{P} = R^{m'}$ (for some positive integer m'), if f is continuously differentiable with respect to both its first and last arguments—so that, by Theorem (I.7.54), Hypothesis (2) certainly holds—and if, for each $u \in \mathscr{U}$, there is a compact subset V_u of V such that $u(t) \in V_u$ for all $t \in I$, then, by (I.3.35 and 46), Hypothesis (4) holds, and we may even choose $\tilde{m}$ to be constant.

7 It is clear that if f and $\mathscr{U}$ satisfy Hypotheses (2)–(4), then, for each $u \in \mathscr{U}$, the function $\hat{F}_u : \mathbf{G} \times I \times \Pi \to R^n$ defined by the formula

8 $$\hat{F}_u(\xi,t,\pi) = f(\xi,u(t),t,\pi), \quad (\xi,t,\pi) \in \mathbf{G} \times I \times \Pi,$$

belongs to the set $\hat{\mathscr{F}}_1$ defined in (IV.4.23).

9 Further, if $\mathscr{U}$ is convex under switching, then evidently the set $\{\hat{F}_u : u \in \mathscr{U}\}$—where $\hat{F}_u$ is defined by (8) for each $u \in \mathscr{U}$— is also convex under switching.

We now consider the following free-time optimal control problem with parameters.

10 *Problem.* Let there be given a function $f : \mathbf{G} \times V \times I \times \Pi \to R^n$ [where $\mathbf{G}$, V, I, and Π are as in (1)] satisfying Hypothesis (2), an open set $A \subset \mathscr{C}^n(I)$ such that $x(t) \in \mathbf{G}$ for all $x \in A$ and $t \in I$, a class $\mathscr{U}$ of functions $u : I \to V$ which is convex under switching and satisfied Hypotheses (3) and (4), continuously differentiable functions $\chi = (\chi^0, \chi^1, \ldots, \chi^{\mu+m}) : (R^n)^k \times \Pi \to R^{\mu+m+1}$ and $\tau = (\tau^1, \ldots, \tau^k) : \Pi \to R^k$, and a subset Π_a of Π which is finitely open in itself and has the property that $\tau(\Pi \cap \operatorname{co} \Pi_a) \subset I^k$. Then find an absolutely continuous function $x_0 \in A$, a function $u_0 \in \mathscr{U}$, and an element $\pi_0 \in \Pi_a$ such that

11 $$\dot{x}_0(t) = f(x_0(t),u_0(t),t,\pi_0) \quad \text{for almost all} \quad t \in I,$$

12 $$\chi^i(x_0(\tau^1(\pi_0)), \ldots, x_0(\tau^k(\pi_0)),\pi_0) \leq 0 \quad \text{for} \quad i = 1, \ldots, \mu,$$

13 $$\chi^i(x_0(\tau^1(\pi_0)), \ldots, x_0(\tau^k(\pi_0)),\pi_0) = 0 \quad \text{for} \quad i = \mu + 1, \ldots, \mu + m,$$

and such that

14 $$\chi^0(x_0(\tau^1(\pi_0)),\ldots,x_0(\tau^k(\pi_0)),\pi_0) \le \chi^0(x(\tau^1(\pi)),\ldots,x(\tau^k(\pi)),\pi)$$

for all $(x,\pi) \in A \times \Pi_a$ with the properties that

(i) x is an absolutely continuous function satisfying the differential equation

15 $$\dot{x}(t) = f(x(t),u(t),t,\pi) \quad \text{for almost all} \quad t \in I$$

for some $u \in \mathscr{U}$,

16 (ii) $\chi^i(x(\tau^1(\pi)),\ldots,x(\tau^k(\pi)),\pi) \le 0 \quad$ for $\quad i = 1,\ldots,\mu$,

17 (iii) $\chi^i(x(\tau^1(\pi)),\ldots,x(\tau^k(\pi)),\pi) = 0 \quad$ for $\quad i = \mu+1,\ldots,\mu+m$.

18 If $(x_0,u_0,\pi_0) \in A \times \mathscr{U} \times \Pi_a$ is a solution of Problem (10), then x_0 will be called an *optimal trajectory*, u_0 an *optimal control*, and π_0 an *optimal parameter*.

We shall use the notation of Chapter IV, with $\mathscr{X} = \mathscr{C}^n(I)$.

Let us set [see (IV.4.5)]

19 $$\hat{\mathscr{W}} = \{\hat{T}_{\hat{x}^*,\hat{F}_u} : u \in \mathscr{U},\ \hat{x}^* \in \hat{\mathscr{C}}^n \text{ and } \hat{x}^* \text{ is constant on } I \times \Pi\},$$

when, for each $u \in \mathscr{U}$, $\hat{F}_u$ is the function in $\hat{\mathscr{F}}_1$ defined by (8) [see (7) and (IV.4.23)] and the operator $\hat{T}_{\hat{x}^*,\hat{F}} : A \times \Pi \to \mathscr{C}^n(I)$ [for each $\hat{x}^* \in \hat{\mathscr{C}}^n$ and each $\hat{F} \in \hat{\mathscr{F}}_1$] is defined by [see (IV.4.29)]

20 $$(\hat{T}_{\hat{x}^*,\hat{F}}(x,\pi))(t) = \hat{x}^*(t,\pi) + \int_{t_1}^{t} \hat{F}(x(s),s,\pi)\,ds, \quad t \in I, \quad x \in A, \quad \pi \in \Pi.$$

Inasmuch as Eq. (15) is clearly equivalent to the equation

21 $$x(t) = (\hat{T}_{\hat{x}^*,\hat{F}_u}(x,\pi))(t) \quad \text{for all} \quad t \in I,$$

where $\hat{x}^*(t,\pi) = x(t_1)$ for all $t \in I$ and $\pi \in \Pi$, Problem (10) has been recast in the form of Problem (IV.2.64). [Recall that, from the results of the Appendix, $\hat{\mathscr{W}} \times \Pi_a \subset \mathscr{S}_0$—see (IV.2.3) and (IV.4.33).]

In order to obtain necessary conditions for Problem (10), we shall appeal to Theorems (IV.2.68) and (IV.5.35).

Thus, suppose that (x_0,u_0,π_0) is a solution of Problem (10) such that x_0 is differentiable at $\tau_0^j = \tau^j(\pi_0)$ for each $j = 1,\ldots,k$. Thus, (11)–(13) hold. For ease of notation, let us suppose that $\tau_0^1 \le \tau_0^2 \le \cdots \le \tau_0^k$. As we have just seen, (x_0,π_0) is then a solution of Problem (IV.2.64), where $\hat{\mathscr{W}}$ is defined by (19), (20), and (8). By Theorem (IV.5.35) and (9), Hypotheses (IV.2.10–12) are satisfied by $\mathscr{W} = \hat{\mathscr{W}} \times \Pi_a$ and by $(\hat{T}_0,\pi_0)$, where the operator

$\hat{T}_0 \in \hat{\mathscr{W}}$ is defined by

22 $$(\hat{T}_0(x,\pi))(t) = x_0(t_1) + \int_{t_1}^{t} f(x(s),u_0(s),s,\pi)\,ds, \quad t \in I, \quad x \in A, \quad \pi \in \Pi.$$

Further, Eq. (11) is clearly equivalent to the operator equation $x_0 = \hat{T}_0(x_0,\pi_0)$, so that $(\hat{T}_0,\pi_0) \in \mathscr{S}_1$. Then, by Theorem (IV.2.68), there exists a nonzero row vector $\alpha = (\alpha^0, \alpha^1, \ldots, \alpha^{\mu+m}) \in \bar{R}^{\mu+1} \times R^m$ such that

23 $$\text{(i)} \sum_{j=1}^{k} \alpha \mathscr{D}_j^0 \chi \delta x_{\xi,u}(\tau_0^j) \le \sum_{j=1}^{k} \alpha \mathscr{D}_j^0 \chi \delta x_{x_0(t_1),u_0}(\tau_0^j) \quad \text{for all} \quad \xi \in R^n \quad \text{and} \quad u \in \mathscr{U},$$

where, for each $\xi \in R^n$ and $u \in \mathscr{U}$, $\delta x_{\xi,u}$ denotes the column-vector-valued function in $\mathscr{C}^n(I)$ defined by [see (IV.2.70) and (IV.4.31)]

24 $$\delta x_{\xi,u}(t) = \int_{t_1}^{t} \mathscr{D}_1 f(x_0(s),u_0(s),s,\pi_0)\delta x_{\xi,u}(s)\,ds + \xi + \int_{t_1}^{t} f(x_0(s),u(s),s,\pi_0)\,ds, \quad t \in I,$$

and $\mathscr{D}_j^0\chi$ (for each j) denotes $\mathscr{D}_j\chi(x_0(\tau_0^1), \ldots, x_0(\tau_0^k),\pi_0)$,

25 $$\text{(ii)} \sum_{i=1}^{\mu} \alpha^i \chi^i(x_0(\tau_0^1), \ldots, x_0(\tau_0^k),\pi_0) = 0,$$

and

26 $$\text{(iii)} \sum_{j=1}^{k} \alpha \mathscr{D}_j^0 \chi[\delta x_{\delta\pi}(\tau_0^j) + \dot{x}_0(\tau_0^j) D^0\tau^j(\delta\pi)] + \alpha D_{k+1}^0 \chi(\delta\pi) \le 0 \quad \text{for all} \quad \delta\pi \in \overline{\Pi_a - \pi_0},$$

where, for each $\delta\pi \in \Pi$, $\delta x_{\delta\pi}$ denotes the column-vector-valued function in $\mathscr{C}^n(I)$ defined by [see (IV.2.72) and (IV.4.32)]

27 $$\delta x_{\delta\pi}(t) = \int_{t_1}^{t} \mathscr{D}_1 f(x_0(s),u_0(s),s,\pi_0)\delta x_{\delta\pi}(s)\,ds + \int_{t_1}^{t} D_4 f(x_0(s),u_0(s),s,\pi_0;\delta\pi)\,ds, \quad t \in I,$$

$D_{k+1}^0\chi(\cdot)$ denotes $D_{k+1}\chi(x_0(\tau_0^1), \ldots, x_0(\tau_0^k),\pi_0;\cdot)$, and $D^0\tau^j(\cdot)$ (for each j) denotes $D\tau^j(\pi_0;\cdot)$.

Arguing as in the proof of Theorem (V.2.34), we can prove the following.

28 THEOREM. *Let (x_0,u_0,π_0) be a solution of Problem (10) such that x_0 is differentiable at $\tau_0^j = \tau^j(\pi_0)$ for each $j = 1, \ldots, k$, and*

suppose that $\tau_0^1 < \tau_0^2 < \cdots < \tau_0^k$. *Then there exist a nonzero row vector* $\alpha = (\alpha^0, \alpha^1, \ldots, \alpha^{\mu+m}) \in \bar{R}_-^{\mu+1} \times R^m$ *and an n* (row)-*vector-valued function* ψ *defined on* $[\tau_0^1,\tau_0^k]$ *such that*

(i) *on each of the intervals* $[\tau_0^1,\tau_0^2), \ldots, [\tau_0^{k-2},\tau_0^{k-1})$, *and* $[\tau_0^{k-1},\tau_0^k]$, $\psi(\cdot)$ *is absolutely continuous and satisfies the linear ordinary differential equation*

29 $$\dot{\psi}(t) = -\psi(t)\mathscr{D}_1 f(x_0(t),u_0(t),t,\pi_0) \quad \text{a.e.},$$

(ii) u_0 *satisfies the maximum condition*

30 $$\int_{\tau_0^1}^{\tau_0^k} \psi(t)f(x_0(t),u_0(t),t,\pi_0)\,dt \geq \int_{\tau_0^1}^{\tau_0^k} \psi(t)f(x_0(t),u(t),t,\pi_0)\,dt \quad \textit{for all} \quad u \in \mathscr{U},$$

(iii) ψ *satisfies the following boundary and jump conditions*:

31 $$\psi(\tau_0^k) = \alpha\mathscr{D}_k^0\chi, \quad \psi(\tau_0^1) = -\alpha\mathscr{D}_1^0\chi,$$

32 $$\psi(\tau_0^j) - \psi(\tau_0^{j-}) = -\alpha\mathscr{D}_j^0\chi \quad \textit{for} \quad j = 2, \ldots, k-1,$$

where $\mathscr{D}_j^0\chi$ *denotes* $\mathscr{D}_j\chi(x_0(\tau_0^1), \ldots, x_0(\tau_0^k),\pi_0)$ *for each j*,

33 (iv) $\alpha^i\chi^i(x_0(\tau_0^1), \ldots, x_0(\tau_0^k),\pi_0) = 0$ *for* $i = 1, \ldots, \mu$, *and*

34 (v) $$\int_{\tau_0^1}^{\tau_0^k} \psi(t)D_4 f(x_0(t),u_0(t),t,\pi_0;\delta\pi)\,dt + \sum_{j=1}^{k} \alpha\mathscr{D}_j^0\chi\dot{x}_0(\tau_0^j)D^0\tau^j(\delta\pi) + \alpha D_{k+1}^0\chi(\delta\pi) \leq 0 \quad \textit{for all} \quad \delta\pi \in \overline{\Pi_a - \pi_0},$$

where $D_{k+1}^0\chi$ *and* $D^0\tau^j$ *denote* $D_{k+1}\chi$ *and* $D\tau^j$, *respectively, evaluated at* $(x_0(\tau_0^1), \ldots, x_0(\tau_0^k),\pi_0)$ *and at* π_0. *If* π_0 *is an internal point of* Π_a, *then* (34) *can be written in the form*

35 $$\int_{\tau_0^1}^{\tau_0^k} \psi(t)D_4 f(x_0(t),u_0(t),t,\pi_0;\delta\pi)\,dt + \sum_{j=1}^{k} \alpha\mathscr{D}_j^0\chi\dot{x}_0(\tau_0^j)D^0\tau^j(\delta\pi) + \alpha D_{k+1}^0\chi(\delta\pi) = 0 \quad \textit{for all} \quad \delta\pi \in \mathscr{P}.$$

The hypothesis in the preceding theorem that $\tau_0^1 < \tau_0^2 < \cdots < \tau_0^k$ was, of course, made only for convenience. Further, if $\tau_0^{i-1} = \tau_0^i$ for some $i = 2, \ldots, k$, then (32) must be modified in an obvious manner, while, in all other respects, the necessary conditions remain in force. If, for some $j' = 1, \ldots, k$, $\tau^{j'}$ is independent of π, i.e., is fixed, then we can delete the hypothesis that x^0 is differentiable at $\tau^{j'}$ [and, of course, $D^0\tau^{j'}$ must be set equal to zero in (34)]. If the function τ is constant on Π_a, so that we have a fixed-time problem, then we can dispense with the hypothesis that x_0 is differentiable at the τ_0^j, and (34) takes the form

36
$$\int_{\tau_0^1}^{\tau_0^k} \psi(t) D_4 f(x_0(t), u_0(t), t, \pi_0; \delta\pi)\, dt + \alpha D_{k+1}^0 \chi(\delta\pi) \leq 0$$
$$\text{for all} \quad \delta\pi \in \overline{\Pi_a - \pi_0}.$$

If $\mathscr{P} = R^{m'}$ (for some positive integer m'), then (34) can be rewritten in the form

37
$$\left[\int_{\tau_0^1}^{\tau_0^k} \psi(t) \mathscr{D}_4 f(x_0(t), u_0(t), t, \pi_0)\, dt + \sum_{j=1}^{k} \alpha \mathscr{D}_j^0 \chi \dot{x}_0(\tau_0^j) \mathscr{D}^0 \tau^j + \alpha \mathscr{D}_{k+1}^0 \chi \right] \delta\pi \leq 0$$
$$0 \quad \text{for all} \quad \delta\pi \in \overline{\Pi_a - \pi_0}$$

(where $\mathscr{D}^0\tau^j$ and $\mathscr{D}_{k+1}^0\chi$ denote the evident Jacobian matrices), and similarly for (35) and (36).

38 If x_0 is not differentiable at $\tau_0^j = \tau^j(\pi_0)$ for some j, then we can still obtain necessary conditions for Problem (10), once we take into account that if (x_0, u_0, π_0) is a solution of Problem (10), then (x_0, u_0) is also a solution of the problem without parameters obtained by fixing π at π_0. This problem is of the form of Problem (V.3.10), and, appealing to Theorem (V.3.75), we may obtain necessary conditions which coincide with those of Theorem (28), with the exception that conclusion (v) should be deleted.

39 If Problem (10) is modified by eliminating either the inequality constraints (16) or the equality constraints (17), or both, then the necessary conditions of Theorem (28) remain in force, but with μ, or m, or both, equal to zero—so that (33) is to be deleted from the necessary conditions if there are no inequality constraints [see (IV.2.63 and 75)].

The necessary conditions of Theorem (28) bear a close resemblance to those of Theorem (V.2.34), which should not be surprising inasmuch as Problem (10) is a generalization of Problem

(V.2.1). In fact, the only essential difference between the two theorems is that relation (V.2.16) is replaced here by the more general relation (34). The extra integral term in (34) reflects the fact that, in Problem (10), the function f may also depend on a parameter.

40 The comments in (V.2.37) and in the second sentence of (V.2.36) regarding the adjoint variable ψ carry over here essentially without change. Further, if we replace Condition (V.2.42) by the following one:

CONDITION. *There is at least one element* $(\xi_1, \ldots, \xi_k, \delta\pi) \in (R^n)^k \times \overline{(\Pi_a - \pi_0)}$ *such that*

$$D^0_{k+1}\chi^i(\delta\pi) + \sum_{j=1}^{k} (\mathscr{D}^0_j \chi^i) \cdot \xi_j \begin{cases} < 0 & \text{for } i = 0 \text{ and each} \\ & i = 1, \ldots, \mu \text{ such that} \\ & \chi^i(x_0(\tau_0^1), \ldots, x_0(\tau^k), \pi_0) = 0, \\ = 0 & \text{for } i = \mu + 1, \ldots, \mu + m, \end{cases}$$

and the m vectors $(\mathscr{D}^0_1\chi^i, \ldots, \mathscr{D}^0_k\chi^i)$, $i = \mu + 1, \ldots, \mu + m$, *are linearly independent*;

then we have obtained a condition which is sufficient to ensure that ψ will not vanish on at least one of the intervals $[\tau_0^i, \tau_0^{i+1})$, $i = 1, \ldots, k - 1$. Finally, if we impose a hypothesis which is essentially the same as Hypothesis (V.2.43), then we can show, as in the proof of Theorem (V.2.44), that the maximum principle in integral form, (30), is equivalent to the following pointwise maximum principle:

41
$$\psi(t)f(x_0(t), u_0(t), t, \pi_0) = \max_{v \in U} \psi(t) f(x_0(t), v, t, \pi_0)$$
$$\text{for almost all} \quad t \in [\tau_0^1, \tau_0^k].$$

In this regard, the comments in (V.2.48–49) are also pertinent here (with evident minor changes).

The reader at this point should review the discussions in (V.1.12) and (V.2.64 and 65) and see how they carry over—with certain obvious modifications—to Problem (10).

42 We point out that the terms $\alpha\mathscr{D}^0_j\chi$ in (34), (35), and (37) may be replaced by expressions involving ψ if we make use of (31) and (32). Also, many of the observations that were made in Section 2 of Chapter V [see, e.g., (V.2.53–63)] regarding relation (V.2.16)

and a Hamiltonian function have analogs here with respect to relations (34)–(37).

Let us now consider the following important special case of Problem (10).

43 *Problem.* Let there be given a function $f: \mathbf{G} \times I \times \Pi \to R^n$ [where $\mathbf{G}$, I, and Π are as in (1)] belonging to the class $\mathscr{F}_1$ [see (IV.4.23)], an open set $A \subset \mathscr{C}^n(I)$ such that $x(t) \in \mathbf{G}$ for all $x \in A$ and $t \in I$, continuously differentiable functions $\chi = (\chi^0, \chi^1, \ldots, \chi^{\mu+m}): (R^n)^k \times \Pi \to R^{\mu+m+1}$ and $\tau = (\tau^1, \ldots, \tau^k): \Pi \to R^k$, and a subset Π_a of Π which is finitely open in itself and is such that $\tau(\Pi \cap \text{co}\, \Pi_a) \subset I^k$. Then find a pair $(x_0, \pi_0) \in A \times \Pi_a$, with x_0 absolutely continuous, such that

44 $$\dot{x}_0(t) = f(x_0(t), t, \pi_0) \quad \text{for almost all} \quad t \in I,$$

such that (12) and (13) hold, and such that (14) holds for all $(x, \pi) \in A \times \Pi_a$ with the properties that (i) x is an absolutely continuous function satisfying the differential equation

45 $$\dot{x}(t) = f(x(t), t, \pi) \quad \text{for almost all} \quad t \in I,$$

and (ii) relations (16) and (17) hold.

Since Problem (43) does not involve any "control" functions u, it is not a true optimal control problem, but is a parameter optimization problem. For this problem, Theorem (28) evidently takes the following special form.

46 THEOREM. *Let* (x_0, π_0) *be a solution of Problem* (43) *such that* x_0 *is differentiable at* $\tau_0^j = \tau^j(\pi_0)$ *for each* $j = 1, \ldots, k$, *and suppose that* $\tau_0^1 < \tau_0^2 < \cdots < \tau_0^k$. *Then there exist a nonzero row vector* $\alpha = (\alpha^0, \alpha^1, \ldots, \alpha^{\mu+m}) \in \bar{R}_-^{\mu+1} \times R^m$ *and an* n *(row) vector-valued function* ψ *defined on* $[\tau_0^1, \tau_0^k]$ *such that conclusions* (i) *and* (iii)–(v) *of Theorem* (28) *hold, except that* (29) *should be replaced by the equation*

47 $$\dot{\psi}(t) = -\psi(t)\mathscr{D}_1 f(x_0(t), t, \pi_0) \quad \text{a.e.},$$

and the integral in (34) *should be replaced by*

$$\int_{\tau_0^1}^{\tau_0^k} \psi(t) D_3 f(x_0(t), t, \pi_0; \delta\pi)\, dt$$

[*and similarly in* (35)].

Virtually all of the remarks following Theorem (28) carry over to Theorem (46) as well.

In certain situations, an optimal control problem *without parameters* may be reduced to a parameter optimization problem of the form of Problem (43), and it is then of interest to compare the necessary conditions from optimal control theory with those of Theorem (46).

For the sake of simplicity, let us consider the fixed-time Problem (V.3.10). Let us make the following additional assumptions on the data of this problem:

48 HYPOTHESIS. *The set* $V = R^r$ *for some positive integer r,* $\mathscr{U}$ *is a subset of* $L_\infty^r(I)$ [*see* (I.6.30)] *which is finitely open in itself, and the function* $f: \mathbf{G} \times R^r \times I \to R^n$ *is continuous, and is, in addition, continuously differentiable with respect to its first and second arguments.*

If we set $\mathscr{P} = \Pi = L_\infty^r(I)$, and define the function $\hat{f}: \mathbf{G} \times I \times \Pi \to R^n$ by the formula

49 $$\hat{f}(\xi,t,u) = f(\xi,u(t),t) \quad \text{for all} \quad (\xi,t) \in \mathbf{G} \times I \quad \text{and} \quad u \in \Pi,$$

then it is not difficult to see, by virtue of (I.3.46) and Theorem (I.7.54), that $\hat{f} \in \mathscr{F}_1$ [see (IV.4.23)].

50 If we now set $\Pi_a = \mathscr{U}$ and set the function τ identically equal to $(\tau^1, \ldots, \tau^k)$—as defined in Problem (V.3.10)—then it is clear that Problem (V.3.10), subject to Hypothesis (48), is of the form of Problem (43), if, in the latter, we replace f by $\hat{f}$. Examining the necessary conditions of Theorem (46) [also see (38)] for this problem, and comparing them with those of Theorem (V.3.75), we find that the two sets of conditions differ only in that the maximum condition (V.3.77) [of Theorem (V.3.75)] takes, in Theorem (46), the form

51 $$\int_{t_1}^{t_2} \psi(t)\mathscr{D}_2 f(x_0(t),u_0(t),t)[u(t) - u_0(t)]\,dt \le 0 \quad \text{for all} \quad u \in \mathscr{U}.$$

Since $\mathscr{U}$ is finitely open in itself, (51) is clearly weaker than (V.3.77); i.e., in this case, Theorem (V.3.75) gives rise to stronger necessary conditions than Theorem (46) gives rise to. In view of the discussion in (IV.2.76), this should not be surprising.

52 Note that if $u_0 \in \operatorname{int} \mathscr{U}$, then (51) is equivalent to the relation

53 $$\psi(t)\mathscr{D}_2 f(x_0(t),u_0(t),t) = 0 \quad \text{for almost all} \quad t \in I,$$

or, in terms of the Hamiltonian function H defined by (V.2.58),

54 $$\mathscr{D}_3 H(\psi(t), x_0(t), u_0(t), t) = 0 \quad \text{for almost all} \quad t \in I.$$

Let us now suppose that $\mathscr{U}$ in addition satisfies the following hypothesis:

55 HYPOTHESIS. *There is a subset U of R^r such that* (i) $u(t) \in U$ *for all $u \in \mathscr{U}$ and every $t \in I$ and* (ii) *every constant function $u: I \to U$ belongs to $\mathscr{U}$.*

Hypothesis (55) is clearly satisfied whenever $\mathscr{U}$ consists of all essentially bounded and measurable (or all piecewise-continuous, or all piecewise-constant) functions $u: I \to U$.

56 Note that if both Hypotheses (48) and (55) hold, then the set U must be finitely open in itself and Hypothesis (2.43) holds. In this case, arguing as in the proof of Theorem (V.2.44), we can show that (51) is equivalent to the following (pointwise) relation:

57 $$\psi(t)\mathscr{D}_2 f(x_0(t), u_0(t), t)v \leq \psi(t)\mathscr{U}_2 f(x_0(t), u_0(t), t)u_0(t)$$
$$\text{for all} \quad v \in U, \quad \text{for almost all} \quad t \in I,$$

which is evidently weaker than the pointwise maximum principle (V.3.85), which also holds here, by Theorem (V.3.84).

58 It is worth pointing out that the weaker maximum principle holds for Problem (V.3.10)—subject to Hypothesis (48)—even if $\mathscr{U}$ is *not* convex under switching. For example, if $\mathscr{U}$ consists of all piecewise-continuous (or, alternately, piecewise-constant) functions $u: I \to U$—where U is some given subset of R^r which is either open or convex—whose points of discontinuity must belong to some given finite subset of I, then $\mathscr{U}$ is not convex under switching, but $\mathscr{U}$ is a subset of $L_\infty^r(I)$ which is finitely open in itself. Indeed, we have the following result.

59 THEOREM. *Let (x_0, u_0) be a solution of Problem* (V.3.10), *modified by replacing the hypothesis that $\mathscr{U}$ is convex under switching by Hypothesis* (48). *Then the necessary conditions of Theorem* (V.3.75) *remain in force, except that the maximum condition* (V.3.77) *must be replaced by* (51). *If also $u_0 \in \operatorname{int} \mathscr{U}$, then* (51) *is equivalent to* (53).

Although we have confined ourselves in this section to the conventional optimization problem, analogous results can be obtained for minimax problems and for problems with vector-valued criteria.

2. A (Second) Unorthodox Fixed-Time Optimal Control Problem

In this section, we shall consider an optimal control problem which is closely related to Problem (V.7.1), and which was already described in (V.7.24). It is unorthodox in a similar manner that Problem (V.7.1) was, in that a differentiability hypothesis on the problem data is replaced by a convexity hypothesis. We shall attack this problem from the viewpoint that was introduced at the end of the preceding section.

The problem we shall consider is the following:

1 *Problem.* Let there be given a continuous function $f:\mathbf{G} \times R^r \times I \to R^n$ (where $\mathbf{G}$ is a given open set in R^n and $I = [t_1, t_2]$ is a given compact interval) which is continuously differentiable with respect to its first and second arguments, an open set $A \subset \mathscr{C}^n(I)$ such that $x(t) \in \mathbf{G}$ for all $x \in A$ and $t \in I$, a set $\mathscr{U} \subset L_\infty^r(I)$ which is finitely open in itself, a continuously differentiable function $\chi = (\chi^0, \chi^1, \ldots, \chi^{\mu+m})\colon (R^n)^k \to R^{\mu+m+1}$, a continuous function $\hat{f} = (\hat{f}^0, \ldots, \hat{f}^\mu)\colon R^n \times R^r \times I \to R^{\mu+1}$ such that (for each $i = 0, \ldots, \mu$ and every $t \in I$) the function $(\xi,v) \to \hat{f}^i(\xi,v,t)\colon R^n \times R^r \to R$ is convex, and a finite subset $I_a = \{\tau^1, \ldots, \tau^k\}$ of I such that $t_1 = \tau^1 < \tau^2 < \cdots < \tau^k = t_2$. Then find an absolutely continuous function $x_0 \in A$ and a function $u_0 \in \mathscr{U}$ such that

2 $$\dot{x}_0(t) = f(x_0(t),u_0(t),t) \quad \text{for almost all} \quad t \in I,$$

3 $$\chi^i(x_0(\tau^1), \ldots ,x_0(\tau^k)) + \int_{t_1}^{t_2} \hat{f}^i(x_0(t),u_0(t),t)\, dt \le 0 \quad \text{for} \quad i = 1, \ldots, \mu,$$

4 $$\chi^i(x_0(\tau^1), \ldots ,x_0(\tau^k)) = 0 \quad \text{for} \quad i = \mu + 1, \ldots, \mu + m,$$

and such that

5 $$\chi^0(x_0(\tau^1), \ldots ,x_0(\tau^k)) + \int_{t_1}^{t_2} \hat{f}^0(x_0(t),u_0(t),t)\, dt \le \chi^0(x(\tau^1), \ldots ,x(\tau^k)) + \int_{t_1}^{t_2} \hat{f}^0(x(t),u(t),t)\, dt$$

for all $(x,u) \in A \times \mathscr{U}$ with x absolutely continuous that satisfy the differential equation

6
$$\dot{x}(t) = f(x(t),u(t),t) \quad \text{a.e. in} \quad I,$$

as well as the constraints

7
$$\chi^i(x(\tau^1),\ldots,x(\tau^k)) + \int_{t_1}^{t_2} \hat{f}^i(x(t),u(t),t)\,dt \le 0 \quad \text{for} \quad i = 1,\ldots,\mu,$$

8
$$\chi^i(x(\tau^1),\ldots,x(\tau^k)) = 0 \quad \text{for} \quad i = \mu + 1,\ldots,\mu + m.$$

For the sake of simplicity in the presentation, we have omitted from Problem (1) restrictions on the phase coordinates of the form

$$\tilde{\chi}^i(x(t),t) \le 0 \quad \text{for all} \quad t \in I_i \quad \text{and each} \quad i = 1,\ldots,\nu.$$

9 If the functions $\hat{f}^i$ are continuously differentiable with respect to their first and second arguments, then, using the procedure outlined in (V.1.12), Problem (1) may be reduced to the problem of Theorem (1.59). If the $\hat{f}^i$ are continuously differentiable with respect to their first argument (but not necessarily their second argument), and if $\mathscr{U}$ is convex under switching, then this same procedure allows us to reduce Problem (1) to Problem (V.3.10). It is just because we make no such differentiability hypotheses on the $\hat{f}^i$ (although we do assume a convexity property) that our problem is unorthodox. If the $\hat{f}^i$ are independent of $v \in R^r$ and if $\mathscr{U}$ is convex under switching, then Problem (1) becomes a special case of Problem (V.7.1).

We shall obtain necessary conditions for Problem (1) by treating it as a problem with parameters and by making use of Theorem (IV.2.34). We begin by reducing Problem (1) to the form of Problem (IV.2.4).

We shall again use the notation of Chapter IV, with $\mathscr{X} = \mathscr{C}^n(I)$. Let us set $\Pi = \mathscr{P} = L^r_\infty(I)$ and $\Pi_a = \mathscr{U}$. For each $\xi \in R^n$, let us define the operator $\hat{T}_\xi : A \times \Pi \to \mathscr{X}$ as follows:

10
$$(\hat{T}_\xi(x,u))(t) = \xi + \int_{t_1}^{t} f(x(s),u(s),s)\,ds, \quad t \in I.$$

It follows from our hypotheses on f [see Theorem (A.12) in the Appendix] that, for each $\xi \in R^n$, $\hat{T}_\xi$ is continuously differentiable (i.e., $\hat{T}_\xi \in \hat{\mathscr{T}}$ for all $\xi \in R^n$) and the Fréchet partial differentials of $\hat{T}_\xi$ with respect to its first and second arguments at any $(x_0,u_0) \in A \times \Pi = \tilde{A}$ are given by the formulas

11 $$(D_1 \hat{T}_\xi(x_0,u_0;\, \delta x))(t) = \int_{t_1}^{t} \mathscr{D}_1 f(x_0(s),u_0(s),s)\delta x(s)\, ds,$$
$$t \in I, \quad \delta x \in \mathscr{C}^n(I),$$

12 $$(D_2 \hat{T}_\xi(x_0,u_0;\, \delta u))(t) = \int_{t_1}^{t} \mathscr{D}_2 f(x_0(s),u_0(s),s)\delta u(s)\, ds,$$
$$t \in I, \quad \delta u \in L_\infty^r(I),$$

independent of ξ.

Let

$$\hat{\mathscr{W}} = \{\hat{T}_\xi : \xi \in R^n\}.$$

Evidently, $\hat{\mathscr{W}}$ is a convex set in $\hat{\mathscr{T}}$. Now set $\tilde{\mathscr{W}} = \hat{\mathscr{W}} \times \mathscr{U}$. Since $\mathscr{U}$ is, by hypothesis, finitely open in itself, $\tilde{\mathscr{W}}$ is also finitely open in itself [see (I.1.40)]. It is easily seen that, for each $\xi \in R^n$ and $u \in \Pi$, $\hat{T}_\xi^u \in \mathscr{V}_2' \subset \mathscr{V}_2$ [see (IV.2.3) and (IV.3.39, 40, and 33)], so that [see (IV.3.41)] Hypotheses (IV.2.10 and 11) hold. Let us further set $\tilde{A}' = \tilde{A}$, and let us define the functions $\hat{\varphi}: \tilde{A} \to R^m$ and the functionals $\hat{\phi}_j^i$ and $\hat{\phi}^i$ $(i = 0, \ldots, \mu; j = 1, 2)$ on $\tilde{A}$ as follows:

13 $$\hat{\varphi}(x,u) = \chi'(x(\tau^1), \ldots, x(\tau^k)),$$
$$\text{where} \quad \chi' = (\chi^{\mu+1}, \ldots, \chi^{\mu+m}), \quad (x,u) \in \tilde{A},$$

14 $$\hat{\phi}_1^i(x,u) = \chi^i(x(\tau^1), \ldots, x(\tau^k)), \quad (x,u) \in \tilde{A}, \quad i = 0, \ldots, \mu,$$

15 $$\hat{\phi}_2^i(x,u) = \int_{t_1}^{t_2} \hat{f}^i(x(t),u(t),t)\, dt, \quad (x,u) \in \tilde{A}, \quad i = 0, \ldots, \mu,$$

16 $$\hat{\phi}^i = \hat{\phi}_1^i + \hat{\phi}_2^i, \quad i = 0, \ldots, \mu.$$

As we indicated in (IV.1.49), it follows from the continuity of χ that $\hat{\varphi}$ and the $\hat{\phi}_1^i$ are continuous, and the continuous differentiability of χ implies that $\hat{\varphi}$ and the $\hat{\phi}_1^i$ are Fréchet differentiable, with Fréchet differentials at any $(x_0,u_0) \in \tilde{A}$ given by (IV.1.50 and 51), but with $D\hat{\varphi}(x_0;\, \delta x)$ and $D\hat{\phi}^i(x_0;\, \delta x)$ to be replaced, respectively, by $D\hat{\varphi}(x_0,u_0;\, \delta x,\delta u)$ and $D\hat{\phi}_1^i(x_0,u_0;\, \delta x,\delta u)$ ($\mathscr{D}_j^0\chi'$ and $\mathscr{D}_j^0\chi^i$ have their usual meanings). Further, by virtue of the convexity and continuity of the functionals $\hat{f}^i$, $\hat{\phi}_2^i$ (for each i) is continuous and convex, and hence [see Corollary (I.7.33)] dually semidifferentiable, with dual semidifferential at any $(x_0,u_0) \in \tilde{A}$ given by the function

$$(\delta x,\delta u) \to \int_{t_1}^{t_2} [\hat{f}^i(x_0(t) + \delta x(t),u_0(t) + \delta u(t),t) - \hat{f}^i(x_0(t),u_0(t),t)]\, dt,$$

$$\delta x \in \mathscr{C}^n(I), \quad \delta u \in L_\infty^r(I),$$

so that [see (16), (I.7.45 and 47), and Theorem (I.7.44)] $\hat{\phi}^i$ (for each i) is continuous and dually semidifferentiable with dual semidifferential at any $(x_0,u_0) \in \tilde{A}$ given by the functional

17 $$D\hat{\phi}^i(x_0,u_0;\, \delta x,\delta u) = \sum_{j=1}^{k} \mathscr{D}_j^0 \chi^i \delta x(\tau^j)$$
$$+ \int_{t_1}^{t_2} [\hat{f}^i(x_0(t) + \delta x(t),u_0(t) + \delta u(t),t) - \hat{f}^i(x_0(t),u_0(t),t)]\, dt,$$
$$\delta x \in \mathscr{C}^n(I), \quad \delta u \in L_\infty^r(I), \quad i = 0, \ldots, \mu.$$

It is now clear that Problem (1) is of the form of Problem (IV.2.4) with the data as described in (IV.2.16)—say with $Z' = \mathscr{Z}' =$ some arbitrary linear topological space and $\hat{\phi}' \equiv 0 \in \mathscr{Z}'$. Let (x_0,u_0) be a solution of Problem (1) and let $\hat{T}_0 = \hat{T}_{x_0(t_1)}$, so that $x_0 = \hat{T}_0(x_0,u_0)$, $(\hat{T}_0,u_0) \in \mathscr{S}_1$, and $(x_0,\hat{T}_0,u_0)$ is a solution of the corresponding Problem (IV.2.4).

As we have seen, Hypotheses (IV.2.10, 11, and 17) hold. Appealing to Theorem (IV.2.34) [see (IV.2.40 and 41)], we conclude that there exist a nonzero vector $\alpha = (\alpha^0, \alpha^1, \ldots, \alpha^{\mu+m}) \in \bar{R}_-^{\mu+1} \times R^m$ and continuous linear functionals Λ^i $(i = 0, \ldots, \mu)$ in $(\mathscr{C}^n(I) \times L_\infty^r(I))^*$ such that [see (10)–(12) and (2)]

18 $$\text{(i)} \sum_{i=\mu+1}^{\mu+m} \sum_{j=1}^{k} \alpha^i \mathscr{D}_j^0 \chi^i \delta x_{\xi,\delta u}(\tau^j) + \sum_{i=0}^{\mu} \alpha^i \Lambda^i(\delta x_{\xi,\delta u},\delta u) \le 0$$
$$\text{for all} \quad \xi \in R^n \quad \text{and} \quad \delta u \in \mathscr{U} - u_0,$$

where, for each $\xi \in R^n$ and $\delta u \in \mathscr{U}$, $\delta x_{\xi,\delta u}$ denotes the column-vector-valued function in $\mathscr{C}^n(I)$ that satisfies the equation

19 $$\delta x_{\xi,\delta u}(t) = \xi + \int_{t_1}^{t} [\mathscr{D}_1 f(x_0(s),u_0(s),s)\delta x_{\xi,\delta u}(s)$$
$$+ \mathscr{D}_2 f(x_0(s),u_0(s),s)\delta u(s)]\, ds, \quad t \in I,$$

20 $$\text{(ii)}\ \Lambda^i(\delta x,\delta u) \le \sum_{j=1}^{k} \mathscr{D}_j^0 \chi^i \delta x(\tau^j)$$
$$+ \int_{t_1}^{t_2} [\hat{f}^i(x_0(t) + \delta x(t),u_0(t) + \delta u(t),t) - \hat{f}^i(x_0(t),u_0(t),t)]\, dt$$
$$\text{for all} \quad \delta x \in \mathscr{C}^n(I) \quad \text{and} \quad \delta u \in L_\infty^r(I), \quad \text{and each} \quad i = 0, \ldots, \mu,$$

21 $$\text{(iii)} \sum_{i=1}^{\mu} \alpha^i \left[\chi^i(x_0(\tau^1), \ldots, x_0(\tau^k)) + \int_{t_1}^{t_2} \hat{f}^i(x_0(t),u_0(t),t)\, dt \right] = 0.$$

By virtue of (I.5.25), Theorems (I.6.14 and 28), and (I.6.20 and 31), we can conclude that there are row-vector-valued functions $\hat{\lambda}_i \in [NBV(I)]^n$, for $i = 0, 1, \ldots, \mu$, and finitely additive real-valued measures $\mathbf{v}_i^j$ defined on all Lebesgue measurable subsets of I, with $\mathbf{v}_i^j(I_0) = 0$ for all subsets I_0 of I of Lebesgue measure zero, for $i = 0, \ldots, \mu$ and $j = 1, \ldots, r$, such that (20) implies that

22 $$\int_{t_1}^{t_2} d\hat{\lambda}_i(t)\delta x(t) + \sum_{j=1}^{r} \int_{t_1}^{t_2} \delta u^j(t)\, d\mathbf{v}_i^j(t) \leq \sum_{j=1}^{k} \mathscr{D}_j^0 \chi^i \delta x(\tau^j)$$
$$+ \int_{t_1}^{t_2} [\hat{f}^i(x_0(t) + \delta x(t),u_0(t) + \delta u(t),t) - \hat{f}^i(x_0(t),u_0(t),t)]\, dt$$
$$\text{for all} \quad \delta x \in \mathscr{C}^n(I) \quad \text{and} \quad \delta u = (\delta u^1, \ldots, \delta u^r) \in L_\infty^r(I),$$
$$\text{and each} \quad i = 0, \ldots, \mu,$$

and such that (18) takes the form

23 $$\sum_{i=\mu+1}^{\mu+m} \alpha^i \sum_{j=1}^{k} \mathscr{D}_j^0 \chi^i \delta x_{\xi,\delta u}(\tau^j) + \sum_{i=0}^{\mu} \alpha^i \int_{t_1}^{t_2} d\hat{\lambda}^i(t)\delta x_{\xi,\delta u}(t)$$
$$+ \sum_{i=0}^{\mu} \alpha^i \sum_{j=1}^{r} \int_{t_1}^{t_2} \delta u^j(t)\, d\mathbf{v}_i^j(t) \leq 0 \quad \text{for all} \quad \xi \in R^n$$
$$\text{and} \quad \delta u = (\delta u^1, \ldots, \delta u^r) \in \mathscr{U} - u_0.$$

Arguing as in Section 7 of Chapter V and in the proof of Lemma (3.100) in the next section, we can prove that there are row-vector-valued functions $\mathscr{D}_1^0 \hat{f}^i \in L_\infty^n(I)$ and $\mathscr{D}_2^0 \hat{f}^i \in L_\infty^r(I)$ ($i = 0, \ldots, \mu$) such that

24 $$\int_{t_1}^{t_2} d\hat{\lambda}^i(t)\delta x(t) = \sum_{j=1}^{k} \mathscr{D}_j^0 \chi^i \delta x(\tau^j) + \int_{t_1}^{t_2} \mathscr{D}_1^0 \hat{f}^i(t)\delta x(t)\, dt$$
$$\text{for all} \quad \delta x \in \mathscr{C}^n(I),$$

25 $$\sum_{j=1}^{r} \int_{t_1}^{t_2} \delta u^j(t)\, d\mathbf{v}_i^j(t) = \int_{t_1}^{t_2} \mathscr{D}_2^0 \hat{f}^i(t)\delta u(t)\, dt$$
$$\text{for all} \quad \delta u = (\delta u^1, \ldots, \delta u^r) \in L_\infty^r(I),$$

and such that, for almost all $t \in I$ and each i, $(\mathscr{D}_1^0 \hat{f}^i(t), \mathscr{D}_2^0 \hat{f}^i(t))$ is a subgradient [see (I.2.29)] for the function $(\xi,v) \to \hat{f}^i(\xi,v,t)$ at $(x_0(t),u_0(t))$.

Hence, (23) can be written in the form

26 $$\sum_{j=1}^{k} \alpha \mathscr{D}_j^0 \chi \delta x_{\xi,\delta u}(\tau^j) + \sum_{i=0}^{\mu} \alpha^i \int_{t_1}^{t_2} [\mathscr{D}_1^0 \hat{f}^i(t) \delta x_{\xi,\delta u}(t) + \mathscr{D}_2^0 \hat{f}^i(t) \delta u(t)]\, dt \leq 0$$

$$\text{for all} \quad \xi \in R^n \quad \text{and} \quad \delta u \in \mathscr{U} - u_0.$$

Performing what by now are routine manipulations on (26) [using (19)], we arrive at the following theorem.

27 THEOREM. *Let (x_0,u_0) be a solution of Problem* (1). *Then there exist a nonzero row vector $\alpha = (\alpha^0, \alpha^1, \ldots, \alpha^{\mu+m}) \in \bar{R}_-^{\mu+1} \times R^m$ and row-vector-valued functions $\mathscr{D}_1^0 \hat{f}^i \in L_\infty^n(I)$, $\mathscr{D}_2^0 \hat{f}^i \in L_\infty^r(I)$ $(i = 0, \ldots, \mu)$, and $\psi \in L_\infty^n(I)$ such that*

(i) *on each of the intervals $[t_1,\tau^2), [\tau^2,\tau^3), \ldots, [\tau^{k-1},t_2]$, $\psi(\cdot)$ is absolutely continuous and satisfies the ordinary differential equation*

28 $$\dot{\psi}(t) = -\psi(t)\mathscr{D}_1 f(x_0(t),u_0(t),t) - \sum_{i=0}^{\mu} \alpha^i \mathscr{D}_1^0 \hat{f}^i(t) \quad \text{a.e.,}$$

29 $$\text{(ii)} \int_{t_1}^{t_2} \left[\psi(t)\mathscr{D}_2 f(x_0(t),u_0(t),t) + \sum_{i=0}^{\mu} \alpha^i \mathscr{D}_2^0 \hat{f}^i(t) \right] \times [u(t) - u_0(t)]\, dt \leq 0$$

$$\textit{for all} \quad u \in \mathscr{U},$$

(iii) *ψ satisfies the boundary conditions*

30 $$\psi(t_1) = -\alpha \mathscr{D}_1^0 \chi, \quad \psi(t_2) = \alpha \mathscr{D}_k^0 \chi,$$

as well as the jump conditions

31 $$\psi(\tau^j) - \psi(\tau^{j-}) = -\alpha \mathscr{D}_j^0 \chi \quad \textit{for} \quad j = 2, \ldots, k-1,$$

where, for each j, $\mathscr{D}_j^0 \chi$ denotes the Jacobian matrix $\mathscr{D}_j \chi$ of χ at $(x_0(\tau^1), \ldots, x_0(\tau^k))$,

32 $$\text{(iv)}\ \alpha^i \left[\chi^i(x_0(\tau^1), \ldots, x_0(\tau^k)) + \int_{t_1}^{t_2} \hat{f}^i(x_0(t),u_0(t),t)\, dt \right] = 0$$

$$\textit{for} \quad i = 1, \ldots, \mu, \quad \textit{and}$$

(v) *for each $i = 0, \ldots, \mu$ and almost all $t \in I$, the vector $(\mathscr{D}_1^0 \hat{f}^i(t), \mathscr{D}_2^0 \hat{f}^i(t))$ in R^{n+r} is a subgradient for the function $(\xi,v) \to \hat{f}^i(\xi,v,t)$ at $(x_0(t),u_0(t))$.*

If also $u_0 \in \text{int } \mathscr{U}$, *then* (29) *is equivalent to the relation*

33
$$\psi(t)\mathscr{D}_2 f(x_0(t),u_0(t),t) + \sum_{i=0}^{\mu} \alpha^i \mathscr{D}_2^0 \hat{f}^i(t) = 0 \quad \text{a.e. } on \quad I.$$

34 We recall [see Lemma (I.7.62)] that if $\hat{f}^i$ is Fréchet differentiable with respect to its first and second arguments at $(x_0(t),u_0(t))$—for some $i = 0, \ldots, \mu$ and some $t \in I$—then $\mathscr{D}_j^0 \hat{f}^i(t) = \mathscr{D}_j \hat{f}^i(x_0(t), u_0(t),t)$ for $j = 1, 2$. In particular, if $\hat{f}^i$ is continuously differentiable with respect to its first and second arguments for each $i = 0, \ldots, \mu$, then we can replace $\mathscr{D}_j^0 \hat{f}^i(t)$ by $\mathscr{D}_j \hat{f}^i(x_0(t),u_0(t),t)$ (for $j = 1, 2$, and each i) in (28) and in (29), and our necessary conditions coincide with those that can be obtained from Theorem (1.59)—if we reformulate our problem as indicated in (9). Note that the maximum conditions (29) and (33) here are of a "local" variety and hence are generally weaker than the more usual "global" maximum principles. Whereas, in the case of Theorem (1.59), we could conclude that the local maximum principle could be strengthened to a global maximum principle if we imposed the additional requirement that $\mathscr{U}$ is convex under switching [see the discussion in (1.50 and 58)], this is not the case for Problem (1), i.e., the local maximum principle is as strong a necessary condition as can here be obtained—even under stronger assumptions for $\mathscr{U}$. Of course, if the $\hat{f}^i$ are continuously differentiable with respect to their first argument, or alternately, if the $\hat{f}^i$ are independent of $v \in R^r$, then (provided that $\mathscr{U}$ is convex under switching) a global maximum principle does hold [see (9) and Theorem (V.7.20)].

35 If the class $\mathscr{U}$ is convex under switching and also satisfies Hypothesis (1.55), then, as was indicated in (1.56), we can show that (29) is equivalent to the relation [compare with (1.57)]

36
$$\left[\psi(t)\mathscr{D}_2 f(x_0(t),u_0(t),t) + \sum_{i=0}^{\mu} \alpha^i \mathscr{D}_2^0 \hat{f}^i(t)\right] v \leq \left[\psi(t)\mathscr{D}_2 f(x_0(t),u_0(t),t) + \sum_{i=0}^{\mu} \alpha^i \mathscr{D}_2^0 \hat{f}^i(t)\right] u_0(t)$$
$$\text{for all} \quad v \in U, \quad \text{for almost all} \quad t \in I.$$

37 If Problem (1) is modified by eliminating either the inequality constraints (7) or the equality constraints (8), or both, then the necessary conditions of Theorem (27) remain in force, but with μ,

or m, or both, equal to zero, so that (32) is to be deleted if there are no inequality constraints [see (IV.2.63 and 75)].

38 Note that the (adjoint) function ψ in Theorem (27) is completely determined by Eqs. (28) and (31), together with either of relations (30), at least in terms of α, x_0, u_0, and the $\mathscr{D}_1^0 \hat{f}^i$. If $\psi \equiv 0$ on I and $\alpha^i = 0$ for $i = 0, \ldots, \mu$, then the basic inequality (29) of Theorem (27) is trivially satisfied, i.e., Theorem (27) yields trivial necessary conditions. If we impose the condition that the m vectors $(\mathscr{D}_1^0\chi^i, \ldots, \mathscr{D}_k^0\chi^i)$, $i = \mu + 1, \ldots, \mu + m$, are linearly independent in $(R^n)^k$, then it is very easy to show—on the basis of (30) and (31), because $\alpha \neq 0$—that it is impossible that $\psi \equiv 0$ on I and that $\alpha^i = 0$ for $i = 0, \ldots, \mu$.

39 If some (or all) of the functions $\chi^0, \ldots, \chi^\mu$ in the problem statement are convex, without necessarily being continuously differentiable, then the necessary conditions of Theorem (27) remain in force, provided that we interpret $(\mathscr{D}_1^0\chi^i, \ldots, \mathscr{D}_k^0\chi^i)$ as a (suitably chosen) subgradient of the function χ^i at $(x_0(\tau^1), \ldots, x_0(\tau^k))$, whenever χ^i is not differentiable at this point (for some $i = 0, \ldots, \mu$). [Compare with (V.6.7).] This can be shown as follows:

40 For the sake of convenience, let us suppose that $\chi^0, \ldots, \chi^\mu$ are all convex. Then the functionals $\hat{\phi}^i$ defined on all of $\mathscr{C}^n(I) \times L_\infty^r(I)$ by (14)–(16) are convex (as well as continuous), and hence dually semidifferentiable [see Corollary (I.7.33)], with dual semidifferential at any $(x_0, u_0) \in \tilde{A}$ given by [compare with (17)]

$$D\hat{\phi}^i(x_0,u_0; \delta x,\delta u) = \hat{\phi}^i(x_0 + \delta x, u_0 + \delta u) - \hat{\phi}^i(x_0,u_0).$$

Hence, when we appeal to Theorem (IV.2.34), relation (20) must be replaced by the inequality

$$\Lambda^i(\delta x,\delta u) \leq \sum_{j=1}^{2} [\hat{\phi}_j^i(x_0 + \delta x, u_0 + \delta u) - \hat{\phi}_j^i(x_0,u_0)]$$

for all $\delta x \in \mathscr{C}^n(I)$ and $\delta u \in L_\infty^r(I)$ and each $i = 0, \ldots, \mu$. If we now appeal to Lemmas (III.4.25) and (I.5.13), we easily conclude that there are linear functionals Λ_1^i and Λ_2^i in $(\mathscr{C}^n(I) \times L_\infty^r(I))^*$ such that $\Lambda^i = \Lambda_1^i + \Lambda_2^i$, and such that

41
$$\Lambda_1^i(\delta x,\delta u) \leq \chi^i(x_0(\tau^1) + \delta x(\tau^1), \ldots, x_0(\tau^k) + \delta x(\tau^k)) - \chi^i(x_0(\tau^1), \ldots, x_0(\tau^k)),$$

$$\Lambda_2^i(\delta x,\delta u) \leq \int_{t_1}^{t_2} [\hat{f}^i(x_0(t) + \delta x(t), u_0(t) + \delta u(t), t) - \hat{f}^i(x_0(t), u_0(t), t)]\, dt$$

for all δx, δu, i. But, in the proof of Theorem (IV.1.64), we showed that, if Λ_1^i satisfies (41) [see (IV.1.60–63)], then

$$\Lambda_1^i(\delta x,\delta u) = \sum_{j=1}^{k} \mathscr{D}_j^0\chi^i\delta x(\tau^j) \quad \text{for all} \quad (\delta x,\delta u),$$

where $(\mathscr{D}_1^0\chi^i, \ldots, \mathscr{D}_k^0\chi^i)$ is a subgradient of χ^i at $(x_0(\tau^1), \ldots, x_0(\tau^k))$. Thus, (20) once again holds, and the remainder of the proof is as before. |||

42 Using Theorem (III.5.16) and Lemma (I.7.62), and arguing much as in the proof of Theorem (V.8.9), we can prove the following: If, the Problem (1), $\mathbf{G} = R^n$, $A = \mathscr{C}^n(I)$, the function f has the form

$$f(\xi,v,t) = \mathbf{A}(t)\xi + \mathbf{B}(t)v, \quad (\xi,v,t) \in R^n \times R^r \times I,$$

where $\mathbf{A}$ and $\mathbf{B}$ are continuous matrix-valued functions, and the functions χ^i are convex for $i = 0, \ldots, \mu$ and affine for $i = \mu + 1, \ldots, \mu + m$, then the necessary conditions of Theorem (27)—with the additional condition $\alpha^0 \neq 0$—are also sufficient for an absolutely continuous function $x_0 \in \mathscr{C}^n(I)$ and a function $u_0 \in \mathscr{U}$ to be optimal, provided that x_0 and u_0 satisfy the problem constraints (2)–(4). [Also, see (35).] This assertion also holds for the modified necessary conditions described in (39). In applying Theorem (III.5.16) here, it is convenient to set $\mathscr{E}' = R^n \times L_\infty^r(I)$ and to define $\mathscr{E}$, etc., in what is then an evident manner.

Finally, we point out that the requirement [in the statement of Problem (1)] that the functions f and $\hat{f}$ are continuous with respect to t can be considerably relaxed.

3. Optimal Control Problems with Mixed Control-Phase Variable Inequality Constraints

In this section, we shall consider a fixed-time optimal control problem of the form of Problem (V.3.1) containing additional constraints of the form

1 $$\tilde{\tilde{\chi}}^i(x(t),u(t),t) \leq 0 \quad \text{for almost all} \quad t \in I_i' \quad \text{and each} \quad i = 1, \ldots, \nu',$$

where $\tilde{\tilde{\chi}}^1, \ldots, \tilde{\tilde{\chi}}^{\nu'}$ are given scalar-valued functions defined on $\mathbf{G} \times V \times I$, and $I_1', \ldots, I_{\nu'}'$ are given Lebesgue measurable subsets of I. Note that if the functions $\tilde{\tilde{\chi}}^i$ are independent of $\xi \in \mathbf{G}$, so that inequalities (1) are of the form

2 $$\tilde{\tilde{\chi}}^i(u(t),t) \leq 0 \quad \text{for almost all} \quad t \in I'_i \quad \text{and each} \quad i = 1, \ldots, \nu',$$

then we may define $\mathscr{U}'$ as the set of all $u \in \mathscr{U}$ which also satisfy (2), and, replacing $\mathscr{U}$ by $\mathscr{U}'$, reduce our problem to the form of Problem (V.3.1), without additional restrictions. Of course, if the functions $\tilde{\tilde{\chi}}^i$ are independent of u, then our problem is immediately of the form of Problem (V.3.1), once we replace $\tilde{\chi}$ by $(\tilde{\chi},\tilde{\tilde{\chi}})$, where $\tilde{\tilde{\chi}} = (\tilde{\tilde{\chi}}^1, \ldots, \tilde{\tilde{\chi}}^{\nu'})$.

In order to obtain necessary conditions for the problem with the general constraints (1), we shall suppose that V is an open set in R^r (for some positive integer r), and that f and $\tilde{\tilde{\chi}}$ are suitably differentiable with respect to their second (as well as their first) arguments. More precisely, we shall suppose that Hypotheses (V.1.16–18) are replaced by the following ones:

3 HYPOTHESIS. *For each* $t \in I$, *the function* $(\xi,v) \to f(\xi,v,t): \mathbf{G} \times V \to R^n$ *is continuously differentiable.*

4 HYPOTHESIS. *For every* $\xi \in \mathbf{G}$ *and each measurable function* $u: I \to V$, *the function* $t \to f(\xi,u(t),t): I \to R^n$ *is measurable.*

5 HYPOTHESIS. *For each compact subset* $\mathbf{G}_c$ *of* $\mathbf{G}$ *and each compact subset* V_c *of* V, *there exists an integrable function* $\tilde{m}: I \to R_+$ *(possibly depending on* $\mathbf{G}_c$ *and on* V_c*) such that*

6 $$|f(\xi,v,t)| + |\mathscr{D}_1 f(\xi,v,t)| + |\mathscr{D}_2 f(\xi,v,t)| \leq \tilde{m}(t)$$
$$\textit{for all} \quad (\xi,v,t) \in \mathbf{G}_c \times V_c \times I.$$

We shall also make the following hypothesis on the set of admissible controls $\mathscr{U}$.

7 HYPOTHESIS. *For every* $u \in \mathscr{U}$, *there is a compact subset* V_u *of* V *such that* $u(t) \in V_u$ *for almost all* $t \in I$. *Further, every* $u \in \mathscr{U}$ *is a measurable function.*

8 Hypothesis (7) clearly implies that $\mathscr{U} \subset L^r_\infty(I)$. If $V = R^r$, then Hypothesis (7) is satisfied if and only if $\mathscr{U} \subset L^r_\infty(I)$. [See (I.3.45).]

9 If f is continuous and continuously differentiable with respect to both its first and second arguments, then it follows from (I.3.46) and Theorem (I.7.54) that Hypotheses (3)–(5) hold, and $\tilde{m}$ may even be taken to be constant in Hypothesis (5). Also note that if f and $\mathscr{U}$ satisfy Hypotheses (3)–(5) and (7), then they satisfy Hypotheses (V.1.16–18) as well.

We can now state the problem to which this section is devoted.

10 *Problem.* Let there be given open sets $\mathbf{G}$ in R^n and V in R^r, a compact interval $I = [t_1, t_2]$, an open interval $\tilde{I} \supset I$, a finite subset $I_a = \{\tau^1, \ldots, \tau^k\}$ of I such that $t_1 = \tau^1 < \tau^2 < \cdots < \tau^k = t_2$, closed subsets $I_1, \ldots, I_\nu$ of I, Lebesgue measurable subsets $I'_1, \ldots, I'_{\nu'}$ of I, an open set A in $\mathscr{C}^n(I)$ such that $x(t) \in \mathbf{G}$ for all $x \in A$ and $t \in I$, and a class $\mathscr{U}$ of functions $u: I \to V$ which (a) is convex under switching, (b) satisfies Hypothesis (7), and (c) is finitely open in itself [as a subset of $L^r_\infty(I)$]. Further, let there be given a function $f: \mathbf{G} \times V \times I \to R^n$ satisfying Hypotheses (3)–(5), continuously differentiable functions $\chi = (\chi^0, \chi^1, \ldots, \chi^{\mu+m})$: $(R^n)^k \to R^{\mu+m+1}$ and $\tilde{\chi} = (\tilde{\chi}^1, \ldots, \tilde{\chi}^\nu)$: $R^n \times \tilde{I} \to R^\nu$ such that $\mathscr{D}_1\tilde{\chi}$ and $\mathscr{D}_2\tilde{\chi}$ are continuously differentiable, and a continuous function $\tilde{\tilde{\chi}} = (\tilde{\tilde{\chi}}^1, \ldots, \tilde{\tilde{\chi}}^{\nu'})$: $\mathbf{G} \times V \times I \to R^{\nu'}$ which is continuously differentiable with respect to its first and second arguments. Then find an absolutely continuous function $x_0 \in A$ and a function $u_0 \in \mathscr{U}$ such that

11 $$\dot{x}_0(t) = f(x_0(t), u_0(t), t) \quad \text{for almost all} \quad t \in I,$$

12 $$\chi^i(x_0(\tau^1), \ldots, x_0(\tau^k)) \leq 0 \quad \text{for} \quad i = 1, \ldots, \mu,$$

13 $$\chi^i(x_0(\tau^1), \ldots, x_0(\tau^k)) = 0 \quad \text{for} \quad i = \mu + 1, \ldots, \mu + m,$$

14 $$\tilde{\chi}^i(x_0(t), t) \leq 0 \quad \text{for all} \quad t \in I_i \quad \text{and each} \quad i = 1, \ldots, \nu,$$

15 $$\tilde{\tilde{\chi}}^i(x_0(t), u_0(t), t) \leq 0 \quad \text{for almost all} \quad t \in I'_i \quad \text{and each} \quad i = 1, \ldots, \nu',$$

and such that

16 $$\chi^0(x_0(\tau^1), \ldots, x_0(\tau^k)) \leq \chi^0(x(\tau^1), \ldots, x(\tau^k))$$

for all absolutely continuous functions $x \in A$, which, for some $u \in \mathscr{U}$, satisfy the differential equation

17 $$\dot{x}(t) = f(x(t), u(t), t) \quad \text{for almost all} \quad t \in I,$$

together with the constraints (12)–(15), with the subzero omitted.

18 Problem (10) differs from Problem (V.3.1) primarily in that it contains the "mixed control-phase variable" inequality constraints (15). In order to obtain meaningful necessary conditions in the presence of such constraints, we are compelled to restrict ourselves to the case where (i) V is an open set in R^r, (ii) f is

differentiable with respect to its second (as well as its first) argument as described in Hypotheses (3) and (5), and (iii) $\mathscr{U}$ is a subset of $L^r_\infty(I)$ which is finitely open in itself, as well as convex under switching. At first sight, the last requirement appears to place a heavy restriction on the classes of admissible controls that we are permitted to consider. However, inequalities (15) may be used to represent additional control constraints (since the functions $\tilde{\tilde{\chi}}^i$ may, as we pointed out earlier, be independent of ξ), and inequalities of the form $\tilde{\tilde{\chi}}(u(t),t) \leq 0$ are the type of control constraints that occur most commonly in concrete problems. On the other hand, if the control variable is constrained to lie in some convex or open subset of V, then such constraints may be handled through the requirement that $\mathscr{U}$ should be finitely open in itself. Very commonly in applications, $V = R^r$ and $\mathscr{U} = L^r_\infty(I)$.

19 The constraints (14) are clearly special cases of the constraints (15), so that, at first sight, it appears redundant to include constraints of the form (14). However, for reasons that will become apparent, it turns out to be convenient to treat these two types of constraints separately.

20 Now suppose that $(x_0,u_0) \in A \times \mathscr{U}$ is a solution of Problem (10). In the notation of Section 2 of Chapter IV, we shall show that, for a suitable choice for $\mathscr{X}$, $\mathscr{P}$, Π, $\tilde{A}'$, $\tilde{\mathscr{W}}$, $\mathscr{Z}_1$, Z_1, $\hat{\varphi}$, $\hat{\varphi}^0$, and $\hat{\phi}_1$, we can identify (x_0,u_0) with a solution of Problem (IV.2.4), that the problem has the form described in (IV.2.16), and that the data of this problem satisfy Hypotheses (IV.2.10–12 and 17). We shall then obtain necessary conditions that x_0 and u_0 must satisfy from Theorem (IV.2.34).

Indeed, let us set $\mathscr{X} = \mathscr{C}^n(I)$, $\mathscr{P} = L^r_\infty(I)$, $\mathscr{Z}' = \mathscr{C}^\nu(I) \times L^{\nu'}_\infty(I)$,

21 $$Z'' = \{z : z = (z^1, \ldots, z^\nu) \in \mathscr{C}^\nu(I),\ z^i(t) \leq 0 \quad \text{for all} \quad t \in I_i \quad \text{and each} \quad i = 1, \ldots, \nu\},$$

22 $$Z''' = \{\zeta : \zeta = (\zeta^1, \ldots, \zeta^{\nu'}) \in L^{\nu'}_\infty(I),\ \zeta^i(t) \leq 0 \quad \text{for almost all} \quad t \in I'_i \quad \text{and each} \quad i = 1, \ldots, \nu'\},$$

$\mathscr{Z}_1 = R^\mu \times \mathscr{Z}'$, $Z' = Z'' \times Z'''$, and $Z_1 = \bar{R}^\mu_- \times Z'$. Further, let

23 $$\Pi = \{u : u \in L^r_\infty(I),\ \overline{\{u(t) : t \in I\}} \subset V\}$$

and let $\tilde{A}' = \tilde{A} = A \times \Pi$, and let us define the functions $\hat{\varphi} : \tilde{A} \to R^m$,

$\hat{\phi}^i:\tilde{A} \to R$ (for $i = 0, \ldots, \mu$), $\hat{\phi}'':\tilde{A} \to \mathscr{C}^{\nu}(I)$, $\hat{\phi}''':\tilde{A} \to L_{\infty}^{\nu'}(I)$, $\hat{\phi}'$: $\tilde{A} \to \mathscr{X}'$, and $\hat{\phi}_1:\tilde{A} \to \mathscr{X}_1$ as follows:

24 $$\hat{\varphi}(x,u) = \chi'(x(\tau^1), \ldots, x(\tau^k)), \quad \text{where} \quad \chi' = (\chi^{\mu+1}, \ldots, \chi^{\mu+m}),$$

25 $$\hat{\phi}^i(x,u) = \chi^i(x(\tau^1), \ldots, x(\tau^k)), \quad i = 0, \ldots, \mu,$$

26 $$(\hat{\phi}''(x,u))(t) = \tilde{\chi}(x(t),t) \quad t \in I,$$

27 $$(\hat{\phi}'''(x,u))(t) = \tilde{\tilde{\chi}}(x(t),u(t),t), \quad t \in I,$$

28 $$\hat{\phi}' = (\hat{\phi}'', \hat{\phi}'''), \quad \hat{\phi}_1 = (\hat{\phi}^1, \ldots, \hat{\phi}^{\mu}, \hat{\phi}').$$

[Note that the function $t \to (\hat{\phi}'''(x,u))(t)$ defined by (27) for each $(x,u) \in \tilde{A}$ is measurable because of the continuity of $\tilde{\tilde{\chi}}$ and the measurability of u, and is essentially bounded in modulus (i.e., is a member of $L_{\infty}^{\nu'}(I)$) by (I.3.46) since the set $\overline{\{x(t):t \in I\}} \times \overline{\{u(t):t \in I\}} \times I$ is a compact subset of $\mathbf{G} \times V \times I$—see (I.3.35 and 45).] We point out that the functions $\hat{\varphi}$, $\hat{\phi}^i$, and $\hat{\phi}''$ are all independent of u.

The preceding definitions are evidently in the spirit of (IV.1.10–14).

29 If $u \in \Pi$, so that $u \in L_{\infty}^r(I)$ and $\overline{\{u(t)\colon t \in I\}}$ is a compact subset of the open set $V \subset R^r$, then it easily follows from the definition of compact sets (I.3.14) that there is an $\varepsilon > 0$ such that $\{\xi\colon \xi \in R^r, |\xi - u(t)| \leq \varepsilon$ for some $t \in I\}$ is contained in V. This immediately implies that Π is open in $L_{\infty}^r(I)$. Note that, by Hypothesis (7) and Lemma (I.3.30), $\mathscr{U} \subset \Pi$.

30 Inasmuch as Z'' and Z''' are both evidently closed convex cones with nonempty interiors, and thus, convex bodies, the same is true of $Z' = Z'' \times Z'''$. [See (I.1.16) and Lemma (I.4.22).]

In order to define $\tilde{\mathscr{W}}$, we first introduce a class of operators $\hat{T} \in \mathscr{V}_2$ [see (IV.4.23, 29, 36, and 5)]. We denote by $\mathbf{I}$ the class of all subsets $\hat{I}$ of I such that $\hat{I}$ is a (possibly empty) finite union of subintervals of I. For each $u' \in \Pi$ and $\hat{I} \in \mathbf{I}$, let $\hat{F}_{u',\hat{I}}$ denote the function from $\mathbf{G} \times I \times \Pi$ into R^n defined by

$$\hat{F}_{u',\hat{I}}(\xi,t,u) = f(\xi,\bar{u}'(t),t), \quad (\xi,t,u) \in \mathbf{G} \times I \times \Pi,$$

where

32 $$\bar{u}'(t) = \begin{cases} u(t) & \text{if} \quad t \in \hat{I}, \\ u'(t) & \text{if} \quad t \in I \quad \text{and} \quad t \notin \hat{I}. \end{cases}$$

Note that $\bar{u}'$ [defined by (32)] is in Π for all $u \in \Pi$ (whatever $u' \in \Pi$ and $\hat{I} \in \mathbf{I}$), and that, since $\mathscr{U}$ is convex under switching,

$\bar{u}' \in \mathscr{U}$ whenever $u \in \mathscr{U}$ and $u' \in \mathscr{U}$. Also note that $\hat{F}_{u',I}$ is independent of u'; indeed, for any $u' \in \Pi$,

33 $$\hat{F}_{u',I}(\xi,t,u) = f(\xi,u(t),t) \quad \text{for all} \quad (\xi,t,u) \in \mathbf{G} \times I \times \Pi.$$

Consequently, we shall simply write $\hat{F}_I$ for $\hat{F}_{u',I}$.

It follows directly from Hypotheses (3)–(5) and (7) that $\hat{F}_{u',\hat{I}} \in \hat{\mathscr{F}}_1$ [see (IV.4.23)] for all $u' \in \Pi$ and $\hat{I} \in \mathbf{I}$, and that, in particular, for all $(\xi,t,u) \in \mathbf{G} \times I \times \Pi$,

34 $$\mathscr{D}_1\hat{F}_I(\xi,t,u) = \mathscr{D}_1 f(\xi,u(t),t),$$

35 $$D_3\hat{F}_I(\xi,t,u;\,\delta u) = \mathscr{D}_2 f(\xi,u(t),t)\delta u(t) \quad \text{for all} \quad \delta u \in L^r_\infty(I).$$

For each $\xi \in R^n$, $u' \in \Pi$, and $\hat{I} \in \mathbf{I}$, we define the operator $\hat{T}_{\xi,u',\hat{I}}: \tilde{A} \to \mathscr{C}^n(I)$ as follows:

36 $$(\hat{T}_{\xi,u',\hat{I}}(x,u))(t) = \xi + \int_{t_1}^{t} \hat{F}_{u',\hat{I}}(x(s),s,u)\,ds, \quad t \in I, \quad (x,u) \in \tilde{A}.$$

If $\hat{I} = I$, then $\hat{T}_{\xi,u',\hat{I}}$ is independent of u', and we simply shall write $\hat{T}_{\xi,I}$. As was pointed out in (IV.4.28 and 33), for every $\xi \in R^n$, $u' \in \Pi$, and $\hat{I} \in \mathbf{I}$, $\hat{T}_{\xi,u',\hat{I}} \in \hat{\mathscr{V}}_2 \subset \hat{\mathscr{V}}$ [see (IV.4.30)] and $(\hat{T}_{\xi,u'\hat{I}},u) \in \mathscr{S}_0$ [see (IV.2.3)] for all $u \in \Pi$. Further, by (IV.4.31 and 32), (34), and (35), for each $\xi \in R^n$ and $(x,u) \in \tilde{A}$,

37 $$(D_1\hat{T}_{\xi,I}(x,u;\,\delta x))(t) = \int_{t_1}^{t} \mathscr{D}_1 f(x(s),u(s),s)\delta x(s)\,ds,$$
$$t \in I, \quad \delta x \in \mathscr{C}^n(I),$$

38 $$(D_2\hat{T}_{\xi,I}(x,u;\,\delta u))(t) = \int_{t_1}^{t} \mathscr{D}_2 f(x(s),u(s),s)\delta u(s)\,ds,$$
$$t \in I, \quad \delta u \in L^r_\infty(I).$$

Now let

39 $$\mathscr{U}' = \left\{u' : u' \in \mathscr{U},\ \operatorname*{ess\,sup}_{t \in I'_i} \tilde{\tilde{\chi}}^i(x_0(t),u'(t),t) < 0 \text{ for each } i = 1,\ldots,v'\right\},$$

and let us define $\tilde{\mathscr{W}}$ as follows [using the notation of (IV.1.9) and (IV.2.3)]:

40 $$\tilde{\mathscr{W}} = \Big\{(\hat{T},u): (\hat{T},u) \in \mathscr{S}_1,\ u \in \mathscr{U},\ \hat{T} = \hat{T}_{\xi,u',\hat{I}}$$
$$\text{for some } \xi \in R^n,\ u' \in \mathscr{U}', \text{ and}$$
$$\hat{I} \in \mathbf{I} \text{ such that } \operatorname*{ess\,sup}_{t \in I'_i} \tilde{\tilde{\chi}}^i(\varphi_0(\hat{T}^u)(t),u'(t),t) < 0$$
$$\text{for each } i = 1,\ldots,v'\Big\}.$$

Note that both $\mathscr{U}'$ and $\tilde{\mathscr{W}}$ depend on x_0. Also, $\mathscr{U}' \subset \mathscr{U} \subset \Pi$ and $\tilde{\mathscr{W}} \subset \hat{\mathscr{V}}_2 \times \Pi$.

41 The data of Problem (IV.2.4) have now been completely specified in terms of the data of Problem (10) and x_0. It is now straightforward to verify that, because (x_0,u_0) is a solution of Problem (10), $(x_0,\hat{T}_{x_0(t_1),I},u_0)$ is a solution of Problem (IV.2.4), which, as is evident, is of the form described in (IV.2.16).

Let us denote the operator $\hat{T}_{x_0(t_1),I}$ by $\hat{T}_0$, so that $(\hat{T}_0,u_0) \in \tilde{\mathscr{W}} \subset \mathscr{S}_1$ and

42 $$x_0 = \hat{T}_0(x_0,u_0) = \hat{T}_0^{u_0}x_0.$$

In addition [see (37) and (38)],

43 $$(D_1\hat{T}_0(x_0,u_0;\delta x))(t) = \int_{t_1}^{t} \mathscr{D}_1 f(x_0(s),u_0(s),s)\delta x(s)\, ds,$$
$$t \in I, \quad \delta x \in \mathscr{C}^n(I),$$

44 $$(D_2\hat{T}_0(x_0,u_0;\delta u))(t) = \int_{t_1}^{t} \mathscr{D}_2 f(x_0(s),u_0(s),s)\delta u(s)\, ds,$$
$$t \in I, \quad \delta u \in L_\infty^r(I).$$

In order to make use of Theorem (IV.2.34), we must verify that Hypotheses (IV.2.10–12 and 17) are satisfied. However, since $\tilde{\mathscr{W}} \subset \hat{\mathscr{V}}_2 \times \Pi$, [see (IV.4.38)] Hypotheses (IV.2.10 and 11) hold. To verify that Hypothesis (IV.2.17) holds, we first point out that the continuity of $\hat{\varphi}, \hat{\phi}^0, \ldots, \hat{\phi}^\mu, \hat{\phi}'', \hat{\phi}'''$, and $\hat{\phi}'$ follow from the continuity of χ, $\tilde{\chi}$, and $\tilde{\tilde{\chi}}$ and from Corollary (I.5.32) and (I.3.38). Further, as we saw in (IV.1.49), the functions $\hat{\varphi}, \hat{\phi}^0, \ldots, \hat{\phi}^\mu$, and $\hat{\phi}''$ are Fréchet differentiable, with Fréchet partial differentials at (x_0,u_0) given by the formulas

45 $$D_1\hat{\varphi}(x_0,u_0;\delta x) = \sum_{j=1}^{k} \mathscr{D}_j^0\chi'\delta x(\tau_j)$$
$$\text{for all} \quad \delta x \in \mathscr{C}^n(I), \quad D_2\hat{\varphi}(x_0,u_0) = 0,$$

46 $$D_1\hat{\phi}^i(x_0,u_0;\delta x) = \sum_{j=1}^{k} \mathscr{D}_j^0\chi^i\delta x(\tau^j)$$
$$\text{for all} \quad \delta x \in \mathscr{C}^n(I) \quad \text{and} \quad D_2\hat{\phi}^i(x_0,u_0) = 0, \quad i = 0, 1, \ldots, \mu,$$

47 $$(D_1\hat{\phi}''(x_0,u_0;\delta x))(t) = \mathscr{D}_1\tilde{\chi}(x_0(t),t)\delta x(t)$$
$$\text{for all} \quad t \in I \quad \text{and} \quad \delta x \in \mathscr{C}^n(I), \quad D_2\hat{\phi}''(x_0,u_0) = 0,$$

where $\mathscr{D}_j^0\chi'$ denotes $\mathscr{D}_j\chi'(x_0(\tau^1), \ldots, x_0(\tau^k))$, and similarly for $\mathscr{D}_j^0\chi^i$. By the same token, $\hat{\phi}'''$ is Fréchet differentiable, with Fréchet partial differentials at (x_0,u_0) given by the formulas

48 $$(D_1\hat{\phi}'''(x_0,u_0\,;\,\delta x))(t) = \mathscr{D}_1\tilde{\tilde{\chi}}(x_0(t),u_0(t),t)\delta x(t), \qquad t \in I, \quad \delta x \in \mathscr{C}^n(I),$$

49 $$(D_2\hat{\phi}'''(x_0,u_0\,;\,\delta u))(t) = \mathscr{D}_2\tilde{\tilde{\chi}}(x_0(t),u_0(t),t)\delta u(t), \qquad t \in I, \quad \delta u \in L_\infty^r(I).$$

Hence, since $\hat{\phi}' = (\hat{\phi}'', \hat{\phi}''')$, Hypothesis (IV.2.17) holds [see (I.7.45 and 49) and Theorem (I.7.44)].

It remains to verify that Hypothesis (IV.2.12) holds. To do this, we first define the subset $\hat{\mathscr{W}}$ of $\hat{\mathscr{V}}_2$ as follows:

50 $$\hat{\mathscr{W}} = \{\hat{T}\colon \hat{T} = \hat{T}_{\xi,u',\hat{I}} \text{ for some } \xi \in R^n,\ u' \in \mathscr{U}' \text{ and } \hat{I} \in \mathbf{I}\}.$$

It is clear that $\hat{\mathscr{W}} \subset \hat{\mathscr{W}} \times \mathscr{U}$. It follows from the convexity under switching of the class $\mathscr{U}$ that $\mathscr{U}'$ and $\{\hat{F}_{u',\hat{I}}\colon u' \in \mathscr{U}', \hat{I} \in \mathbf{I}\}$ are both also convex under switching. Then, by Theorem (IV.5.35), Hypothesis (IV.2.12) holds when $\hat{\mathscr{W}}$ is replaced by $(\hat{\mathscr{W}} \times \mathscr{U})$ and π_0 by u_0.

51 Let us show that Hypothesis (IV.2.12) holds with $\hat{\mathscr{W}}$ itself (and with $\pi_0 = u_0$). Thus, let $\{(\hat{T}_1,u_1), \ldots ,(\hat{T}_\rho,u_\rho)\}$ be an arbitrary finite subset of $\hat{\mathscr{W}}$ and let, for each $j = 1, \ldots, \rho$, $\hat{T}_j = \hat{T}_{\xi_j,u_j',\hat{I}_j}$ where $\xi_j \in R^n$, $u_j' \in \mathscr{U}'$, and $\hat{I}_j \in \mathbf{I}$. Since $\hat{\mathscr{W}} \subset (\hat{\mathscr{W}} \times \mathscr{U})$, and because Hypothesis (IV.2.12) holds when $\hat{\mathscr{W}}$ is replaced by $(\hat{\mathscr{W}} \times \mathscr{U})$, there are a subset $\hat{\mathscr{W}}_1$ of $(\hat{\mathscr{W}} \times \mathscr{U})$ and a number $\varepsilon_1 > 0$ such that Conclusions A and B of Hypothesis (IV.2.12) hold—with $\hat{\mathscr{W}}_0$, π, π_i, and ε_0 replaced by $\hat{\mathscr{W}}_1$, u, u_i, and ε_1, respectively. Further [see (IV.5.37) in Theorem (IV.5.35)], $\hat{\mathscr{W}}_1 \subset (\hat{\mathscr{W}}_1 \times \Pi)$, where [see (IV.5.13 and 14)]

$$\hat{\mathscr{W}}_1 \subset \{\hat{T}_{\hat{x}^*,\hat{F}}\colon \hat{x}^* \equiv \xi,\ \xi \in \text{co}\,\{x_0(t_1),\xi_1, \ldots ,\xi_\rho\}, \quad \hat{F} \in \text{Ch}\,\{\hat{F}_I,\hat{F}_{u_1',\hat{I}_1}, \ldots ,\hat{F}_{u_\rho',\hat{I}_\rho}\}\}.$$

But if $\hat{F} \in \text{Ch}\,\{\hat{F}_I,\hat{F}_{u_1',\hat{I}_1}, \ldots ,\hat{F}_{u_\rho',\hat{I}_\rho}\}$, then $\hat{F} = \hat{F}_{u',\hat{I}}$ for some $u' \in \text{Ch}\,\{u_1', \ldots ,u_\rho'\}$ and some $\hat{I} \in \mathbf{I}$, so that

52 $$\hat{\mathscr{W}}_1 \subset \{\hat{T}_{\xi,u',\hat{I}}\colon \xi \in R^n,\ u' \in \text{Ch}\,\{u_1', \ldots ,u_\rho'\},\ \hat{I} \in \mathbf{I}\}.$$

Further, since $u'_j \in \mathscr{U}'$ for $j = 1, \ldots, \rho$, [see (39)] there is an $\eta_0 < 0$ such that

$$\operatorname*{ess\,sup}_{t \in I'_i} \tilde{\tilde{\chi}}^i(x_0(t), u'_j(t), t) \leq \eta_0 \quad \text{for each} \quad i = 1, \ldots, \nu'$$
$$\text{and} \quad j = 1, \ldots, \rho,$$

which means that, for every $u' \in \text{Ch}\,\{u'_1, \ldots, u'_\rho\}$,

$$\tilde{\tilde{\chi}}^i(x_0(t), u'(t), t) \leq \eta_0 \quad \text{for almost all} \quad t \in I'_1$$
$$\text{and each} \quad i = 1, \ldots, \nu'.$$

Further, because $\mathscr{U}' \subset \Pi$, [see (23)] there is a compact subset V_c of V such that

$$\{u'(t) : t \in I, u' \in \text{Ch}\,\{u'_1, \ldots, u'_\rho\}\} \subset V_c.$$

Hence, by Corollary (I.5.32), there is an $\tilde{\varepsilon} > 0$ such that

53 $$\operatorname*{ess\,sup}_{t \in I'_i} \tilde{\tilde{\chi}}^i(x(t), u'(t), t) \leq \frac{\eta_0}{2} < 0 \quad \text{for each} \quad i = 1, \ldots, \nu',$$
$$\text{every} \quad u' \in \text{Ch}\,\{u'_1, \ldots, u'_\rho\}, \quad \text{and every} \quad x \in \mathscr{C}^n(I)$$
$$\text{such that} \quad \|x - x_0\| < \tilde{\varepsilon}.$$

Since Conclusions A and B of Hypothesis (IV.2.12) hold with $\mathscr{W}$, $\mathscr{W}_0$, π, π_i, and ε_0 replaced by $(\hat{\mathscr{W}} \times \mathscr{U})$, $\tilde{\mathscr{W}}_1$, u, u_i, and ε_1, respectively, there are a neighborhood $\hat{N}_0$ of 0 in $\hat{\mathscr{T}}$ and a number $\varepsilon_2 \in (0, \varepsilon_1)$ such that $\|\varphi_0(\hat{T}^u) - \varphi_0(\hat{T}_0^{u_0})\| < \tilde{\varepsilon}$ whenever (i) $(\hat{T}, u) \in (\tilde{\mathscr{W}}_1 \cap \mathscr{S}_1)$, (ii) $\hat{T} \in \hat{T}_0 + \hat{N}_0$, and (iii) $u = u_0 + \sum_1^\rho \lambda^j(u_j - u_0)$ with $0 \leq \lambda^j \leq \varepsilon_2$ for each $j = 1, \ldots, \rho$. Let us denote by $\tilde{\mathscr{W}}_0$ the set of all $(\hat{T}, u)$ which satisfy conditions (i)–(iii) of the preceding sentence. Note that $\tilde{\mathscr{W}}_0 \subset \tilde{\mathscr{W}}_1 \subset (\hat{\mathscr{W}}_1 \times \mathscr{U})$ and that $\tilde{\mathscr{W}}_0 \subset \mathscr{S}_1$. Since $\varphi_0(\hat{T}_0^{u_0}) = x_0$ [see (42)], it now follows from (40), (50), (52), and (53) that $\tilde{\mathscr{W}}_0 \subset \tilde{\mathscr{W}}$. Let $\hat{N}_1$ be a neighborhood of 0 in $\hat{\mathscr{T}}$ such that $\hat{N}_1 + \hat{N}_1 \subset \hat{N}_0$, and let $\varepsilon_0 \in (0, \varepsilon_2)$ be such that $\sum_1^\rho \lambda^j(\hat{T}_j - \hat{T}_0) \in \hat{N}_1$ whenever $0 \leq \lambda^j \leq \varepsilon_0$ for each $j = 1, \ldots, \rho$ [see Corollaries (I.4.3 and 4)]. It is evident that Condition A of Hypothesis (IV.2.12) now holds, with π and π_i replaced by u and u_i, respectively. Further, since Condition B in this Hypothesis holds with π_i replaced by u_i and $\tilde{\mathscr{W}}_0$ by $\tilde{\mathscr{W}}_1$, and, because $\tilde{\mathscr{W}}_1 \subset (\hat{\mathscr{W}}_1 \times \mathscr{U})$, we conclude that, for every neighborhood $\hat{N}$ of 0 in

$\hat{\mathscr{T}}$, there is a continuous function $\hat{\gamma}$: $\{\lambda: \lambda = (\lambda^1, \ldots, \lambda^\rho) \in R^\rho$, $0 \leq \lambda^i \leq \varepsilon_0$ for $i = 1, \ldots, \rho\} \to \hat{N} \cap \hat{N}_1$ such that, for every $\lambda = (\lambda^1, \ldots, \lambda^\rho) \in R^\rho$ satisfying $0 \leq \lambda^j \leq \varepsilon_0$ for each j,

54 $$\text{(i) } \hat{T}_0 + \sum_{j=1}^{\rho} \lambda^j(\hat{T}_j - \hat{T}_0) + \hat{\gamma}(\lambda^1, \ldots, \lambda^\rho) \in (\hat{T}_0 + \hat{N}_1 + \hat{N}_1) \subset (\hat{T}_0 + \hat{N}_0),$$

55 $$\text{(ii) } u_0 + \sum_{j=1}^{\rho} \lambda^j(u_j - u_0) \in \mathscr{U},$$

and (iii) $(\hat{T}_\lambda, u_\lambda) \in (\mathscr{S}_1 \cap \tilde{\mathscr{W}}_1)$, where $\hat{T}_\lambda$ is the operator in (54) and u_λ is the function in (55). But this means that $(\hat{T}_\lambda, u_\lambda) \in (\mathscr{S}_1 \cap \tilde{\mathscr{W}}_0)$ whenever $0 \leq \lambda^j \leq \varepsilon_0$ for each j, i.e., Hypothesis (IV.2.12) is satisfied by $\tilde{\mathscr{W}}$ itself.

56 In the preceding arguments—beginning in (51)—we never made use of the fact that $(\hat{T}_j, u_j) \in \tilde{\mathscr{W}}$ for each $j = 1, \ldots, \rho$ beyond the fact that $(\hat{T}_j, u_j) \in (\hat{\mathscr{W}} \times \mathscr{U})$(for each j). Consequently, all our conclusions remain in effect even under the weaker requirement that $(\hat{T}_j, u_j) \in (\hat{\mathscr{W}} \times \mathscr{U})$ for each j. Let us use this to demonstrate that

57 $$(\hat{\mathscr{W}} \times \mathscr{U}) \subset (\hat{T}_0, u_0) + \overline{\text{co co}}\, (\tilde{\mathscr{W}} - (\hat{T}_0, u_0)).$$

Indeed, let $(\hat{T}_1, u_1) \in (\hat{\mathscr{W}} \times \mathscr{U})$ be arbitrary. By what we just said, we may repeat all of the arguments, begining in (51), with $\rho = 1$ and the indicated $(\hat{T}_1, u_1)$. In particular, if $\hat{N}_2$ is an arbitrary neighborhood of 0 in $\hat{\mathscr{T}}$, we can conclude that there are a number $\lambda^1 \in (0,1)$ and an element $(\hat{T}'_1, \tilde{u}_1) \in \tilde{\mathscr{W}}$ such that $\hat{T}'_1 \in \hat{T}_0 + \lambda^1(\hat{T}_1 - \hat{T}_0) + \lambda^1 \hat{N}_2$ such that $\tilde{u}_1 = u_0 + \lambda^1(u_1 - u_0)$, i.e., such that $(1/\lambda^1)(\hat{T}'_1 - \hat{T}_0, \tilde{u}_1 - u_0) \in [(\hat{T}_1 - \hat{T}_0, u_1 - u_0) + \hat{N}_2 \times \{0\}] \cap \text{co co}\, (\tilde{\mathscr{W}} - (\hat{T}_0, u_0))$, which implies that $(\hat{T}_1 - \hat{T}_0, u_1 - u_0) \in \overline{\text{co co}}\, (\tilde{\mathscr{W}} - (\hat{T}_0, u_0))$. But this means that (57) holds.

If we now appeal to Theorem (IV.2.34), (IV.2.51), and Corollaries (IV.2.46 and 47), taking into account (42)–(50), (57), (36), (31), and (32), and make use of (I.5.25), (I.6.8, 19, and 31), and Theorems (I.6.14, 22, and 32) in order to obtain a representation for $(Z')^*$, we arrive at the following conclusion:

58 There exist a row vector $\alpha = (\alpha^0, \alpha^1, \ldots, \alpha^{\mu+m}) \in \bar{R}_-^{\mu+1} \times R^m$, a row-vector-valued function $\tilde{\lambda} = (\tilde{\lambda}^1, \ldots, \tilde{\lambda}^\nu) \in (NBV(I))^\nu$, and finitely additive real-valued measures $\mathbf{v}^1, \ldots, \mathbf{v}^{\nu'}$ defined on all

Lebesgue measurable subsets of I, such that

$$\text{(i)}\ \|\tilde{\lambda}\| + |\alpha| + \sum_{i=1}^{\nu'} |\mathbf{v}^i(I)| > 0,$$

59
$$\text{(ii)}\ \sum_{j=1}^{k} \alpha \mathscr{D}_j^0 \chi \delta x_{\xi,u''}(\tau^j) + \int_{t_1}^{t_2} d\tilde{\lambda}(t) \mathscr{D}_1 \tilde{\chi}(x_0(t),t) \delta x_{\xi,u''}(t) + \sum_{i=1}^{\nu'} \int_{t_1}^{t_2} \mathscr{D}_1 \tilde{\tilde{\chi}}^i(x_0(t),u_0(t),t) \delta x_{\xi,u''}(t)\, d\mathbf{v}^i(t)$$
$$\leq \sum_{j=1}^{k} \alpha \mathscr{D}_j^0 \chi \delta x_{x_0(t_1),u_0}(\tau^j) + \int_{t_1}^{t_2} d\tilde{\lambda}(t) \mathscr{D}_1 \tilde{\chi}(x_0(t),t) \delta x_{x_0(t_1),u_0}(t) + \sum_{i=1}^{\nu'} \int_{t_1}^{t_2} \mathscr{D}_1 \tilde{\tilde{\chi}}^i(x_0(t),u_0(t),t) \delta x_{x_0(t_1),u_0}(t)\, d\mathbf{v}^i(t)$$

for all $\xi \in R^n$ and $u'' \in \mathscr{U}''$

where

60
$$\mathscr{U}'' = \{u'' : u'' \in \text{Ch}\,\{u_0, u'\} \text{ for some } u' \in \mathscr{U}'\}$$

and, for each $\xi \in R^n$ and $u'' \in \mathscr{U}''$, $\delta x_{\xi,u''}$ denotes the column-vector-valued function in $\mathscr{C}^n(I)$ defined by

61
$$\delta x_{\xi,u''}(t) = \int_{t_1}^{t} \mathscr{D}_1 f(x_0(s),u_0(s),s) \delta x_{\xi,u''}(s)\, ds + \xi + \int_{t_1}^{t} f(x_0(s),u''(s),s)\, ds, \quad t \in I,$$

and $\mathscr{D}_j^0 \chi$ (for each j) denotes $\mathscr{D}_j \chi(x_0(\tau^1), \ldots, x_0(\tau^k))$,

62
$$\text{(iii)}\ \sum_{i=1}^{\mu} \alpha^i \chi^i(x_0(\tau^1), \ldots, x_0(\tau^k)) = \int_{t_1}^{t_2} d\tilde{\lambda}(t) \tilde{\chi}(x_0(t),t) + \sum_{i=1}^{\nu'} \int_{t_1}^{t_2} \tilde{\tilde{\chi}}^i(x_0(t),u_0(t),t)\, d\mathbf{v}^i(t) = 0,$$

63
$$\text{(iv)}\ \sum_{j=1}^{k} \alpha \mathscr{D}_j^0 \chi \delta x_{\delta u}(\tau^j) + \int_{t_1}^{t_2} d\tilde{\lambda}(t) \mathscr{D}_1 \tilde{\chi}(x_0(t),t) \delta x_{\delta u}(t) + \sum_{i=1}^{\nu'} \int_{t_1}^{t_2} [\mathscr{D}_1 \tilde{\tilde{\chi}}^i(x_0(t),u_0(t),t) \delta x_{\delta u}(t) + \mathscr{D}_2 \tilde{\tilde{\chi}}^i(x_0(t),u_0(t),t) \delta u(t)]\, d\mathbf{v}^i(t) \leq 0$$

for all $\delta u \in \mathscr{U} - u_0$,

where, for each $\delta u \in L_\infty^r(I)$, $\delta x_{\delta u}$ denotes the column-vector-valued function in $\mathscr{C}^n(I)$ defined by

64 $$\delta x_{\delta u}(t) = \int_{t_1}^{t} \mathscr{D}_1 f(x_0(s),u_0(s),s)\delta x_{\delta u}(s)\,ds + \int_{t_1}^{t} \mathscr{D}_2 f(x_0(s),u_0(s),s)\delta u(s)\,ds, \quad t \in I,$$

(v) for each $i = 1, \ldots, \nu$, $\tilde{\lambda}^i$ is nonincreasing on I and is constant on every subinterval of I which does not meet I_i, and

(vi) for each $i = 1, \ldots, \nu'$, $\mathbf{v}^i(\hat{I}) \leq 0$ for every Lebesgue measurable subset $\hat{I}$ of I, and $\mathbf{v}^i(\hat{I}) = 0$ if the set $(\hat{I} \cap I'_i)$ has Lebesgue measure zero.

Now conclusions (iii), (v), and (vi), together with relations (12), (14), and (15), and the fact that $\alpha^i \leq 0$ for $i = 1, \ldots, \mu$, imply that

65 $$\alpha^i \chi^i(x_0(\tau^1), \ldots, x_0(\tau^k)) = 0 \quad \text{for} \quad i = 1, \ldots, \mu,$$

66 $$\int_{t_1}^{t_2} d\tilde{\lambda}^i(t)\tilde{\chi}^i(x_0(t),t) = 0 \quad \text{for} \quad i = 1, \ldots, \nu,$$

67 $$\int_{t_1}^{t_2} \tilde{\tilde{\chi}}^i(x_0(t),u_0(t),t)\,d\mathbf{v}^i(t) = 0 \quad \text{for} \quad i = 1, \ldots, \nu'.$$

But (66), (14), and (v) imply that, for each $i = 1, \ldots, \nu$, $\tilde{\lambda}^i$ is constant on every subinterval of I which does not meet the closed set I_i^0 defined as follows:

68 $$I_i^0 = \{t : t \in I_i, \tilde{\chi}^i(x_0(t),t) = 0\}, \quad i = 1, \ldots, \nu.$$

Further, (67), (15), and (vi) imply that, for each $i = 1, \ldots, \nu'$, and each $\theta > 0$, $\mathbf{v}^i(I_i'^{\theta}) = 0$, where

69 $$I_i'^{\theta} = \{t : t \in I'_i, \tilde{\tilde{\chi}}^i(x_0(t),u_0(t),t) < -\theta\}.$$

Note that, since $\mathbf{v}^i$ is not necessarily countably additive, it does *not* follow from this that $\mathbf{v}^i(\{t : t \in I'_i, \tilde{\tilde{\chi}}^i(x_0(t),u_0(t),t) < 0\}) = 0$.

As we saw in the derivation of formula (V.2.21), $\delta x_{\xi,u''}$ may be written in the form

70 $$\delta x_{\xi,u''}(t) = \Phi(t)\xi + \Phi(t)\int_{t_1}^{t} \Phi^{-1}(s) f(x_0(s),u''(s),s)\,ds, \quad t \in I,$$

where Φ is the absolutely continuous $(n \times n)$-matrix-valued function defined on I that satisfies

71 $$\dot{\Phi}(t) = \mathscr{D}_1 f(x_0(t),u_0(t),t)\Phi(t)$$

$$\text{a.e. on} \quad I, \quad \Phi(t_1) = \text{the identity matrix.}$$

By the same token,

72 $$\delta x_{\delta u}(t) = \Phi(t) \int_{t_1}^{t} \Phi^{-1}(s)\mathscr{D}_2 f(x_0(s),u_0(s),s)\delta u(s)\, ds, \quad t \in I.$$

Consequently, (59) can be rewritten in the form

73 $$\sum_{j=1}^{k} \alpha \mathscr{D}_j^0 \chi \Phi(\tau^j) \left\{\xi + \int_{t_1}^{\tau^j} \Phi^{-1}(s)\delta f_{u''}^0(s)\, ds\right\} + \int_{t_1}^{t_2} d\tilde{\lambda}(t)\psi_1(t)\left\{\xi + \int_{t_1}^{t} \Phi^{-1}(s)\delta f_{u''}^0(s)\, ds\right\} + \sum_{i=1}^{v'} \int_{t_1}^{t_2} \psi_2^i(t)\left\{\xi + \int_{t_1}^{t} \Phi^{-1}(s)\delta f_{u''}^0(s)\, ds\right\} d\mathbf{v}^i(t) \le 0$$

for all $\xi \in R^n$ and $u'' \in \mathscr{U}''$,

where, for every $t \in I$,

74 $$\psi_1(t) = \mathscr{D}_1\tilde{\chi}(x_0(t),t)\Phi(t) \quad \text{and} \quad \psi_2^i(t) = \mathscr{D}_1\tilde{\tilde{\chi}}^i(x_0(t),u_0(t),t)\Phi(t)$$

for $i = 1, \ldots, v'$,

and, for each $s \in I$ and every $u'' \in \mathscr{U}''$,

75 $$\delta f_{u''}^0(s) = f(x_0(s),u''(s),s) - f(x_0(s),u_0(s),s).$$

Similarly, (63) can rewritten in the form

76 $$\sum_{j=1}^{k} \alpha \mathscr{D}_j^0 \chi \Phi(\tau^j) \int_{t_1}^{\tau^j} \Phi^{-1}(s)\mathscr{D}_2^0 f(s)\delta u(s)\, ds + \int_{t_1}^{t_2} d\tilde{\lambda}(t)\left(\psi_1(t)\int_{t_1}^{t} \Phi^{-1}(s)\mathscr{D}_2^0 f(s)\delta u(s)\, ds\right) + \sum_{i=1}^{v'} \int_{t_1}^{t_2}\left[\psi_2^i(t)\int_{t_1}^{t} \Phi^{-1}(s)\mathscr{D}_2^0 f(s)\delta u(s)\, ds + \mathscr{D}_2\tilde{\tilde{\chi}}^i(x_0(t),u_0(t),t)\delta u(t)\right] d\mathbf{v}^i(t) \le 0$$

for all $\delta u \in \mathscr{U} - u_0$,

where, for ease of notation, we have set

77 $$\mathscr{D}_2^0 f(s) = \mathscr{D}_2 f(x_0(s),u_0(s),s) \quad s \in I.$$

Since $u_0 \in \mathscr{U}''$ and $\delta f_{u_0}^0(s) \equiv 0$, (73) implies that

78 $$\sum_{j=1}^{k} \alpha \mathscr{D}_j^0 \chi \Phi(\tau^j) + \int_{t_1}^{t_2} d\tilde{\lambda}(t)\psi_1(t) + \sum_{i=1}^{v'} \int_{t_1}^{t_2} \psi_2^i(t)\, d\mathbf{v}^i(t) = 0.$$

Further, since (73) holds for $\xi = 0$, we have

79 $$\int_{t_1}^{t_2} \psi(s)\delta f^0_{u''}(s)\,ds + \int_{t_1}^{t_2} d\tilde{\lambda}(t)\left(\psi_1(t)\int_{t_1}^{t} \Phi^{-1}(s)\delta f^0_{u''}(s)\,ds\right)$$
$$+ \sum_{i=1}^{v'} \int_{t_1}^{t_2} \psi_2^i(t)\int_{t_1}^{t} \Phi^{-1}(s)\delta f^0_{u''}(s)\,ds\,d\mathbf{v}^i(t) \le 0 \quad \text{for all} \quad u'' \in \mathscr{U}'',$$

where the function ψ is defined by

80 $$\psi(s) = \sum_{j=1}^{k} \alpha\mathscr{D}_j^0\chi\Phi(\tau^j)\Phi^{-1}(s) \quad \text{for} \quad \tau^{i-1} \le s < \tau^i$$
$$\text{and each} \quad i = 2, \ldots, k,$$

81 $$\psi(\tau^k) = \psi(t_2) = \alpha\mathscr{D}_k^0\chi.$$

Let us define the functions $\tilde{\tilde{\psi}}$, $\hat{\psi}$, $\bar{\psi}$, and $\tilde{\psi}$ on I as follows:

82 $$\tilde{\tilde{\psi}}(s) = \sum_{i=1}^{v'} \int_{s}^{t_2} \psi_2^i(t)\,d\mathbf{v}^i(t)\Phi^{-1}(s), \quad s \in I,$$

83 $$\hat{\psi}(s) = \int_{s}^{t_2} d\tilde{\lambda}(t)\psi_1(t)\Phi^{-1}(s), \quad s \in I,$$

84 $$\bar{\psi} = \psi + \hat{\psi},$$

85 $$\tilde{\psi}(s) = \bar{\psi}(s) + \tilde{\lambda}(s)\mathscr{D}_1\tilde{\chi}(x_0(s),s), \quad s \in I.$$

Note that the functions ψ, ψ_1, $\delta f^0_{u''}$, $\hat{\psi}$, $\bar{\psi}$, and $\tilde{\psi}$ are defined here as in Section 3 of Chapter V [see (V.3.23, 27–29, 32, 33, 36)].

Interchanging the order of integration in each of the two double integrals in (76), we can now rewrite this relation as follows:

86 $$\int_{t_1}^{t_2} [\tilde{\psi}(t) - \tilde{\lambda}(t)\mathscr{D}_1\tilde{\chi}(x_0(t),t) + \tilde{\tilde{\psi}}(t)]\mathscr{D}_2 f(x_0(t),u_0(t),t)\delta u(t)\,dt$$
$$+ \sum_{i=1}^{v'} \int_{t_1}^{t_2} \mathscr{D}_2\tilde{\tilde{\chi}}^i(x_0(t),u_0(t),t)\delta u(t)\,d\mathbf{v}^i(t) \le 0 \quad \text{for all} \quad \delta u \in \mathscr{U} - u_0.$$

Similarly, if we interchange the order of integration in each of the two double integrals in (79), this relation can be rewritten in the form

87 $$\int_{t_1}^{t_2} [\tilde{\psi}(t) - \tilde{\lambda}(t)\mathscr{D}_1\tilde{\chi}(x_0(t),t) + \tilde{\tilde{\psi}}(t)]f(x_0(t),u_0(t),t)\,dt$$
$$\ge \int_{t_1}^{t_2} [\tilde{\psi}(t) - \tilde{\lambda}(t)\mathscr{D}_1\tilde{\chi}(x_0(t),t) + \tilde{\tilde{\psi}}(t)]f(x_0(t),u(t),t)\,dt$$
$$\text{for all} \quad u \in \mathscr{U}''.$$

Further, as we saw in Section 3 of Chapter V, $\tilde{\psi}$ is absolutely continuous on each of the intervals $[t_1,\tau^2)$, $[\tau^2,\tau^3)$, $\ldots$, $[\tau^{k-1},t_2]$, and, on each of these intervals, satisfies Eqs. (V.3.38 and 39), i.e.,

88 $$\dot{\tilde{\psi}}(t) = -\tilde{\psi}(t)\mathscr{D}_1 f(x_0(t),u_0(t),t) + \tilde{\lambda}(t)\mathscr{D}_1 p(x_0(t),u_0(t),t) \quad \text{a.e.,}$$

where

89 $$p(\xi,v,t) = \mathscr{D}_1\tilde{\chi}(\xi,t)f(\xi,v,t) + \mathscr{D}_2\tilde{\chi}(\xi,t) \quad \text{for all} \quad (\xi,v,t) \in \mathbf{G} \times V \times I,$$

and, in addition,

90 $$\tilde{\psi}(\tau^j) - \tilde{\psi}(\tau^{j-}) = -\alpha\mathscr{D}_j^0\chi \quad \text{for} \quad j = 2, \ldots, k-1,$$

91 $$\tilde{\psi}(t_2) = \alpha\mathscr{D}_k^0\chi.$$

Finally, it follows from (82)–(85), (80), (71), and (78), since $\tau^1 = t_1$, that

92 $$\tilde{\psi}(t_1) - \tilde{\lambda}(t_1)\mathscr{D}_1\tilde{\chi}(x_0(t_1),t_1) + \tilde{\tilde{\psi}}(t_1) = -\alpha\mathscr{D}_1^0\chi.$$

Thus, we have proved the following theorem:

93 THEOREM. *Let* (x_0,u_0) *be a solution of Problem* (10). *Then there exist a row vector* $\alpha = (\alpha^0, \alpha^1, \ldots, \alpha^{\mu+m}) \in \bar{R}_-^{\mu+1} \times R^m$, *finitely additive real-valued measures* $\mathbf{v}^1, \ldots, \mathbf{v}^{\nu'}$ *defined on all Lebesgue measurable subsets of* I, *and row-vector-valued functions* $\tilde{\lambda} = (\tilde{\lambda}^1, \ldots, \tilde{\lambda}^\nu) \in (NBV(I))^\nu$, $\tilde{\psi} \in L_\infty^n(I)$, *and* $\tilde{\tilde{\psi}} \in L_\infty^n(I)$ *such that*

(i) $\|\tilde{\lambda}\| + |\alpha| + \sum_{i=1}^{\nu'} |\mathbf{v}^i(I)| > 0$,

(ii) $\tilde{\tilde{\psi}}$ *is defined by* (82), (74), *and* (71),

(iii) *on each of the intervals* $[t_1,\tau^2)$, $[\tau^2,\tau^3)$, $\ldots$, $[\tau^{k-1},t_2]$, $\tilde{\psi}$ *is absolutely continuous and satisfies the linear inhomogeneous ordinary differential equation* (88), *where* p *is defined by* (89),

(iv) u_0 *satisfies the maximum principle in integral form* (87), *where* $\mathscr{U}''$ *is defined by* (60) *and* (39),

(v) $\tilde{\psi}$ *satisfies the transversality conditions* (91) *and* (92) *as well as the jump conditions* (90), *where, for each* j, $\mathscr{D}_j^0\chi$ *denotes* $\mathscr{D}_j\chi(x_0(\tau^1), \ldots, x_0(\tau^k))$,

(vi) *relation* (86) *holds*,

(vii) *Eqs.* (65) *hold*,

(viii) *for each* $i = 1, \ldots, \nu$, $\tilde{\lambda}^i$ *is nonincreasing on* I *and is constant on every subinterval of* I *which does not meet the closed subset* I_i^0 *of* I *defined by* (68),

(ix) *for each* $i = 1, \ldots, v'$, $\mathbf{v}^i(\hat{I}) \leq 0$ *for every Lebesgue measurable subset* $\hat{I}$ *of* I, *and* $\mathbf{v}^i(\hat{I}) = 0$ *if the set* $(\hat{I} \cap I'_i)$ *has Lebesgue measure zero, and also, for each* $\theta > 0$, $\mathbf{v}^i(I'^{\theta}_i) = 0$, *where the set* I'^{θ}_i *is defined by* (69), *and* (67) *holds.*

94 The necessary conditions of Theorem (93) differ from those of Theorem (V.3.44) in the following respects—all of which reflect the presence here of the additional mixed control-phase variable constraint—(i) there is an additional adjoint variable $\tilde{\tilde{\psi}}$, defined in terms of the nonpositive finitely additive measures $\mathbf{v}^i$, which appears both in the maximum condition (87) and in the transversality condition (92), (ii) the maximum condition (87) is with respect to all u in the subset $\mathscr{U}''$ of $\mathscr{U}$ rather than with respect to all $u \in \mathscr{U}$, and (iii) there is an additional inequality, (86).

95 Consider the following subset of $\mathscr{U}$:

96 $$\mathscr{U}''' = \{u\colon u \in \mathscr{U}, \tilde{\tilde{\chi}}^i(x_0(t),u(t),t) \leq 0 \quad \text{for almost all} \quad t \in I'_i \quad \text{and each} \quad i = 1, \ldots, v'\}.$$

Evidently, $\mathscr{U}' \subset \mathscr{U}'''$ and $u_0 \in \mathscr{U}'''$, and $\mathscr{U}'''$ is convex under switching, so that also $\mathscr{U}'' \subset \mathscr{U}'''$. If $\mathscr{U}'$ is dense in $\mathscr{U}'''$ [in $L^r_\infty(I)$], as is often the case in applications, then it follows from Hypotheses (3)–(5) and (7) and the Lebesgue dominated convergence theorem that the maximum principle (87) holds not only for all $u \in \mathscr{U}''$, but also for all $u \in \mathscr{U}'''$.

97 If Problem (10) is modified by eliminating the phase inequality constraint (14), then the necessary conditions of Theorem (93) remain in force, but with $\tilde{\lambda} \equiv 0$. On the other hand, if either (or both) the equality constraints (13), or the inequality constraints (12), are eliminated in Problem (10), then m or μ (or both) should be set equal to zero in Theorem (93), and Eqs. (65) must be deleted if there are no inequality constraints [see (IV.2.75)]. If $k = 2$—so that $\tau^1 = t_1$ and $\tau^2 = t_2$, and the constraints (12) and (13), as well as the cost functional appearing in (16), do not depend on any "intermediate" values of x—then (90) must be deleted from the necessary conditions, and it follows that $\tilde{\psi}$ is absolutely continuous on I.

The reader should review the remarks in (V.3.46 and 48) which are pertinent here also with evident minor modifications.

We shall now show that if the set $\mathscr{U}$ and the functions $\tilde{\tilde{\chi}}^i$ are, in a certain sense, "well-behaved" at the pair (x_0,u_0), then the additional adjoint variable $\tilde{\tilde{\psi}}$ is absolutely continuous, and the sum $(\tilde{\psi} + \tilde{\tilde{\psi}})$ even satisfies a linear inhomogeneous differential equation similar to Eq. (88) on each of the intervals $[t_1, \tau^2)$, etc.

Indeed, let us make the following hypothesis.

98 HYPOTHESIS. *There is a function* $u^* \in \mathscr{U}$ *such that*

99 $$\operatorname*{ess\,sup}_{t \in I'_i} [\tilde{\tilde{\chi}}^i(x_0(t),u_0(t),t) + \mathscr{D}_2\tilde{\tilde{\chi}}^i(x_0(t),u_0(t),t)(u^*(t) - u_0(t))] < 0 \quad \textit{for} \quad i = 1, \ldots, \nu'.$$

Hypothesis (98) essentially asserts that the constraints (15) are compatible with one another "near (x_0,u_0), to first order, in u," as well as with the constraint $u \in \mathscr{U}$.

We first prove the following lemma.

100 LEMMA. *If* (x_0,u_0) *is a solution of Problem* (10) *such that Hypothesis* (98) *is satisfied, and if* $\mathscr{U}$ *is strongly convex under switching* [*see* (IV.5.15)], *then the measures* $\mathbf{v}^i$ *in Theorem* (93) *are countably additive and absolutely continuous* (*with respect to Lebesgue measure*).

Proof. If Hypothesis (98) holds, there evidently is a number $\tilde{\varepsilon} > 0$ such that

$$\tilde{\tilde{\chi}}^i(x_0(t),u_0(t),t) + \mathscr{D}_2\tilde{\tilde{\chi}}^i(x_0(t),u_0(t),t)(u^*(t) - u_0(t)) + \xi^i \le 0$$

for almost all $t \in I'_i$, each $i = 1, \ldots, \nu'$, and every $\xi = (\xi^1, \ldots, \xi^{\nu'}) \in R^{\nu'}$ satisfying $|\xi| \le \tilde{\varepsilon}$. Let us arbitrarily fix a Lebesgue measurable subset $\hat{I}$ of I, and let $\mathbf{v}(\hat{I}) = (\mathbf{v}^1(\hat{I}), \ldots, \mathbf{v}^{\nu'}(\hat{I}))$, so that $\mathbf{v}(\hat{I}) \in R^{\nu'}$. It follows at once from (67), (15), and conclusion (ix) in Theorem (93) that

$$\int_{\hat{I}} \tilde{\tilde{\chi}}^i(x_0(t),u_0(t),t)\, d\mathbf{v}^i(t) = 0 \quad \text{for} \quad i = 1, \ldots, \nu'.$$

If $\mathbf{v}(\hat{I}) \ne 0$, let us set $\hat{\xi} = -\tilde{\varepsilon}\mathbf{v}(\hat{I})/|\mathbf{v}(\hat{I})|$. We can then conclude, on the basis of what we have just shown, by virtue of (86) and the strong convexity under switching of $\mathscr{U}$, that

$$\begin{aligned} \tilde{\varepsilon}|\mathbf{v}(\hat{I})| &= -\sum_{i=1}^{\nu'} \hat{\xi}^i \int_{\hat{I}} d\mathbf{v}^i(t) \\ &\le \sum_{i=1}^{\nu'} \int_{\hat{I}} \mathscr{D}_2\tilde{\tilde{\chi}}^i(x_0(t),u_0(t),t)[u^*(t) - u_0(t)]\, d\mathbf{v}^i(t) \\ &\le -\int_{\hat{I}} [\tilde{\psi}(t) - \tilde{\lambda}(t)\mathscr{D}_1\tilde{\chi}(x_0(t),t) \\ &\quad + \tilde{\tilde{\psi}}(t)]\mathscr{D}_2 f(x_0(t),u_0(t),t)(u^*(t) - u_0(t))\, dt. \end{aligned}$$

Since $\int_{\hat{I}} z(t)\,dt \to 0$ as meas $(\hat{I}) \to 0$ for any fixed Lebesgue integrable function $z: I \to R$, we can conclude that $|\mathbf{v}(\hat{I})| \to 0$ as meas $(\hat{I}) \to 0$, or that

$$\mathbf{v}^i(\hat{I}) \to 0 \quad \text{as} \quad \text{meas}\,(\hat{I}) \to 0 \quad \text{for each} \quad i = 1, \ldots, v'.$$

Let us show that the last relation (which is the assertion that each $\mathbf{v}^i$ is absolutely continuous) implies that each $\mathbf{v}^i$ is countably additive. Indeed, if $\tilde{I} = \bigcup_{j=1}^{\infty} \tilde{I}_j$ where the $\tilde{I}_j$ are disjoint Lebesgue measurable subsets of I, then meas $(\bigcup_{j=k}^{\infty} \tilde{I}_j) \to 0$ as $k \to \infty$, so that $\mathbf{v}^i(\bigcup_{j=k}^{\infty} \tilde{I}_j) \to 0$ as $k \to \infty$, which (since $\mathbf{v}^i$ is finitely additive) means that

$$\sum_{j=1}^{k} \mathbf{v}^i(\tilde{I}_j) = \mathbf{v}^i\left(\sum_{j=1}^{k} \tilde{I}_j\right) \to \mathbf{v}^i(\tilde{I}) \quad \text{as} \quad k \to \infty,$$

$$\text{i.e.,} \quad \mathbf{v}^i \quad \text{is countably additive.} \quad |||$$

101 Now if $\mathbf{v}^i$ is absolutely continuous with respect to Lebesgue measure (as well as countably additive), then, by the Radon-Nikodým Theorem [see, e.g., reference 7 of Chapter I, Theorem 5.1 on page 181], there is a function $\boldsymbol{\mu}^i \in L_1(I)$ such that

$$\mathbf{v}^i(\hat{I}) = \int_{\hat{I}} \boldsymbol{\mu}^i(t)\,dt$$

for every Lebesgue measurable subset $\hat{I}$ of I. Since $\mathbf{v}^i$ is non-positive, $\boldsymbol{\mu}^i(t) \leq 0$ a.e. on I, and, because $\mathbf{v}^i(J_i) = 0$, where $J_i = \{t: t \in I, t \notin I'_i\}$, $\boldsymbol{\mu}^i(t) = 0$ a.e. on J_i. In this case also (see, e.g., reference 4 of Chapter I, Corollary III.10.5 on page 179)

$$\int_{\hat{I}} \mathbf{F}(t)\,d\mathbf{v}^i(t) = \int_{\hat{I}} \mathbf{F}(t)\boldsymbol{\mu}^i(t)\,dt$$

for every Lebesgue measurable subset $\hat{I}$ of I and every function $\mathbf{F} \in L_\infty(I)$, so that, in particular, (82) can be rewritten as [see (74)]

$$\tilde{\tilde{\psi}}(s) = \sum_{i=1}^{v'} \int_s^{t_2} \mathscr{D}_1\tilde{\tilde{\chi}}^i(x_0(t),u_0(t),t)\Phi(t)\boldsymbol{\mu}^i(t)\,dt\,\Phi^{-1}(s), \quad s \in I,$$

which means that $\tilde{\tilde{\psi}}$ is absolutely continuous on I and, by virtue of (71), satisfies the differential equation [also see (V.2.30 and 20)]

$$\dot{\tilde{\tilde{\psi}}}(t) = -\tilde{\tilde{\psi}}(t)\mathscr{D}_1 f(x_0(t),u_0(t),t) - \sum_{i=1}^{v'} \boldsymbol{\mu}^i(t)\mathscr{D}_1\tilde{\tilde{\chi}}^i(x_0(t),u_0(t),t)$$

$$\text{a.e. on} \quad I.$$

Consequently, if we denote $(\tilde{\psi} + \tilde{\tilde{\psi}})$ by $\tilde{\boldsymbol{\psi}}$, it follows that $\tilde{\boldsymbol{\psi}}$ is absolutely continuous on each of the intervals $[t_1,\tau^2)$, etc., and, on each of these intervals, satisfies the equation [see (88)]

102 $$\dot{\tilde{\boldsymbol{\psi}}}(t) = -\tilde{\boldsymbol{\psi}}(t)\mathscr{D}_1 f(x_0(t),u_0(t),t) - \boldsymbol{\mu}(t)\mathscr{D}_1\tilde{\tilde{\chi}}(x_0(t),u_0(t),t) + \tilde{\lambda}(t)\mathscr{D}_1 p(x_0(t),u_0(t),t) \quad \text{a.e.,}$$

[where $\boldsymbol{\mu} = (\boldsymbol{\mu}^1, \ldots, \boldsymbol{\mu}^{\nu'})$], as well as the conditions [see (90)–(92)]

103 $$\tilde{\boldsymbol{\psi}}(\tau^j) + \tilde{\boldsymbol{\psi}}(\tau^{j-}) = -\alpha\mathscr{D}_j^0\chi \quad \text{for} \quad j = 2, \ldots, k-1,$$

104 $$\tilde{\boldsymbol{\psi}}(t_2) = \alpha\mathscr{D}_k^0\chi,$$

105 $$\tilde{\boldsymbol{\psi}}(t_1) - \tilde{\lambda}(t_1)\mathscr{D}_1\tilde{\chi}(x_0(t_1),t_1) = -\alpha\mathscr{D}_1^0\chi.$$

Further, the maximum principle (87) can be rewritten in the form

106 $$\int_{t_1}^{t_2} [\tilde{\psi}(t) - \tilde{\lambda}(t)\mathscr{D}_1\tilde{\chi}(x_0(t),t)] f(x_0(t),u_0(t),t)\,dt \geq \int_{t_1}^{t_2} [\tilde{\psi}(t) - \tilde{\lambda}(t)\mathscr{D}_1\tilde{\chi}(x_0(t),t)] f(x_0(t),u(t),t)\,dt \quad \text{for all} \quad u \in \mathscr{U}'',$$

and (86) takes the form

107 $$\int_{t_1}^{t_2} \{[\tilde{\psi}(t) - \tilde{\lambda}(t)\mathscr{D}_1\tilde{\chi}(x_0(t),t)]\mathscr{D}_2 f(x_0(t),u_0(t),t) + \boldsymbol{\mu}(t)\mathscr{D}_2\tilde{\tilde{\chi}}(x_0(t),u_0(t),t)\}\delta u(t)\,dt \leq 0 \quad \text{for all} \quad \delta u \in \mathscr{U} - u_0.$$

Finally, it follows from (67) and (15) that, for each $i = 1, \ldots, \nu'$, $\boldsymbol{\mu}^i(t)\tilde{\tilde{\chi}}^i(x_0(t),u_0(t),t) = 0$ a.e. on I_i'.

Summarizing all the preceding, we have the following theorem.

108 THEOREM. *Let* (x_0,u_0) *be a solution of Problem* (10) *such that Hypothesis* (98) *holds and suppose that* $\mathscr{U}$ *is strongly convex under switching. Then there exist a row vector* $\alpha = (\alpha^0, \alpha^1, \ldots, \alpha^{\mu+m}) \in \bar{R}_-^{\mu+1} \times R^m$ *and row-vector-valued functions* $\tilde{\lambda} = (\tilde{\lambda}^1, \ldots, \tilde{\lambda}^\nu) \in [NBV(I)]^\nu$, $\boldsymbol{\mu} = (\boldsymbol{\mu}^1, \ldots, \boldsymbol{\mu}^{\nu'}) \in L_1^{\nu'}(I)$, *and* $\tilde{\boldsymbol{\psi}} \in L_\infty^n(I)$ *such that*

(i) $\|\tilde{\lambda}\| + \|\boldsymbol{\mu}\| + |\alpha| > 0$,

(ii) *on each of the intervals* $[t_1,\tau^2), [\tau^2,\tau^3), \ldots, [\tau^{k-1},t_2]$, $\tilde{\boldsymbol{\psi}}$ *is absolutely continuous and satisfies the linear inhomogeneous ordinary differential equation* (102), *where* p *is given by* (89),

(iii) u_0 *satisfies the maximum principle in integral form* (106), *where* $\mathscr{U}''$ *is defined by* (60) *and* (39),

(iv) $\tilde{\boldsymbol{\psi}}$ *satisfies the transversality conditions* (104) *and* (105), *as well as the jump conditions* (103), *where, for each* j, $\mathscr{D}_j^0\chi$ *denotes* $\mathscr{D}_j\chi(x_0(\tau^1), \ldots, x_0(\tau^k))$,

(v) *relation* (107) *holds,*

(vi) *Eqs.* (65) *hold,*

(vii) *for each* $i = 1, \ldots, \nu$, $\tilde{\lambda}^i$ *is nonincreasing on* I *and is constant on every subinterval of* I *which does not meet the closed subset* I_i^0 *of* I *defined by* (68), *and*

(viii) *for each* $i = 1, \ldots, \nu'$, $\boldsymbol{\mu}^i(t) \leq 0$ *a.e. on* I, $\boldsymbol{\mu}^i(t) = 0$ *for almost all* $t \in \{s\colon s \in I,\ s \notin I_i'\}$, *and* $\boldsymbol{\mu}^i(t)\tilde{\tilde{\chi}}^i(x_0(t),u_0(t),t) = 0$ *a.e. on* I_i'.

Note that if $u_0 \in \mathscr{U}$, as must be the case if $\mathscr{U}$ is open, then relation (107) takes the form

109 $$[\tilde{\boldsymbol{\psi}}(t) - \tilde{\lambda}(t)\mathscr{D}_1\tilde{\chi}(x_0(t),t)]\mathscr{D}_2 f(x_0(t),u_0(t),t) + \boldsymbol{\mu}(t)\mathscr{D}_2\tilde{\tilde{\chi}}(x_0(t),u_0(t),t) = 0 \quad \text{a.e. on} \quad I.$$

As before, if $\mathscr{U}'$ is dense in $\mathscr{U}'''$ [defined in (96)], then (106) holds for all $u \in \mathscr{U}'''$.

If $\boldsymbol{\mu} \equiv 0$ on I, then the necessary conditions of Theorem (108) differ from those of Theorem (V.3.44) only in that the maximum principle (106) is with respect to all $u \in \mathscr{U}''$ rather than with respect to all $u \in \mathscr{U}$, and in that (107) holds.

Corresponding to Lemma (V.3.68), we have the following result concerning the nontriviality of our maximum principle.

110 LEMMA. *Let* (x_0,u_0) *be a solution of Problem* (10), *where* $\mathscr{U}$ *is strongly convex under switching, and suppose that Hypothesis* (98) *as well as Conditions* (V.3.52 *and* 54) *are satisfied. Then the functions* $\tilde{\boldsymbol{\psi}}$ *and* $\tilde{\lambda}$ *in Theorem* (108) *have the property that*

111 $$\tilde{\boldsymbol{\psi}}(t) - \tilde{\lambda}(t)\mathscr{D}_1\tilde{\chi}(x_0(t),t) \neq 0 \quad \textit{for all} \quad t$$

in a subset of I *of positive measure.*

Proof. In exactly the same way that we proved Lemma (V.3.68), we can show that, under the hypotheses of the present lemma, (111) holds if $|\alpha| = \|\boldsymbol{\mu}\| = 0$ or if $\alpha \neq 0$. This leaves the case where $\alpha = 0$ and $\|\boldsymbol{\mu}\| \neq 0$. We argue by contradiction, and suppose that $\tilde{\boldsymbol{\psi}}(t) - \tilde{\lambda}(t)\mathscr{D}_1\tilde{\chi}(x_0(t),t) = 0$ a.e. on I. Then (107) takes the form

112 $$\sum_{i=1}^{\nu'} \int_{t_1}^{t_2} \boldsymbol{\mu}^i(t)\mathscr{D}_2\tilde{\tilde{\chi}}^i(x_0(t),u_0(t),t)[u(t) - u_0(t)]\, dt \leq 0 \quad \text{for all} \quad u \in \mathscr{U}.$$

Since $\boldsymbol{\mu}^j(t) \neq 0$ for t in some subset of I of positive measure for at least one $j = 1, \ldots, \nu'$ (because $\|\boldsymbol{\mu}\| \neq 0$), and because (112) must hold in particular for $u = u^*$, where u^* satisfies (99), then, once we take into account conclusion (viii) of Theorem (108), we immediately obtain a contradiction. |||

113 If Problem (10) is modified by eliminating the phase inequality constraints (14), then, in Lemma (110), Conditions (V.3.52 and 54) should be replaced by Condition (V.3.82).

Let us now turn to an investigation of when the maximum principle in integral form, (106), can be written as a "pointwise" maximum principle. In fact, we shall show that a pointwise maximum principle holds whenever the following hypothesis is satisfied:

114 HYPOTHESIS. *The class $\mathscr{U}$ is strongly convex under switching, and there is a subset U of V such that* (i) *$u(t) \in U$ for all $u \in \mathscr{U}$ and every $t \in I$, and* (ii) *every constant function $u: I \to U$ belongs to $\mathscr{U}$.*

Note that if V, f, and $\mathscr{U}$ satisfy the requirements of Problem (10) [in particular that $V \subset R^r$ and that Hypothesis (3) holds] as well as Hypothesis (114), then Hypothesis (V.2.43) holds.

In correspondence with Theorem (V.3.70), we have the following result.

115 THEOREM. *Let (x_0, u_0) be a solution of Problem* (10) *such that Hypotheses* (98) *and* (114) *hold. Then the maximum principle in integral form* (106) *in Theorem* (108) *is equivalent to the following pointwise maximum principle*:

116 $$[\tilde{\boldsymbol{\psi}}(t) - \tilde{\lambda}(t)\mathscr{D}_1\tilde{\chi}(x_0(t),t)]f(x_0(t),u_0(t),t) = \sup_{v \in \omega(x_0(t),t)} [\tilde{\boldsymbol{\psi}}(t) - \tilde{\lambda}(t)\mathscr{D}_1\tilde{\chi}(x_0(t),t)]f(x_0(t),v,t) \quad \textit{for almost all} \quad t \in I,$$

where, for each $(\xi,t) \in \mathbf{G} \times I$,

117 $$\omega(\xi,t) = \{v: v \in U, \tilde{\tilde{\chi}}^i(\xi,v,t) < 0 \textit{ for all } i \textit{ such that } t \in I_i'\},$$

and relation (107) *is equivalent to the following pointwise condition*:

118 $$\{[\tilde{\boldsymbol{\psi}}(t) - \tilde{\lambda}(t)\mathscr{D}_1\tilde{\chi}(x_0(t),t)]\mathscr{D}_2 f(x_0(t),u_0(t),t) + \boldsymbol{\mu}(t)\mathscr{D}_2\tilde{\tilde{\chi}}(x_0(t),u_0(t),t)\}[v - u_0(t)] \le 0 \quad \textit{for all} \quad v \in U, \quad \textit{for almost all} \quad t \in I.$$

Proof. Arguing essentially as in the proof of Theorem (V.2.44), taking into account that almost every point of I_i' (or of the complement of I_i'), for each i, is a *point of density* of this set [$\hat{t}$ is said to be a point of density of a Lebesgue measurable subset $\hat{I}$ of I if

$$\lim_{\text{meas } J \to 0} \frac{\text{meas } (J \cap \hat{I})}{\text{meas } J} = 1,$$

where J is an arbitrary subinterval of I that contains $\hat{t}$]—this is a general property of Lebesgue measurable subsets of R—we can show that (106) implies that relation (116) holds with = replaced by $\geq$. To show that equality holds in (116), we point out that it follows from (15) and Hypothesis (98), since $\mathscr{U}$ is finitely open in itself, that, for almost all $t \in I$, $[u_0(t) + \varepsilon(u^*(t) - u_0(t))] \in \omega(x_0(t),t)$ for all $\varepsilon > 0$ sufficiently small, which, by virtue of Hypothesis (3), implies that equality in (116) holds.

By an analogous argument, we can show that (107) implies that (118) holds.

It is obvious that (116) and (118) imply, respectively, that (108) and (107) hold. |||

119 Corollary. *If* (x_0,u_0) *is a solution of Problem* (10) *such that Hypotheses* (98) *and* (114) *hold, and if*

120 $$\{v\colon v \in U, \tilde{\tilde{\chi}}^i(x_0(t),v,t) \leq 0 \textit{ for all } i = 1, \ldots, \nu' \textit{ such that } t \in I_i'\} \subset \overline{\omega(x_0(t),t)} \quad \textit{for almost all} \quad t \in I,$$

then relation (116) *in Theorem* (115) *may be replaced by the following stronger maximum principle*:

121 $$[\tilde{\psi}(t) - \tilde{\lambda}(t)\mathscr{D}_1\tilde{\chi}(x_0(t),t)]f(x_0(t),u_0(t),t) = \max_{\substack{v \in U \\ \tilde{\tilde{\chi}}^i(x_0(t),v,t) \leq 0 \textit{ for all} \\ i=1,\ldots,\nu' \textit{ such that } t \in I_i'}} [\tilde{\psi}(t) - \tilde{\lambda}(t)\mathscr{D}_1\tilde{\chi}(x_0(t),t)]f(x_0(t),v,t)$$

for almost all $t \in I$.

The corollary follows at once from a standard continuity argument.

122 In case the functions $\tilde{\tilde{\chi}}^i$, $i = 1, \ldots, \nu'$, are independent of ξ, so that Problem (10) becomes a special case of Problem (V.3.1), then the necessary conditions of Theorems (108) and (115) and Corollary (119)—subject to the corresponding additional required hypotheses—essentially coincide with those of Theorems (V.3.44 and 70). [Also see (V.2.48).] However, we point out that the necessary conditions of Chapter V were derived under weaker hypotheses than those that we had to introduce in this section in

order to be able to treat problems with mixed control-phase variable constraints.

123 Note that if Hypothesis (114) is satisfied, then, since $\mathscr{U}$ is finitely open in itself, U also is (as a subset of R^r). Let us consider the maximum condition (121) for some fixed $t \in I$ (for which it is valid). In this case, $u_0(t)$ is a solution of the problem of minimizing the function $\boldsymbol{\phi}^0 : U \to R$ defined by

$$\boldsymbol{\phi}^0(v) = -[\tilde{\boldsymbol{\psi}}(t) - \tilde{\lambda}(t)\mathscr{D}_1\tilde{\chi}(x_0(t),t)]f(x_0(t),v,t), \quad v \in U$$

subject to the constraints $\boldsymbol{\phi}^i(v) \leq 0$ for all $i = 1, \ldots, \nu'$ such that $t \in I_i'$, where the functions $\boldsymbol{\phi}^i : U \to R$ are defined by

$$\boldsymbol{\phi}^i(v) = \tilde{\tilde{\chi}}^i(x_0(t),v,t), \quad v \in U, \quad i = 1, \ldots, \nu'.$$

Our hypotheses on f and on $\tilde{\tilde{\chi}}$ ensure that the functions $\boldsymbol{\phi}^i$, $i = 0, \ldots, \nu'$, are continuously differentiable. Thus, $u_0(t)$ is a solution of a problem which is of the form of Problem (II.1.16)—with $\mathscr{E} = U$, $\mathscr{Y} = R^r$, $\phi^0 = \boldsymbol{\phi}^0$, and $\phi^i = \boldsymbol{\phi}^{j_i}$ for $i = 1, \ldots, \mu$, and some suitable subset $\{j_1, \ldots, j_\mu\}$ of $\{1, \ldots, \nu'\}$—with Hypotheses (II.1.2 and 17) both satisfied. Then, by Theorem (II.1.18), there exist nonpositive numbers $\beta^0, \beta^1, \ldots, \beta^{\nu'}$, not all zero, such that $\beta^i = 0$ (for $i > 0$) if $t \notin I_i'$ or if $\boldsymbol{\phi}^i(u_0(t)) = \tilde{\tilde{\chi}}^i(x_0(t),u_0(t),t) < 0$, and such that

$$\begin{aligned}\sum_{i=0}^{\mu} \beta^i D\boldsymbol{\phi}^i(u_0(t); v - u_0(t)) &= \Big\{-\beta^0[\tilde{\boldsymbol{\psi}}(t) - \tilde{\lambda}(t)\mathscr{D}_1\tilde{\chi}(x_0(t),t)]\mathscr{D}_2 f(x_0(t),u_0(t),t) \\ &\quad + \sum_{i=1}^{\nu'} \beta^i \mathscr{D}_2\tilde{\tilde{\chi}}^i(x_0(t),u_0(t),t)\Big\}(v - u_0(t)) \leq 0 \\ &\qquad \text{for all} \quad v \in U.\end{aligned}$$

But, by relation (118) and conclusion (viii) of Theorem (108), we conclude that, for almost all $t \in I$, the numbers $-1, \boldsymbol{\mu}^1(t), \ldots, \boldsymbol{\mu}^{\nu'}(t)$ satisfy the conditions we have just listed for $\beta^0, \beta^1, \ldots, \beta^{\nu'}$, respectively. Thus [subject to the hypotheses of Corollary (119)], for almost every $t \in I$, the numbers $\boldsymbol{\mu}^i(t)$ may be viewed as "Lagrange multipliers" for the constrained maximization problem in U defined by (121).

The reader should at this point refer back to the remarks in (V.2.48 and 49), which, with some minor changes, carry over here.

124 Arguing as in Section 4 of Chapter V, we can also develop necessary conditions for the minimax problem which differs from Problem (10) in that the minimization, instead of being defined by inequality (16), is defined by inequality (V.4.6), where $\tilde{\chi}^0$ and I_0 have the same properties as $\tilde{\chi}^i$ and I_i for $i > 0$. The resultant necessary conditions differ only slightly from those already obtained [much as those of Theorem (V.4.11) differ from those of Theorem (V.3.44)]. The case where $\tilde{\chi}^0$ depends on u as well as on x will be the subject of the next section.

125 Finally, we point out that most of the results in Sections 5–8 of Chapter V, as well as in Sections 1 and 2 of this chapter, carry over to Problem (10), provided that certain evident modifications are made.

4. Minimax Optimal Control Problems of Mixed Control-Phase Variable Type

In this section, we shall consider a minimax variant of the problem considered in the preceding section. Specifically, we shall find necessary conditions for optimality for the following problem.

1 *Problem.* Let there be given sets $\mathbf{G}$, V, I, $\tilde{I}$, I_a, $I_1, \ldots, I_\nu$, A, and $\mathscr{U}$ as described in Problem (3.10), together with a Lebesgue measurable subset I_0' of I and a nonempty compact set $\mathbf{K}$ in $R^{\nu'}$. Let there also be given functions f, $\chi = (\chi^1, \ldots, \chi^{\mu+m})\colon (R^m)^k \to R^{\mu+m}$, $\tilde{\chi}$, and $\tilde{\tilde{\chi}}$ with the same properties as in Problem (3.10). Then find an absolutely continuous function $x_0 \in A$ and a function $u_0 \in \mathscr{U}$ such that (3.11–14) hold, and such that

2 $$\operatorname*{ess\,sup}_{t\in I_0'} \max_{\xi\in\mathbf{K}} [\xi \cdot \tilde{\tilde{\chi}}(x_0(t),u_0(t),t)] \leq \operatorname*{ess\,sup}_{t\in I_0'} \max_{\xi\in\mathbf{K}} [\xi \cdot \tilde{\tilde{\chi}}(x(t),u(t),t)]$$

for all pairs $(x,u) \in A \times \mathscr{U}$, with x absolutely continuous, that satisfy (3.11–14), with the subzero omitted.

3 If $\nu' = 1$ and $\mathbf{K} = \{1\}$, then Problem (1) reduces to minimizing the functional

$$\operatorname*{ess\,sup}_{t\in I_0'} \tilde{\tilde{\chi}}(x(t),u(t),t)$$

subject to the constraints indicated in the problem statement. If, in addition [but also see (V.4.12)], $\tilde{\tilde{\chi}}$ is independent of $v \in V$ (and if I_0' is closed), then Problem (1) becomes a special case of Problem

(V.4.1). In contrast to Problem (V.4.1), we impose stronger requirements here on V, f, and $\mathscr{U}$ in order to be able to consider the case where $\tilde{\tilde{\tilde{\chi}}}$ also depends on v [see (3.18)].

If $\mathbf{K} = \{\xi : \xi \in R^{\nu'}, |\xi| = 1\}$, then (2) is clearly equivalent to

4 $$\operatorname*{ess\,sup}_{t \in I'_0} |\tilde{\tilde{\tilde{\chi}}}(x_0(t),u_0(t),t)| \leq \operatorname*{ess\,sup}_{t \in I'_0} |\tilde{\tilde{\tilde{\chi}}}(x(t),u(t),t)|.$$

5 If $\tilde{\tilde{\tilde{\chi}}}$ is independent of $\xi \in R^n$, then our problem consists in finding what is in some loose sense a control which is best in terms of "minimum control effort," where effort is defined as $\operatorname*{ess\,sup}_{t \in I'_0} \max_{\xi \in \mathbf{K}} \xi \cdot \tilde{\tilde{\tilde{\chi}}}(u(t),t)$. For example, if $\nu' = r$ and $\tilde{\tilde{\tilde{\chi}}}(\xi,v,t) = v$ for all $(\xi,v,t) \in \mathbf{G} \times V \times I$, and if $\mathbf{K} = \{(1,0,\ldots,0), (0,1,0,\ldots,0), \ldots, (0,0,\ldots,0,1), (-1,0,\ldots,0), \ldots, (0,0,\ldots,0,-1)\}$, then control "effort" means $\operatorname*{ess\,sup}_{t \in I'_0} \max_{1 \leq i \leq r} |u^i(t)|$.

Merely for the sake of simplicity, we have omitted from Problem (1) additional constraints of the form of (3.15).

We shall obtain necessary conditions for Problem (1) by appealing to Theorem (IV.2.80), using arguments very similar to those of the preceding section.

6 We begin by rewriting (2) in terms of functionals which belong to $(L_\infty^{\nu'}(I))^*$. Let $\mathscr{M}$ denote the set of all finitely additive nonnegative real-valued measures $\mathbf{v}$ which are defined on all Lebesgue measurable subsets of I such that (i) $\mathbf{v}(\hat{I}) = 0$ for every Lebesgue measurable subset $\hat{I}$ of I whose intersection with I'_0 has Lesbesgue measure zero, and (ii) $\mathbf{v}(I) = \mathbf{v}(I'_0) = 1$. For every $\xi \in R^{\nu'}$ and each $\mathbf{v} \in \mathscr{M}$, define the linear functional $\ell_{\xi,\mathbf{v}}$ on $L_\infty^{\nu'}(I)$ as follows:

7 $$\ell_{\xi,\mathbf{v}}(z) = \int_I \xi \cdot z(t)\, d\mathbf{v}(t), \quad z \in L_\infty^{\nu'}(I).$$

It is easily seen [see Theorem (I.6.28)] that $\ell_{\xi,\mathbf{v}} \in (L_\infty^{\nu'}(I))^*$ for all $\xi \in R^{\nu'}$ and $\mathbf{v} \in \mathscr{M}$. Now set

8 $$\mathscr{L} = \{\ell_{\xi,\mathbf{v}} : \xi \in \mathbf{K}, \mathbf{v} \in \mathscr{M}\}.$$

Then it is not difficult to see that

9 $$\operatorname*{ess\,sup}_{t \in I'_0} \max_{\xi \in \mathbf{K}} [\xi \cdot z(t)] = \sup_{\ell \in \mathscr{L}} \ell(z) \quad \text{for all} \quad z \in L_\infty^{\nu'}(I),$$

so that (2) can equivalently be written as

10 $$\sup_{\ell \in \mathscr{L}} \ell \circ \hat{\phi}'''(x_0,u_0) \leq \sup_{\ell \in \mathscr{L}} \ell \circ \hat{\phi}'''(x,u),$$

where the function $\hat{\phi}''' : A \times \Pi \to L_\infty^{\nu'}(I)$ is defined by (3.23 and 27).

11 It is a direct consequence of (9) that if $\mathscr{L}$ is defined by (8), then Hypothesis (III.1.22), where $\mathscr{Z}_2 = L_\infty^{\nu'}(I)$, is satisfied. Also, if there is a vector $\xi_0 \in R^{\nu'}$ such that $\xi \cdot \xi_0 = 1$ for all $\xi \in \mathbf{K}$, then Hypothesis (III.3.17) holds, inasmuch as we can choose the function which is identically equal to ξ_0 on I for ζ_0. In this latter case [see (III.3.18)], Hypothesis (III.1.21) also holds.

12 Now suppose that $(x_0,u_0) \in A \times \mathscr{U}$ is a solution of Problem (1). In the notation of Section 2 of Chapter IV, we shall show that, for a suitable choice for $\mathscr{X}$, $\mathscr{P}$, Π, $\tilde{A}'$, $\tilde{\mathscr{W}}'$, $\mathscr{Z}_1$, Z_1, $\hat{\varphi}$, $\hat{\phi}_1$, and $\hat{\phi}_2$, and with $\mathscr{Z}_2 = L_\infty^{\nu'}(I)$ and $\mathscr{L} \subset \mathscr{Z}_2^*$ defined by (7) and (8), we can identify (x_0,u_0) with a solution of Problem (IV.2.79), that the problem has the form described in (IV.2.16), and that the problem data satisfy Hypotheses (IV.2.10–12) and (IV.1.72)—with x_0 replaced by (x_0,u_0) in the latter. As we have already seen, Hypotheses (III.1.22) holds. We shall then obtain necessary conditions that x_0 and u_0 must satisfy from Theorem (IV.2.80).

13 Indeed, let us set $\mathscr{X} = \mathscr{C}^n(I)$, $\mathscr{P} = L_\infty^r(I)$, $\mathscr{Z}' = \mathscr{C}^\nu(I)$,

$$Z' = \{z : z = (z^1, \ldots, z^\nu) \in \mathscr{C}^\nu(I),\ z^i(t) \leq 0 \quad \text{for all} \quad t \in I_i \quad \text{and each} \quad i = 1, \ldots, \nu\},$$

$\mathscr{Z}_1 = R^\mu \times \mathscr{Z}'$, and $Z_1 = \bar{R}_-^\mu \times Z'$. Further, let Π be given by (3.23), let $\tilde{A}' = \tilde{A} = A \times \Pi$, and let us define the functions $\hat{\varphi} : \tilde{A} \to R^m$, $\hat{\phi}^i : \tilde{A} \to R$ (for $i = 1, \ldots, \mu$), $\hat{\phi}'' : \tilde{A} \to \mathscr{C}^\nu(I)$, and $\hat{\phi}''' : \tilde{A} \to L_\infty^{\nu'}(I)$ by (3.24–27). Finally, let

14 $$\hat{\phi}_1 = (\hat{\phi}^1, \ldots, \hat{\phi}^\mu, \hat{\phi}''), \quad \hat{\phi}_2 = \hat{\phi}'''.$$

[see the parenthetical remark following (3.28), as well as the remarks in (3.29)]. Note that the functions $\hat{\varphi}$, $\hat{\phi}^i$, $\hat{\phi}''$, and $\hat{\phi}_1$ are all independent of u, and that Z' is a closed, convex cone in $\mathscr{Z}'$ with a nonempty interior and thus a convex body.

We now define the functions $\hat{F}_{u',\hat{I}} : \mathbf{G} \times I \times \Pi \to R^n$ and the operators $\hat{T}_{\xi,u',\hat{I}} : \tilde{A} \to \mathscr{C}^n(I)$—for each $u' \in \Pi$, $\hat{I} \in \mathbf{I}$, and $\xi \in R^n$—as in Section 3 [see (3.31–38)]. Further, let [compare with (3.39)]

15 $$\mathscr{U}' = \left\{u' : u' \in \mathscr{U}, \operatorname*{ess\,sup}_{t \in I_0'} \max_{\xi \in \mathbf{K}} [\xi \cdot \tilde{\tilde{\chi}}(x_0(t),u'(t),t)] < \kappa_0\right\},$$

16 where [see (9)]

$$\kappa_0 = \operatorname*{ess\,sup}_{t \in I_0'} \max_{\xi \in \mathbf{K}} [\xi \cdot \tilde{\tilde{\chi}}(x_0(t),u_0(t),t)] = \sup_{\ell \in \mathscr{L}} \ell \circ \hat{\phi}_2(x_0,u_0),$$

and let us define $\tilde{\mathscr{W}}$ as follows [using the notation of (IV.1.9) and (IV.2.3)]:

17 $$\tilde{\mathscr{W}} = \Big\{(\hat{T},u) : (\hat{T},u) \in \mathscr{S}_1, u \in \mathscr{U}, \hat{T} = \hat{T}_{\xi,u',\hat{I}} \text{ for some } \xi \in R^n, u' \in \mathscr{U}', \text{ and } \hat{I} \in \mathbf{I} \text{ such that} \Big(\operatorname*{ess\,sup}_{t \in I_0'} \max_{\xi \in \mathbf{K}} [\xi \cdot \tilde{\tilde{\chi}}(\varphi_0(\hat{T}^u)(t),u'(t),t)] \Big) < \kappa_0 \Big\}$$

[compare with (3.40)]. Note that both $\mathscr{U}'$ and $\tilde{\mathscr{W}}$ depend on x_0 and u_0. Also, $\mathscr{U}' \subset \mathscr{U} \subset \Pi$, and $\tilde{\mathscr{W}} \subset (\hat{\mathscr{V}}_2 \times \Pi)$.

18 The data of Problem (IV.2.79) have now been completely specified in terms of the data of Problem (1) and of x_0 and u_0. It is now straightforward to verify that, because (x_0,u_0) is a solution of Problem (1), $(x_0,\hat{T}_{x_0(t_1),I},u_0)$ is a solution of Problem (IV.2.79), which, as is evident, is of the form described in (IV.2.16).

If we denote the operator $\hat{T}_{x_0(t_1),I}$ by $\hat{T}_0$, so that $(\hat{T}_0,u_0) \in \tilde{\mathscr{W}} \subset \mathscr{S}_1$, then (3.42–44) hold.

In order to make use of Theorem (IV.2.80), we must verify that Hypotheses (IV.2.10–12) and (IV.1.72)—with x_0 replaced by $\tilde{x}_0 = (x_0, u_0)$—hold. However, as we saw in Section 3, our problem data satisfy Hypotheses (IV.2.10, 11, and 17), and $\hat{\phi}''' = \hat{\phi}_2$ is continuous and Fréchet differentiable, so that Hypothesis (IV.1.72) holds [see (IV.1.73)]. Further, the Fréchet partial differentials of $\hat{\varphi}, \hat{\phi}^1, \ldots, \hat{\phi}^\mu, \hat{\phi}''$, and $\hat{\phi}'''$ are given by (3.45–49). Arguing almost exactly as in Section 3, we can also show that $\tilde{\mathscr{W}}$, as defined in this section by (17), satisfies Hypothesis (IV.2.12) and that (3.57) holds, where $\hat{\mathscr{W}}$ is given by (3.50). [In this regard, it is important to note that the function $(\xi_1,v,t) \to \max_{\xi \in \mathbf{K}} \xi \cdot \tilde{\tilde{\chi}}(\xi_1,v,t): \mathbf{G} \times V \times I \to R$ is continuous.]

If we now appeal to Theorem (IV.2.80), taking into account (11) and the remarks in (IV.2.84) as well as (3.31, 32, 36, 42–50, and 57), and make use of (I.5.25), (I.6.8, 19, and 31) and Theorems (I.6.14, 22, and 28) in order to obtain representations for $(Z')^*$ and $\mathscr{Z}_2^*$, we arrive at the following conclusion:

19 Let (x_0,u_0) be a solution of Problem (1) and suppose that either $\kappa_0 \neq 0$ [see (16)] or that the following hypothesis holds.

20 HYPOTHESIS. *There is a vector* $\xi_0 \in R^{\nu'}$ *such that* $\max_{\xi \in \mathbf{K}} \xi \cdot \xi_0 < 0$. Then there exist a row vector $\alpha = (\alpha^1, \ldots, \alpha^{\mu+m}) \in \bar{R}_-^{\mu} \times R^m$, a row-vector-valued function $\tilde{\lambda} = (\tilde{\lambda}^1, \ldots, \tilde{\lambda}^{\nu}) \in (NBV(I))^{\nu}$, and finitely additive real-valued measures $\mathbf{v}^1, \ldots, \mathbf{v}^{\nu'}$, defined on all Lebesgue measurable subsets of I, such that (for each $i = 1, \ldots, \nu'$) $\mathbf{v}^i(I_0) = 0$ for any subset I_0 of I of Lebesgue measure zero, with the following properties: (i) if the measures $\mathbf{v}^1, \ldots, \mathbf{v}^{\nu'}$ all vanish identically, then $\|\tilde{\lambda}\| + |\alpha| > 0$, (ii) relation (3.59) holds, where $\mathscr{U}''$ is given by (3.60) and (15), the function $\delta x_{\xi,u''}$ (for each $\xi \in R^n$ and $u'' \in \mathscr{U}''$) is given by (3.61), and $\mathscr{D}_j^0\chi$ (for each j) denotes $\mathscr{D}_j\chi(x_0(\tau^1), \ldots, x_0(\tau^k))$,

21 $$\text{(iii)} \sum_{i=1}^{\mu} \alpha^i \chi^i(x_0(\tau^1), \ldots, x_0(\tau^k)) = 0,$$

$$\text{(iv)} \int_{t_1}^{t_2} d\tilde{\lambda}(t)\tilde{\chi}(x_0(t),t) = 0,$$

22 $$\text{(v)} \sum_{i=1}^{\nu'} \int_{t_1}^{t_2} z^i(t)\, d\mathbf{v}^i(t) \geq \sum_{i=1}^{\nu'} \int_{t_1}^{t_2} \tilde{\tilde{\chi}}^i(x_0(t),u_0(t),t)\, d\mathbf{v}^i(t)$$

for all functions $z = (z^1, \ldots, z^{\nu'}) \in L_{\infty}^{\nu'}(I)$ that satisfy the inequality [see (9) and (III.1.17)]

23 $$\xi \cdot z(t) \leq \kappa_0 \quad \text{for all} \quad \xi \in \mathbf{K}, \quad \text{for almost all} \quad t \in I_0',$$

(vi) relation (3.63) holds, where, for each $\delta u \in L_{\infty}^r(I)$, the function $\delta x_{\delta u}$ is given by (3.64), and

(vii) for each $i = 1, \ldots, \nu$, $\tilde{\lambda}^i$ is nonincreasing on I and is constant on any subinterval of I which does not meet I_i.

Arguing almost exactly as in the derivation of Theorem (3.93), we can now prove the following theorem.

24 THEOREM. *Let* (x_0,u_0) *be a solution of Problem* (1) *such that either* κ_0 [*given by* (16)] *does not vanish, or such that Hypothesis* (20) *holds. Then there exist a row vector* $\alpha = (\alpha^1, \ldots, \alpha^{\mu+m}) \in \bar{R}_-^{\mu} \times R^m$, *finitely additive real-valued measures* $\mathbf{v}^1, \ldots, \mathbf{v}^{\nu'}$ *defined on all Lebesgue measurable subsets of* I, *and row-vector valued functions* $\tilde{\lambda} = (\tilde{\lambda}^1, \ldots, \tilde{\lambda}^{\nu}) \in [NBV(I)]^{\nu}$, $\tilde{\psi} \in L_{\infty}^n(I)$, *and* $\tilde{\tilde{\psi}} \in L_{\infty}^n(I)$ *such that* (i) *if the measures* $\mathbf{v}^i$ $(i = 1, \ldots, \nu')$ *all vanish identically, then* $\|\tilde{\lambda}\| + |\alpha| > 0$, (ii) *conclusions* (ii)–(viii) *of Theorem* (3.93) *hold, except that* (3.39) *is to be replaced by* (15), (iii) *for each* $i = 1, \ldots, \nu'$, $\mathbf{v}^i(I_0) = 0$ *for each subset* I_0 *of* I *of Lebesgue measure zero, and* (iv) (22) *holds for all* $z \in L_{\infty}^{\nu'}(I)$ *that satisfy* (23).

25 Theorem (24) differs from Theorem (3.93) essentially only in that conclusion (ix) of the latter is replaced by conclusions (iii) and (iv) in the former. Depending on the nature of the set **K**, conclusion (iv) of Theorem (24) may be put in a more concrete form. For example, if **K** satisfies the following Hypothesis [which is evidently stronger than Hypothesis (20)]:

26 HYPOTHESIS. *There is a vector* $\xi_0 \in R^{\nu'}$ *such that* $\xi \cdot \xi_0 = 1$ *for all* $\xi \in \mathbf{K}$;

so that [see (11)] Hypothesis (III.3.17) holds, then conclusion (iv) of Theorem (24) takes the following form [see (IV.2.84), Corollary (IV.1.78), (III.1.17), and (9)]:

27 $$\text{(iva)} \ \sum_{i=1}^{\nu'} \int_{t_1}^{t_2} z^i(t)\, d\mathbf{v}^i(t) \geq 0$$

for all $z = (z^1, \ldots, z^{\nu'}) \in L_\infty^{\nu'}(I)$ such that

28 $$\xi \cdot z(t) \leq 0 \quad \text{for all} \quad \xi \in \mathbf{K}, \quad \text{for almost all} \quad t \in I'_0,$$

and

29 $$\text{(ivb)} \ \sum_{i=1}^{\nu'} \int_{t_1}^{t_2} \tilde{\tilde{\chi}}^i(x_0(t),u_0(t),t)\, d\mathbf{v}^i(t)$$
$$= \left(\sum_{i=1}^{\nu'} \xi_0^i \mathbf{v}^i(I) \right) \cdot \left(\operatorname*{ess\,sup}_{t \in I'_0} \max_{\xi \in \mathbf{K}} [\xi \cdot \tilde{\tilde{\chi}}(x_0(t),u_0(t),t)] \right).$$

If, in particular, **K** is the finite set

$$\{(1,0,\ldots,0),(0,1,0,\ldots,0),\ldots,(0,\ldots,0,1)\},$$

so that (2) can be written as

$$\operatorname*{ess\,sup}_{t \in I'_0} \max_{1 \leq i \leq \nu'} \tilde{\tilde{\chi}}^i(x_0(t),u_0(t),t) \leq \operatorname*{ess\,sup}_{t \in I'_0} \max_{1 \leq i \leq \nu'} \tilde{\tilde{\chi}}^i(x(t),u(t),t),$$

then (iva) and (ivb) are easily seen to imply that condition (ix) of Theorem (3.93) holds, provided that I'_i (for each i) is replaced by I'_0, and, in (67) and (69), $\tilde{\tilde{\chi}}^i(x_0(t),u_0(t),t)$ is replaced by

$$[\tilde{\tilde{\chi}}^i(x_0(t),u_0(t),t) - \kappa_0].$$

The necessary conditions of Theorem (24) differ from those of Theorem (V.4.11) much as those of Theorem (3.93) differ from those of Theorem (V.3.44)—see (3.94). The remarks in (3.95 and 97) carry over here, provided that an evident minor modification is made in (3.96)—see (15).

Let us now show, much as was done in Section 3, that if $\mathscr{U}$ and $\tilde{\tilde{\chi}}$ are "well-behaved" at (x_0,u_0), then the additional adjoint variable $\tilde{\tilde{\psi}}$ is absolutely continuous, and the sum $(\tilde{\psi} + \tilde{\tilde{\psi}})$ satisfies a differential equation similar to Eq. (3.88) on each of the intervals $[t_1,\tau^2)$, etc.

We thus make the following hypothesis [compare with Hypothesis (3.98)].

30 HYPOTHESIS. *There is a function* $u^* \in \mathscr{U}$ *such that*

$$\operatorname*{ess\,sup}_{t\in I_0'} \max_{\xi\in\mathbf{K}} \{\xi \cdot [\tilde{\tilde{\chi}}(x_0(t),u_0(t),t) + \mathscr{D}_2\tilde{\tilde{\chi}}(x_0(t),u_0(t),t)(u^*(t) - u_0(t))]\} < \kappa_0$$

[*where* κ_0 *is given by* (16)].

Hypothesis (30) essentially asserts that "to first-order, in u," Problem (1) is well-posed with respect to the functional being minimized. Note that if Hypothesis (30) is satisfied and if $\kappa_0 \le 0$, then evidently Hypothesis (20) holds.

The following lemma is analogous to Lemma (3.100).

31 LEMMA. *If* (x_0,u_0) *is a solution of Problem* (1) *such that Hypothesis* (30) *is satisfied, and if* $\mathscr{U}$ *is strongly convex under switching* [*see* (IV.5.15)], *then the measures* $\mathbf{v}^i$ *in Theorem* (24) *are countably additive and absolutely continuous (with respect to Lebesgue measure).*

Proof. If Hypothesis (30) holds, then, since the set $\mathbf{K}$ is compact, there is a number $\tilde{\varepsilon} > 0$ such that

$$\max_{\xi\in\mathbf{K}} \{\xi \cdot [\tilde{\tilde{\chi}}(x_0(t),u_0(t),t) + \mathscr{D}_2\tilde{\tilde{\chi}}(x_0(t),u_0(t),t)(u^*(t) - u_0(t)) + \xi_1]\} \le \kappa_0$$

for almost all $t \in I_0'$ and all $\xi_1 \in R^{\nu'}$ with $|\xi_1| \le \tilde{\varepsilon}$. Let us arbitrarily fix a Lebesgue measurable subset $\hat{I}$ of I, and let $\mathbf{v}(\hat{I}) = (\mathbf{v}^1(\hat{I}), \ldots, \mathbf{v}^{\nu'}(\hat{I}))$, so that $\mathbf{v}(\hat{I}) \in R^{\nu'}$. If $\mathbf{v}(\hat{I}) \neq 0$, let us set $\hat{\xi}_1 = -\tilde{\varepsilon}\mathbf{v}(\hat{I})/|\mathbf{v}(\hat{I})|$. It now follows from the last conclusion of Theorem (24) that

$$\begin{aligned}\tilde{\varepsilon}|\mathbf{v}(\hat{I})| &= -\sum_{i=1}^{\nu'} \hat{\xi}_1^i \int_{\hat{I}} d\mathbf{v}^i(t) \\ &\le \sum_{i=1}^{\nu'} \int_{\hat{I}} \mathscr{D}_2\tilde{\tilde{\chi}}^i(x_0(t),u_0(t),t)(u^*(t) - u_0(t))\, d\mathbf{v}^i(t).\end{aligned}$$

If we now make use of (3.86) and the strong convexity under switching of the set $\mathscr{U}$, we conclude that

$$\tilde{\varepsilon}|\mathbf{v}(\hat{I})| \leq -\int_{\hat{I}} [\tilde{\psi}(t) - \tilde{\lambda}(t)\mathscr{D}_1\tilde{\chi}(x_0(t),t) + \tilde{\tilde{\psi}}(t)]\mathscr{D}_2 f(x_0(t),u_0(t),t)(u^*(t) - u_0(t))\,dt.$$

The remainder of the proof is as in the proof Lemma (3.100). |||

As was pointed out in (3.101), if the measures $\mathbf{v}^i$ are absolutely continuous w.r. to Lebesgue measure, then there are functions $\boldsymbol{\mu}^i \in L_1(I)$ such that, for each i,

$$\mathbf{v}^i(\hat{I}) = \int_{\hat{I}} \boldsymbol{\mu}^i(t)\,dt \quad \text{and} \quad \int_{\hat{I}} \mathbf{F}(t)\,d\mathbf{v}^i(t) = \int_{\hat{I}} \mathbf{F}(t)\boldsymbol{\mu}^i(t)\,dt$$

for every Lebesgue measurable subset $\hat{I}$ of I and every $\mathbf{F} \in L_\infty(I)$, as a result of which we can conclude that the function $\boldsymbol{\psi} = \tilde{\psi} + \tilde{\tilde{\psi}}$ is absolutely continuous on each of the intervals $[t_1,\tau^2)$, etc., and, on each of these intervals, satisfies Equation (3.102) [where $\boldsymbol{\mu} = (\boldsymbol{\mu}^1, \ldots, \boldsymbol{\mu}^{v'})$], as well as (3.103–107). In addition, (22)—which holds for all $z \in L_\infty^{v'}(I)$ that satisfy (23)—takes the form

32 $$\int_{t_1}^{t_2} \boldsymbol{\mu}(t)z(t)\,dt \geq \int_{t_1}^{t_2} \boldsymbol{\mu}(t)\tilde{\tilde{\chi}}(x_0(t),u_0(t),t)\,dt.$$

33 Let us show that the assertion that (32) holds for all $z \in L_\infty^{v'}(I)$ which satisfy (23) implies that $\boldsymbol{\mu}$ has some special properties. Let us set

34 $$\mathbf{Z}_0 = \{\xi_1 : \xi_1 \in R^{v'}, \xi \cdot \xi_1 \leq \kappa_0 \text{ for all } \xi \in \mathbf{K}\}.$$

It is evident that $\mathbf{Z}_0$ is a convex set in $R^{v'}$ and that $\tilde{\tilde{\chi}}(x_0(t),u_0(t),t) \in \mathbf{Z}_0$ for almost all $t \in I_0'$. Further, let

35 $$\tilde{I}_0' = \left\{t : t \in I_0', \max_{\xi \in \mathbf{K}} [\xi \cdot \tilde{\tilde{\chi}}(x_0(t),u_0(t),t)] = \kappa_0\right\},$$

and let J_0 denote the complement of $\tilde{I}_0'$ with respect to I. Clearly, $\tilde{\tilde{\chi}}(x_0(t),u_0(t),t)$ is a boundary point [see (I.3.6)] of $\mathbf{Z}_0$ for every $t \in \tilde{I}_0'$, and $\tilde{\tilde{\chi}}(x_0(t),u_0(t),t) \in \operatorname{int} \mathbf{Z}_0$ for almost every $t \in J_0 \cap I_0'$. We shall show that $\boldsymbol{\mu}(t) = 0$ for almost all $t \in J_0$, and that

36 $$\boldsymbol{\mu}(t)\xi_1 \geq \boldsymbol{\mu}(t)\tilde{\tilde{\chi}}(x_0(t),u_0(t),t) \quad \text{for all} \quad \xi_1 \in \mathbf{Z}_0, \qquad \text{for almost all} \quad t \in \tilde{I}_0'.$$

We shall argue by contradiction. Let $J_0' = J_0 \cap I_0'$, and let $J_0'' = \{t : t \in J_0, t \notin I_0'\}$. If the set $\tilde{J}_0'' = \{t : t \in J_0'', \boldsymbol{\mu}(t) \neq 0\}$ has positive Lebesgue measure, we define the function $z \in L_\infty^{\nu'}(I)$ as follows:

$$z(t) = \begin{cases} \tilde{\tilde{\chi}}(x_0(t),u_0(t),t) - \boldsymbol{\mu}(t)/|\boldsymbol{\mu}(t)| & \text{if } t \in \tilde{J}_0'', \\ \tilde{\tilde{\chi}}(x_0(t),u_0(t),t) & \text{if } t \in I, \quad t \notin \tilde{J}_0'', \end{cases}$$

in which case z satisfies (23) but not (32), which is impossible. Thus $\boldsymbol{\mu}(t) = 0$ for almost all $t \in J_0''$. On the other hand, if the set $\tilde{J}_0' = \{t: t \in J_0', \boldsymbol{\mu}(t) \neq 0\}$ has positive Lebesgue measure, then, for some positive integer n_0, the set

$$\hat{J}_0' = \left\{t: t \in J_0', \boldsymbol{\mu}(t) \neq 0, \max_{\xi \in \mathbf{K}} [\xi \cdot \tilde{\tilde{\chi}}(x_0(t),u_0(t),t)] < (\kappa_0 - 1/n_0)\right\}$$

has positive measure. Let $\varepsilon_0 > 0$ be such that $\max_{\xi \in \mathbf{K}} \xi \cdot \xi_1 \leq 1/n_0$ for all $\xi_1 \in R^{\nu'}$ with $|\xi_1| \leq \varepsilon_0$. Then the function $z \in L_\infty^{\nu'}(I)$ defined by

$$z(t) = \begin{cases} \tilde{\tilde{\chi}}(x_0(t),u_0(t),t) - \varepsilon_0 \boldsymbol{\mu}(t)/|\boldsymbol{\mu}(t)| & \text{if } t \in \hat{J}_0', \\ \tilde{\tilde{\chi}}(x_0(t),u_0(t),t) & \text{if } t \notin \hat{J}_0', \quad t \in I, \end{cases}$$

satisfies (23) but not (32), which is again a contradiction. Consequently, $\boldsymbol{\mu}(t) = 0$ for almost all $t \in (J_0' \cup J_0'') = J_0$.

It remains for us to show that (36) holds. Again, suppose the contrary, so that the set

$$\hat{I}_0 = \{t: t \in \tilde{I}_0, \boldsymbol{\mu}(t)\tilde{\tilde{\chi}}(x_0(t),u_0(t),t) > \boldsymbol{\mu}(t)\xi_1 \text{ for some } \xi_1 \in \mathbf{Z}_0\}$$

has positive Lebesgue measure. Let $\{\bar{\xi}_1,\bar{\xi}_2,\ldots\}$ be a countable dense subset of $\mathbf{Z}_0$. Then, since

$$\hat{I}_0 = \bigcup_{j=1}^{\infty} \{t: t \in \tilde{I}_0, \boldsymbol{\mu}(t)\tilde{\tilde{\chi}}(x_0(t),u_0(t),t) > \boldsymbol{\mu}(t)\bar{\xi}_j\},$$

the set $\{t: t \in \tilde{I}_0, \boldsymbol{\mu}(t)\tilde{\tilde{\chi}}(x_0(t),u_0(t),t) > \boldsymbol{\mu}(t)\bar{\xi}_{j_0}\}$ has positive Lebesque measure for some integer $j_0 = 1, 2, \ldots$. Let us denote this latter set by $\hat{I}_{j_0}$. Therefore, the function $z \in L_\infty^{\nu'}(I)$ defined by

$$z(t) = \begin{cases} \bar{\xi}_{j_0} & \text{if } t \in \hat{I}_{j_0}, \\ \tilde{\tilde{\chi}}(x_0(t),u_0(t),t) & \text{if } t \in I, \quad t \notin \hat{I}_{j_0}, \end{cases}$$

satisfies (23) but not (32), which is a contradiction. Thus, (36) holds.

We may now summarize our results as follows:

37 THEOREM. *Let (x_0,u_0) be a solution of Problem* (1) *such that Hypothesis* (30) *holds, and suppose that $\mathscr{U}$ is convex under switching. Then there exist a row vector $\alpha = (\alpha^1, \ldots, \alpha^{\mu+m}) \in \bar{R}_-^{\mu} \times R^m$ and*

row-vector-valued functions $\tilde{\lambda} = (\tilde{\lambda}^1, \ldots, \tilde{\lambda}^\nu) \in [NBV(I)]^\nu$, $\boldsymbol{\mu} = (\boldsymbol{\mu}^1, \ldots, \boldsymbol{\mu}^{\nu'}) \in L_1^{\nu'}(I)$, *and* $\tilde{\boldsymbol{\psi}} \in L_\infty^n(I)$ *such that conclusions* (i)–(vii) *of Theorem* (3.108) *hold—except that* (3.39) *is to be replaced by* (15)*—and such that* $\boldsymbol{\mu}(t) = 0$ *for almost all* $t \in I$ *which do not belong to the set* $\tilde{I}_0'$ *defined by* (35) *and* (16), *and* (36) *holds.*

If $u_0 \in \operatorname{int} \mathscr{U}$, as must be the case if $\mathscr{U}$ is open, then (3.107) takes the form of (3.109).

It is easily seen that Lemma (3.110) concerning the nontriviality of the maximum principle [also see (3.113)], as well as Theorem (3.115) and its Corollary (3.119) on the pointwise maximum principle, carry over here with evident minor modifications. For example, relation (3.117) in Theorem (3.115) should here be replaced by

$$\omega(\xi,t) = \begin{cases} \left\{ v \colon v \in U, \max\limits_{\xi_1 \in \mathbf{K}} [\xi_1 \cdot \tilde{\tilde{\chi}}(\xi,v,t)] < \kappa_0 \right\} & \text{if} \quad t \in I_0', \\ U \quad \text{if} \quad t \notin I_0'. \end{cases}$$

Finally, the comments in (3.123) regarding the interpretation of $\boldsymbol{\mu}^1(t), \ldots, \boldsymbol{\mu}^{\nu'}(t)$ as Lagrange multipliers in the pointwise maximum principle also carry over with minor changes. In this regard, we point out that here $u_0(t)$ is a solution of a problem which is of the form of Problem (III.1.2)—with the equality constraint omitted—rather than of Problem (II.1.16), as was the case in (3.123), and that we must make use of Theorem (III.3.34), rather than Theorem (II.1.18).

CHAPTER VII

Miscellaneous Optimal Control Problems

IN THE PRECEDING two chapters, we have been concerned exclusively with optimal control problems with restrictions in the form of ordinary differential equations. However, the machinery which we developed in Chapter IV allows us to consider much more general classes of restrictions, such as operator equations in terms of Volterra-type operators [see (IV.3.3)].

We shall illustrate the way in which we can generalize the ordinary differential equation restrictions by considering two generalizations of such equations: functional differential equations (in Section 1)—including ordinary differential equations with retardations as a special case (in Section 2), and Volterra integral equations (in Section 3). In Section 4, we shall study optimal control problems with restrictions in the form of difference equations.

Although it is possible to investigate essentially all of the classes of optimal control problems which we considered in Chapters V and VI, with the ordinary differential equations replaced by the just-described types of equations, we shall confine ourselves to a relatively simple problem which is analogous to Problem (V.3.10). We do this because our aim in this chapter is merely to *illustrate* the way in which the results of Chapters V and VI can be extended. The reader who wishes to find necessary conditions for problems other than the simple ones which we shall consider, should be able to do so by bringing to bear the machinery and techniques which we have provided for him.

Finally, we shall not touch on sufficient conditions at all in this chapter. But again, the reader should be able to extend the sufficiency conditions obtained in Chapters V and VI to the problems discussed in the sequel.

1. Optimal Control with Functional Differential Equations

In this section, we shall consider a variant of Problem (V.3.10) in which the ordinary differential equation (V.3.7) is replaced by a functional differential equation. We shall suppose that this functional differential equation is defined in terms of some given n-vector-valued function g whose domain is $A \times V \times I$, where I is a given compact interval $[t_1,t_2]$, V is a given arbitrary set, and A is a given open set in $\mathscr{C}^n(I)$. As in Chapters V and VI, we shall also suppose that we are given a class $\mathscr{U}$ of functions $u:I \to V$. We shall suppose that $\mathscr{U}$ is convex under switching.

In order to obtain necessary conditions for our optimal control problem, we shall have to make certain assumptions on $\mathscr{U}$ and on the character of the function g [which are analogous to Hypotheses (V.1.16–18)]. Indeed, we shall suppose that g and $\mathscr{U}$ satisfy the following hypotheses.

1 HYPOTHESIS. *For each $(v,t) \in V \times I$, the function $x \to g(x,v,t)$: $A \to R^n$ is continuously differentiable.*

2 HYPOTHESIS. *For every $x \in A$ and each $u \in \mathscr{U}$, the function $t \to g(x,u(t),t)$: $I \to R^n$ is measurable.*

3 HYPOTHESIS. *For each $u \in \mathscr{U}$ and each $x_1 \in A$, there exist an integrable function $\tilde{m}:I \to R_+$ and a number $\zeta > 0$ (both possibly depending on x_1 and on u) such that*

4 $$|g(x,u(t),t)| + \|D_1 g(x,u(t),t)\| \le \tilde{m}(t) \quad \textit{for all} \quad t \in I$$
$$\textit{and all} \quad x \in A \quad \textit{such that} \quad \|x - x_1\| \le \zeta.$$

In addition, we shall make the following "causality" assumption concerning g.

5 HYPOTHESIS. *If $x_1 \in A$, $x_2 \in A$, and $t \in I$ satisfy the relations $x_1(s) = x_2(s)$ for all $s \in [t_1,t]$, then $g(x_1,v,t) = g(x_2,v,t)$ for all $v \in V$.*

If t represents time, Hypothesis 5 asserts that $g(x,v,t)$—for each $v \in V$—depends only on the "past" and "present" values of x, i.e., only on the values of the function x at times "no later" than t.

6 Note that if V is a topological space and g is continuous, then Hypothesis (2) holds so long as each $u \in \mathscr{U}$ is measurable. If also g is continuously differentiable with respect to its first argument [see (I.7.51)]—so that Hypothesis (1) certainly holds—and if, for each $u \in \mathscr{U}$, there is a compact subset V_u of V such that $u(t) \in V_u$ for all $t \in I$, then it easily follows from (I.3.46) and Corollary (I.5.32) that Hypothesis (3) holds, and we may even choose $\tilde{m}$ to be constant.

7 It is clear that if g and $\mathscr{U}$ satisfy Hypotheses (1)–(3) and (5), then, for each $u \in \mathscr{U}$, the function $G_u: A \times I \to R^n$ defined by the formula

8 $$G_u(x,t) = g(x,u(t),t), \quad (x,t) \in A \times I,$$

belongs to the set $\mathscr{G}$ defined in (IV.3.9).

9 Further, if $\mathscr{U}$ is convex under switching [see (IV.5.14)], then the set $\{G_u: u \in \mathscr{U}\}$—where G_u is defined by (8) for each $u \in \mathscr{U}$—is evidently also convex under switching.

Let us consider the following problem:

10 *Problem.* Let there be given a compact interval $I = [t_1, t_2]$, an arbitrary set V, an open set $A \subset \mathscr{C}^n(I)$, a function $g: A \times V \times I \to R^n$ which satisfies Hypotheses (1) and (5), a class $\mathscr{U}$ of functions $u: I \to V$ which is convex under switching and satisfies Hypotheses (2) and (3), a subset A' of $\mathscr{C}^n(I)$ which is finitely open in itself, a number $t_3 \in (t_1,t_2)$, and a continuously differentiable function $\chi = (\chi^0, \chi^1, \ldots, \chi^{\mu+m}): (R^n)^3 \to R^{\mu+m+1}$. Then find functions $x_0 \in A$, $x_0^* \in A'$, and $u_0 \in \mathscr{U}$ such that $(x_0 - x_0^*)$ is absolutely continuous, such that

11 $$\frac{d}{dt}(x_0 - x_0^*)(t) = g(x_0,u_0(t),t) \quad \text{for almost all} \quad t \in I,$$

12 $$x_0(t_1) = x_0^*(t_1),$$

13 $$\chi^i(x_0(t_1),x_0(t_2),x_0(t_3)) \leq 0 \quad \text{for} \quad i = 1, \ldots, \mu,$$

14 $$\chi^i(x_0(t_1),x_0(t_2),x_0(t_3)) = 0 \quad \text{for} \quad i = \mu + 1, \ldots, \mu + m,$$

and such that

15 $$\chi^0(x_0(t_1),x_0(t_2),x_0(t_3)) \leq \chi^0(x(t_1),x(t_2),x(t_3))$$

for all functions $x \in A$ with the properties that—for some $x^* \in A'$ and some $u \in \mathscr{U}$—$(x - x^*)$ is absolutely continuous and satisfies the differential equation

16 $$\frac{d}{dt}(x - x^*)(t) = g(x,u(t),t) \quad \text{for almost all} \quad t \in I,$$

with initial condition

17 $$x(t_1) = x^*(t_1),$$

and x satisfies the constraints

18 $$\chi^i(x(t_1),x(t_2),x(t_3)) \leq 0 \quad \text{for} \quad i = 1, \ldots, \mu,$$

19 $$\chi^i(x(t_1),x(t_2),x(t_3)) = 0 \quad \text{for} \quad i = \mu + 1, \ldots, \mu + m.$$

Problem (10) differs from Problem (V.3.10) in that we have (for the sake of simplicity) set $k = 3$, and have replaced the ordinary differential equations (V.3.2 and 7) by the functional differential equations (11) and (16) together with (12) and (17). If A' consists of all *constant* functions in $\mathscr{C}^n(I)$, and if g has the form

20 $$g(x,v,t) = f(x(t),v,t) \quad \text{for all} \quad (x,v,t) \in A \times V \times I,$$

then the functional differential equation restriction of this section reduces to the ordinary differential equation restriction of Problem (V.3.10).

The reader should here review (V.1.12) in order to see what kinds of problems may be reformulated so as to take on the form of Problem (10).

21 If the functions $x^* \in A'$ all have the property that $x^*(t) = x^*(t_3)$ for every $t \in [t_3,t_2]$, and if g is such that $g(x,v,t) = 0$ for all $(x,v,t) \in A \times V \times [t_1,t_3)$, then Problem (10) consists in finding functions $x_0 \in A$, $x_0^* \in A'$, and $u_0 \in \mathscr{U}$ such that (i) x_0 is absolutely continuous on $[t_3,t_2]$, (ii) x_0 satisfies the initial condition

22 $$x_0(t) = x_0^*(t) \quad \text{for all} \quad t \in [t_1,t_3],$$

(iii) x_0 satisfies the functional differential equation

23 $$\dot{x}_0(t) = g(x_0,u_0(t),t) \quad \text{a.e. on} \quad [t_3,t_2],$$

(iv) x_0 satisfies the boundary value constraints (13) and (14), and (v) relation (15) holds for all x which (for some $u \in \mathscr{U}$ and $x^* \in A'$) satisfy the conditions (i)–(iv), with the subzeroes omitted.

Equation (23) differs from an ordinary differential equation in that $\dot{x}_0(t)$ depends on the entire "past history" of x_0 rather than only on the "present value" of x_0. Since $\dot{x}_0(t)$ does not depend on "future" values of x_0, Eq. (23) is often referred to as a functional differential equation of retarded type. Note that, in contrast to ordinary differential equations, the initial condition (22) on x_0 is on an "initial interval" $[t_1,t_3]$ rather than at a single initial value of t.

In this and the next section, we shall use the notation of Chapter IV, with $\mathscr{X} = \mathscr{C}^n(I)$.

Let us set

$$\mathscr{W} = \{T_{x^*,G_u}: u \in \mathscr{U}, x^* \in A'\},$$

where, for each $u \in \mathscr{U}$, G_u is the function in $\mathscr{G}$ defined by (8) [see (7) and (IV.3.9)], and the operator $T_{x^*,G}: A \to \mathscr{C}^n(I)$ [for each $x^* \in \mathscr{C}^n(I)$ and $G \in \mathscr{G}$] is defined by

24 $$(T_{x^*,G}x)(t) = x^*(t) + \int_{t_1}^{t} G(x,s)\, ds, \quad t \in I, \quad x \in A.$$

Then, inasmuch as Eqs. (16) and (17) are clearly equivalent to the equation

$$x(t) = (T_{x^*,G_u}x)(t) \quad \text{for all} \quad t \in I,$$

Problem (10) has been recast in the form of Problem (IV.1.15), with $k = 3$ and $\tau^1 = t_1$, $\tau^2 = t_3$, $\tau^3 = t_2$, but with the constraints (IV.1.19) omitted. [Recall that, from the results of the Appendix, $\mathscr{W} \subset \mathscr{T}^u \subset \mathscr{T}_0$—see (IV.1.2) and (IV.3.7 and 15).]

In order to obtain necessary conditions for Problem (10), we shall appeal to Theorems (IV.5.34) and (IV.1.53).

25 Thus, let $(x_0, x_0^*, u_0) \in A \times A' \times \mathscr{U}$ be a solution of Problem (10), so that (11)–(14) hold. Let us denote $T_{x_0^*,G_{u_0}}$ simply by T_0, so that $T_0 \in \mathscr{W}$ and

26 $$(T_0x)(t) = x_0^*(t) + \int_{t_1}^{t} g(x,u_0(s),s)\, ds, \quad t \in I, \quad x \in A,$$

and Eqs. (11) and (12) are clearly equivalent to the operator equation $x_0 = T_0x_0$, which means that $T_0 \in \mathscr{T}_1$. Then, by Theorem (IV.5.34) and (9), Hypotheses (IV.1.21–23) hold. Appealing to Theorem (IV.1.53) and to (IV.1.58), we conclude that there exists a nonzero row vector $\alpha = (\alpha^0, \alpha^1, \ldots, \alpha^{\mu+m}) \in R^{\mu+m+1}$ such that

27 (i) $\alpha^i \le 0$ for $i = 0, \ldots, \mu$,

28 (ii) $$\sum_{j=1}^{3} \alpha \mathscr{D}_j^0 \chi\, \delta x_{x^*,u}(t_j) \le \sum_{j=1}^{3} \alpha \mathscr{D}_j^0 \chi\, \delta x_{x_0^*,u_0}(t_j)$$

for all $x^* \in A'$ and $u \in \mathscr{U}$,

where, for each $x^* \in A'$ and $u \in \mathscr{U}$, $\delta x_{x^*,u}$ denotes the column-vector-valued function in $\mathscr{C}^n(I)$ defined by [see (IV.1.55) and

(IV.3.17)]

29 $$\delta x_{x^*,u}(t) = \int_{t_1}^{t} D_1 g(x_0,u_0(s),s;\, \delta x_{x^*,u})\, ds + x^*(t) + \int_{t_1}^{t} g(x_0,u(s),s)\, ds, \quad t \in I,$$

and $\mathscr{D}_j^0\chi$ $(j = 1, 2, 3)$ denotes $\mathscr{D}_j\chi(x_0(t_1),x_0(t_2),x_0(t_3))$, and

30 (iii) $\sum_{i=1}^{\mu} \alpha^i \chi^i(x_0(t_1),x_0(t_2),x_0(t_3)) = 0.$

It follows from (30), (27), and (13) that

31 $$\alpha^i \chi^i(x_0(t_1),x_0(t_2),x_0(t_3)) = 0 \quad \text{for} \quad i = 1, \ldots, \mu.$$

Now Eq. (29) is clearly equivalent to the linear inhomogeneous functional differential equation

32 $$\frac{d}{dt}(\delta x_{x^*,u} - x^*)(t) = D_1 g(x_0,u_0(t),t;\, \delta x_{x^*,u} - x^*) + g(x_0,u(t),t) + D_1 g(x_0,u_0(t),t;\, x^*) \quad \text{a.e. in} \quad I,$$

with initial condition

33 $$\delta x_{x^*,u}(t_1) = x^*(t_1).$$

34 For each $t \in I$, the function $\delta x \to D_1 g(x_0,u_0(t),t;\, \delta x)$ belongs to $\mathbf{B}(\mathscr{C}^n(I),R^n)$ [see (I.5.1)]. Hence, by Corollary (I.6.16) [also see (I.6.21)], there is an $(n \times n)$-matrix-valued function Λ defined on $I \times I$ such that

35 (i) $D_1 g(x_0,u_0(t),t;\, \delta x) = \int_{t_1}^{t_2} d_1\Lambda(s,t)\delta x(s)$

for all $\delta x \in \mathscr{C}^n(I)$ and $t \in I$,

where $d_1\Lambda$ denotes that the first argument of Λ is to be taken as the independent variable for the Stieltjes integration, and (ii) the functions $s \to \lambda^{ij}(s,t)$, for every $t \in I$—where λ^{ij} denotes the i, j-th element of Λ—belong to $NBV(I)$. Further, it easily follows from Hypothesis (5) that $\Lambda(s,t) = 0$ for all $s \in [t,t_2]$ and each $t \in I$, so that (35) can be rewritten in the form

36 $$D_1 g(x_0,u_0(t),t;\, \delta x) = \int_{t_1}^{t} d_1\Lambda(s,t)\delta x(s), \quad \delta x \in \mathscr{C}^n(I), \quad t \in I.$$

Further, it follows from Hypothesis (3) and the last conclusion in Corollary (I.6.16) that there is an integrable function $\tilde{m}: I \to R_+$ such that

37 $$TV\lambda^{ij}(\cdot,t) \leq \tilde{m}(t) \quad \text{for all} \quad t \in I \quad \text{and each} \quad i, j = 1, \ldots, n,$$

which at once implies [since $\lambda^{ij}(t_2) = 0$ for all i, j] that

38 $$|\lambda^{ij}(s,t)| \leq \tilde{m}(t) \quad \text{for all} \quad (s,t) \in I \times I \quad \text{and each} \quad i, j = 1, \ldots, n.$$

Finally, it follows from Hypotheses (1) and (2) that, for each $\delta x \in \mathscr{C}^n(I)$, the function $t \to D_1 g(x_0,u_0(t),t; \delta x): I \to R^n$ is measurable, by virtue of which we can conclude [using (36) with suitable functions δx] that the functions $t \to \lambda^{ij}(s,t): I \to R$ are measurable (for every $s \in I$ and each i, j). Inasmuch as $\lambda^{ij}(\cdot,t)$ is in $NBV(I)$ for every $t \in I$ (and each i, j), this implies, as is easily seen, that the functions $(s,t) \to \lambda^{ij}(s,t): I \times I \to R$ are measurable (for each i and j).

Setting $\delta x_{x^*,u} - x^* = \delta\tilde{x}_{x^*,u}$, we can now rewrite (32) and (33) in the form

39 $$\frac{d}{dt}(\delta\tilde{x}_{x^*,u}(t)) = \int_{t_1}^{t} d_1\Lambda(s,t)\delta\tilde{x}_{x^*,u}(s) + \int_{t_1}^{t} d_1\Lambda(s,t)x^*(s) + g(x_0,u(t),t) \quad \text{a.e. in} \quad I,$$

40 $$\delta\tilde{x}_{x^*,u}(t_1) = 0.$$

Appealing to Theorem (A.70) in the Appendix, we conclude that the (unique) solution of (39) and (40) is given by

$$\begin{aligned}\delta\tilde{x}_{x^*,u}(t) &= \delta x_{x^*,u}(t) - x^*(t) \\ &= \int_{t_1}^{t} \Gamma(t; s)\Big[g(x_0,u(s),s) + \int_{t_1}^{s} d_1\Lambda(\tau,s)x^*(\tau)\Big]\, ds,\end{aligned}$$

or

41 $$\delta x_{x^*,u}(t) = x^*(t) + \int_{t_1}^{t} \Gamma(t; s)g(x_0,u(s),s)\, ds + \int_{t_1}^{t} \Gamma(t; s)\int_{t_1}^{s} d_1\Lambda(\tau,s)x^*(\tau)\, ds, \quad t \in I,$$

where, for each $t \in I$, $\Gamma(t;\cdot)$ is the unique $(n \times n)$-matrix-valued function defined on I all of whose elements are functions of bounded variation and which satisfies the equations

42 $$\Gamma(t; s) + \int_{s}^{t} \Gamma(t; \zeta)\Lambda(s,\zeta)\, d\zeta = \text{the identity matrix for all } s \in [t_1,t],$$

43 $$\Gamma(t; s) = 0 \quad \text{for all} \quad s \in (t,t_2].$$

Note that (42) implies that

44 $$\Gamma(t;\, t) = \text{the identity matrix for all } t \in I.$$

Substituting (41) into (28), and taking (43) into account, we obtain that

45 $$\sum_{j=1}^{3} \alpha\mathscr{D}_j^0\chi\delta x(t_j) + \sum_{j=2}^{3} \alpha\mathscr{D}_j^0\chi \left\{\int_{t_1}^{t_2} \Gamma(t_j;\, t)[g(x_0,u(t),t) - g(x_0,u_0(t),t)]\, dt + \int_{t_1}^{t_2} \Gamma(t_j;\, t) \int_{t_1}^{t} d_1\Lambda(s,t)\delta x(s)\, dt\right\} \leq 0$$

$$\text{for all} \quad \delta x \in A' - x_0^* \quad \text{and} \quad u \in \mathscr{U}.$$

Since $x_0^* \in A'$, (45) implies that

46 $$\int_{t_1}^{t_2} \psi(t)g(x_0,u_0(t),t)\, dt \geq \int_{t_1}^{t_2} \psi(t)g(x_0,u(t),t)\, dt \quad \text{for all} \quad u \in \mathscr{U},$$

where ψ is the n (row)-vector-valued function defined on I which is given by the formulas

47 $$\begin{cases} \psi(t) = \displaystyle\sum_{j=2}^{3} \alpha\mathscr{D}_j^0\chi\Gamma(t_j;\, t) & \text{for} \quad t \in I \quad \text{and} \quad t \neq t_3, \\ \psi(t_3) = \alpha\mathscr{D}_2^0\chi\Gamma(t_2;\, t_3). \end{cases}$$

Note that, inasmuch as the elements of the matrix-valued functions $\Gamma(t_2;\cdot)$ and $\Gamma(t_3;\cdot)$ are of bounded variation, ψ is also of bounded variation, and hence bounded on I.

It follows from (42)–(44) and (47) that

48 $$\psi(t) + \int_{t}^{t_2} \psi(s)\Lambda(t,s)\, ds = \begin{cases} \alpha\mathscr{D}_2^0\chi + \alpha\mathscr{D}_3^0\chi & \text{if} \quad t_1 \leq t < t_3, \\ \alpha\mathscr{D}_2^0\chi & \text{if} \quad t_3 \leq t \leq t_2. \end{cases}$$

Since $u_0 \in \mathscr{U}$, (45) implies that

49 $$\sum_{j=1}^{3} \alpha\mathscr{D}_j^0\chi\delta x(t_j) + \int_{t_1}^{t_2} \psi(t) \int_{t_1}^{t} d_1\Lambda(s,t)\delta x(s)\, dt \leq 0$$

$$\text{for all} \quad \delta x \in A' - x_0^*.$$

Applying the unsymmetric Fubini theorem† to interchange the

† See R. H. Cameron and W. T. Martin, An unsymmetric Fubini theorem, *Bull. Amer. Math. Soc.* **47** (1941), 121–125.

order of integration in the double integral in (49), we obtain that

$$\sum_{j=1}^{3} \alpha \mathscr{D}_j^0 \chi \delta x(t_j) + \int_{t_1}^{t_2} d\left(\int_s^{t_2} \psi(t)\Lambda(s,t)\,dt\right)\delta x(s) \leq 0$$

$$\text{for all} \quad \delta x \in A' - x_0^*,$$

which, by virtue of (48), may be rewritten as follows:

50 $$\int_{t_1}^{t_2} d\hat{\psi}(t)[x^*(t) - x_0^*(t)] \geq 0 \quad \text{for all} \quad x^* \in A',$$

where the function $\hat{\psi}: I \to R^n$ is given by the formula

51 $$\hat{\psi}(t) = \begin{cases} \psi(t_1) + \alpha \mathscr{D}_1^0 \chi & \text{for} \quad t = t_1, \\ \psi(t) & \text{for} \quad t_1 < t < t_2, \\ 0 & \text{for} \quad t = t_2. \end{cases}$$

We may now summarize our results as follows:

52 THEOREM. *Let* (x_0, x_0^*, u_0) *be a solution of Problem* (10), *and let* Λ *be the* $(n \times n)$*-matrix-valued function defined on* $I \times I$ *which satisfies* (35) *and which has the property that the functions* $s \to \lambda^{ij}(s,t)$ *are in* $NBV(I)$ *for every* $t \in I$ *and each* $i, j = 1, \ldots, n$ (λ^{ij} *denotes the* i, j*-th element of* Λ). *Then there exist a nonzero row vector* $\alpha = (\alpha^0, \alpha^1, \ldots, \alpha^{\mu+m}) \in \bar{R}_-^{\mu-1} \times R^m$ *and* n *(row)-vector-valued functions* ψ *and* $\hat{\psi}$ *defined and of bounded variation on* I *such that*

(i) ψ *satisfies the "adjoint" equation* (48), *and* $\hat{\psi}$ *is given by* (51),

(ii) u_0 *satisfies the maximum principle* (46),

(iii) *relation* (50) *holds*, and

(iv) *Eqs.* (31) *hold*,

where, in (48) *and* (51), $\mathscr{D}_j^0 \chi$ $(j = 1, 2, 3)$ *denotes*

$$\mathscr{D}_j \chi(x_0(t_1), x_0(t_2), x_0(t_3)).$$

53 Note that (48) implies that $\psi(t_2) = \alpha \mathscr{D}_2^0 \chi$. As we pointed out earlier, if A' consists of all constant functions in $\mathscr{C}^n(I)$ and if g is of the form of (20)—where A and f are as in Problem (V.3.10)—then our problem becomes a particular case of Problem (V.3.10). In this case also

$$\Lambda(s,t) = \begin{cases} -\mathscr{D}_1 f(x_0(t), u_0(t), t) & \text{if} \quad s < t, \\ 0 & \text{if} \quad s \geq t, \end{cases}$$

and it is not hard to see that the necessary conditions of Theorem (52) are the same as those obtained from Theorem (V.3.75).

54 It also follows easily from (48), by virtue of (38) and the boundedness of ψ, that the "adjoint variable" ψ is continuous from the left at t_2. Further, since $\Lambda(\cdot,s)$ is, for each $s \in I$, continuous from the right at every point of (t_1,t_2), we can conclude, on the basis of (48) and the Lebesgue dominated convergence theorem, that ψ is continuous from the right at each point of (t_1,t_2). However, we cannot conclude here, as was true in the case of ordinary differential equation restrictions, that ψ is necessarily continuous in $[t_1,t_3) \cup (t_3,t_2]$. Consequently [see (51)], $\hat{\psi} \in [NBV(I)]^n$.

55 Let us more closely examine the special case of Problem (10) which was described in (21). In this case, $D_1 g(x_0,u_0(t),t) = 0$ for all $t \in [t_1,t_3)$; therefore [see (35)], $\Lambda(t,s) = 0$ for all $s \in [t_1,t_3)$, for every $t \in I$. Hence, (48) can here be rewritten as

56 $$\psi(t) = -\int_{t_3}^{t_2} \psi(s)\Lambda(t,s)\,ds + \alpha\mathscr{D}_2^0\chi + \alpha\mathscr{D}_3^0\chi \quad \text{if} \quad t_1 \leq t < t_3,$$

57 $$\psi(t) = -\int_{t}^{t_2} \psi(s)\Lambda(t,s)\,ds + \alpha\mathscr{D}_2^0\chi \quad \text{if} \quad t_3 \leq t \leq t_2.$$

Also, since $\delta x(t) = \delta x(t_3)$ for all $t \in [t_3,t_2]$ and every $\delta x \in A' - x_0^*$, relation (50), by virtue of (51), can here be written in the form

58 $$\int_{t_1}^{t_3} d\tilde{\psi}(t)[x^*(t) - x_0^*(t)] \geq 0 \quad \text{for all} \quad x^* \in A',$$

where the function $\tilde{\psi}:[t_1,t_3] \to R^n$ is defined as follows:

59 $$\tilde{\psi}(t) = \begin{cases} \psi(t_1) + \alpha\mathscr{D}_1^0\chi & \text{for} \quad t = t_1, \\ \psi(t) & \text{for} \quad t_1 < t < t_3, \\ 0 & \text{for} \quad t = t_3. \end{cases}$$

Note that $\tilde{\psi} \in [NBV(I_1)]^n$, where $I_1 = [t_1, t_3]$. Let us denote by A_1' the set of all functions $x_1: I_1 \to R^n$ which are the restriction to I_1 of a function $x \in A'$. Evidently, $A_1' \subset \mathscr{C}^n(I_1)$. If A_1' is open in $\mathscr{C}^n(I_1)$, then, by Lemma (I.5.14) and Theorem (I.6.14), (58) implies that $\tilde{\psi} \equiv 0$ on I_1, i.e., that $\psi(t) = 0$ for $t_1 < t < t_3$ and that $\psi(t_1) = -\alpha\mathscr{D}_1^0\chi$, which, by (56), is equivalent to the equations

60 $$\int_{t_3}^{t_2} \psi(s)\Lambda(t,s)\,ds = \alpha\mathscr{D}_2^0\chi + \alpha\mathscr{D}_3^0\chi \quad \text{for all} \quad t \in (t_1,t_3),$$

61 $$\int_{t_3}^{t_2} \psi(s)\Lambda(t_1,s)\,ds = \sum_{i=1}^{3} \alpha\mathscr{D}_i^0\chi.$$

Finally, if g is as indicated in (21), then the lower limit in the integrals in the maximum principle (46) can clearly be changed to t_3.

Thus, we have shown the following.

62 THEOREM. *Let* (x_0, x_0^*, u_0) *be a solution of Problem* (10), *where* A' *and* g *are as indicated in* (21). *Then there exist a nonzero row vector* $\alpha = (\alpha^0, \alpha^1, \ldots, \alpha^{\mu+m}) \in \bar{R}_-^{\mu+1} \times R^m$ *and an* n *(row)-vector-valued function* ψ *defined and of bounded variation on* I *such that*

(i) ψ *satisfies Eqs.* (56) *and* (57), *where* Λ *is as described in the statement of Theorem* (52),

63 (ii) $\int_{t_3}^{t_2} \psi(t) g(x_0, u_0(t), t)\, dt \geq \int_{t_3}^{t_2} \psi(t) g(x_0, u(t), t)\, dt$ *for all* $u \in \mathscr{U}$,

(iii) *relation* (58) *holds, where* $\tilde{\psi}$ *is given by* (59), *and*

(iv) *Eqs.* (31) *hold,*

where in (56), (57), *and* (59), $\mathscr{D}_j^0 \chi$ $(j = 1, 2, 3)$ *denotes*

$$\mathscr{D}_j \chi(x_0(t_1), x_0(t_2), x_0(t_3)).$$

If also A' *is such that the set* $\{\tilde{x}: \tilde{x} \in \mathscr{C}^n(I_1),\ \tilde{x}(t) = x^*(t)$ *for all* $t \in I_1$ *and for some* $x^* \in A'\}$ *is open in* $\mathscr{C}^n(I_1)$, *where* $I_1 = [t_1, t_3]$, *then Eqs.* (56) *and* (58) *may be replaced by Eqs.* (60) *and* (61).

64 Returning to the general case of Problem (10), we point out that relation (50) may take on various concrete forms, depending on the nature of A'. We just saw one example of this, when the functions in A' are all constant for $t \geq t_3$. Alternately, if A' consists of all constant functions in $\mathscr{C}^n(I)$ which take on their values in some given set $\mathbf{G}_0$ in R^n ($\mathbf{G}_0$ must be finitely open in itself, since A' is), then (50) takes the form [by virtue of (51)]

65 $$(\psi(t_1) + \alpha \mathscr{D}_1^0 \chi)(\xi - \xi_0) \leq 0 \quad \text{for all} \quad \xi \in \mathbf{G}_0,$$

where $x_0^*(t) \equiv \xi_0$. If $\xi_0 \in \operatorname{int} \mathbf{G}_0$ (as must be the case if $\mathbf{G}_0$ is open), then (65) is obviously equivalent to the equation

66 $$\psi(t_1) = -\alpha \mathscr{D}_1^0 \chi.$$

If A' consists of a single function x_0^*, then (50) conveys no useful information.

Let us now turn to the question of sufficient conditions in order that the set $\{t: t \in I,\ \psi(t) \neq 0\}$ have positive (Lebesgue) measure, i.e., in order that the maximum condition (46) be nontrivial. Let us consider the following conditions.

67 CONDITION. *There is a vector* $(\xi_1, \xi_2, \xi_3) \in (R^n)^3$ *such that*

68 $$\sum_{j=1}^{3} (\mathscr{D}_j^0 \chi^i) \cdot \xi_j \begin{cases} < 0 & \textit{for} \quad i = 0 \quad \textit{and each} \quad i = 1, \ldots, \mu \\ & \quad \textit{such that } \chi^i(x_0(t_1), x_0(t_2), x_0(t_3)) = 0, \\ = 0 & \textit{for} \quad i = \mu + 1, \ldots, \mu + m, \end{cases}$$

and the m vectors $(\mathscr{D}_1^0\chi^i,\mathscr{D}_2^0\chi^i,\mathscr{D}_3^0\chi^i)$, $i = \mu + 1, \ldots, \mu + m$, *are linearly independent in* $(R^n)^3$.

69 CONDITION. $x_0^*(t_1) \in \operatorname{int}\{x^*(t_1): x^* \in A'\}$.

70 Condition (67) is a compatibility condition which is analogous to Condition (V.3.82). Note that Condition (67) is satisfied whenever the vectors $(\mathscr{D}_1^0\chi^i,\mathscr{D}_2^0\chi^i,\mathscr{D}_3^0\chi^i)$, as i ranges over 0 and over those integers from 1 to $\mu + m$ for which $\chi^i(x_0(t_1),x_0(t_2),x_0(t_3)) = 0$, are linearly independent in $(R^n)^3$.

71 LEMMA. *Let* (x_0,x_0^*,u_0) *be a solution of Problem* (10) *such that Conditions* (67) *and* (69) *are satisfied. Then the set* $\{t: t \in I, \psi(t) \neq 0\}$, *where* ψ *is the function whose existence is asserted in Theorem* (52), *has positive Lebesgue measure.*

Proof. We shall argue by contradiction. Thus suppose that $\psi = 0$ a.e. on I. Since [see (54)] ψ is continuous from the left at t_2 and continuous from the right at every point of (t_1,t_2), $\psi(t) = 0$ for all $t \in (t_1,t_2]$, so that, by (48), $\alpha\mathscr{D}_2^0\chi = \alpha\mathscr{D}_3^0\chi = 0$, as a consequence of which, again by (48), $\psi(t_1) = 0$, i.e., $\psi \equiv 0$. Then, (50) and (51) imply that

72 $$\alpha\mathscr{D}_1^0\chi[x^*(t_1) - x_0^*(t_1)] \leq 0 \quad \text{for all} \quad x^* \in A'.$$

Let $(\xi_1,\xi_2,\xi_3) \in (R^n)^3$ be as indicated in Condition (67). Replacing (ξ_1,ξ_2,ξ_3) by a sufficiently small positive multiple thereof, if necessary, we may assume, by Condition (69), that $\xi_1 = x_1^*(t_1) - x_0^*(t_1)$ for some $x_1^* \in A'$. We then have

$$\sum_{j=1}^{3}\sum_{i=0}^{\mu+m} \alpha^i(\mathscr{D}_j^0\chi^i)\cdot\xi_j = (\alpha\mathscr{D}_1^0\chi,\alpha\mathscr{D}_2^0\chi,\alpha\mathscr{D}_3^0\chi)\cdot(\xi_1,\xi_2,\xi_3)$$
$$= \alpha\mathscr{D}_1^0\chi\cdot\xi_1 \leq 0,$$

which, by (68), the nonpositivity of $\alpha^0, \alpha^1, \ldots, \alpha^\mu$, and (31), implies that $\alpha^i = 0$ for each $i = 0, \ldots, \mu$. But Condition (69) and (72) imply that $\alpha\mathscr{D}_1^0\chi = 0$. Thus, we have that $\alpha\mathscr{D}_j^0\chi = 0$ for $j = 1$, 2, 3, $\alpha \neq 0$, and $\alpha^i = 0$ for $i = 0, \ldots, \mu$; i.e.,

$$\sum_{i=\mu+1}^{\mu+m} \alpha^i(\mathscr{D}_1^0\chi^i,\mathscr{D}_2^0\chi^i,\mathscr{D}_3^0\chi^i) = 0,$$

$$\alpha^i \neq 0 \quad \text{for some} \quad i = \mu + 1, \ldots, \mu + m,$$

which contradicts the last assertion in Condition (67). |||

We point out that, as is easily seen, Lemma (71) remains in force if we replace the hypothesis that Conditions (67) and (69) hold by the hypothesis that the following condition holds.

73 CONDITION. *There is a vector* $(\xi_2,\xi_3) \in (R^n)^2$ *such that*

$$\sum_{j=2}^{3} (\mathscr{D}_j^0\chi^i) \cdot \xi_j \begin{cases} < 0 \quad \text{for} \quad i = 0 \quad \text{and each} \quad i = 1, \ldots, \mu \\ \quad \text{such that} \quad \chi^i(x_0(t_1),x_0(t_2),x_0(t_3)) = 0, \\ = 0 \quad \text{for} \quad i = \mu + 1, \ldots, \mu + m, \end{cases}$$

and the m vectors $(\mathscr{D}_2^0\chi^i,\mathscr{D}_3^0\chi^i)$, $i = \mu + 1, \ldots, \mu + m$, *are linearly independent in* $(R^n)^2$.

In case the data of Problem (10) are as described in (21), it is of interest to find sufficient conditions in order that the set $\{t: t_3 \leq t \leq t_2, \psi(t) \neq 0\}$ have positive Lebesgue measure [see (63)].

74 CONDITION. $(x_0^*(t_1),x_0^*(t_3)) \in \text{int}\,\{(x^*(t_1),x^*(t_3)): x^* \in A'\}$.

75 LEMMA. *Let* (x_0,x_0^*,u_0) *be a solution of Problem* (10), *where* A' *and* g *are as indicated in* (21), *and suppose that Conditions* (67) *and* (74) *are satisfied. Then the set* $\{t:\ t_3 \leq t \leq t_2,\ \psi(t) \neq 0\}$, *where* ψ *is the function whose existence is asserted in Theorem* (62), *has positive Lebesgue measure.*

Proof. We again argue by contradiction and suppose that $\psi(t) = 0$ for almost all $t \in [t_3,t_2]$. As we saw in the proof of Lemma (71), this means that $\psi \equiv 0$ on $[t_3,t_2]$. By (57), we then conclude that $\alpha\mathscr{D}_2^0\chi = 0$, and, by (56), we obtain that $\psi(t) \equiv \alpha\mathscr{D}_3^0\chi$ on $[t_1,t_3)$. Then (58) and (59) imply that

$$\alpha\mathscr{D}_1^0\chi\delta x(t_1) + \alpha\mathscr{D}_3^0\chi\delta x(t_3) \leq 0 \quad \text{for all} \quad \delta x \in A' - x_0^*,$$

which, by Condition (74), implies that $\alpha\mathscr{D}_1^0\chi = \alpha\mathscr{D}_3^0\chi = 0$. The remainder of the proof parallels that of Lemma (71). |||

In order to obtain a "pointwise" maximum principle in place of (46), we shall make the following hypothesis, which is analogous to Hypothesis (V.2.43).

76 HYPOTHESIS. *There are a subset* U *of* V *and a subset* I^* *of* I *whose complement with respect to* I *has Lebesgue measure zero such that* (i) $u(t) \in U$ *for all* $u \in \mathscr{U}$ *and every* $t \in I$, (ii) *every constant function* $u: I \to U$ *belongs to* $\mathscr{U}$, *and either* (iiia) *for every* $x \in A$ *and* $v \in U$, *the function* $t \to g(x,v,t)\colon I^* \to R^n$ *is continuous, or* (iiib) *for every* $x \in A$, *there is a finite or countable subset* $\tilde{U}$ *of* U *such that* $g(x,\tilde{U},t)$ *is dense in* $g(x,U,t)$ *for all* $t \in I^*$.

In essentially the same way that we proved Theorem (V.2.44), we can now prove the following.

77 THEOREM. *Let* (x_0,x_0^*,u_0) *be a solution of Problem* (10) *such that Hypothesis* (76) *holds. Then the maximum condition* (46) *of Theorem* (52) *is equivalent to the "pointwise" maximum principle*

78 $$\psi(t)g(x_0,u_0(t),t) = \max_{v\in U} \psi(t)g(x_0,v,t) \quad \textit{for almost all} \quad t \in I.$$

In this regard, the remarks in (V.2.48–49) carry over here with certain obvious minor changes.

79 If Problem (10) is modified by eliminating either the inequality constraints (13), (18) or the equality constraints (14), (19) (or both), then the necessary conditions of Theorems (52) and (62) remain in force, but with μ or m (or both) equal to zero, so that (31) is to be deleted from the necessary conditions if there are no inequality constraints [see (IV.1.58 and 92)].

2. Optimal Control with Differential Equations with Retardations

In this section, we shall particularize the results of the preceding section to the case where the functional differential equation is an ordinary differential equation with retardations. Here, g and A' will be as described in (1.21).

More precisely, we shall suppose that the set $A \subset \mathscr{C}^n(I)$ and the function g are as follows:

1 $$A = \{x: x \in \mathscr{C}^n(I),\, x(t) \in \mathbf{G} \text{ for all } t \in I\},$$

2 $$g(x,v,t) = \begin{cases} 0 \quad \text{if} \quad t \in [t_1,t_3), \quad \text{for all} \quad (x,v) \in A \times V, \\ f(x(t-\omega_1),\ldots,x(t-\omega_k),v,t) \quad \text{if} \quad t \in [t_3,t_2], \\ \qquad\qquad \text{for all} \quad (x,v) \in A \times V, \end{cases}$$

where $\mathbf{G}$ is some given open set in R^n, f is a given function from $(\mathbf{G})^k \times V \times [t_3,t_2]$ into R^n, and the ω_i are real numbers such that $0 = \omega_1 < \omega_2 < \cdots < \omega_k = t_3 - t_1$.

As usual, we shall have to make certain hypotheses concerning f and $\mathscr{U}$ which are similar to Hypotheses (V.1.16–18).

3 HYPOTHESIS. *For each* $(v,t) \in V \times [t_3,t_2]$, *the function* $\xi \to f(\xi,v,t)\colon (\mathbf{G})^k \to R^n$ *is continuously differentiable.*

4 HYPOTHESIS. *For every* $\xi \in (\mathbf{G})^k$ *and each* $u \in \mathscr{U}$, *the function* $t \to f(\xi,u(t),t)$: $[t_3,t_2] \to R^n$ *is measurable.*

5 HYPOTHESIS. *For each* $u \in \mathscr{U}$ *and each compact subset* $\mathbf{G}_c$ *of* $\mathbf{G}$, *there exists an integrable function* $\tilde{m}: I \to R_+$ *(possibly depending on* u *and on* $\mathbf{G}_c$*) such that*

6 $$|f(\xi,u(t),t)| + \sum_{j=1}^{k} |\mathscr{D}_j f(\xi,u(t),t)| \leq \tilde{m}(t)$$

$$\textit{for every} \quad \xi \in (\mathbf{G}_c)^k \quad \textit{and every} \quad t \in [t_3,t_2].$$

7 The remarks in (V.1.20), which pertain to Hypotheses (V.1.16–18), have obvious analogs here with regard to Hypotheses (3)–(5). It is also easy to see that the set A defined by (1) is open in $\mathscr{C}^n(I)$—since $\mathbf{G}$ is open in R^n—and that $\mathscr{U}$ and the function $g: A \times V \times I \to R^n$ defined by (2) satisfy Hypotheses (1.1–3 and 5) whenever $\mathscr{U}$ and f satisfy Hypotheses (3)–(5). In connection with this,

8 $$D_1 g(x,v,t;\delta x) = \sum_{j=1}^{k} \mathscr{D}_j f(x(t-\omega_1),\ldots,x(t-\omega_k),v,t)\delta x(t-\omega_j)$$

$$\text{for all} \quad \delta x \in \mathscr{C}^n(I), \quad \text{and each} \quad (x,v,t) \in A \times V \times [t_3,t_2],$$

9 $$D_1 g(x,v,t;\delta x) = 0 \quad \text{for all} \quad \delta x \in \mathscr{C}^n(I)$$

$$\text{and each} \quad (x,v,t) \in A \times V \times [t_1,t_3).$$

Let us now state our problem precisely.

10 *Problem.* Let there be given a compact interval $I = [t_1, t_2]$ and a number $t_3 \in (t_1,t_2)$, an arbitrary set V, an open set $\mathbf{G} \subset R^n$, a function $f: (\mathbf{G})^k \times V \times [t_3,t_2] \to R^n$ which satisfies Hypothesis (3), a class $\mathscr{U}$ of functions $u: I \to V$ which is convex under switching and satisfies Hypotheses (4) and (5), real numbers $\omega_1, \ldots, \omega_k$ such that $0 = \omega_1 < \omega_2 < \cdots < \omega_k = t_3 - t_1$, a subset A_1' of $\mathscr{C}^n([t_1,t_3])$ which is finitely open in itself, and a continuously differentiable function $\chi = (\chi^0, \chi^1, \ldots, \chi^{\mu+m}): (R^n)^3 \to R^{\mu+m+1}$. Then find functions $x_0 \in A$ [where A is given by (1)], $\tilde{x}_0 \in A_1'$, and $u_0 \in \mathscr{U}$ such that x_0 is absolutely continuous on $[t_3,t_2]$ and satisfies the equations

11 $$\dot{x}_0(t) = f(x_0(t-\omega_1),\ldots,x_0(t-\omega_k),u_0(t),t)$$

$$\text{for almost all} \quad t \in [t_3,t_2],$$

12 $$x_0(t) = \tilde{x}_0(t) \quad \text{for all} \quad t \in [t_1,t_3],$$

together with relations (1.13 and 14), and such that (1.15) holds for every function $x \in A$ which (i) is absolutely continuous on $[t_3,t_2]$, (ii) satisfies—for some $\tilde{x} \in A'_1$ and some $u \in \mathscr{U}$—the equations

13 $$\dot{x}(t) = f(x(t - \omega_1), \ldots, x(t - \omega_k), u(t), t) \quad \text{for almost all} \quad t \in [t_3,t_2],$$

14 $$x(t) = \tilde{x}(t) \quad \text{for all} \quad t \in [t_1,t_3],$$

and (iii) satisfies (1.18 and 19).

It is now easily seen that, if we define the function g by (2) and the set $A' \subset \mathscr{C}^n(I)$ by

15 $$A' = \{x^*: x^* \in \mathscr{C}^n(I),\ x^*(t) = \tilde{x}(t) \text{ for some } \tilde{x} \in A'_1 \text{ and all } t \in [t_1,t_3],\ x^*(t) = x^*(t_3) \text{ for all } t \in [t_3,t_2]\},$$

then Problem (10) has been reduced to the form of Problem (1.10), with g and A' as in (1.21). Note that if $k = 1$ and χ is independent of its first argument, then Problem (10) reduces to a special case of Problem (V.3.10).

Let us now appeal to Theorem (1.62) in order to obtain necessary conditions that a solution $(x_0,\tilde{x}_0,u_0)$ of Problem (10) must satisfy. By (8), (9), and (1.34), it follows directly that, for every $t \in [t_1,t_2]$,

16 $$\Lambda(t,s) = \begin{cases} 0 & \text{for} \quad t_1 \le s \le t, \\ -\sum_{j=1}^{\ell} \mathscr{D}_j^0 f(s) & \text{for} \quad t + \omega_\ell < s \le t + \omega_{\ell+1} \\ & \quad \text{and each} \quad \ell = 1, \ldots, k-1, \\ & \quad \text{except when} \quad (s,t) = (t_3,t_1), \\ -\sum_{j=1}^{k} \mathscr{D}_j^0 f(s) & \text{for} \quad t + \omega_k < s \le t_2 \\ & \quad \text{and for} \quad (s,t) = (t_3,t_1), \end{cases}$$

where, for each $j = 1, \ldots, k$,

17 $$\mathscr{D}_j^0 f(s) = \begin{cases} 0 & \text{for} \quad t_1 \le s < t_3, \\ \mathscr{D}_j f(x_0(s - \omega_1), \ldots, x_0(s - \omega_k), u_0(s), s) & \\ & \quad \text{for} \quad t_3 \le s \le t_2, \\ 0 & \text{for} \quad s > t_2. \end{cases}$$

Consequently, (1.56 and 57) here take the form

18 $$\psi(t) = \sum_{j=1}^{k} \int_{t+\omega_j}^{t_2} \psi(s)\mathcal{D}_j^0 f(s)\, ds + \alpha\mathcal{D}_2^0\chi + \begin{cases} 0 & \text{for } \; t_3 \leq t \leq t_2, \\ \alpha\mathcal{D}_3^0\chi & \text{for } \; t_1 \leq t < t_3, \end{cases}$$

where we shall agree that

19 $$\psi(t) = 0 \quad \text{for} \quad t > t_2.$$

Hence, ψ is absolutely continuous both on the interval $[t_1,t_3)$ and on the interval $[t_3,t_2]$, and, on each of these intervals, satisfies the linear "advanced" differential equation

20 $$\dot{\psi}(t) = -\sum_{j=1}^{k} \psi(t + \omega_j)\mathcal{D}_j^0 f(t + \omega_j) \quad \text{for almost all} \quad t.$$

In addition,

21 $$\psi(t_3) - \psi(t_3^-) = -\alpha\mathcal{D}_3^0\chi,$$

22 $$\psi(t_2) = \alpha\mathcal{D}_2^0\chi.$$

Therefore, (1.58) takes the following form [see (1.59) and (15)]:

23 $$\alpha\mathcal{D}_1^0\chi[\tilde{x}(t_1) - \tilde{x}_0(t_1)] + [\psi(t_3) + \alpha\mathcal{D}_3^0\chi][\tilde{x}(t_3) - \tilde{x}_0(t_3)] + \sum_{j=1}^{k} \int_{t_1}^{t_3} \psi(t + \omega_j)\mathcal{D}_j^0 f(t + \omega_j)[\tilde{x}(t) - \tilde{x}_0(t)]\, dt \leq 0 \quad \text{for all} \quad \tilde{x} \in A_1'.$$

If A_1' is open, then relations (1.60 and 61), which here take the form

24 $$0 = \sum_{j=1}^{k} \int_{t+\omega_j}^{t_2} \psi(s)\mathcal{D}_j^0 f(s)\, ds + \alpha\mathcal{D}_2^0\chi + \alpha\mathcal{D}_3^0\chi \quad \text{for all} \quad t \in (t_1,t_3),$$

25 $$0 = \sum_{j=1}^{k} \int_{t_3}^{t_2} \psi(s)\mathcal{D}_j^0 f(s)\, ds + \sum_{i=1}^{3} \alpha\mathcal{D}_i^0\chi,$$

hold, which implies, by virtue of (18) and (21), that $\psi(t) = 0$ for all $t \in (t_1,t_3)$ and that

26 $$\psi(t_3) = -\alpha\mathcal{D}_3^0\chi.$$

Thus, $\dot{\psi}(t) = 0$ for all $t \in (t_1,t_3)$, which, by (20), means that

27 $$\sum_{j=1}^{k} \psi(t + \omega_j)\mathcal{D}_j^0 f(t + \omega_j) = 0 \quad \text{for almost all} \quad t \in [t_1,t_3].$$

Passing to the limit in (24) as $t \to t_1^+$, and comparing with (25), we obtain

28 $$\alpha \mathscr{D}_1^0 \chi = 0.$$

The maximum principles (1.63) and (1.78) take the forms, respectively,

29 $$\int_{t_3}^{t_2} \psi(t) f(x_0(t - \omega_1), \ldots, x_0(t - \omega_k), u_0(t), t)\, dt$$
$$\geq \int_{t_3}^{t_2} \psi(t) f(x_0(t - \omega_1), \ldots, x_0(t - \omega_k), u(t), t)\, dt \quad \text{for all} \quad u \in \mathscr{U},$$

30 $$\psi(t) f(x_0(t - \omega_1), \ldots, x_0(t - \omega_k), u_0(t), t)$$
$$= \max_{v \in U} \psi(t) f(x_0(t - \omega_1), \ldots, x_0(t - \omega_k), v, t)$$
$$\text{for almost all} \quad t \in [t_3, t_2].$$

Thus we have proved the following.

31 THEOREM. *Let* $(x_0, \tilde{x}_0, u_0)$ *be a solution of Problem* (10). *Then there exist a nonzero vector* $\alpha = (\alpha^0, \alpha^1, \ldots, \alpha^{\mu+m}) \in \bar{R}_-^{\mu+1} \times R^m$ *and an n (row)-vector-valued function* ψ *defined on* $[t_1, t_2 + \omega_k]$ *satisfying* (19) *such that*

(i) ψ *is absolutely continuous on* $[t_3, t_2]$, *and, on this interval, satisfies the linear advanced differential equation* (20), *where* $\mathscr{D}_j^0 f$ *(for each j) is given by* (17),

(ii) ψ *satisfies the transversality condition* (22),

(iii) u_0 *satisfies the maximum principle in integral form* (29),

(iv) *relation* (23) *holds, and*

(v) *Eqs.* (1.31) *are satisfied*,

where, in (22) *and* (23), $\mathscr{D}_j^0 \chi$ *(for* $j = 1, 2, 3$) *denotes* $\mathscr{D}_j \chi(x_0(t_1), x_0(t_2), x_0(t_3))$.

If A_1' *is open, then* (23) *may be replaced by* (26)–(28).

Note that, inasmuch as $\mathscr{D}_j^0 f(t) = 0$ for $t \notin [t_3, t_2]$, the values of $\psi(t)$ for $t \notin [t_3, t_2]$ are of no importance in the necessary conditions of Theorem (31).

32 By Lemma (1.75), if Condition (1.67) is satisfied, and if

$$(\tilde{x}_0(t_1), \tilde{x}_0(t_3)) \in \text{int}\,\{(\tilde{x}(t_1), \tilde{x}(t_3)) : \tilde{x} \in A_1'\},$$

then we can assert that the "adjoint variable" ψ in Theorem (31) does not vanish on some subset of $[t_3, t_2]$ of positive Lebesgue measure. Further, by Theorem (1.77), if f satisfies a hypothesis which differs in an obvious minor manner from Hypothesis

(V.2.43), then the maximum principle in integral form (29) is equivalent to the "pointwise" maximum principle (30). [Also see (V.2.48–49).]

33 If A'_1 is open, then the function χ is ordinarily independent of its first argument, i.e., the equality and inequality constraints in the problem statement, as well as the cost functional, then do not ordinarily depend on $x(t_1)$. Indeed, if A'_1 is open, then the problem is not really well-posed if χ depends on its first argument. Of course, if χ is independent of ξ_1, then (28) must hold (for every α).

If $k = 1$, then the necessary conditions of Theorem (31) coincide with those obtained from Theorem (V.3.75).

34 The results of this section may easily be generalized to the case where the "retardations" $\omega_1, \ldots, \omega_k$ are functions of t, so long as these functions are continuous and take their values in $[0, t_3 - t_1]$.

35 Also of interest is the case where Eq. (11) in Problem (10) is replaced by an equation of the form

36 $$\dot{x}_0(t) = f(x_0(t-\omega_1), \ldots, x_0(t-\omega_k), u_0(t-\omega_1), \ldots, u_0(t-\omega_k), t) \quad \text{a.e. in} \quad [t_3, t_2]$$

[and similarly for Eq. (13)], where f is a given function from $(\mathbf{G})^k \times (V)^k \times [t_3, t_2] \to R^n$. Thus, we here permit retardations (in the differential equation) not only for the state variable x, but also for the control variable u. This problem cannot be investigated in the same manner as the problems which we have considered up until now in this chapter and in the preceding two chapters. The reason for this is the following. To each $u \in \mathcal{U}$, let us correspond the function $\mathbf{u} : I \to V^k$ defined by

$$\mathbf{u}(t) = (u(t-\omega_1), u(t-\omega_2), \ldots, u(t-\omega_k)).$$

Then it is not generally true that the set of all such functions $\mathbf{u}$ (as u ranges over $\mathcal{U}$) is convex under switching (even if $\mathcal{U}$ is). Consequently, in order to obtain necessary conditions for this problem, it is necessary to sharpen the results on convexity under switching of Section 5 of Chapter IV. The easiest case is the one in which there is a number $\omega_0 > 0$ such that $t_2 - t_3 = \ell_0\omega_0$ and $\omega_i = \ell_i\omega_0$ for each $i = 2, \ldots, k$, where $\ell_0, \ell_2, \ldots, \ell_k$ are all positive integers (recall that $\omega_1 = 0$). In this situation, it is only necessary to observe that the "chattering lemma" (IV.5.3) can be strengthened by adding the conclusion that [for any $\omega_0 > 0$ such that $(t_2 - t_3)/\omega_0$ is a positive integer] the sets J_j (or equivalently the points s_j) in the proof of the lemma can be chosen in such a way that meas $J_j = \omega_0/\tilde{k}$ for every j, for some suitably large positive integer $\tilde{k}$ (which generally depends on ε_0 as well as on the H_i). If we correspondingly strengthen the remaining results in Section 5 of Chapter IV, we can obtain necessary conditions for the problem with retardations in the controls. For the more general case, where a number ω_0 with the previously indicated properties does not exist, we can also obtain necessary conditions by using an additional limiting argument.

The remarks in (1.79) carry over here essentially without change to Theorem (31).

3. Optimal Control with Volterra Integral Equations

In the first two sections of this chapter, we considered generalizations of the ordinary differential equation constraints to more general differential equations. In this section we shall consider restrictions which are not differential equations, but rather are Volterra integral equations.

Indeed, we shall consider a variant of Problem (V.3.10) in which the ordinary differential equation (V.3.7) is replaced by a Volterra integral equation which is defined in terms of some given n-vector-valued function f whose domain is $\mathbf{G} \times V \times I \times I$, where, as usual, $\mathbf{G}$ is a given open set in R^n, V is a given arbitrary set, and I is a given compact interval $[t_1,t_2]$. As always, we shall also suppose that we are given a class $\mathscr{U}$ of functions $u: I \to V$, where $\mathscr{U}$ is convex under switching.

The assumptions which we shall make on f and on $\mathscr{U}$ in order to obtain necessary conditions are contained in the following hypotheses. [Compare with Hypotheses (V.1.16–18).]

1 HYPOTHESIS. *For each $(v,s) \in V \times I$, the function $(\xi,t) \to f(\xi,v,s,t)$: $\mathbf{G} \times I \to R^n$ is continuous and continuously differentiable with respect to its first argument.*

2 HYPOTHESIS. *For every $(\xi,t) \in \mathbf{G} \times I$ and each $u \in \mathscr{U}$, the function $s \to f(\xi,u(s),s,t)$: $I \to R^n$ is measurable.*

3 HYPOTHESIS. *For each $u \in \mathscr{U}$ and each compact subset $\mathbf{G}_c$ of $\mathbf{G}$, there exists an integrable function $\tilde{m}: I \to R_+$ (possibly depending*

on u and on $\mathbf{G}_c$) such that

4 $$|f(\xi,u(s),s,t)| + |\mathscr{D}_1 f(\xi,u(s),s,t)| \leq \tilde{m}(s) \quad \textit{for all} \quad (\xi,s,t) \in \mathbf{G}_c \times I \times I.$$

5 If V is a topological space and f is continuous, then Hypothesis (2) holds so long as each $u \in \mathscr{U}$ is measurable. If also f is continuously differentiable with respect to its first argument—so that Hypothesis (1) certainly holds—and if, for each $u \in \mathscr{U}$, there is a compact subset V_u of V such that $u(t) \in V_u$ for all $t \in I$, then, by (I.3.46), Hypothesis (3) holds and we may even choose $\tilde{m}$ to be constant.

6 It is easily seen that if f and $\mathscr{U}$ satisfy Hypotheses (1)–(3), then, for each $u \in \mathscr{U}$, the function $F_u{:}\mathbf{G} \times I \times I \to R^n$ defined by the formula

7 $$F_u(\xi,s,t) = f(\xi,u(s),s,t), \quad (\xi,s,t) \in \mathbf{G} \times I \times I,$$

belongs to the set $\mathscr{F}$ defined in (IV.3.22).

8 Further, if $\mathscr{U}$ is convex under switching, then evidently the set $\{F_u{:}\ u \in \mathscr{U}\}$—where F_u is defined by (7) for each $u \in \mathscr{U}$—is also convex under switching [see the parenthetical remark preceding Lemma (IV.5.27)].

We shall consider the following problem.

9 *Problem.* Let there be given a compact interval $I = [t_1, t_2]$, an arbitrary set V, an open set $\mathbf{G} \subset R^n$, a function $f{:}\mathbf{G} \times V \times I \times I \to R^n$ which satisfies Hypothesis (1), an open set $A \subset \mathscr{C}^n(I)$ such that $x(t) \in \mathbf{G}$ for all $x \in A$ and $t \in I$, a class $\mathscr{U}$ of functions $u{:}I \to V$ which is convex under switching and satisfies Hypotheses (2) and (3), a subset A' of $\mathscr{C}^n(I)$ which is finitely open in itself, and a continuously differentiable function $\chi = (\chi^0, \chi^1, \ldots, \chi^{\mu+m})$: $R^n \times R^n \to R^{\mu+m+1}$. Then find functions $x^0 \in A$, $x_0^* \in A'$, and $u_0 \in \mathscr{U}$ such that

10 $$x_0(t) = x_0^*(t) + \int_{t_1}^{t} f(x_0(s),u_0(s),s,t)\, ds \quad \text{for all} \quad t \in I,$$

such that

11 $$\chi^i(x_0(t_1),x_0(t_2)) \leq 0 \quad \text{for} \quad i = 1, \ldots, \mu,$$

12 $$\chi^i(x_0(t_1),x_0(t_2)) = 0 \quad \text{for} \quad i = \mu + 1, \ldots, \mu + m,$$

and such that

13 $$\chi^0(x_0(t_1),x_0(t_2)) \leq \chi^0(x(t_1),x(t_2))$$

for all functions $x \in A$ which—for some $x^* \in A'$ and $u \in \mathscr{U}$—satisfy the relations

14 $$x(t) = x^*(t) + \int_{t_1}^{t} f(x(s),u(s),s,t)\, ds \quad \text{for all} \quad t \in I,$$

15 $$\chi^i(x(t_1),x(t_2)) \leq 0 \quad \text{for} \quad i = 1, \ldots, \mu,$$

16 $$\chi^i(x(t_1),x(t_2)) = 0 \quad \text{for} \quad i = \mu + 1, \ldots, \mu + m.$$

Problem (9) differs from Problem (V.3.10) in that we have (for the sake of simplicity) set $k = 2$ and have replaced the ordinary differential equations (V.3.2 and 7) by the Volterra integral equations (10) and (14).

The reader should here review (V.1.12) in order to see what kinds of problems may be reformulated so as to take on the form of Problem (9).

If the function f in Problem (9) is independent of its last argument, and if A' consists of all constant functions in $\mathscr{C}^n(I)$, then Eq. (10) reduces to the ordinary differential equation

$$\dot{x}_0(t) = f(x_0(t),u_0(t),t) \quad \text{a.e. in} \quad I,$$

[and similarly for (14)], and our problem reduces to a special case of Problem (V.3.10).

We shall use the notation of Chapter IV, with $\mathscr{X} = \mathscr{C}^n(I)$.

We set

$$\mathscr{W} = \{T_{x^*,F_u} : u \in \mathscr{U},\, x^* \in A'\},$$

where, for each $u \in \mathscr{U}$, F_u is the function in $\mathscr{F}$ defined by (7) [see (6) and (IV.3.22)], and the operator $T_{x^*,F} : A \to \mathscr{C}^n(I)$ [for each $x^* \in \mathscr{C}^n(I)$ and $F \in \mathscr{F}$] is defined by

17 $$(T_{x^*,F}x)(t) = x^*(t) + \int_{t_1}^{t} F(x(s),s,t)\, ds, \quad t \in I, \quad x \in A.$$

Then, inasmuch as Eq. (14) is clearly equivalent to the equation

$$x(t) = (T_{x^*,F_u}x)(t) \quad \text{for all} \quad t \in I,$$

Problem (9) has been recast in the form of Problem (IV.1.15), with $k = 2$ and $\tau^1 = t_1$ and $\tau^2 = t_2$, but with the constraints (IV.1.19) omitted. [Recall that, from the results of the Appendix, $\mathscr{W} \subset \mathscr{T}^u \subset \mathscr{T}_0$—see (IV.1.2) and (IV.3.7 and 28).]

In order to obtain necessary conditions for Problem (9), we shall appeal to Theorems (IV.1.53) and (IV.5.34).

18 Thus, let $(x_0,x_0^*,u_0) \in A \times A' \times \mathscr{U}$ be a solution of Problem (9), so that (10)–(12) hold. Set $T_{x_0^*,F_{u_0}}$ equal to T_0, so that

19 $$(T_0x)(t) = x_0^*(t) + \int_{t_1}^{t} f(x(s),u_0(s),s,t)\,ds, \quad t \in I, \quad x \in A,$$

and Eq. (10) is clearly equivalent to $x_0 = T_0x_0$, which means that $T_0 \in \mathscr{T}_1$. Then, by Theorem (IV.5.34) and (8), Hypotheses (IV.1.21–23) hold. Appealing to Theorem (IV.1.53) and to (IV.1.58), we conclude that there exists a nonzero row vector $\alpha = (\alpha^0, \alpha^1, \ldots, \alpha^{\mu+m}) \in R^{\mu+m+1}$ such that

20 (i) $\alpha^i \leq 0$ for $i = 0, \ldots, \mu$,

21 (ii) $\sum_{j=1}^{2} \alpha\mathscr{D}_j^0\chi\delta x_{x^*,u}(t_j) \leq \sum_{j=1}^{2} \alpha\mathscr{D}_j^0\chi\delta x_{x_0^*,u_0}(t_j)$

for all $x^* \in A'$ and $u \in \mathscr{U}$,

where, for each $x^* \in A'$ and $u \in \mathscr{U}$, $\delta x_{x^*,u}$ denotes the column-vector-valued function in $\mathscr{C}^n(I)$ defined by [see (IV.1.55), (IV.3.30), (17), and (7)]

22 $$\delta x_{x^*,u}(t) = \int_{t_1}^{t} \mathscr{D}_1 f(x_0(s),u_0(s),s,t)\delta x_{x^*,u}(s)\,ds + x^*(t) + \int_{t_1}^{t} f(x_0(s),u(s),s,t)\,ds, \quad t \in I,$$

and $\mathscr{D}_j^0\chi$ $(j = 1, 2)$ denotes $\mathscr{D}_j\chi(x_0(t_1),x_0(t_2))$,

23 (iii) $\sum_{i=1}^{\mu} \alpha^i\chi^i(x_0(t_1),x_0(t_2)) = 0.$

It follows from (23), (20), and (11) that

24 $$\alpha^i\chi^i(x_0(t_1),x_0(t_2)) = 0 \quad \text{for} \quad i = 1, \ldots, \mu.$$

Now Eq. (22) is evidently a linear, inhomogeneous Volterra integral equation with kernel $\mathbf{K}(s,t)$ defined by

25 $$\mathbf{K}(s,t) = \mathscr{D}_1 f(x_0(s),u_0(s),s,t), \quad s,t \in I.$$

Let $\mathbf{K}^*(s,t)$ be the resolvent kernel of $\mathbf{K}$ [see Theorem (A.65) in the Appendix], so that the function $s \to \mathbf{K}^*(s,t)$ is integrable over I (for each $t \in I$), and the solution of Eq. (22) can be written in the form

26 $$\delta x_{x^*,u}(t) = x^*(t) + \int_{t_1}^{t} f(x_0(s),u(s),s,t)\,ds + \int_{t_1}^{t} \mathbf{K}^*(s,t)x^*(s)\,ds + \int_{t_1}^{t} \mathbf{K}^*(s,t) \int_{t_1}^{s} f(x_0(\tau),u(\tau),\tau,s)\,d\tau\,ds, \quad t \in I,$$

and Eq. (21) can be rewritten as follows:

27 $$\alpha\mathscr{D}_1^0\chi\delta x^*(t_1) + \alpha\mathscr{D}_2^0\chi\left[\delta x^*(t_2) + \int_{t_1}^{t_2}\mathbf{K}^*(s,t_2)\delta x^*(s)\,ds\right]$$
$$+ \alpha\mathscr{D}_2^0\chi\left[\int_{t_1}^{t_2}\delta f_u^0(s,t_2)\,ds + \int_{t_1}^{t_2}\mathbf{K}^*(s,t_2)\int_{t_1}^{s}\delta f_u^0(\tau,s)\,d\tau\,ds\right] \leq 0$$
$$\text{for all} \quad \delta x^* \in A' - x_0^* \quad \text{and} \quad u \in \mathscr{U},$$

where, for each $u \in \mathscr{U}$,

28 $$\delta f_u^0(s,t) = f(x_0(s),u(s),s,t) - f(x_0(s),u_0(s),s,t), \quad s,t \in I.$$

Since $\delta f_{u_0}^0 \equiv 0$ and $u_0 \in \mathscr{U}$, (27) implies that

29 $$\alpha\mathscr{D}_1^0\chi\delta x^*(t_1) + \alpha\mathscr{D}_2^0\chi\left[\delta x^*(t_2) + \int_{t_1}^{t_2}\mathbf{K}^*(s,t_2)\delta x^*(s)\,ds\right] \leq 0$$
$$\text{for all} \quad \delta x^* \in A' - x_0^*.$$

Since $x_0^* \in A'$, (27) also implies, once we interchange the order of integration in the double integral, that

30 $$\alpha\mathscr{D}_2^0\chi\left\{\int_{t_1}^{t_2}\left[f(x_0(s),u_0(s),s,t_2) + \int_{s}^{t_2}\mathbf{K}^*(\tau,t_2)f(x_0(s),u_0(s),s,\tau)\,d\tau\right]ds\right\}$$
$$\geq \alpha\mathscr{D}_2^0\chi\left\{\int_{t_1}^{t_2}\left[f(x_0(s),u(s),s,t_2) + \int_{s}^{t_2}\mathbf{K}^*(\tau,t_2)f(x_0(s),u(s),s,\tau)\,d\tau\right]ds\right\}$$
$$\text{for all} \quad u \in \mathscr{U}.$$

Thus we have shown the following.

31 THEOREM. *Let* (x_0,x_0^*,u_0) *be a solution of Problem* (9), *and let* $\mathbf{K}^*$ *be the resolvent kernel of the kernel* $\mathbf{K}$ *defined by* (25). *Then there exists a nonzero row vector* $\alpha = (\alpha^0, \alpha^1, \ldots, \alpha^{\mu+m}) \in \bar{R}_-^{\mu+1} \times R^m$ *such that*

(i) u_0 *satisfies the maximum principle in integral form* (30),
(ii) *relation* (29) *holds, and*
(iii) *Eqs.* (24) *hold,*

where, in (29) *and* (30), $\mathscr{D}_j^0\chi$ $(j = 1, 2)$ *denotes* $\mathscr{D}_j\chi(x_0(t_1),x_0(t_2))$.

32 Note that the necessary conditions of Theorem (31) are not expressed in terms of an "adjoint variable," as was the case in all of our preceding problems. This is because there does not appear to be any useful way to characterize an adjoint variable for Problem (9).

In case f is independent of its last argument, then, as is easily seen,

$$\mathbf{K}^*(s,t) = \Phi(t)\Phi^{-1}(s)\mathscr{D}_1 f(x_0(s),u_0(s),s),$$

where Φ is the $(n \times n)$-matrix-valued function defined on I that satisfies (V.2.22 and 20)—so that (V.2.30) holds. In this case, relations (29) and (30) can be rewritten as follows:

33 $$\alpha\mathscr{D}_1^0\chi\delta x^*(t_1) + \psi(t_2)\delta x^*(t_2) + \int_{t_1}^{t_2} \psi(s)\mathscr{D}_1 f(x_0(s),u_0(s),s)\delta x^*(s)\, ds \leq 0 \quad \text{for all} \quad \delta x^* \in A' - x_0^*,$$

34 $$\int_{t_1}^{t_2} \left[\psi(t_2) + \int_s^{t_2} \psi(\tau)\mathscr{D}_1 f(x_0(\tau),u_0(\tau),\tau)\, d\tau\right] f(x_0(s),u_0(s),s)\, ds \geq \int_{t_1}^{t_2} \left[\psi(t_2) + \int_s^{t_2} \psi(\tau)\mathscr{D}_1 f(x_0(\tau),u_0(\tau),\tau)\, d\tau\right] f(x_0(s),u(s),s)\, ds \quad \text{for all} \quad u \in \mathscr{U},$$

where $\psi(t) = \alpha\mathscr{D}_2^0\chi\Phi(t_2)\Phi^{-1}(t)$ for all $t \in I$. Thus, ψ is absolutely continuous on I, and, by virtue of Eqs. (V.2.30 and 20), satisfies the ordinary differential equation

35 $$\dot{\psi}(t) = -\psi(t)\mathscr{D}_1 f(x_0(t),u_0(t),t) \quad \text{a.e. on} \quad I,$$

as well as the boundary condition

$$\psi(t_2) = \alpha\mathscr{D}_2^0\chi.$$

Using (35) in (34), we conclude that

$$\int_{t_1}^{t_2} \psi(t)f(x_0(t),u_0(t),t)\, dt \qquad \int_{t_1}^{t_2} \psi(t)f(x_0(t),u(t),t)\, dt \quad \text{for all} \quad u \in \mathscr{U}.$$

Further, it follows from (35) and (33) that Eqs. (1.50 and 51) hold. If also A' consists of all constant functions in $\mathscr{C}^n(I)$—so that Problem (9) reduces to a special case of Problem (V.3.10)—then

(1.50 and 51) imply that

$$[\psi(t_1) + \alpha \mathscr{D}_1^0 \chi]\xi \leq 0 \quad \text{for all} \quad \xi \in R^n,$$

which can only hold if

$$\psi(t_1) = -\alpha \mathscr{D}_1^0 \chi,$$

and our necessary conditions coincide with those that can be obtained from Theorem (V.3.75).

If $\alpha \mathscr{D}_2^0 \chi = 0$, then the necessary conditions of Theorem (31) are trivial. In order to find sufficient conditions for $\alpha \mathscr{D}_2^0 \chi$ not to vanish, we introduce the following compatibility condition [compare with Conditions (1.67 and 73) and (V.3.82)]:

36 CONDITION. *There is a vector* $(\xi_1,\xi_2) \in R^n \times R^n$ *such that*

37
$$\sum_{j=1}^{2} (\mathscr{D}_j^0 \chi^i) \cdot \xi_j \begin{cases} < 0 & \text{for } \ i = 0 \ \text{ and each } \ i = 1, \ldots, \mu \\ & \text{such that } \ \chi^i(x_0(t_1),x_0(t_2)) = 0, \\ = 0 & \text{for } \ i = \mu + 1, \ldots, \mu + m, \end{cases}$$

and the m vectors $(\mathscr{D}_1^0 \chi^i, \mathscr{D}_2^0 \chi^i)$, $i = \mu + 1, \ldots, \mu + m$, *are linearly independent in* $R^n \times R^n$.

Then we have the following result. [Compare with Lemmas (1.71 and 75).]

38 LEMMA. *Let* (x_0, x_0^*, u_0) *be a solution of Problem* (9) *such that Conditions* (36) *and* (1.69) *hold. Then* $\alpha \mathscr{D}_2^0 \chi \neq 0$, *where* α *is the vector whose existence is asserted in Theorem* (31).

Proof. We argue by contradiction and suppose that $\alpha \mathscr{D}_2^0 \chi = 0$. Hence, by (29), $\alpha \mathscr{D}_1^0 \chi [x^*(t_1) - x_0^*(t_1)] \leq 0$ for all $x^* \in A'$, which implies, by virtue of Condition (1.69), that $\alpha \mathscr{D}_1^0 \chi = 0$, i.e., $0 = \sum_{i=0}^{\mu+m} \alpha^i \mathscr{D}_j^0 \chi^i$ for $j = 1$ and 2. Let (ξ_1,ξ_2) be the vector whose existence is asserted in Condition (36). Evidently,

39
$$0 = \sum_{j=1}^{2} \left(\sum_{i=0}^{\mu+m} \alpha^i \mathscr{D}_j^0 \chi^i \right) \cdot \xi_j = \sum_{i=0}^{\mu+m} \alpha^i \sum_{j=1}^{2} (\mathscr{D}_j^0 \chi^i) \cdot \xi_j.$$

It follows from (39), (37), (20), and (24) that $\alpha^i = 0$ for $i = 0, \ldots, \mu$, so that

$$\sum_{i=\mu+1}^{\mu+m} \alpha^i (\mathscr{D}_1^0 \chi^i, \mathscr{D}_2^0 \chi^i) = 0 \quad \text{in} \quad (R^n)^2,$$

with at least one of the numbers $\alpha^{\mu+1}, \ldots, \alpha^{\mu+m}$ different from zero. But this contradicts the last assertion of Condition (36). |||

Let us now turn to the question of when the maximum principle in integral form (30) can be written in "pointwise" form. To this end, we introduce the following hypothesis [compare with Hypothesis (V.2.43)].

40 HYPOTHESIS. *There are a subset* U *of* V *and a subset* I^* *of* I *whose complement with respect to* I *has Lebesgue measure zero such that* (i) $u(t) \in U$ *for all* $u \in \mathscr{U}$ *and every* $t \in I$, (ii) *every constant function* $u: I \to U$ *belongs to* $\mathscr{U}$, *and either* (iiia) *for every* $v \in U$, *the function* $(\xi,s,t) \to f(\xi,v,s,t): \mathbf{G} \times I^* \times I \to R^n$ *is continuous, or* (iiib) U *is a topological space with a countable* (*or finite*) *dense subset* $\tilde{U}$, *and, for every* $(\xi,s) \in \mathbf{G} \times I^*$, *the function* $(v,t) \to f(\xi,v,s,t): U \times I \to R^n$ *is continuous.*

Note that if condition (iiib) of Hypothesis (40) holds, then $f(\xi,\tilde{U},s,t)$ is dense in $f(\xi,U,s,t)$ for each $(\xi,s,t) \in \mathbf{G} \times I^* \times I$.

41 THEOREM. *Let* (x_0,x_0^*,u_0) *be a solution of Problem* (9) *such that Hypothesis* (40) *holds. Then the maximum condition* (30) *in Theorem* (31) *is equivalent to the pointwise maximum principle*

42 $$\alpha\mathscr{D}_2^0\chi\left[f(x_0(t),u_0(t),t,t_2) + \int_t^{t_2} \mathbf{K}^*(s,t_2)f(x_0(t),u_0(t),t,s)\,ds\right]$$
$$= \max_{v\in U}\left\{\alpha\mathscr{D}_2^0\chi\left[f(x_0(t),v,t,t_2) + \int_t^{t_2} \mathbf{K}^*(s,t_2)f(x_0(t),v,t,s)\,ds\right]\right\}$$
$$\text{for almost all} \quad t \in I.$$

Proof. Let $\mathbf{f}$ denote the function from $\mathbf{G} \times V \times I$ into R defined by

$$\mathbf{f}(\xi,v,t) = \alpha\mathscr{D}_2^0\chi\left[f(\xi,v,t,t_2) + \int_t^{t_2} \mathbf{K}^*(s,t_2)f(\xi,v,t,s)\,ds\right].$$

It is easily seen [using Corollary (I.5.32)] that, under the hypotheses of our theorem, Hypotheses (V.1.16–18) and (V.2.43) are satisfied—with f replaced by $\mathbf{f}$, and R^n by R. Inasmuch as (30) and (42) can be rewritten in the forms

$$\int_{t_1}^{t_2} \mathbf{f}(x_0(s),u_0(s),s)\,ds \geq \int_{t_1}^{t_2} \mathbf{f}(x_0(s),u(s),s)\,ds \quad \text{for all} \quad u \in \mathscr{U},$$

$$\mathbf{f}(x_0(t),u_0(t),t) = \max_{v\in U} \mathbf{f}(x_0(t),v,t) \quad \text{for almost all} \quad t \in I,$$

respectively, our theorem now follows essentially in the same way as Theorem (V.2.44). |||

The observations in (V.2.48–49) are relevant here also.

The remarks in (1.79) carry over, with evident minor modifications, to Theorem (31).

4. Discrete Optimal Control Problems

In this section, we shall concern ourselves with so-called discrete, or finite-dimensional, optimal control problems. Such problems are characterized by constraints which have the form of difference equations. More precisely, we shall suppose that x is a function from some finite set of integers $I = \{1, 2, \ldots, k\}$ into R^n, and we shall suppose that x is constrained by an equation of the form

1 $$x(j + 1) = f(x(j),u(j),j), \quad j = 1, \ldots, k - 1,$$

where f is a given function from $\mathbf{G} \times V \times I'$ into R^n, where $I' = \{1, 2, \ldots, k - 1\}$, $\mathbf{G}$ is a given open set in R^n, and V is an arbitrary set. In Eq. (1), u is some function from I' into V.

In analogy with the "continuous" optimal control problems which we have considered in the earlier sections of this chapter, as well as in the two preceding chapters, we shall refer to x as a *state variable* (or *phase vector*), and to u as a *control* variable. Evidently, if we approximate the differential equation (V.1.2) (which characterized the optimal control problems discussed in Chapters V and VI) by a finite-difference equation, we arrive at an equation of the form of (1).

Corresponding to Hypothesis (V.1.16), we shall make the following hypothesis on the function f.

2 HYPOTHESIS. *For each* $(v,j) \in V \times I'$, *the function* $\xi - f(\xi,v,j)$: $\mathbf{G} \to R^n$ *is continuously differentiable.*

We then consider the following "discrete" optimal control problem.

3 *Problem.* Let there be given an open set $\mathbf{G}$ in R^n, an arbitrary set V, a positive integer k, a function $f: \mathbf{G} \times V \times I' \to R^n$ satisfying Hypothesis (2) (where I' denotes $\{1, 2, \ldots, k - 1\}$), a class $\mathscr{U}$ of functions $u: I' \to V$, and a continuously differentiable function $\chi = (\chi^0, \chi^1, \ldots, \chi^{\mu+m})$: $(R^n)^k \to R^{\mu+m+1}$. Then find functions $x_0: I \to \mathbf{G}$ (where I denotes $\{1, \ldots, k\}$) and $u_0 \in \mathscr{U}$

such that

4 $$x_0(j + 1) = f(x_0(j),u_0(j),j) \quad \text{for} \quad j = 1, \ldots, k - 1,$$

5 $$\chi^i(x_0(1), \ldots ,x_0(k)) \leq 0 \quad \text{for} \quad i = 1, \ldots, \mu,$$

6 $$\chi^i(x_0(1), \ldots ,x_0(k)) = 0 \quad \text{for} \quad i = \mu + 1, \ldots, \mu + m,$$

and such that

7 $$\chi^0(x_0(1), \ldots ,x_0(k)) \leq \chi^0(x(1), \ldots ,x(k))$$

for all functions $x:I \to \mathbf{G}$ satisfying the difference equation

8 $$x(j + 1) = f(x(j),u(j),j), \quad j = 1, \ldots, k - 1,$$

for some $u \in \mathscr{U}$, together with the constraints

9 $$\chi^i(x(1), \ldots ,x(k)) \leq 0 \quad \text{for} \quad i = 1, \ldots, \mu,$$

10 $$\chi^i(x(1), \ldots ,x(k)) = 0 \quad \text{for} \quad i = \mu + 1, \ldots, \mu + m.$$

11 It is evident that Problem (3) may be viewed as a "discrete form" of Problem (V.3.10) [or, for that matter, of Problem (V.3.1)]. However, note that here we have not imposed convexity under switching as a requirement on $\mathscr{U}$. Indeed, in order to obtain a maximum principle for Problem (3) which is analogous to that of Theorem (V.3.75), we shall have to impose a stronger requirement on $\mathscr{U}$ than convexity under switching.

For each $u \in \mathscr{U}$, let us define the function $F_u:\mathbf{G} \times I' \to R^n$ as follows:

12 $$F_u(\xi,j) = f(\xi,u(j),j), \quad \xi \in \mathbf{G}, \quad j \in I'.$$

We then make the following hypothesis:

13 HYPOTHESIS. *The set* $\{F_u\colon u \in \mathscr{U}\}$ [*see* (12)] *is finitely open in itself.*

This is, of course, a rather strong hypothesis, analogous to that discussed in Section 5 of Chapter V. However, it turns out that, in the case of discrete optimal control problems, convexity under switching is not a strong enough condition in order to obtain a global maximum principle of the form of Theorem (V.3.75). As we shall see later in this section, it is possible to obtain a "local" maximum principle—analogous to relation (VI.1.51) of Theorem (VI.1.59)—under alternate, essentially weaker, requirements on the set $\mathscr{U}$.

14 Note that if V is a subset of a linear vector space $\mathbf{V}$, if $\mathscr{U}$ is finitely open in itself (as a subset of the linear vector space of all function $u:I' \to \mathbf{V}$), and if f has the form

$$f(\xi,v,j) = g_1(\xi,j) + g_2(\xi,j)v, \quad (\xi,v,j) \in \mathbf{G} \times V \times I',$$

where g_1 is some given function from $\mathbf{G} \times I'$ into R^n and $g_2(\xi,j)$—for each $(\xi,j) \in \mathbf{G} \times I'$—is a linear transformation from $\mathbf{V}$ into R^n, then Hypothesis (13) is satisfied.

15 Let $\mathscr{X}$ denote the linear vector space of all functions $x:I \to R^n$. We define a norm on $\mathscr{X}$ through the relation

$$\|x\| = \max_{1 \le j \le k} |x(j)|.$$

It is easily seen that, with this norm, $\mathscr{X}$ becomes a Banach space. Let $A = \{x\colon x \in \mathscr{X},\ x(j) \in \mathbf{G} \text{ for each } j \in I\}$. It is immediately seen that A is open in $\mathscr{X}$. For each pair $(\xi,u) \in \mathbf{G} \times \mathscr{U}$, let us define the operator $T_{\xi,u}\colon A \to \mathscr{X}$ as follows:

16 $$(T_{\xi,u}x)(j) = \begin{cases} \xi & \text{for } \ j = 1, \\ f(x(j-1),u(j-1),j-1) & \text{for } \ j = 2,\dots,k, \end{cases} \qquad x \in A.$$

Finally, let

17 $$\mathscr{W} = \{T_{\xi,u}\colon \xi \in \mathbf{G},\ u \in \mathscr{U}\}.$$

It easily follows from Hypothesis (2) that each element of $\mathscr{W}$ is continuously differentiable, i.e., $\mathscr{W}$ is a subset of the linear vector space $\mathscr{T}$ of all operators $T\colon A \to \mathscr{X}$ which are continuously differentiable.

18 Note that if Hypothesis (13) is satisfied, then the set $\mathscr{W}$ defined by (17) and (16) is finitely open in itself.

It is evident that the difference equation (8) is equivalent to the operator equation [see (16)]

$$x = T_{x(1),u}x.$$

It is also evident that the set $\mathscr{W}$ defined by (17) is a subset of $\mathscr{T}_0$ [see (IV.1.2)]. Let us define real-valued functions $\hat{\varphi}^1, \dots, \hat{\varphi}^m$, $\hat{\phi}^0, \dots, \hat{\phi}^\mu$ on A as follows:

19 $$\hat{\varphi}^i(x) = \chi^{\mu+i}(x(1),\dots,x(k)), \quad x \in A, \quad i = 1,\dots,m,$$

20 $$\hat{\phi}^i(x) = \chi^i(x(1),\dots,x(k)), \quad x \in A, \quad i = 0,\dots,\mu.$$

Further, let $\hat{\varphi} = (\hat{\varphi}^1, \dots, \hat{\varphi}^m)$ and $\hat{\phi}_1 = (\hat{\phi}^1, \dots, \hat{\phi}^\mu)$. It is now evident that Problem (3) is of the form of Problem (IV.1.3), with $\mathscr{X}$,

A, $\mathscr{W}$, and $\mathscr{T}$ as defined in (15), $A' = A$, $\mathscr{Z}_1 = R^\mu$, and $Z_1 = \bar{R}^\mu_-$.

Now suppose that (x_0,u_0) is a solution of Problem (3) subject to Hypothesis (13), so that (4)–(6) hold. Denote $T_{x_0(1),u_0}$ by T_0, so that $T_0 \in (\mathscr{W} \cap \mathscr{T}_1)$ and $x_0 = T_0 x_0$. Let us make use of Theorem (IV.1.30) in order to obtain necessary conditions that x_0 and u_0 must satisfy. Inasmuch as $\mathscr{W}$ is finitely open in itself [see (18)], we need only verify that Hypotheses (IV.1.21, 22, and 28) hold [with $\hat{\phi}'$ deleted in Hypothesis (IV.1.28)]. But Hypothesis (IV.1.21) is trivially seen to hold, and the continuous differentiability of χ implies that Hypothesis (IV.1.28) holds [see (IV.1.29)]. In fact, $\hat{\varphi}$ and the $\hat{\phi}^i$ are Fréchet differentiable with Fréchet differentials at any $x \in A$ given by the formulas

21 $$D\hat{\varphi}(x;\delta x) = \sum_{j=1}^{k} \mathscr{D}_j \chi'(x(1),\ldots,x(k))\delta x(j), \quad x \in A, \quad \delta x \in \mathscr{X},$$

22 $$D\hat{\phi}^i(x;\delta x) = \sum_{j=1}^{k} \mathscr{D}_j \chi^i(x(1),\ldots,x(k))\delta x(j),$$
$$x \in A, \quad \delta x \in \mathscr{X}, \quad i = 0,\ldots,\mu,$$

where $\chi' = (\chi^{\mu+1},\ldots,\chi^{\mu+m})$. With regard to Hypothesis (IV.1.22), we note that [by Hypothesis (2)] each $T \in \text{co } \mathscr{W}$—say $T = \sum_{i=1}^{\nu} \beta^i T_{\xi_i,u_i}$, where (for each i) $\beta^i \geq 0$, $\xi_i \in \mathbf{G}$, and $u_i \in \mathscr{U}$, and $\sum_1^\nu \beta^i = 1$—is Fréchet differentiable, with Fréchet differential (at any $x \in A$) given by the formula

23 $$(DT(x;\delta x))(j) = \begin{cases} 0 \quad \text{for} \quad j = 1 \\ \displaystyle\sum_{i=1}^{\nu} \beta^i \mathscr{D}_1 f(x(j-1),u_i(j-1),j-1)\delta x(j-1) \\ \qquad \text{for} \quad j = 2,\ldots,k, \quad \text{for all} \quad \delta x \in \mathscr{X}. \end{cases}$$

Thus, the equation $\delta x - DT(x;\delta x) = \bar{x}$ (as an equation for δx), for each fixed $x \in A$ and $\bar{x} \in \mathscr{X}$, is equivalent to the linear difference equation

24 $$\delta x(j) = \sum_{i=1}^{\nu} \beta^i \mathscr{D}_1 f(x(j-1),u_i(j-1),j-1)\delta x(j-1) + \bar{x}(j).$$
$$j = 2,\ldots,k,$$

with initial value

25 $$\delta x(1) = \bar{x}(1).$$

Since Eqs. (24) and (25) evidently have a unique solution in $\mathscr{X}$ for every $\bar{x} \in \mathscr{X}$, Hypothesis (IV.1.22) is satisfied.

Hence, by Theorem (IV.1.30), using (16), (17) and (20)–(23), we conclude that there exists a nonzero row vector $\alpha = (\alpha^0, \alpha^1, \ldots, \alpha^{\mu+m}) \in R^{\mu+m+1}$ such that

26 $$\text{(i) } \alpha^i \le 0 \quad \text{for} \quad i = 0, \ldots, \mu,$$

27 $$\text{(ii) } \sum_{j=1}^{k} \alpha \mathscr{D}_j^0 \chi \delta x_{\xi,u}(j) \le \sum_{j=1}^{k} \alpha \mathscr{D}_j^0 \chi \delta x_{x_0(1),u_0}(j)$$
$$\text{for all} \quad \xi \in \mathbf{G} \quad \text{and} \quad u \in \mathscr{U},$$

where, for each $\xi \in R^n$ and $u \in \mathscr{U}$, $\delta x_{\xi,u}$ denotes the column-vector-valued function in $\mathscr{X}$ defined by

28 $$\delta x_{\xi,u}(j) = \mathscr{D}_1 f(x_0(j-1), u_0(j-1), j-1)\delta x_{\xi,u}(j-1) + f(x_0(j-1), u(j-1), j-1),$$
$$j = 2, \ldots, k,$$

29 $$\delta x_{\xi,u}(1) = \xi,$$

and $\mathscr{D}_j^0 \chi$ (for each j) denotes $\mathscr{D}_j \chi(x_0(1), \ldots, x_0(k))$, and

30 $$\text{(iii) } \sum_{i=1}^{\mu} \alpha^i \chi^i(x_0(1), \ldots, x_0(k)) = 0.$$

Now it follows from (30), (26), and (5) that

31 $$\alpha^i \chi^i(x_0(1), \ldots, x_0(k)) = 0 \quad \text{for} \quad i = 1, \ldots, \mu.$$

Further, the solution of Eqs. (28) and (29) is (as is easily shown by an induction argument) given by the formula†

32 $$\delta x_{\xi,u}(j) = \prod_{\ell=1}^{j-1} \mathscr{D}^0 f(\ell)\xi + \sum_{s=1}^{j-2} \left[\prod_{\ell=s+1}^{j-1} \mathscr{D}^0 f(\ell) \right] f(x_0(s), u(s), s) + f(x_0(j-1), u(j-1), j-1) \quad \text{for} \quad j = 2, \ldots, k,$$

[the second summand in the right-hand side of (32) should be deleted for $j = 2$], where, for ease of notation, we have written

33 $$\mathscr{D}^0 f(j) = \mathscr{D}_1 f(x_0(j), u_0(j), j), \quad j = 1, \ldots, k-1.$$

† In (32), and in the sequel, products of the form $\prod_{\ell=s+1}^{j-1} \mathscr{D}^0 f(\ell)$ are to be understood with the order such that the index decreases from left to right.

Thus, (27) may be rewritten in the form

34 $$\alpha\mathscr{D}_1^0\chi(\xi - x_0(1)) + \sum_{j=2}^{k} \alpha\mathscr{D}_j^0\chi\left[\prod_{\ell=1}^{j-1} \mathscr{D}^0 f(\ell)\right](\xi - x_0(1)) + \sum_{j=3}^{k} \alpha\mathscr{D}_j^0\chi \sum_{s=1}^{j-2}\left[\prod_{\ell=s+1}^{j-1} \mathscr{D}^0 f(\ell)\right]\delta f_u(s) + \sum_{j=2}^{k} \alpha\mathscr{D}_j^0\chi\delta f_u(j-1) \leq 0$$

for all $\xi \in \mathbf{G}$ and $u \in \mathscr{U}$,

where, for each $u \in \mathscr{U}$ and $j = 1, \ldots, k-1$, we have set

35 $$\delta f_u(j) = f(x_0(j),u(j),j) - f(x_0(j),u_0(j),j).$$

Since $u_0 \in \mathscr{U}$ and $\delta f_{u_0} = 0$, (34) implies in particular that

36 $$\left(\alpha\mathscr{D}_1^0\chi + \sum_{j=2}^{k} \alpha\mathscr{D}_j^0\chi\left[\prod_{\ell=1}^{j-1} \mathscr{D}^0 f(\ell)\right]\right)(\xi - x_0(1)) \leq 0$$

for all $\xi \in \mathbf{G}$.

Since $\mathbf{G}$ is open and $x_0(1) \in \mathbf{G}$, (36) can hold only if

37 $$\alpha\mathscr{D}_1^0\chi + \sum_{j=2}^{k} \alpha\mathscr{D}_j^0\chi\left[\prod_{\ell=1}^{j-1} \mathscr{D}^0 f(\ell)\right] = 0.$$

Also, since $x_0(1) \in \mathbf{G}$, (34) implies that

38 $$\sum_{s=3}^{k} \alpha\mathscr{D}_s^0\chi \sum_{j=1}^{s-2}\left[\prod_{\ell=j+1}^{s-1} \mathscr{D}^0 f(\ell)\right]\delta f_u(j) + \sum_{j=1}^{k-1} \alpha\mathscr{D}_{j+1}^0\chi\delta f_u(j) \leq 0$$

for all $u \in \mathscr{U}$.

Interchanging the order of summation in the double sum in (38), and setting

39 $$\psi(j) = \sum_{s=j+2}^{k} \alpha\mathscr{D}_s^0\chi\left[\prod_{\ell=j+1}^{s-1} \mathscr{D}^0 f(\ell)\right] + \alpha\mathscr{D}_{j+1}^0\chi$$

for $j = 0, 1, \ldots, k-2$,

40 $$\psi(k-1) = \alpha\mathscr{D}_k^0\chi,$$

we obtain that

41 $$\sum_{j=1}^{k-1} \psi(j) f(x_0(j),u_0(j),j) \geq \sum_{j=1}^{k-1} \psi(j) f(x_0(j),u(j),j)$$

for all $u \in \mathscr{U}$.

It follows from (39) and (37) that

42 $$\psi(0) = 0,$$

and that [see (33)]

43 $$\psi(j-1) = \psi(j)\mathscr{D}_1 f(x_0(j),u_0(j),j) + \alpha\mathscr{D}_j^0\chi, \quad j = 1,\ldots,k-1.$$

Thus, we have shown the following:

44 THEOREM. *Let (x_0,u_0) be a solution of Problem* (3) *subject to Hypothesis* (13). *Then there exist a nonzero row vector $\alpha = (\alpha^0, \alpha^1, \ldots, \alpha^{\mu+m}) \in \bar{R}_-^{\mu+1} \times R^m$ and an n (row)-vector-valued function ψ defined on $\{0, 1, \ldots, k-1\}$ such that*

(i) *ψ satisfies the linear inhomogeneous difference equation* (43),

(ii) *u_0 satisfies the maximum condition* (41),

(iii) *ψ satisfies the boundary conditions* (40) *and* (42), *and*

(iv) *relations* (31) *hold,*

where, in (43) *and* (40), $\mathscr{D}_j^0\chi = \mathscr{D}_j\chi(x_0(1),\ldots,x_0(k))$ *for each* j.

45 Theorem (44) is evidently a "discrete analog" of Theorem (V.3.75). The function ψ in this theorem may be viewed as an adjoint variable, and Eq. (43)—which [together with (40)] uniquely determines ψ, in terms of α, x_0, and u_0—may be viewed as an adjoint difference equation. Note that if the functions χ^i depend only on their first and last arguments—i.e., if the problem constraints (5), (6), (9), and (10) and the cost functional appearing in (7) only involve the initial and final values of x—so that $\mathscr{D}_j^0\chi = 0$ for $j = 2, \ldots, k-1$, then the adjoint equation (43) is homogeneous for $j > 1$. Relation (41) is a maximum principle in summed form; below we shall see that, under suitable hypotheses, it is equivalent to a "pointwise" maximum principle. Finally, Eqs. (40) and (42) play the role of transversality conditions.

46 If Problem (3) is modified by eliminating either the inequality constraints (5), (9) or the equality constraints (6), (10), then the necessary conditions of Theorem (44) remain in force, but with μ or m (or both) equal to zero, so that (31) is to be deleted if there are no inequality constraints [see (IV.1.58 and 92)].

As usual, it is of interest to find sufficient conditions under which $\psi \not\equiv 0$ on $\{1, \ldots, k-1\}$, for otherwise the necessary conditions of Theorem (44) are trivial. We thus invoke the following compatibility condition [compare with Condition (V.3.82)].

47 CONDITION. *There exists a vector* $(\xi_1, \ldots, \xi_k) \in (R^n)^k$ *such that*

48
$$\sum_{j=1}^{k} (\mathscr{D}_j^0 \chi^i) \cdot \xi_j \begin{cases} < 0 & \textit{for} \quad i = 0 \quad \textit{and each} \quad i = 1, \ldots, \mu \\ & \quad \textit{such that } \chi^i(x_0(1), \ldots, x_0(k)) = 0, \\ = 0 & \textit{for} \quad i = \mu + 1, \ldots, \mu + m, \end{cases}$$

and the m vectors $(\mathscr{D}_1^0 \chi^i, \ldots, \mathscr{D}_k^0 \chi^i)$, $i = \mu + 1, \ldots, \mu + m$, *are linearly independent in* $(R^n)^k$.

49 LEMMA. *If* (x_0, u_0) *is a solution of Problem* (3) *such that Hypothesis* (13) *and Condition* (47) *hold, then at least one of the vectors* $\psi(1), \ldots, \psi(k-1)$ *in Theorem* (44) *does not vanish.*

Proof. We argue by contradiction, and suppose that $\psi(j) = 0$ for $j = 1, \ldots, k-1$. It then follows from (40), (42), and (43) that $\alpha \mathscr{D}_j^0 \chi = 0$ for $j = 1, \ldots, k$. If $(\xi_1, \ldots, \xi_k)$ is the vector whose existence is asserted in Condition (47), this implies, by virtue of (26), (31), and (48), that $\alpha^i = 0$ for $i = 0, \ldots, \mu$. Hence, the numbers $\alpha^{\mu+1}, \ldots, \alpha^{\mu+m}$ are not all zero, and $\sum_{i=\mu+1}^{\mu+m} \alpha^i \mathscr{D}_j^0 \chi^i = 0$ for $j = 1, \ldots, k$ which contradicts the last assertion in Condition (47). |||

To obtain a "pointwise" maximum principle, we make the following hypothesis:

50 HYPOTHESIS. *There exist subsets* $U_1, \ldots, U_{k-1}$ *of* V *such that* $\mathscr{U} = \{u \colon u(j) \in U_j$ for $j = 1, \ldots, k-1\}$.

51 THEOREM. *If* (x_0, u_0) *is a solution of Problem* (3) *such that Hypotheses* (13) *and* (50) *hold, then the maximum condition* (41) *in Theorem* (44) *is equivalent to the condition*

52
$$\psi(j) f(x_0(j), u_0(j), j) = \max_{v \in U_j} \psi(j) f(x_0(j), v, j)$$
$$\textit{for each} \quad j = 1, \ldots, k-1.$$

Proof. It is obvious that (52) implies (41). We prove the converse by a contradiction argument, and thus suppose that, for some $\hat{j} = 1, \ldots, k-1$ and some $\hat{v} \in U_{\hat{j}}$,

$$\psi(\hat{j}) f(x_0(\hat{j}), u_0(\hat{j}), \hat{j}) < \psi(\hat{j}) f(x_0(\hat{j}), \hat{v}, \hat{j}).$$

If we define the function $\tilde{u} \colon I' \to V$ by the formula

$$\tilde{u}(j) = \begin{cases} u_0(j) & \text{for} \quad j = 1, \ldots, k-1, \quad j \neq \hat{j}, \\ \hat{v} & \text{for} \quad j = \hat{j}, \end{cases}$$

then $\tilde{u} \in \mathscr{U}$ and

$$\sum_{j=1}^{k-1} \psi(j)f(x_0(j),u_0(j),j) < \sum_{j=1}^{k-1} \psi(j)f(x_0(j),\tilde{u}(j),j),$$

contradicting (41). |||

53 It often turns out that the data of Problem (3) do not satisfy Hypothesis (13), but that, by a suitable reformulation of the problem, this hypothesis can be made to hold. This reformulation typically consists of an enlargement of the set of "admissible" phase-trajectories x, without a corresponding reduction of the minimum value of the "cost" function. Let us illustrate this with the following example. Suppose that we replace Hypothesis (13) by the following one:

54 HYPOTHESIS. *For each* $(\xi_1, \ldots, \xi_{k-1}) \in \mathbf{G}^{k-1}$, *the set*

$$\{(f(\xi_1,u(1),1), f(\xi_2,u(2),2), \ldots, f(\xi_{k-1},u(k-1), k-1): u \in \mathscr{U}\}$$

is convex [*in* $(R^n)^{k-1}$].

Note that if the set $\{F_u: u \in \mathscr{U}\}$—see (12)—is convex, then Hypothesis (54) holds. Also, if Hypothesis (50) holds, and if the sets $f(\xi,U_j,j)$ are convex for each $\xi \in \mathbf{G}$ and $j \in I'$, then Hypothesis (54) holds.

55 We then modify Problem (3) as follows:

1. We "enlarge" V by replacing it by the set $\mathbf{V} = \bigcup_{\nu=1}^{\infty} (V^\nu \times R^\nu)$.

2. We "enlarge" $\mathscr{U}$ by replacing it by the set of all functions $(\mathbf{u},\boldsymbol{\beta})$, where $\mathbf{u} \in \mathscr{U}^\nu$ and $\boldsymbol{\beta}$ is a constant function from I' into $\{\beta: \beta = (\beta^1, \ldots, \beta^\nu) \in \bar{R}^\nu_+, \sum_{i=1}^{\nu} \beta^i = 1\}$, ν being an arbitrary positive integer.

3. We redefine f as follows (denoting the "new" f by $\mathbf{f}$):

$$\mathbf{f}(\xi,v_1, \ldots, v_\nu,\beta^1, \ldots, \beta^\nu,j) = \sum_{i=1}^{\nu} \beta^i f(\xi,v_i,j), \; \xi \in \mathbf{G},$$

$$(v_1, \ldots, v_\nu,\beta^1, \ldots, \beta^\nu) \in \mathbf{V}, \quad j \in I'.$$

In this reformulated form, Problem (3) evidently satisfies Hypothesis (13). Suppose that $(\tilde{x}_0,u_{0,1}, \ldots, u_{0,\nu},\boldsymbol{\beta}_0^1, \ldots, \boldsymbol{\beta}_0^\nu)$ is a solution of this new problem, so that

$$\tilde{x}_0(j+1) = \sum_{i=1}^{\nu} \boldsymbol{\beta}_0^i f(\tilde{x}_0(j),u_{0,i}(j),j) \quad \text{for} \quad j = 1, \ldots, k-1.$$

By Hypothesis (54), there is a function $\tilde{u}_0 \in \mathscr{U}$ such that

$$\sum_{i=1}^{\nu} \boldsymbol{\beta}_0^i f(\tilde{x}_0(j),u_{0,i}(j),j) = f(\tilde{x}_0(j),\tilde{u}_0(j),j) \quad \text{for} \quad j = 1,\ldots,k-1,$$

so that

$$\tilde{x}_0(j+1) = f(\tilde{x}_0(j),\tilde{u}_0(j),j) \quad \text{for} \quad j = 1,\ldots,k-1.$$

From this it follows at once that $(\tilde{x}_0,\tilde{u}_0)$ is a solution of Problem (3)—in its original form. Hence, if (x_0,u_0) is any solution of Problem (3) (in its original form), then $(x_0,u_0,1)$ is a solution of Problem (3) in its "enlarged" form. Inasmuch as the enlarged problem satisfies Hypothesis (13), $(x_0,u_0,1)$ then satisfies the necessary conditions of Theorem (44)—modified for the enlarged problem data. But, when $\nu = 1$ (and $\boldsymbol{\beta} \equiv 1$), these modified conditions revert to their original form. Thus, we have proved the following:

56 THEOREM. *Let* (x_0,u_0) *be a solution of Problem* (3) *subject to Hypothesis* (54). *Then the conclusions of Theorem* (44) *hold.*

57 In many cases of interest, the "cost" function in Problem (3) is not expressed in terms of a function $\chi^0(x(1),\ldots,x(k))$, but rather is given in terms of a sum of the following kind:

$$\sum_{j=1}^{k-1} \hat{f}^0(x(j),u(j),j) + \chi^0(x(1),\ldots,x(k)),$$

where $\hat{f}^0$ is a scalar-valued function defined on $\mathbf{G} \times V \times I'$, satisfying a hypothesis analogous to Hypothesis (2). In this case, one may adjoin an additional phase coordinate x^0 [compare with (V.1.12)] to the phase vector x, constrained by the equation

$$x^0(j+1) = x^0(j) + \hat{f}^0(x(j),u(j),j), \quad j = 1,\ldots,k-1,$$
$$x^0(1) = 0,$$

and then reformulate the problem—much as was done in (V.1.12)—so that it takes the form of Problem (3).

Of course, in order to be able to appeal to Theorem (44) or to Theorem (56), in order to obtain necessary conditions that solutions of the just-described problem satisfy, the "augmented" function $(\hat{f}^0,f)$, together with $\mathscr{U}$, must satisfy Hypothesis (13) or Hypothesis (54). Even if $(\hat{f}^0,f)$ does not satisfy either Hypothesis (54) or (13), it is sometimes possible to reformulate the problem, by "enlarging" the set of admissible controls (without reducing the minimum cost) in such a way that one or the other of these hypotheses holds for the "enlarged" problem. For example, we may replace V by $V \times R$, $\mathscr{U}$ by $\mathscr{U} \times \{\theta : \theta \text{ is a function from } I' \text{ into } \bar{R}_+\}$, and redefine $\hat{f}^0$ and f as follows (the "new" $\hat{f}^0$ and f will be denoted by $\hat{\mathbf{f}}^0$ and $\mathbf{f}$, respectively):

$$\mathbf{f}(\xi,v,\rho,j) = f(\xi,v,j),$$
$$\hat{\mathbf{f}}^0(\xi,v,\rho,j) = \hat{f}^0(\xi,v,j) + \rho$$

for all $\xi \in \mathbf{G}$, $v \in V$, $\rho \in R$, and $j \in I'$. It is easily seen that (x_0,u_0) is a solution of the original problem if and only if $(x_0,u_0,0)$ is a solution of the reformulated problem. But it is clearly easier for $(\hat{\mathbf{f}}^0,\mathbf{f})$—with the "augmented" $\mathscr{U}$—to satisfy either Hypothesis (54) or (13) than it is for $(\hat{f}^0,f)$—with $\mathscr{U}$—to do so. For example, if the (original) set $\mathscr{U}$ satisfies Hypothesis (50), and if, in addition, the following hypothesis holds:

58 HYPOTHESIS. *For each $j \in I'$, each $\xi \in \mathbf{G}$, each $\beta \in [0,1]$, and each pair (v_1,v_2) in $(U_j)^2$, there is an element $v_3 \in U_j$ (possibly depending on ξ as well as on j, β, v_1, and v_2) such that*

59 $$\beta f(\xi,v_1,j) + (1-\beta)f(\xi,v_2,j) = f(\xi,v_3,j),$$

60 $$\beta\hat{f}^0(\xi,v_1,j) + (1-\beta)\hat{f}^0(\xi,v_2,j) \geq \hat{f}^0(\xi,v_3,j);$$

then $(\hat{\mathbf{f}}^0,\mathbf{f})$ and the augmented $\mathscr{U}$ satisfy Hypothesis (54), while $(\hat{f}^0,f)$ and $\mathscr{U}$ generally do not [unless (60) can be made to hold with an equality sign].

Assertions similar to those in the preceding paragraph can be made with regard to modifying the inequality constraints (5), (9) and the equality constraints (6), (10).

We now turn to a variant of the discrete optimal control problem. For this problem, we shall not impose the strong convexity assumption that we were forced to invoke for Problem (3). Instead, we shall make certain additional, relatively mild, assumptions concerning V, $\mathscr{U}$, and the dependence of f on its second argument. We shall then treat the optimal control problem as a parametric optimization problem—much as was done in the last part of Section 1 in Chapter VI—and shall obtain necessary conditions in the form of a local maximum principle, which is analogous to Theorem (VI.1.59).

We shall henceforth suppose that V is an open set in some Banach space $\mathscr{R}$. Here, Hypothesis (2) will be replaced by the following stronger hypothesis.

61 HYPOTHESIS. *For each $j \in I'$, the function $(\xi,v) \to f(\xi,v,j)$: $\mathbf{G} \times V \to R^n$ is continuously differentiable.*

We then consider the following problem.

62 *Problem.* Let there be given an open set $\mathbf{G}$ in R^n, an open set V in a Banach space $\mathscr{R}$, a positive integer k, a function $f:\mathbf{G} \times V \times I' \to R^n$ satisfying Hypothesis (61) (where I' denotes $\{1, 2, \ldots, k-1\}$), a class $\mathscr{U}$ of functions $u:I' \to V$ which is finitely open in itself, and a continuously differentiable function $\chi = (\chi^0, \chi^1, \ldots, \chi^{\mu+m})$: $(R^n)^k \times (V)^{k-1} \to R^{\mu+m+1}$. Then find functions $x_0:I \to \mathbf{G}$ (where I denotes $\{1, \ldots, k\}$) and $u_0 \in \mathscr{U}$ such that

63 $$x_0(j+1) = f(x_0(j),u_0(j),j) \quad \text{for} \quad j = 1, \ldots, k-1,$$

64 $$\chi^i(x_0(1), \ldots, x_0(k), u_0(1), \ldots, u_0(k-1)) \leq 0 \quad \text{for} \quad i = 1, \ldots, \mu,$$

65 $$\chi^i(x_0(1), \ldots, x_0(k), u_0(1), \ldots, u_0(k-1)) = 0$$
$$\text{for} \quad i = \mu+1, \ldots, \mu+m,$$

and such that

66 $$\chi^0(x_0(1), \ldots, x_0(k), u_0(1), \ldots, u_0(k-1)) \leq \chi^0(x(1), \ldots, x(k), u(1), \ldots, u(k-1))$$

for all pairs (x,u), where x is a function from I into $\mathbf{G}$ and u is a function in $\mathscr{U}$ such that the following relations hold:

67 $$x(j+1) = f(x(j),u(j),j) \quad \text{for} \quad j = 1, \ldots, k-1,$$

68 $$\chi^i(x(1), \ldots, x(k), u(1), \ldots, u(k-1)) \leq 0 \quad \text{for} \quad i = 1, \ldots, \mu,$$

69 $$\chi^i(x(1), \ldots, x(k), u(1), \ldots, u(k-1)) = 0$$
$$\text{for} \quad i = \mu+1, \ldots, \mu+m.$$

70 Problem (62) differs from Problem (3) in that V is now an open set in a Banach space (rather than an arbitrary set), in that we require f to be continuously differentiable jointly in both its first and second (rather than only its first) arguments, in that we require $\mathscr{U}$ to be finitely open in itself, and in that the constraint functionals χ^i are now allowed to depend on the control, as well as the phase, variable. Because of the last feature, Problem (62)

bears some resemblance to Problem (VI.3.10). Of course, some of the functions χ^i for $i > 0$ may be independent of their first k arguments, in which case the corresponding problem constraints represent inequality or equality constraints on the allowed values of the control variable.

The difference equation restriction (63), (67) may evidently be incorporated in the constraint (65), (69). However, we find it convenient to treat the former restriction separately because its special form gives rise to a particular form for the necessary conditions.

71 Let us define the Banach space $\mathscr{X}$ and its open subset A as described in (15). Let $\mathscr{P}_0$ denote the linear vector space of all functions $u: I' \to \mathscr{R}$. We define a norm on $\mathscr{P}_0$ through the relation

$$\|u\| = \max_{1 \le j \le k-1} \|u(j)\|.$$

It is easily seen that, with this norm, $\mathscr{P}_0$ becomes a Banach space. Let $\Pi_0 = \{u: u \in \mathscr{P}_0,\ u(j) \in V \text{ for each } j \in I'\}$. It is immediately seen that Π_0 is open in $\mathscr{P}_0$. Finally, let $\mathscr{P} = \mathscr{P}_0 \times R^n$ and $\Pi = \Pi_0 \times \mathbf{G}$, so that Π is an open set in the Banach space $\mathscr{P}$ [see (I.4.38)], and let $\tilde{A} = A \times \Pi$.

We now define the operator $\hat{T}_0: \tilde{A} \to \mathscr{X}$ as follows:

72 $$(\hat{T}_0(x,u,\xi))(j) = \begin{cases} \xi & \text{for } j = 1, \\ f(x(j-1), u(j-1), j-1) & \text{for } j = 2, \dots, k, \end{cases}$$
$$x \in A,\ u \in \Pi_0,\ \xi \in \mathbf{G}.$$

It easily follows from Hypothesis (61) that $\hat{T}_0$ is continuously differentiable (i.e., $\hat{T}_0 \in \hat{\mathscr{T}}$, the space of all continuously differentiable operators from $\tilde{A}$ into $\mathscr{X}$), and that the Fréchet differential of $\hat{T}_0$ at any $(x,u,\xi) \in \tilde{A}$ is given by the formula [see (I.7.51 and 60)]

73 $$(D\hat{T}_0(x,u,\xi;\ \delta x,\delta u,\delta\xi))(j)$$
$$= \begin{cases} \delta\xi \quad \text{for } j = 1 \\ \mathscr{D}_1 f(x(j-1), u(j-1), j-1)\delta x(j-1) \\ \quad + D_2 f(x(j-1), u(j-1), j-1;\ \delta u(j-1)) \\ \qquad\qquad \text{for } j = 2, \dots, k, \end{cases}$$

for all $\delta x \in \mathscr{X}$, $\delta u \in \mathscr{P}_0$, and $\delta\xi \in R^n$.

74 Let us denote $\mathscr{U} \times \mathbf{G}$ by Π_a, so that $\Pi_a \subset \Pi$ and Π_a is finitely open in itself [see (I.1.40) and (I.4.16)], and let $\tilde{\mathscr{W}} = \hat{\mathscr{W}} \times \Pi_a$, where $\hat{\mathscr{W}}$ consists of the single operator $\hat{T}_0$. It is immediately verifiable that Hypotheses (IV.2.10 and 11) hold and that $\tilde{\mathscr{W}} \subset \mathscr{S}_0$. Also [see (I.1.40)], $\tilde{\mathscr{W}}$ is evidently finitely open in itself.

Clearly, the difference equation (67) is equivalent to the operator equation

75 $$x = \hat{T}_0(x,u,x(1)).$$

Let us define real-valued functions $\hat{\varphi}^1, \ldots, \hat{\varphi}^m, \hat{\phi}^0, \ldots, \hat{\phi}^\mu$ on $\tilde{A}$ as follows:

76 $$\hat{\varphi}^i(x,u,\xi) = \chi^{\mu+i}(x(1),\ldots,x(k),u(1),\ldots,u(k-1)),$$
$$(x,u,\xi) \in \tilde{A}, \quad i = 1,\ldots,m,$$

77 $$\hat{\phi}^i(x,u,\xi) = \chi^i(x(1),\ldots,x(k),u(1),\ldots,u(k-1)),$$
$$(x,u,\xi) \in \tilde{A}, \quad i = 0,1,\ldots,\mu.$$

Further, let $\hat{\varphi} = (\hat{\varphi}^1, \ldots, \hat{\varphi}^m)$ and $\hat{\phi}_1 = (\hat{\phi}^1, \ldots, \hat{\phi}^\mu)$. We can now assert that Problem (62) is of the form of Problem (IV.2.4), with $\mathscr{X}$, A, $\mathscr{P}$, Π, $\tilde{A}$, $\tilde{\mathscr{T}}$, and $\tilde{\mathscr{W}}$ as described in (71) and (74), $\tilde{A}' = \tilde{A}$, $\mathscr{Z}_1 = R^\mu$, and $Z_1 = \bar{R}^\mu_-$.

Now suppose that (x_0,u_0) is a solution of Problem (62), so that (63)–(65) hold, and $x_0 = \hat{T}_0(x_0,u_0,x_0(1))$. Let us make use of Theorem (IV.2.34) in order to obtain necessary conditions that x_0 and u_0 must satisfy. As we have seen, Hypotheses (IV.2.10 and 11) hold, and $\tilde{\mathscr{W}}$ is finitely open in itself. Further, because of the continuous differentiability of χ, Hypothesis (IV.2.17)—with $\hat{\phi}'$ deleted—holds [see (IV.2.18)]. In fact, $\hat{\varphi}$ and $\hat{\phi}^i$ are Fréchet differentiable, with Fréchet partial differentials given by the formulas [for each $(x,u,\xi) \in \tilde{A}$]

78 $$D_1\hat{\varphi}(x,u,\xi;\delta x) = \sum_{j=1}^{k} \mathscr{D}_j\chi'(x(1),\ldots,x(k),u(1),\ldots,u(k-1))\delta x(j), \quad \delta x \in \mathscr{X},$$

79 $$D_2\hat{\varphi}(x,u,\xi;\delta u) = \sum_{j=1}^{k-1} D_{k+j}\chi'(x(1),\ldots,x(k),u(1),\ldots,u(k-1);\delta u(j)), \quad \delta u \in \mathscr{P}_0,$$

80 $$\mathscr{D}_3\hat{\varphi}(x,u,\xi) = 0,$$

81 $$D_1\hat{\phi}^i(x,u,\xi;\delta x) = \sum_{j=1}^{k} \mathscr{D}_j\chi^i(x(1),\ldots,x(k),u(1),\ldots,u(k-1))\delta x(j), \quad \delta x \in \mathscr{X}, \quad i = 0,\ldots,\mu,$$

82 $$D_2\hat{\phi}^i(x,u,\xi;\delta u) = \sum_{j=1}^{k-1} D_{k+j}\chi^i(x(1),\ldots,x(k),u(1),\ldots,u(k-1);\delta u(j)), \quad \delta u \in \mathscr{P}_0, \quad i = 0,\ldots,\mu,$$

83 $$\mathscr{D}_3\hat{\phi}^i(x,u,\xi) = 0, \quad i = 0,\ldots,\mu,$$

where $\chi' = (\chi^{\mu+1},\ldots,\chi^{\mu+m})$.

Hence, by Theorem (IV.2.34) and its Corollary (IV.2.47), using (73) and (78)–(83), we conclude that there exists a nonzero row vector $\alpha = (\alpha^0, \alpha^1, \ldots, \alpha^{\mu+m}) \in R^{\mu+m+1}$ such that

84 (i) $\alpha^i \leq 0$ for $i = 0,\ldots,\mu$,

85 (ii) $$\sum_{j=1}^{k} \alpha\mathscr{D}_j^0\chi\delta x_{\delta\xi,\delta u}(j) + \sum_{j=1}^{k-1} \alpha D_{k+j}^0\chi(\delta u(j)) \leq 0$$

for all $\delta\xi \in \mathbf{G} - x_0(1)$ and $\delta u \in \mathscr{U} - u_0$,

where, for each $\delta\xi \in R^n$ and $\delta u \in \mathscr{P}_0, \delta x_{\delta\xi,\delta u}$ denotes the column-vector-valued function in $\mathscr{X}$ defined by

86 $$\delta x_{\delta\xi,\delta u}(j) = \mathscr{D}_1 f(x_0(j-1), u_0(j-1), j-1)\delta x_{\delta\xi,\delta u}(j-1) + D_2 f(x_0(j-1), u_0(j-1), j-1; \delta u(j-1)), \quad j = 2,\ldots,k,$$

87 $$\delta x_{\delta\xi,\delta u}(1) = \delta\xi,$$

$\mathscr{D}_j^0\chi$ (for each j) denotes $\mathscr{D}_j\chi(x_0(1),\ldots,x_0(k),u_0(1),\ldots,u_0(k-1))$, and $D_{k+j}^0\chi(\cdot)$ (for each j) denotes

$$D_{k+j}\chi(x_0(1),\ldots,x_0(k),u_0(1),\ldots,u_0(k-1);\cdot),$$

and

88 (iii) $$\sum_{i=1}^{\mu} \alpha^i\chi^i(x_0(1),\ldots,x_0(k),u_0(1),\ldots,u_0(k-1)) = 0.$$

Proceeding as in the proof of Theorem (44), we can now prove the following theorem:

89 THEOREM. *Let* (x_0,u_0) *be a solution of Problem* (62). *Then there exist a nonzero row vector* $\alpha = (\alpha^0, \alpha^1, \ldots, \alpha^{\mu+m}) \in \bar{R}_-^{\mu+1} \times R^m$

and an n (row)-vector-valued function ψ defined on $\{0, 1, \ldots, k-1\}$ such that

(i) *ψ satisfies the linear inhomogeneous difference equation* (43),

(ii) *u_0 satisfies the inequality*

90 $$\sum_{j=1}^{k-1} (\psi(j)D_2 f(x_0(j),u_0(j),j) + \alpha D^0_{k+j}\chi)(u(j) - u_0(j)) \leq 0$$

$$\textit{for all} \quad u \in \mathscr{U},$$

(iii) *ψ satisfies the boundary conditions* (40) *and* (42), *and*

91 (iv) $\alpha^i \chi^i(x_0(1), \ldots, x_0(k), u_0(1), \ldots, u_0(k-1)) = 0$

$$\textit{for} \quad i = 1, \ldots, \mu,$$

where (for each j) $\mathscr{D}^0_j\chi$ denotes

$$\mathscr{D}_j\chi(x_0(1), \ldots, x_0(k), u_0(1), \ldots, u_0(k-1))$$

and $D^0_{k+j}\chi$ denotes

$$D_{k+j}\chi(x_0(1), \ldots, x_0(k), u_0(1), \ldots, u_0(k-1)).$$

92 The basic difference between Theorems (44) and (89) lies in the different forms—(41) and (90)—that the maximum conditions take. The "extra" term in (90) of course arises from the fact that, in Problems (62), we allow χ to depend on the control variable (as well as on the phase variable). If we suppose that χ is independent of its last $k - 1$ arguments (i.e., is independent of the control), so that $D^0_{k+j}\chi = 0$ for each j, then it is seen at once that (90) is essentially a "local" maximum principle, in contrast to (41), which is a "global" maximum principle. This weakening of the maximum principle is the price we pay for dispensing with our convexity hypothesis (13). The reader should find it interesting to compare our results here with the discussion in (VI.1.50 and 58). In fact, Theorem (89) may be viewed as a discrete "analog" of Theorem (VI.1.59).

93 Note that if $u_0 \in \operatorname{int} \mathscr{U}$, then (90) takes the form

94 $$\sum_{j=1}^{k-1} \{\psi(j)D_2 f(x_0(j),u_0(j),j) + \alpha D^0_{k+j}\chi\} = 0.$$

The observations in (45) regarding ψ carry over here, essentially without change, as do the usual remarks in (46). Further, more general "cost" functions may be handled as described in (57).

95 It is easy to see, as in the proof of Theorem (51), that if Hypothesis (50) holds for Problem (62) (in which case the sets U_j must be finitely open in themselves), then the maximum condition (90) in Theorem (89), which is of course in summed form, is equivalent to the following "pointwise" maximum condition:

96 $$(\psi(j)D_2f(x_0(j),u_0(j),j) + \alpha D^0_{k+j}\chi)(v - u_0(j)) \leq 0 \quad \text{for all } v \in U_j \text{ and each } j = 1, \ldots, k-1$$

[compare with (52)]. If $u_0(j) \in \operatorname{int} U_j$ for some j, then (96) implies that (for each j)

97 $$\psi(j)D_2f(x_0(j),u_0(j),j) + \alpha D^0_{k+j}\chi = 0.$$

As a sufficient condition for $\psi \not\equiv 0$ in Theorem (89), we introduce the following compatibility condition [compare with Condition (47)].

98 CONDITION. *There exist a vector* $(\xi_1, \ldots, \xi_k) \in (R^n)^k$ *and a function* $u_1 \in \mathscr{U}$ *such that*

99 $$\sum_{j=1}^{k} (\mathscr{D}^0_j\chi^i)\cdot\xi_j + \sum_{j=1}^{k-1} D^0_{k+j}\chi^i(u_1(j) - u_0(j)) \begin{cases} < 0 & \text{for } i = 0 \text{ and each } i = 1, \ldots, \mu \\ & \text{such that } \chi^i(x_0(1), \ldots, x_0(k), \\ & \quad u_0(1), \ldots, u_0(k-1)) = 0, \\ = 0 & \text{for } i = \mu + 1, \ldots, \mu + m, \end{cases}$$

and the m elements

$$(\mathscr{D}^0_1\chi^i, \ldots, \mathscr{D}^0_k\chi^i, D^0_{k+1}\chi^i, \ldots, D^0_{2k-1}\chi^i),\ i = \mu + 1, \ldots, \mu + m,$$

are linearly independent as elements of $(R^n)^k \times (\mathscr{R}^*)^{k-1}$ [*see* (I.5.4)].

100 LEMMA. *If* (x_0,u_0) *is a solution of Problem* (62) *such that* $u_0 \in$ int $\mathscr{U}$ *and such that Condition* (98) *holds, then at least one of the vectors* $\psi(1), \ldots, \psi(k-1)$ *in Theorem* (89) *does not vanish.*

Proof. We argue by contradiction, and thus suppose that $\psi(j) = 0$ for $j = 1, \ldots, k-1$. It follows from (40), (42), and (43) that then $\alpha\mathscr{D}^0_j\chi = 0$ for $j = 1, \ldots, k$. If $(\xi_1, \ldots, \xi_k) \in (R^n)^k$ and $u_1 \in \mathscr{U}$ are the vector and function whose existence is asserted in Condition (98), this implies, by virtue of (84), (90), (91), and (99), that $\alpha^i = 0$ for $i = 0, \ldots, \mu$. Hence, the numbers $\alpha^{\mu+1}, \ldots, \alpha^{\mu+m}$

are not all zero, $\sum_{i=\mu+1}^{\mu+m} \alpha^i \mathscr{D}_j^0 \chi^i = 0$ for $j = 1, \ldots, k$, and, by (90),

101 $$\sum_{i=\mu+1}^{\mu+m} \alpha^i \sum_{j=1}^{k-1} D_{k+j}^0 \chi^i(u(j) - u_0(j)) \leq 0 \quad \text{for all} \quad u \in \mathscr{U}.$$

Since $u_0 \in \operatorname{int} \mathscr{U}$, (101) implies that $\sum_{i=\mu+1}^{\mu+m} \alpha^i D_{k+j}^0 \chi^i = 0$ for each $j = 1, \ldots, k-1$, contradicting the last assertion in Condition (98). |||

For consistency with the problems considered in the first three sections of this chapter, as well as in the preceding two chapters, we have only considered the case where the range of the phase variable x is contained in R^n (for some positive integer n). However, it is not difficult to see that essentially all of the results of this section remain in force if we allow x to take on its values in some arbitrary Banach space, where **G** is a given open set in this space.

Volterra-Type Operators

THIS APPENDIX IS devoted to a study of Volterra-type operators which were introduced in (IV.3.3). In particular, we shall prove some existence, continuation, uniqueness, and continuous dependence theorems for equations defined in terms of such operators. These theorems were used in Chapters IV–VII. We shall also show that the operators in the classes $\mathscr{V}_1$ and $\mathscr{V}_2$ [see (IV.3.38 and 39)], which arise in optimal control problems with restrictions in the form of ordinary—or, more generally, functional—differential equations or of Volterra integral equations, are Volterra-type operators. Analogous assertions for operators defined in terms of parameters will also be proved. Finally, we shall obtain some representation theorems for solutions of linear ordinary differential equations, linear functional differential equations, and linear Volterra integral equations. These representation theorems were used in Chapters IV–VII in our derivations of necessary conditions.

Throughout this appendix, we shall use the notation and terminology introduced in Chapter IV, and the reader should review (IV.1.1 and 2), (IV.2.1–3), (IV.3.1–7), and (IV.4.1–5) in order to remind himself of them. Here, of course, $\mathscr{X} = \mathscr{C}^n(I)$, where $I = [t_1, t_2]$.

Let us begin by showing that the classes $\mathscr{V}_1$ and $\mathscr{V}_2$ defined in (IV.3.38 and 39) are included in $\mathscr{V}$.

1 THEOREM. *Let the operator* $T_{x^*,G}: A \to \mathscr{C}^n(I)$ *be defined by* (IV.3.16), *where* x^* *is some given function in* $\mathscr{C}^n(I)$ *and* G *is some given function in* $\mathscr{G}$ [*see* (IV.3.9)]. Then $T_{x^*,G} \in \mathscr{V}$, *and its Fréchet differential at any* $x \in A$ *is given by* (IV.3.17).

Proof. We first prove that $T_{x^*,G}$ is continuously differentiable, i.e., that $T_{x^*,G} \in \mathscr{T}$. For ease of notation, let us henceforth write T for $T_{x^*,G}$. In order to show that T is Fréchet differentiable at any $x \in A$, with differential given by (IV.3.17), it is clearly sufficient to prove that

2 $$\int_{t_1}^{t} \frac{G^i(x+\delta x, s) - G^i(x,s) - D_1 G^i(x,s;\delta x)}{\|\delta x\|}\, ds \xrightarrow[\|\delta x\| \to 0]{} 0$$

$$\text{for} \quad i = 1, \ldots, n,$$

and that the convergence in (2) is uniform with respect to $t \in I$. But, by Theorem (I.7.43), if $\|\delta x\|$ is sufficiently small, then

3 $$\left| \int_{t_1}^{t} \frac{G^i(x+\delta x, s) - G^i(x,s) - D_1 G^i(x,s;\delta x)}{\|\delta x\|}\, ds \right|$$
$$\leq \int_{t_1}^{t} \frac{|D_1 G^i(x+\theta\delta x, s;\delta x) - D_1 G^i(x,s;\delta x)|}{\|\delta x\|}\, ds$$
$$\leq \int_{t_1}^{t_2} \|D_1 G^i(x+\theta\delta x, s;\cdot) - D_1 G^i(x,s;\cdot)\|\, ds,$$
$$\text{for} \quad i = 1, \ldots, n \quad \text{and all} \quad t \in I,$$

where θ is some number in $[0,1]$ which may depend on δx, s, and i (as well as on x). But, by virtue of (IV.3.10 and 13) and the Lebesgue dominated convergence theorem, the last term in (3) tends to 0 as $\|\delta x\| \to 0$, which means that (2) holds, uniformly in t. A similar argument shows that T is continuously differentiable.

Let us now show that T is of Volterra type. It is evident that T is of fixed initial value and [by virtue of (IV.3.12)] also causal. To show that T is locally compact, we fix $x_1 \in A$. Let ζ and $\tilde{m}$ be as indicated in (IV.3.13). Then if $y = Tx$ for some $x \in A$ with $\|x - x_1\| \leq \zeta$, we have that

$$|y(t'') - y(t')| \leq |x^*(t'') - x^*(t')| + \int_{t'}^{t''} \tilde{m}(s)\, ds$$
$$\text{for all} \quad t', t'' \quad \text{with} \quad t_1 \leq t' < t'' \leq t_2,$$

from which it follows at once that the set of functions

4 $$\{Tx \colon x \in A, \|x - x_1\| \leq \zeta\}$$

is equicontinuous. Since $(Tx)(t_1) = x^*(t_1)$ for all $x \in A$, this implies that the set (4) is, as a subset of $\mathscr{C}^n(I)$, bounded, i.e., [see (IV.3.1)] belongs to $\mathscr{K}$. Thus, T is locally compact and of Volterra type.

It is easily seen that, for each $x \in A$, the function $(\delta x, s) \to D_1 G(x,s;\delta x)$: $\mathscr{C}^n(I) \times I \to R^n$ has properties analogous to

(IV.3.10–13) of G. We can then show as above that $DT(x;\cdot)$ is of Volterra type. Thus, $T \in \mathscr{V}$. |||

5 THEOREM. *Let the operator* $T_{x^*,F}: A \to \mathscr{C}^n(I)$ *be defined by* (IV.3.29), *where* x^* *is some given function in* $\mathscr{C}^n(I)$ *and* F *is some given function in* $\mathscr{F}$ [*see* (IV.3.22)]. *Then* $T_{x^*,F} \in \mathscr{V}$, *and its Fréchet differential at any* $x \in A$ *is given by* (IV.3.30).

Proof. The proof of Theorem (5) is almost the same as that of Theorem (1). For illustration, we shall carry out the proof that $T_{X^*,F}$ is locally compact. Thus, let $x_1 \in A$ be fixed but arbitrary. Since $\{x_1(s): s \in I\}$ is a compact subset of the open set $\mathbf{G}$ [see Theorem (I.3.26)], there is a number $\zeta > 0$ such that the set $\{\xi: \xi \in R^N, |\xi - x_1(s)| \leq \zeta$ for some $s \in I\}$, which we shall denote by $\mathbf{G}_c$, is contained in $\mathbf{G}$. It is easily seen that $\mathbf{G}_c$ is compact in R^n.

If $x \in A$ is such that $\|x - x_1\| \leq \zeta$, then clearly $x(s) \in \mathbf{G}_c$ for all $s \in I$. Thus, if y belongs to the set

6 $$\{T_{x^*,F}x: x \in A, \|x - x_1\| \leq \zeta\},$$

and $\tilde{m}$ is the function described in (IV.3.26), then

$$\begin{aligned} |y(t'') - y(t')| \leq\; & |x^*(t'') - x^*(t')| \\ & + \int_{t_1}^{t_2} \max_{\xi \in \mathbf{G}_c} |F(\xi,s,t'') - F(\xi,s,t')|\, ds \\ & + \int_{t'}^{t''} \tilde{m}(s)\, ds \quad \text{for all} \quad t', t'' \quad \text{such that} \quad t_1 \leq t' < t'' \leq t_2. \end{aligned}$$

Taking into account (IV.3.24 and 26) and Corollary (I.5.32), and appealing to the Lebesgue dominated convergence theorem, we conclude that the set of functions (6) is equicontinuous. Since $(T_{x^*,F}x)(t_1) = x^*(t_1)$ for all $x \in A$, this means that the set (6) is also a bounded subset of $\mathscr{C}^n(I)$, i.e., the set belongs to $\mathscr{K}$. |||

We now turn to the operators in $\hat{\mathscr{T}}$ in order to show that the sets $\hat{\mathscr{V}}_1$ and $\hat{\mathscr{V}}_2$ defined in (IV.4.35 and 36) are contained in $\hat{\mathscr{V}}$.

THEOREM. *Let the operator* $\hat{T}_{\hat{x}^*,\hat{G}}: \tilde{A} \to \mathscr{C}^n(I)$ *be defined by* (IV.4.15), *where* $\hat{x}^*$ *is some given function in* $\hat{\mathscr{C}}^n$ *and* $\hat{G}$ *is some given function in* $\hat{\mathscr{G}}$ [*see* (IV.4.7)]. *Then* $\hat{T}_{\hat{x}^*,\hat{G}} \in \hat{\mathscr{V}}$, *and its Fréchet partial differentials at any* $(x,\pi) \in \tilde{A}$ *are given by* (IV.4.16 *and* 17).

Proof. Arguing essentially as in the first part of the proof of Theorem (1), using (I.7.53), we can show that $\hat{T}_{\hat{x}^*,\hat{G}}$—which we shall simply denote by $\hat{T}$—is continuously differentiable, i.e., belongs to $\hat{\mathscr{T}}$, and that (IV.4.16 and 17) hold. Further, since the function $(x,s) \to \hat{G}(x,s,\pi)$ evidently belongs to $\mathscr{G}$ for each $\pi \in \Pi$, it follows at once from Theorem (1) that $\hat{T}^{\pi} \in \mathscr{V}$ for each $\pi \in \Pi$.

It only remains to show that $\hat{T}$ is locally compact [see (IV.4.1)]. Thus, let us arbitrarily fix $\pi_1 \in \Pi$ and the finite subset $\{\pi'_1, \ldots, \pi'_\rho\}$ of $\mathscr{P}$. We must show that there exists an $\varepsilon_1 > 0$ with the property that, for each $x_1 \in A$, there is a $\zeta_1 > 0$ such that the set

8 $$\Big\{\hat{T}(x,\pi)\colon x \in A, \pi \in \Pi, \pi = \pi_1 + \sum_{i=1}^{\rho} \lambda^i \pi'_i, 0 \le \lambda^i \le \varepsilon_1 \text{ for each } i = 1, \ldots, \rho, \|x - x_1\| \le \zeta_1\Big\}$$

is in $\mathscr{K}$. Let us fix $x_1 \in A$, and let ζ_1, ζ_2, and $\tilde{m}$ be as indicated in (IV.4.11). Let $\varepsilon_1 > 0$ be such that (i) $\|\sum_{i=1}^{\rho} \lambda^i \pi'_i\| \le \zeta_2$ and (ii) $(\pi_1 + \sum_1^{\rho} \lambda^i \pi'_i) \in \Pi$ whenever $0 \le \lambda^i \le \varepsilon_1$ for each $i = 1, \ldots, \rho$ [see Corollary (I.4.3)]. Note that ε_1 is independent of x_1, since ζ_2 is. It follows from Corollary (I.5.32) that, for each $\varepsilon > 0$, there is a $\delta > 0$ such that

9 $$|\hat{x}^*(t'',\pi) - \hat{x}^*(t',\pi)| < \frac{\varepsilon}{2}$$

$$\text{whenever} \quad t_1 \le t' < t'' \le t_2, t'' - t' < \delta, \quad \text{and}$$

$$\pi = \pi_1 + \sum_1^{\rho} \lambda^i \pi'_i \quad \text{with} \quad 0 \le \lambda^i \le \varepsilon_1 \quad \text{for each} \quad i.$$

Without loss of generality, we may suppose that, in addition,

10 $$\int_{t'}^{t''} \tilde{m}(s)\, ds < \frac{\varepsilon}{2} \quad \text{whenever} \quad t_1 \le t' < t'' \le t_2 \quad \text{and} \quad t'' - t' < \delta.$$

Making use of (IV.4.12 and 15), (9), and (10), we directly conclude that the functions of the set (8) are equicontinuous. Further, since $(\hat{T}(x,\pi))(t_1) = \hat{x}^*(t_1,\pi)$ for all $(x,\pi) \in \tilde{A}$, it follows from Theorem (I.3.26) that the set $\{y(t_1)\colon y \in \text{the set (8)}\}$ is bounded in R^n, which allows us to conclude that the set (8) is also bounded [as a subset of $\mathscr{C}^n(I)$], i.e., that this set belongs to $\mathscr{K}$. |||

In almost the same way as we proved Theorem (7), we can prove the following two theorems.

11 THEOREM. *Let the operator* $\hat{T}_{\hat{x}^*,\hat{F}}\colon \tilde{A} \to \mathscr{C}^n(I)$ *be defined by* (IV.4.30), *where* $\hat{x}^*$ *is some given function in* $\hat{\mathscr{C}}^n$ *and* $\hat{F}$ *is some given function in* $\hat{\mathscr{F}}_1$ [*see* (IV.4.23)]. *Then* $\hat{T}_{\hat{x}^*,\hat{F}} \in \hat{\mathscr{V}}$, *and its Fréchet partial differentials at any* $(x,\pi) \in \tilde{A}$ *are given by* (IV.4.31 *and* 32).

12 THEOREM. *Let there be given an open set* $\mathbf{G} \subset R^n$, *a compact interval* $I = [t_1,t_2]$, *a continuous function* $f:\mathbf{G} \times R^r \times I \to R^n$ *which is continuously differentiable with respect to its first and second arguments, and an open set* $A \subset \mathscr{C}^n(I)$ *such that* $x(t) \in \mathbf{G}$ *for all* $x \in A$ *and* $t \in I$. *For some fixed* $\xi \in R^n$, *let the operator* $\hat{T}_\xi : A \times L^r_\infty(I) \to \mathscr{C}^n(I)$ *be defined by* (VI.2.10). *Then* $\hat{T}_\xi$ *is continuously differentiable, and its Fréchet partial differentials at any* $(x_0,u_0) \in A \times L^r_\infty(I)$ *are given by* (VI.2.11 *and* 12).

Let us now turn to some properties which all Volterra-type operators share. We begin with a local existence theorem.

13 THEOREM. *Let* T *be a Volterra-type operator from* A *into* $\mathscr{C}^n(I)$ *such that* $x_1(t_1) = (Tx_1)(t_1)$ *for some* $x_1 \in A$. *Then the equation* $x = Tx$ *has at least one local solution* [*see* (IV.3.4)].

Proof. Let $\zeta > 0$ be such that the set $\mathbf{N} = \{x\colon x \in \mathscr{C}^n(I), \|x - x_1\| \leq \zeta\}$ has the properties that $\mathbf{N} \subset A$ and $T(\mathbf{N}) \in \mathscr{K}$ [see (IV.3.2)]. Let $\delta > 0$ be such that $t \in [t_1, t_1 + \delta]$ implies that $|x_1(t) - x_1(t_1)| < \zeta/2$ and that $|(Tx)(t) - (Tx)(t_1)| < \zeta/2$ for all $x \in \mathbf{N}$.

Let us denote $[t_1, t_1 + \delta]$ by $\tilde{I}$, the restriction of x_1 to $\tilde{I}$ by $\tilde{x}_1$, and $\{\tilde{x}\colon \tilde{x} \in \mathscr{C}^n(\tilde{I}), \tilde{x}(t_1) = x_1(t_1), \|\tilde{x} - \tilde{x}_1\| \leq \zeta\}$ by $\tilde{A}$. For each $\tilde{x} \in \tilde{A}$, let us define the function $\tilde{x}' \in \mathbf{N}$ as follows:

$$\tilde{x}'(t) = \begin{cases} \tilde{x}(t) & \text{for } t \in \tilde{I}, \\ x_1(t) + \tilde{x}(t_1 + \delta) - x_1(t + \delta) & \text{for } t_1 + \delta \leq t \leq t_2. \end{cases}$$

Further, define the operator $\tilde{T}:\tilde{A} \to \mathscr{C}^n(\tilde{I})$ by the relation $(\tilde{T}\tilde{x})(t) = (T\tilde{x}')(t)$, $t \in \tilde{I}$. Now $\tilde{T}$ is continuous because T is, and since also $T(\mathbf{N}) \in \mathscr{K}$, $\tilde{T}(\tilde{A})$ is an equicontinuous family of functions in $\tilde{A}$. Further, since T is causal, $(\tilde{T}\tilde{x})(t_1) = (T\tilde{x}')(t_1) = (Tx_1)(t_1) = x_1(t_1)$ for all $\tilde{x} \in \tilde{A}$, which implies that $\tilde{T}(\tilde{A})$ is bounded in $\mathscr{C}^n(\tilde{I})$, so that $\tilde{T}(\tilde{A})$ is conditionally compact [see (IV.3.1)]. Recalling the definition of δ, and taking into account that $(\tilde{T}\tilde{x})(t_1) = x_1(t_1)$ for all $\tilde{x} \in \tilde{A}$, we conclude that $\tilde{T}(\tilde{A}) \subset \tilde{A}$. Since $\tilde{A}$ is a closed, convex subset of $\mathscr{C}^n(\tilde{I})$, $\overline{\text{co}}\,(\tilde{T}(\tilde{A}))$ is a compact, convex subset of $\tilde{A}$ which $\tilde{T}$ maps into itself. By Theorem (I.5.38), there is an $\tilde{x}_0 \in \tilde{A}$ such that $\tilde{x}_0 = \tilde{T}\tilde{x}_0$. But this means that $\tilde{x}'_0(t) = (T\tilde{x}'_0)(t)$ for all $t \in \tilde{I} = [t_1, t_1 + \delta]$, i.e., that $\tilde{x}'_0$ is a local solution of the equation $x = Tx$. |||

14 COROLLARY. *Let* T *be a continuous, causal, locally compact operator from* A *into* $\mathscr{C}^n(I)$ *such that* $x_1(t_1) = (Tx_1)(t_1)$ *for some*

$x_1 \in A$. Then the equation $x = Tx$ has at least one local solution x_0 such that $x_0(t_1) = x_1(t_1)$.

Proof. Examining the proof of Theorem (13), one can easily convince oneself that Corollary (14) holds. |||

15 THEOREM. *If T is as in Theorem (13), then every local solution of the equation $x = Tx$ which is not a solution can be continued [see (IV.3.6)].*

Proof. Suppose that $x_0 \in A$ is such that $x_0(t) = (Tx_0)(t)$ for $t_1 \leq t \leq \hat{t}$, where $\hat{t} \in (t_1,t_2)$, but that x_0 does not satisfy the equation $x = Tx$ on any interval $[t_1,\tilde{t}]$ with $\tilde{t} > \hat{t}$. Let $\zeta > 0$ be such that the set $\mathbf{N} = \{x: x \in \mathscr{C}^n(I), \|x - x_0\| \leq \zeta\}$ has the properties that $\mathbf{N} \subset A$ and $T(\mathbf{N}) \in \mathscr{K}$. Let us denote $[\hat{t},t_2]$ by $\hat{I}$, the restriction to $\hat{I}$ of x_0 by $\hat{x}_0$, and the set $\{\hat{x}: \hat{x} \in \mathscr{C}^n(\hat{I}), \|\hat{x} - \hat{x}_0\| < \zeta\}$ by $\hat{A}$. Clearly, $\hat{A}$ is open in $\mathscr{C}^n(\hat{I})$ and $\hat{x}_0 \in \hat{A}$.

For each $\hat{x} \in \hat{A}$, let us define the function $\hat{x}' \in \mathbf{N} \subset A$ as follows:

$$\hat{x}'(t) = \begin{cases} x_0(t) + \hat{x}(\hat{t}) - x_0(\hat{t}) & \text{for} \quad t_1 \leq t \leq \hat{t}, \\ \hat{x}(t) \quad \text{for} \quad t \in \hat{I}. \end{cases}$$

Further, define the operator $\hat{T}: \hat{A} \to \mathscr{C}^n(\hat{I})$ by the relation $(\hat{T}\hat{x})(t) = (T\hat{x}')(t)$, $t \in \hat{I}$. Because T is continuous, causal, and locally compact, it easily follows that $\hat{T}$ also is. Appealing to Corollary (14)—but with t_1, I, A, x_1, and T replaced by $\hat{t}$, $\hat{I}$, $\hat{A}$, $\hat{x}_0$, and $\hat{T}$, respectively—we conclude that there is a function $\hat{x}_1 \in \hat{A}$ which is a local solution of the equation $\hat{x} = \hat{T}\hat{x}$ and satisfies $\hat{x}_1(\hat{t}) = x_0(\hat{t})$. But this implies that $\hat{x}_1' \in A$ is a continuation of x_0. |||

16 If $x_0 \in A$ is a local solution of the equation $x = Tx$ (where $T \in \mathscr{T}$) satisfying this equation on $[t_1,\hat{t}]$ for some $\hat{t} \in (t_1,t_2)$, but not satisfying it on $[t_1,\tilde{t}]$ for any $\tilde{t} \in (\hat{t},t_2]$, then we shall say that a function $\bar{x}:[t_1,\bar{t}) \to R^n$, where $\hat{t} < \bar{t} \leq t_2$, is a *maximal continuation* of x_0 if (i) $x_0(t) = \bar{x}(t)$ for all $t \in [t_1,\hat{t}]$, (ii) for each $t \in [t_1,\bar{t})$, there is a function $\bar{x}_t \in A$ such that $\bar{x}_t(s) = \bar{x}(s) = (T\bar{x}_t)(s)$ for all $s \in [t_1,t]$, and either (iiia) there is no function $x' \in A$ such that $x'(t) = \bar{x}(t)$ for all $t \in [t_1,\bar{t})$, or (iiib) $\bar{t} = t_2$ and $\bar{x}$ is the restriction to $[t_1,t_2)$ of a solution $x' \in A$ of the equation $x = Tx$.

17 If alternative (iiib) holds, then we shall also refer to x' as a maximal continuation of x_0. Any function $\bar{x}:[t_1,\bar{t}) \to R^n$—where $\bar{t} \in (t_1,t_2]$—satisfying conditions (ii) and (iiia) in (16) will be referred to as a maximal continuation of the equation $x = Tx$.

For convenience, we shall also refer to solutions of this equation as maximal continuations of the equation.

Note that a maximal continuation is always a continuous function.

18 THEOREM. *If T is as in Theorem* (13) *and $x_0 \in A$ is a local solution (but not a solution) of the equation $x = Tx$, then x_0 has at least one maximal continuation.*

Proof. Let x_0 be a local solution of $x = Tx$ satisfying this equation on $[t_1,\hat{t}]$ for some $\hat{t} \in (t_1,t_2)$, and suppose that there is no solution x_0' of this equation such that $x_0'(t) = x_0(t)$ for all $t \in [t_1,\hat{t}]$. Let **P** denote the set of all pairs $(\bar{x},\bar{t})$ where $\bar{t} \in (\hat{t},t_2]$ and $\bar{x}$ is a function from $[t_1,\bar{t})$ into R^n which satisfies conditions (i) and (ii) in (16). Define the following partial ordering on $\mathbf{P}$:$(\bar{x}_1,\bar{t}_1) \prec (\bar{x}_2,\bar{t}_2)$ if $\bar{t}_1 \leq \bar{t}_2$ and $\bar{x}_1(t) = \bar{x}_2(t)$ for all $t \in [t_1,\bar{t}_1)$. Our desired conclusion now follows at once from Theorem (15) and Zorn's Lemma [see Reference 4 of Chapter I, Theorem 7, p. 6].

If we suppose that $A = \mathscr{C}^n(I)$ and that the operator T is affine [see (I.2.16)], then we can considerably sharpen our preceding results, as is seen in the following theorem.

19 THEOREM. *If L is a linear Volterra-type operator from $\mathscr{C}^n(I)$ into itself, then, for every $\tilde{x} \in \mathscr{C}^n(I)$, the equation*

20 $$x = Lx + \tilde{x}$$

has exactly one solution in $\mathscr{C}^n(I)$, which is also the only maximal continuation of this equation. Further, there is a linear operator R:$\mathscr{C}^n(I) \to \mathscr{C}^n(I)$, also of Volterra type and depending only on L (and not on $\tilde{x}$), such that the solution x of Eq. (20) *is given by the formula*

21 $$x = \tilde{x} + R\tilde{x}.$$

Proof. Let us fix $\tilde{x} \in \mathscr{C}^n(I)$. We define the operator T:$\mathscr{C}^n(I) \to \mathscr{C}^n(I)$ as follows:

22 $$Tx = Lx + \tilde{x}.$$

Evidently, T is affine, and, since L is of Volterra type, T also is. By Theorem (13), Eq. (20) has at least one local solution, say x_0. By Theorem (18), x_0 has at least one maximal continuation, say $\bar{x}$, defined on $[t_1,\bar{t})$, for some $\bar{t} \in (t_1,t_2]$. Recall that $\bar{x}$ is continuous.

We shall first prove that $\bar{x}$ and $\bar{t}$ satisfy alternative (iiib) in (16). We shall argue by contradiction, and thus suppose that $\bar{x}$ and $\bar{t}$ satisfy alternative (iiia) in (16).

Let us denote $[t_1,\bar{t})$ by $\tilde{I}$, and let, for each $t \in \tilde{I}$, $\bar{x}_t$ denote the function in $\mathscr{C}^n(I)$ which satisfies the relations $\bar{x}_t(s) = \bar{x}(s)$ for $s \in [t_1,t]$ and $\bar{x}_t(s) = \bar{x}(t)$ for $s \in [t,t_2]$. Let us show that there is an increasing sequence $\{\tau_i\} \subset \tilde{I}$ such that $\tau_i \to \bar{t}$ and $|\bar{x}(\tau_i)| \to \infty$ as $i \to \infty$. Indeed, in the contrary case, the functions $\bar{x}_t$, for $t \in \tilde{I}$, are uniformly bounded, so that—by virtue of the local compactness and linearity of L—$\{T\bar{x}_t : t \in \tilde{I}\} \in \mathscr{K}$. But, as is easily seen, this implies—because $\bar{x}(s) = \bar{x}_t(s) = (T\bar{x}_t)(s)$ for all $s \in [t_1,t]$ and every $t \in \tilde{I}$—that $\bar{x}(t) \to \xi_0$ as $t \to \bar{t}^-$, for some $\xi_0 \in R^n$. Considering the function $x' \in \mathscr{C}^n(I)$ defined by $x'(t) = \bar{x}(t)$ for $t \in \tilde{I}$ and $x'(t) = \xi_0$ for $t \in [\bar{t},t_2]$, we obtain a contradiction to our hypothesis that alternative (iiia) holds, so that a sequence $\{\tau_i\} \subset \tilde{I}$ as previously described does exist.

Since L is linear and locally compact, L is also a compact operator, i.e., if A_0 is any bounded set in $\mathscr{C}^n(I)$, then $L(A_0) \in \mathscr{K}$. Thus, [see Theorem (I.5.31)] there is a $\delta \in (0, \bar{t} - t_1)$ such that

23 $$|(Lx)(t'') - (Lx)(t')| < \tfrac{1}{4}$$

$$\text{whenever} \quad \|x\| \le 1,\ t', t'' \in I, \quad |t'' - t'| < \delta.$$

Let $\beta = \max\{|\bar{x}(t)| : t_1 \le t \le \bar{t} - \delta/2\}$, let $\gamma = 8\|\bar{x}\| + 2\beta + 1$, and let $\hat{I} = \{t : t \in \tilde{I}, |\bar{x}(t)| = \gamma\}$. By what we have shown, $\hat{I}$ is not empty. Finally, let $\tilde{t} = \min\{t : t \in \hat{I}\}$.

Let $x_1 \in \mathscr{C}^n(I)$ be defined by $x_1(t) = \gamma^{-1}\bar{x}(t)$ for $t_1 \le t \le \tilde{t}$ and $x_1(t) = \gamma^{-1}\bar{x}(\tilde{t})$ for $\tilde{t} \le t \le t_2$. Clearly, $\|x_1\| = 1$, $|x_1(\bar{t} - \delta/2)| < 1/2$, $|x_1(\tilde{t})| = 1$, and $t_1 < \bar{t} - \delta/2 < \tilde{t} \le \bar{t} \le t_2$. Hence,

24 $$|x_1(\tilde{t}) - x_1(\bar{t} - \delta/2)| > \tfrac{1}{2},$$

and, by (23),

$$|(Lx_1)(\tilde{t}) - (Lx_1)(\bar{t} - \delta/2)| < \tfrac{1}{4}.$$

But it follows from the definition of $\bar{x}$ that

$$x_1(t) = (Lx_1)(t) + \gamma^{-1}\tilde{x}(t) \quad \text{for all} \quad t \in [t_1,\tilde{t}],$$

so that

$$\begin{aligned} |x_1(\tilde{t}) - x_1(\bar{t} - \delta/2)| &\le |(Lx_1)(\tilde{t}) - (Lx_1)(\bar{t} - \delta/2)| \\ &\quad + \gamma^{-1}|\tilde{x}(\tilde{t}) - \tilde{x}(\bar{t} - \delta/2)| \\ &< \frac{1}{4} + \frac{2\|\tilde{x}\|}{8\|\tilde{x}\| + 2\beta + 1} < \frac{1}{2}, \end{aligned}$$

contradicting (24). Hence, $\bar{x}$ and $\bar{t}$ satisfy alternative (iiib) in (16), and Eq. (20) has at least one solution, x'. To show that x' is the unique solution of Eq. (20), we first prove the following lemma.

25 LEMMA. *Let L be a linear Volterra-type operator from $\mathscr{C}^n(I)$ into $\mathscr{C}^n(I)$. Then $x \equiv 0$ is the only solution of the equation $x = Lx$.*

Proof. We first point out that, since L is linear and of fixed initial value, $(Lx)(t_1) = 0$ for all $x \in \mathscr{C}^n(I)$. We now argue by contradiction, and thus suppose that $Lx_1 = x_1$ for some $x_1 \in \mathscr{C}^n(I)$, where $x_1 \neq 0$. Let $I' = \{t: t \in I,\ x_1(s) = 0 \text{ for all } s \in [t_1,t]\}$. Clearly, $t_1 \in I'$ and $I' \neq I$. Let $t' = \max \{t: t \in I'\}$, and let $\delta \in (0, t_2 - t')$ be such that (23) holds. Replacing x_1 by $\zeta \cdot x_1$ with a suitable $\zeta > 0$, if necessary, we may suppose that $\max \{|x_1(t)|: t_1 \leq t \leq t' + \delta/2\} = 1$. Let $t'' \in (t', t' + \delta/2]$ be such that $|x_1(t'')| = 1$, so that

26 $$|x_1(t') - x_1(t'')| = 1, |t'' - t'| \leq \delta/2 < \delta,$$

and let $x_2 \in \mathscr{C}^n(I)$ be defined by $x_2(t) = x_1(t)$ for $t_1 \leq t \leq t' + \delta/2$ and $x_2(t) = x_1(t' + \delta/2)$ for $t' + \delta/2 \leq t \leq t_2$. Then $\|x_2\| = 1$ and $(Lx_2)(t) = (Lx_1)(t) = x_1(t)$ for all $t \in [t_1, t' + \delta/2]$, and (26) contradicts (23), completing the proof of the lemma.

But Lemma (25) and the linearity of L immediately imply that x' is the unique solution of Eq. (20).

It only remains to prove that there is a linear Volterra-type operator $R:\mathscr{C}^n(I) \to \mathscr{C}^n(I)$ (which is independent of $\tilde{x}$) such that

27 $$x' = \tilde{x} + R\tilde{x}.$$

If E denotes the identity operator on $\mathscr{C}^n(I)$, then we have shown that the continuous linear operator $(E - L):\mathscr{C}^n(I) \to \mathscr{C}^n(I)$ is one-to-one and onto. Hence [see Reference 4 of Chapter I, Theorem 2, p. 57], $(E - L)$ is a homeomorphism. Let us denote the linear continuous operator $[(E - L)^{-1} - E]$ by R, so that x' is a solution of Eq. (20) if and only if it satisfies (27). It only remains to show that R is of Volterra type. It is obvious that R is of fixed initial value (because L is). Since $R = L(E - L)^{-1}$ and L is locally compact, we at once conclude that R is locally compact. Finally, if we fix $t \in (t_1,t_2]$, and use the existence and uniqueness result already obtained in this proof—but with I replaced by $[t_1,t]$—we at once conclude that $(E - L)^{-1}$, and hence also R, are causal. |||

28 Corollary. *If L is a linear, Volterra-type operator from $\mathscr{C}^n(I)$ into itself, then $(E - L)$ is a linear homeomorphism of $\mathscr{C}^n(I)$ onto itself [E denotes the identity operator on $\mathscr{C}^n(I)$].*

Let us now turn to the question of uniqueness for solutions of the equation $x = Tx$ when T is not necessarily affine. Although one can prove quite general uniqueness theorems for Volterra-type operators, we shall content ourselves with showing that the operators in $\mathscr{V}_1$ and $\mathscr{V}_2$ have the uniqueness property [see (IV.3.7)].

29 Theorem. *Every operator T in $\mathscr{V}_1$ [see (IV.3.38)] has the uniqueness property, i.e., belongs to $\mathscr{T}^u$.*

Proof. We shall argue by contradiction, and thus suppose that there are functions $x_1, x_2 \in A$ and a number $\bar{t} \in (t_1,t_2]$ such that $x_j(t) = (Tx_j)(t)$ for all $t \in [t_1,\bar{t}]$ and $j = 1$, 2, and such that $x_1(\bar{t}) \neq x_2(\bar{t})$. Let $G \in \mathscr{G}$ and $x^* \in \mathscr{C}^n(I)$ be such that $T = T_{x^*,G}$ [see (IV.3.15)]. Let $\tilde{m} \in L_1(I)$ and $\zeta > 0$ be such that (i) $x \in A$ whenever $\|x - x_1\| \leq \zeta$ and (ii) $\|D_1 G(x,s;\cdot)\| \leq \tilde{m}(s)$ for all $s \in I$ and all $x \in A$ such that $\|x - x_1\| \leq \zeta$ [see (IV.3.13)]. Since $x_j(t_1) = (Tx_j)(t_1) = x^*(t_1)$ for $j = 1$ and 2, we may assume, without loss of generality, that $|x_2(t) - x_1(t)| \leq \zeta$ for all $t \in [t_1,\bar{t}]$. For each $t \in [t_1,\bar{t}]$, define the function $x_t \in A$ as follows:

$$x_t(s) = \begin{cases} x_2(s) & \text{for} \quad t_1 \leq s \leq t, \\ x_2(t) + x_1(s) - x_1(t) & \text{for} \quad t \leq s \leq t_2. \end{cases}$$

Note that $\|x_t - x_1\| = \max_{t_1 \leq s \leq t} |x_2(s) - x_1(s)| \leq \zeta$ for all $t \in [t_1,\bar{t}]$. Since G is causal, $G(x_2,t) = G(x_t,t)$ for all $t \in [t_1,\bar{t}]$. Appealing to Theorem (I.7.42), we obtain that

$$\begin{aligned} |x_2(t) - x_1(t)| &= |(Tx_2)(t) - (Tx_1)(t)| \\ &= \left| \int_{t_1}^t [G(x_2,s) - G(x_1,s)]\, ds \right| \\ &\leq \int_{t_1}^t |G(x_2,s) - G(x_1,s)|\, ds \\ &= \int_{t_1}^t |G(x_s,s) - G(x_1,s)|\, ds \\ &\leq \int_{t_1}^t \tilde{m}(s)[\max_{t_1 \leq \sigma \leq s} |x_2(\sigma) - x_1(\sigma)|]\, ds \\ &\qquad \text{for all} \quad t \in [t_1,\bar{t}]. \end{aligned}$$

Let the function $y \in \mathscr{C}(I)$ be defined by $y(t) = |x_2(t) - x_1(t)|$, $t \in I$.

Thus, for every $\varepsilon > 0$,

$$y(t) < \varepsilon + \int_{t_1}^{t} \tilde{m}(s)[\max_{t_1 \le \sigma \le s} y(\sigma)]\, ds, \quad t \in [t_1,\bar{t}].$$

For each $\varepsilon > 0$, let y_ε denote the function $t \to \varepsilon \exp\{\int_{t_1}^{t} \tilde{m}(s)\, ds\}$: $I \to R_+$. Evidently,

$$\begin{aligned} y_\varepsilon(t) &= \varepsilon + \int_{t_1}^{t} \tilde{m}(s)\, y_\varepsilon(s)\, ds \\ &= \varepsilon + \int_{t_1}^{t} \tilde{m}(s)[\max_{t_1 \le \sigma \le s} y_\varepsilon(\sigma)]\, d\sigma, \quad t \in I, \end{aligned}$$

so that

30 $$y_\varepsilon(t) - y(t) > \int_{t_1}^{t} \tilde{m}(s)\{[\max_{t_1 \le \sigma \le s} y_\varepsilon(\sigma)] - [\max_{t_1 \le \sigma \le s} y(\sigma)]\}\, ds$$

$$\text{for each} \quad \varepsilon > 0 \quad \text{and} \quad t \in [t_1,\bar{t}].$$

In particular, $y_\varepsilon(t_1) - y(t_1) > 0$ for all $\varepsilon > 0$. Let us show that $y_\varepsilon(t) - y(t) > 0$ for all $\varepsilon > 0$ and all $t \in [t_1,\bar{t}]$. Indeed, in the contrary case, for some $\varepsilon_1 > 0$ and some $\tilde{t} \in [t_1,\bar{t}]$, we would have that $y_{\varepsilon_1}(t) - y(t) > 0$ for all $t \in [t_1,\tilde{t})$ and that $y_{\varepsilon_1}(\tilde{t}) = y(\tilde{t})$, contradicting (30) with $\varepsilon = \varepsilon_1$ and $t = \tilde{t}$. Hence,

$$|x_2(t) - x_1(t)| = y(t) < y_\varepsilon(t) \le \varepsilon \exp\left\{\int_{t_1}^{\bar{t}} \tilde{m}(s)\, ds\right\}$$

$$\text{for all} \quad t \in [t_1,\bar{t}] \quad \text{and all} \quad \varepsilon > 0,$$

which implies that $x_1(t) = x_2(t)$ for all $t \in [t_1,\bar{t}]$. But this contradicts the relation $x_1(\bar{t}) \ne x_2(\bar{t})$. |||

31 THEOREM. *Every operator T in $\mathscr{V}_2$ [see (IV.3.39)] has the uniqueness property, i.e., belongs to $\mathscr{T}^u$.*

Proof. We again argue by contradiction, and thus suppose that $x_1, x_2 \in A$ and $\bar{t} \in (t_1,t_2]$ are such that $x_j(t) = (Tx_j)(t)$ for all $t \in [t_1,\bar{t}]$ and $j = 1, 2$, and such that $x_1(\bar{t}) \ne x_2(\bar{t})$. Let $F \in \mathscr{F}$ and $x^* \in \mathscr{C}^n(I)$ be such that $T = T_{x^*,F}$ [see (IV.3.28)]. Let $\zeta > 0$ be such that the compact set $\mathbf{G}_c = \{\xi: |\xi - x_1(t)| \le \zeta$ for some $t \in I\}$ is contained in the open set $\mathbf{G}$ in R^n which defines A [see (IV.3.22)]. Let $\tilde{m} \in L_1(I)$ be such that $|\mathscr{D}_1 F(\xi,s,t)| \le \tilde{m}(s)$ for all $\xi \in \mathbf{G}_c$ and $(s,t) \in I \times I$ [see (IV.3.26)]. Since $x_j(t_1) = (Tx_j)(t_1) = x^*(t_1)$ for $j = 1, 2$, we may assume, without loss of generality, that $|x_2(t) - x_1(t)| \le \zeta$ for all $t \in [t_1,\bar{t}]$. Appealing to Theorem

(I.7.42) and (I.6.3), we conclude that

$$\begin{aligned}|x_2(t) - x_1(t)| &= |(Tx_2)(t) - (Tx_1)(t)| \\ &\le \int_{t_1}^{t} |F(x_2(s),s,t) - F(x_1(s),s,t)|\, ds \\ &\le \int_{t_1}^{t} \tilde{m}(s)|x_2(s) - x_1(s)|\, ds \\ &\text{for all} \quad t \in [t_1,\bar{t}].\end{aligned}$$

If we define the functions y and y_ε (for each $\varepsilon > 0$) as in the proof of Theorem (29), we conclude that

$$y_\varepsilon(t) - y(t) > \int_{t_1}^{t} \tilde{m}(s)[y_\varepsilon(s) - y(s)]\, ds \quad \text{for all} \quad t \in [t_1,\bar{t}].$$

Arguing as in the last lines of the preceding proof, we can show that this implies that $x_1(\bar{t}) = x_2(\bar{t})$, contradicting our hypothesis. |||

We now turn to the question of the continuous dependence of a fixed point of a Volterra-type operator on the operator.

32 THEOREM. *Let $T_0 \in \mathscr{T}_1$ be a Volterra-type operator with the uniqueness property, and let x_0 denote the fixed point of T_0 in A. Then, for every regular subset* [*(see* (IV.3.43)] *$\mathscr{W}_1$ of $\mathscr{T}^u \cap \mathscr{V}$, and every $\varepsilon > 0$, there is a neighborhood N of 0 in $\mathscr{T}$* [*in the topology of pointwise convergence—see* (I.6.33)]*—possibly depending on $\mathscr{W}_1$ as well as on ε and on T_0—such that* (i) $(T_0 + N) \cap \mathscr{W}_1 \subset \mathscr{T}_1$ *and* (ii) $\|x - x_0\| < \varepsilon$ *for all $x \in A$ which are fixed points of some $T \in (T_0 + N) \cap \mathscr{W}_1$.*

In essence, Theorem (32) asserts that if $T_0 \in \mathscr{T}_1$ is of Volterra type and has the uniqueness property, and if $\mathscr{W}_1$ is a regular subset of $\mathscr{T}^u \cap \mathscr{V}$, then every $T \in \mathscr{W}_1$ which is sufficiently close to T_0 will be in $\mathscr{T}_1$, with its fixed point arbitrarily near the fixed point of T_0.

Proof. Let $\mathscr{W}_1$ be any regular subset of $\mathscr{T}^u \cap \mathscr{V}$, and let $\zeta > 0$ such that

33 $$\{Tx: x \in A, \|x - x_0\| \le \zeta, T \in \mathscr{W}_1\} \in \mathscr{K}.$$

Without loss of generality—since A is open—we shall also assume that $\{x: x \in \mathscr{C}^n(I), \|x - x_0\| \le \zeta\} \subset A$. Further, let $\{\tilde{x}_i\}$, $i = 1, 2, \ldots$, be a countable dense subset of A. (See Reference 8 of

Chapter I, Problem 2, p. 94, for a sketch of a proof that such a set exists.)

If $T \in \mathscr{W}_1$, and if the function $\bar{x}:[t_1,\bar{t}) \to R^n$ is a maximal continuation of the equation $x = Tx$ satisfying alternative (iiia) in (16) as well as the inequality $|\bar{x}(t_1) - x_0(t_1)| \leq \zeta$, then there must be a $\tilde{t} \in [t_1,\bar{t})$ such that $|\bar{x}(\tilde{t}) - x_0(\tilde{t})| = \zeta$. [This may be shown by arguing essentially as in the first part of the proof of Theorem (19).]

We now argue by contradiction. Supposing the theorem is false, we conclude, on the basis of the preceding paragraph as well as (16) and Theorems (13) and (18), that there exist a regular subset $\mathscr{W}_1$ of $\mathscr{T}^u \cap \mathscr{V}$, positive numbers ε and ζ such that $\varepsilon \leq \zeta$ and (33) holds, and sequences $\{x_i\} \subset A$, $\{T_i\} \subset \mathscr{W}_1$, and $\{\tau_i\} \subset I$ such that, for every $i = 1, 2, \ldots$,

34
$$\begin{gathered} \tau_i > t_1 \quad \text{and} \quad |x_i(t) - x_0(t)| < \varepsilon \quad \text{for all} \quad t \in [t_1,\tau_i), \\ |x_i(\tau_i) - x_0(\tau_i)| = \varepsilon, \\ x_i(t) = (T_i x_i)(t) \quad \text{for all} \quad t \in [t_1,\tau_i], \end{gathered}$$

35
$$\|T_i\tilde{x}_j - T_0\tilde{x}_j\| < \frac{\varepsilon}{i} \quad \text{for each} \quad j = 1, \ldots, i.$$

Now for each $i = 1, 2, \ldots$, define the function $x_i' \in \mathscr{C}^n(I)$ as follows:

36
$$x_i'(t) = \begin{cases} x_i(t) & \text{for} \quad t_1 \leq t \leq \tau_i, \\ x_0(t) + x_i(\tau_i) - x_0(\tau_i) & \text{for} \quad \tau_i \leq t \leq t_2. \end{cases}$$

Evidently, for each i, $\|x_i' - x_0\| = \varepsilon \leq \zeta$, so that $x_i' \in A$. Since each T_i is causal, we have

37
$$x_i'(t) = x_i(t) = (T_i x_i)(t) = (T_i x_i')(t) \quad \text{for all} \quad t \in [t_1,\tau_i] \quad \text{and every} \quad i = 1, 2, \ldots.$$

By (33), we can conclude that $\{T_i x_i' : i = 1, 2, \ldots\} \in \mathscr{K}$, from which it follows [see (36) and (37)] that $\{x_i' : i = 1, 2, \ldots\} \in \mathscr{K}$. By Theorem (I.4.43), there is a subsequence of the sequence $\{x_i'\}$ which converges to some element $\tilde{x} \in \mathscr{C}^n(I)$. Without loss of generality, we shall assume that $x_i' \to \tilde{x}$ as $i \to \infty$, i.e., that

38
$$\|x_i' - \tilde{x}\| \xrightarrow[i \to \infty]{} 0.$$

Evidently, $\|\tilde{x} - x_0\| \le \zeta$, so that $\tilde{x} \in A$. By the same token, we may assume that $\tau_i \to \tilde{\tau}$ as $i \to \infty$ for some $\tilde{\tau} \in I$. Since the T_i, $i = 0, 1, 2, \ldots$, are of fixed initial value, we have that $x_i(t_1) = (T_i x_i)(t_1) = (T_i \tilde{x}_j)(t_1)$ for each i and j, and it thus follows from (35) and (36) that $x_i'(t_1) \to x_0(t_1)$ as $i \to \infty$. Since the functions x_i', $i = 1, 2, \ldots$, are equicontinuous, this implies, by virtue of (34), that $\tilde{\tau} > t_1$.

It also follows from (35) that $T_i \tilde{x}_j \to T_0 \tilde{x}_j$ as $i \to \infty$ for each $j = 1, 2, \ldots$. Since the $\tilde{x}_j$ are dense in A and [see (IV.3.43)] the T_i are equicontinuous, this is easily seen to imply that $T_i x \to T_0 x$ as $i \to \infty$ for every $x \in A$, and, in particular, that

39 $$\|T_i \tilde{x} - T_0 \tilde{x}\| \xrightarrow[i \to \infty]{} 0.$$

Now, for each $i = 1, 2, \ldots$, and every $t \in [t_1, \tau_i]$, we have [see (37)]

40 $$\begin{aligned} |(T_0\tilde{x})(t) - \tilde{x}(t)| &\le |(T_0\tilde{x})(t) - (T_i\tilde{x})(t)| \\ &\quad + |(T_i\tilde{x})(t) - (T_i x_i)(t)| + |(T_i x_i)(t) - \tilde{x}(t)| \\ &\le \|T_0\tilde{x} - T_i\tilde{x}\| + \|T_i\tilde{x} - T_i x_i'\| + \|x_i' - \tilde{x}\|. \end{aligned}$$

It now follows from (38)–(40), by virtue of the equicontinuity of the T_i, that $\tilde{x}(t) = (T_0\tilde{x})(t)$ for all $t \in [t_1, \tilde{\tau})$. Since T_0 has the uniqueness property by hypothesis, this means that $x_0(t) = \tilde{x}(t)$ for all $t \in [t_1, \tilde{\tau}]$, which, because of (38), implies that $x_i'(\tilde{\tau}) \to x_0(\tilde{\tau})$ as $i \to \infty$. But, since $\tau_i \to \tilde{\tau}$ and the x_i' are equicontinuous, this contradicts (34) and (36). |||

We now turn to some special properties of the solutions of *linear* ordinary differential equations, linear Volterra integral equations, and linear functional differential equations.

Let us begin with the linear homogeneous ordinary differential equation

41 $$\dot{x}(t) = \mathbf{A}(t)x(t), \quad t_1 \le t \le t_2,$$

where $\mathbf{A}$ is an integrable $(n \times n)$-matrix-valued function defined on $I = [t_1, t_2]$. By a solution of (41), we mean an absolutely continuous function $x: I \to R^n$ which satisfies this equation for almost all $t \in I$. If $\tilde{t} \in I$ and $\xi \in R^n$ are fixed (but arbitrary), then x is a solution of (41) such that $x(\tilde{t}) = \xi$ if and only if

42 $$x(t) = \xi + \int_{\tilde{t}}^{t} \mathbf{A}(s)x(s)\, ds, \quad t \in I.$$

Let us suppose that $\tilde{t} \in (t_1, t_2)$. Denote $[t_1, \tilde{t}]$ by I' and $[\tilde{t}, t_2]$ by I''. It is easily seen that x satisfies (42) if and only if there are functions $x' \in \mathscr{C}^n(I')$ and $x'' \in \mathscr{C}^n(I'')$ such that

43 $$x'(t) = \xi - \int_{t_1}^{t} \mathbf{A}(\tilde{t} + t_1 - s)x'(s)\,ds, \quad t \in I',$$

44 $$x''(t) = \xi + \int_{\tilde{t}}^{t} \mathbf{A}(s)x''(s)\,ds, \quad t \in I'',$$

45 $$x(t) = \begin{cases} x'(\tilde{t} + t_1 - t) & \text{for} \quad t \in I', \\ x''(t) & \text{for} \quad t \in I''. \end{cases}$$

Since the operators $L' : \mathscr{C}^n(I') \to \mathscr{C}^n(I')$ and $L'' : \mathscr{C}^n(I'') \to \mathscr{C}^n(I'')$ defined by

$$(L'x')(t) = -\int_{t_1}^{t} \mathbf{A}(\tilde{t} + t_1 - s)x'(s)\,ds, \quad t \in I', \quad x' \in \mathscr{C}^n(I'),$$

$$(L''x'')(t) = \int_{\tilde{t}}^{t} \mathbf{A}(s)x''(s)\,ds, \quad t \in I'', \quad x'' \in \mathscr{C}^n(I''),$$

are evidently linear and in $\mathscr{V}'_2$ [see (IV.3.40 and 33)], and hence of Volterra type [see Theorem (5)], it follows from Theorem (19) that Eq. (42) has a unique solution (for each ξ and $\tilde{t}$). The preceding arguments may be applied in a simplified manner to arrive at the same conclusion if $\tilde{t} = t_1$ or t_2.

For each $i = 1, \ldots, n$, let x_i denote the solution of Eq. (41) that satisfies the initial condition $x_i(t_1) = e_i$, where e_i denotes the vector in R^n whose i-th coordinate is equal to one, and all of whose other coordinates are equal to zero. Let $\Phi(t)$ denote the $(n \times n)$-matrix-valued function defined on I whose i-th column coincides with x_i. Then evidently

46 $$\dot{\Phi}(t) = \mathbf{A}(t)\Phi(t) \quad \text{a.e. on} \quad I, \quad \Phi(t_1) = \text{the identity matrix}.$$

The matrix-valued function $\Phi(t)$ is called the fundamental matrix of solutions of Eq. (41).

47 THEOREM. *Let $\Phi(t)$ be the fundamental matrix of solutions of Eq.* (41), *i.e., let $\Phi(t)$ be the absolutely continuous $(n \times n)$-matrix-valued function defined on I which satisfies* (46). *Then the matrix $\Phi(t)$ is nonsingular for every $t \in I$, and Φ^{-1} is absolutely continuous and satisfies the equation*

48 $$\frac{d}{dt}(\Phi^{-1}(t)) = -\Phi^{-1}(t)\mathbf{A}(t) \quad \text{a.e. } on \quad I,$$
$$\Phi^{-1}(t_1) = \textit{the identity matrix}.$$

Also, for every $\tilde{t} \in I$ *and* $\xi \in R^n$, *there is a unique solution of Eq.* (41) *that satisfies the relation* $x(\tilde{t}) = \xi$, *which is given by the formula* $x(t) = \Phi(t)\Phi^{-1}(\tilde{t})\xi$.

Proof. The assertion concerning the existence and uniqueness of the solution of Eq. (41) follows from the preceding discussion, and the given formula for this solution may be verified by direct substitution. It only remains to prove that $\Phi(t)$ is nonsingular for every $t \in I$ and that (48) holds.

Suppose that $\Phi(\bar{t})$ is singular for some $\bar{t} \in I$. Then, for some nonzero $\xi_0 \in R^n$, we have $\Phi(\bar{t})\xi_0 = 0$. Thus, the function $\tilde{x}: I \to R^n$ defined by $\tilde{x}(t) = \Phi(t)\xi_0$ satisfies Eq. (41) as well as $\tilde{x}(\bar{t}) = 0$, and $\tilde{x}(t_1) = \xi_0 \neq 0$. But the function $\hat{x}: I \to R^n$ defined by $\hat{x}(t) = 0$ for all $t \in I$ also satisfies Eq. (41), and $\hat{x}(\bar{t}) = 0$, which contradicts the uniqueness result already obtained.

To verify that (48) holds, we simply differentiate the identity

$$\Phi(t)\Phi^{-1}(t) \equiv \text{the identity matrix},$$

and obtain

$$\begin{aligned} 0 &= \dot{\Phi}(t)\Phi^{-1}(t) + \Phi(t)\frac{d}{dt}(\Phi^{-1}(t)) \\ &= \mathbf{A}(t) + \Phi(t)\frac{d}{dt}(\Phi^{-1}(t)) \quad \text{a.e. on} \quad I, \end{aligned}$$

from which (48) follows at once. |||

49 COROLLARY. *If the absolutely continuous function* $x: I \to R^n$ *satsifies Eq.* (41) *and* $x(\tilde{t}) = 0$ *for some* $\tilde{t} \in I$, *then* $x(t) = 0$ *for all* $t \in I$.

Now consider the linear inhomogeneous ordinary differential equation

50 $$\dot{x}(t) = \mathbf{A}(t)x(t) + b(t), \quad t_1 \leq t \leq t_2,$$

where $\mathbf{A}$ is as before, and b is an integrable function from I into R^n.

51 THEOREM. *For every* $\tilde{t} \in I$ *and* $\xi \in R^n$, *there is a unique solution of Eq.* (50) *that satisfies the relation* $x(\tilde{t}) = \xi$ *which is given by the formula*

52 $$x(t) = \Phi(t)\left[\Phi^{-1}(\tilde{t})\xi + \int_{\tilde{t}}^{t} \Phi^{-1}(s)b(s)\,ds\right], \quad t_1 \leq t \leq t_2,$$

where Φ *is the fundamental matrix of solutions of Eq.* (41)—*see* (46).

Proof. Clearly, x is a solution of Eq. (50) satisfying $x(\tilde{t}) = \xi$ if and only if

$$x(t) = \int_{\tilde{t}}^{t} \mathbf{A}(s)x(s)\,ds + \int_{\tilde{t}}^{t} b(s)\,ds + \xi \quad \text{for} \quad t_1 \leq t \leq t_2.$$

The existence and uniqueness of such a solution now follow from Theorem (19), as in the proof of Theorem (47). Formula (52) may be verified by direct substitution. |||

Formula (52) is often referred to as the variations of parameters (or variations of constants) formula.

We now pass over to linear Volterra integral equations. Indeed, let us consider the integral equation [for a function $x \in L_\infty^n(I)$]

53 $$x(t) = x^*(t) + \int_{t_1}^{t} \mathbf{K}(s,t)x(s)\,ds, \quad t_1 \le t \le t_2,$$

where x^* is some given function in $L_\infty^n(I)$ and the kernel $\mathbf{K}$ is a given measurable $(n \times n)$-matrix-valued function defined on $I \times I$. We shall suppose that there is a function $\tilde{m} \in L_1(I)$ such that

54 $$|\mathbf{K}(s,t)| \le \tilde{m}(s) \quad \text{for all} \quad (s,t) \in I \times I.$$

Let us consider solving Eq. (53) by the method of successive approximations. Let us set $x_1 = x^*$ and define the functions x_i $(i > 1)$ inductively through the formula

55 $$x_{i+1}(t) = x^*(t) + \int_{t_1}^{t} \mathbf{K}(s,t)x_i(s)\,ds, \quad t_1 \le t \le t_2, \quad i = 1, 2, \ldots .$$

It is clear that $x_1 \in L_\infty^n(I)$, and, by induction, it easily follows that $x_i \in L_\infty^n(I)$ for each $i = 1, 2, \ldots$. We shall show that, for each $t \in I$, $\lim_{i\to\infty} x_i(t)$ exists, and that the limiting function is a solution of Eq. (53). We shall also give a representation theorem for solutions of this equation which is similar to Theorem (51).

We define iterated kernels $\mathbf{K}_i$, $i = 1, 2, \ldots$, inductively as follows [each $\mathbf{K}_i$ is an $(n \times n)$-matrix-valued function defined on $I \times I$]:

56 $$\mathbf{K}_1(s,t) = \mathbf{K}(s,t),$$

57 $$\mathbf{K}_{i+1}(s,t) = \int_{s}^{t} \mathbf{K}(\sigma,t)\mathbf{K}_i(s,\sigma)\,d\sigma, \quad i = 1, 2, \ldots .$$

Let us show that the integrals in (57) are all finite (it is clear, by an induction argument, that the integrands are measurable for each i). Indeed, we shall show that

58 $$|\mathbf{K}_i(s,t)| \le \frac{[\mathbf{k}(s,t)]^{i-1}}{(i-1)!}\,\tilde{m}(s) \quad \text{whenever} \quad t_1 \le s \le t \le t_2,$$
$$\text{for each} \quad i = 2, 3, \ldots,$$

where

59 $$\mathbf{k}(s,t) = \int_{s}^{t} \tilde{m}(\sigma)\,d\sigma \quad \text{for all} \quad (s,t) \in I \times I \quad \text{with} \quad s \le t.$$

To see this, observe that (for any s,t with $t_1 \leq s \leq t \leq t_2$)

$$\begin{aligned}|\mathbf{K}_2(s,t)| &= \left|\int_s^t \mathbf{K}(\sigma,t)\mathbf{K}(s,\sigma)\, d\sigma\right| \\ &\leq \int_s^t |\mathbf{K}(\sigma,t)|\, |\mathbf{K}(s,\sigma)|\, d\sigma \leq \left(\int_s^t \tilde{m}(\sigma)\, d\sigma\right) \tilde{m}(s) \\ &= \mathbf{k}(s,t)\tilde{m}(s).\end{aligned}$$

Proceeding by induction, we suppose that (58) holds for some fixed i. Then

$$\begin{aligned}|\mathbf{K}_{i+1}(s,t)| &\leq \int_s^t |\mathbf{K}(\sigma,t)|\, |\mathbf{K}_i(s,\sigma)|\, d\sigma \\ &\leq \frac{1}{(i-1)!}\int_s^t \tilde{m}(\sigma)[\mathbf{k}(s,\sigma)]^{i-1}\, d\sigma\, \tilde{m}(s) \\ &= \frac{1}{(i-1)!}\int_s^t \mathscr{D}_2\mathbf{k}(s,\sigma)[\mathbf{k}(s,\sigma)]^{i-1}\, d\sigma\, \tilde{m}(s) \\ &= \frac{[\mathbf{k}(s,t)]^i}{i!}\tilde{m}(s),\end{aligned}$$

as was to be shown.

It is clear that [see (55)]

$$x_{i+1}(t) - x_i(t) = \int_{t_1}^t \mathbf{K}(s,t)[x_i(s) - x_{i-1}(s)]\, ds, \quad t_1 \leq t \leq t_2,$$
$$i = 2, 3, \ldots .$$

Using an induction argument, it is straightforward to show that also

60 $$x_{i+1}(t) - x_i(t) = \int_{t_1}^t \mathbf{K}_i(s,t)x^*(s)\, ds, \quad t_1 \leq t \leq t_2, \quad i = 1, 2, \ldots,$$

so that, by (58), for every $t \in I$,

$$\begin{aligned}|x_{i+1}(t) - x_i(t)| &\leq \int_{t_1}^t |\mathbf{K}_i(s,t)|\, |x^*(s)|\, ds \\ &\leq \int_{t_1}^t \frac{[\mathbf{k}(s,t)]^{i-1}}{(i-1)!}\tilde{m}(s)|x^*(s)|\, ds\end{aligned}$$
$$\text{for i} = 2, 3, \ldots .$$

But, for $t_1 \leq s \leq t \leq t_2$,

61 $$0 \leq \mathbf{k}(s,t) = \int_s^t \tilde{m}(\sigma)\, d\sigma \leq \int_{t_1}^{t_2} \tilde{m}(\sigma)\, d\sigma.$$

Hence, denoting $\int_{t_1}^{t_2} \tilde{m}(\sigma)\, d\sigma$ by α_0 and $\int_{t_1}^{t_2} \tilde{m}(s)|x^*(s)|\, ds$ by α_1, we obtain that

$$|x_{i+1}(t) - x_i(t)| \leq \frac{(\alpha_0)^{i-1}}{(i-1)!}\alpha_1$$

$$\text{for all} \quad t \in I \quad \text{and} \quad i = 2, 3, \ldots .$$

Since the series

$$\sum_{i=1}^{\infty} \frac{(\alpha_0)^i}{i!}\alpha_1$$

evidently converges, this implies, by virtue of the Weierstrass **M**-test, that the series

$$\sum_{i=1}^{\infty} [x_{i+1}(t) - x_i(t)]$$

converges uniformly in I, i.e., that there is a function $\tilde{x}: I \to R^n$ such that

$$x_i(t) \xrightarrow[i \to \infty]{} \tilde{x}(t) \text{ uniformly in } I.$$

Since each x_i is measurable, $\tilde{x}$ also is, and it directly follows from the preceding [recall that $x_1 = x^* \in L_\infty^n(I)$] that the functions x_i are uniformly bounded, which implies that $\tilde{x} \in L_\infty^n(I)$. Passing to the limit in (55) as $i \to \infty$, and using the Lebesgue dominated convergence theorem, we conclude that

62 $$\tilde{x}(t) = x^*(t) + \int_{t_1}^{t} \mathbf{K}(s,t)\tilde{x}(s)\, ds, \quad t_1 \leq t \leq t_2,$$

so that $\tilde{x}$ is a solution of Eq. (53), satisfying this equation for *all* $t \in I$.

It follows from (58) and (61) that, for all s,t with $t_1 \leq s \leq t \leq t_2$, the series

$$\sum_{i=1}^{\infty} \mathbf{K}_i(s,t)$$

converges—say, to a function $\mathbf{K}^*(s,t)$. Using (58), (60), (61), and the Lebesgue dominated convergence theorem, we conclude, after a passage to the limit, that

63 $$\tilde{x}(t) = x^*(t) + \int_{t_1}^{t} \mathbf{K}^*(s,t)x^*(s)\, ds, \quad t_1 \leq t \leq t_2.$$

The $(n \times n)$-matrix-valued function $\mathbf{K}^*$ is commonly referred to as the *resolvent kernel* of $\mathbf{K}$.

Let us show that $\tilde{x}$ is the unique solution of Eq. (53). This is clearly equivalent to showing that the identically zero function is the only solution of the homogeneous equation

64 $$x(t) = \int_{t_1}^{t} \mathbf{K}(s,t)x(s)\,ds, \quad t \in I.$$

We argue by contradiction, and suppose that $x \in L_\infty^n(I)$ satisfies (64) for *all* $t \in I$, where $x \not\equiv 0$. Let $t' = \max\{t: t \in I, x(s) = 0$ a.e. on $[t_1,t]\}$. It is evidently impossible that $t' = t_2$. Let $t'' \in (t',t_2)$ be such that $\int_{t'}^{t''} \tilde{m}(s)\,ds \leq 1/2$. Let $\beta = \operatorname{ess\,sup}\{|x(t)|: t' \leq t \leq t''\}$, so that $\beta > 0$, and let $t''' \in [t',t'']$ be such that $|x(t''')| > \beta/2$. Then [see (54) and (64)]

$$\begin{aligned}\beta/2 < |x(t''')| &= \left|\int_{t_1}^{t'''} \mathbf{K}(s,t)x(s)\,ds\right| = \left|\int_{t'}^{t'''} \mathbf{K}(s,t)x(s)\,ds\right| \\ &\leq \int_{t'}^{t''} |\mathbf{K}(s,t)|\,|x(s)|\,ds \leq \int_{t'}^{t''} \tilde{m}(s)\beta\,ds \leq \beta/2,\end{aligned}$$

which is absurd.

Thus, we have proved the following theorem.

65 THEOREM. *Let there be given a function $x^* \in L_\infty^n(I)$ and a measurable $(n \times n)$-matrix-valued function $\mathbf{K}$ defined on $I \times I$. Suppose that there is a function $\tilde{m} \in L_1(I)$ such that $|\mathbf{K}(s,t)| \leq \tilde{m}(s)$ for all $(s,t) \in I \times I$. Then there is a unique function $x \in L_\infty^n(I)$ which satisfies the integral equation*

$$x(t) = x^*(t) + \int_{t_1}^{t} \mathbf{K}(s,t)x(s)\,ds$$

for all $t \in I$, which is given by the formula

$$x(t) = x^*(t) + \int_{t_1}^{t} \mathbf{K}^*(s,t)x^*(s)\,ds, \quad t \in I,$$

where the "resolvent kernel" $\mathbf{K}^*$ is given by the formula

$$\mathbf{K}^*(s,t) = \sum_{i=1}^{\infty} \mathbf{K}_i(s,t)$$

(this series converges for all s,t with $t_1 \leq s \leq t \leq t_2$), where $\mathbf{K}_1 = \mathbf{K}$ and

$$\mathbf{K}_{i+1}(s,t) = \int_{s}^{t} \mathbf{K}(\sigma,t)\mathbf{K}_i(s,\sigma)\,d\sigma \quad \textit{for all} \quad (s,t) \in I \times I$$
$$\textit{and each} \quad i = 1, 2, \ldots .$$

66 We close this appendix with an investigation of the linear functional differential equation

67 $$\dot{x}(t) = \int_{t_1}^{t} d_1\Lambda(s,t)x(s) + b(t), \quad t \in I,$$

where $x \in \mathscr{C}^n(I)$ is to be absolutely continuous, b is a given function in $L_1^n(I)$, and Λ is a given measurable $(n \times n)$-matrix-valued function defined on $I \times I$ satisfying $\Lambda(s,t) = 0$ for $s \geq t$ whose elements λ^{ij} are such that (i) for each $t \in I$, the function $s \to \lambda^{ij}(s,t)$ is in $NBV(I)$ and (ii) there is a function $\tilde{m} \in L_1(I)$ such that

$$TV\lambda^{ij}(\cdot,t) \leq \tilde{m}(t) \quad \text{for all} \quad t \in I$$
$$\text{and each} \quad i, j = 1, \ldots, n.$$

The symbol d_1 in (67) denotes that the first argument of Λ is to be taken as the independent variable for the Stieltjes integration.

For each fixed $t \in I$, let us consider the integral equation

68 $$\Gamma(s) + \int_{s}^{t} \Gamma(\zeta)\Lambda(s,\zeta)\, d\zeta = \text{the identity matrix}, \quad s \in [t_1,t],$$

for the $(n \times n)$-matrix-valued function Γ defined on $[t_1,t]$. After making some evident variable changes, we can show that, if we denote by $\mathbf{\Gamma}$ the matrix-valued function $s \to \Gamma^T(t - s)$ (where T denotes transpose), then each column of $\mathbf{\Gamma}$ satisfies an equation which is of the form of Eq. (53). Appealing to Theorem (65), we can then conclude that there is a unique solution of Eq. 68 in $L_\infty^{n^2}([t_1,t])$, for each $t \in I$. To emphasize that Γ depends on t, let us write $\Gamma(t; s)$ in place of $\Gamma(s)$, and, for convenience, let us define $\Gamma(t; s) = 0$ for $s \in (t,t_2)$ for every $t \in I$.

69 We point out that, because of our hypotheses on Λ, for each $t \in I$, the function $s \to \Gamma(t; s): I \to R^{n^2}$ is of bounded variation on I. This may easily be verified from (68).

We then have the following representation theorem for the solutions of Eq. (67).

70 THEOREM. *Let* Λ *be a function as described in* (66). *Then, for each* $\xi \in R^n$, *Eq.* (67) *has a unique solution satisfying the initial condition* $x(t_1) = \xi$ *which is given by the formula*

71 $$x(t) = \Gamma(t; t_1)\xi + \int_{t_1}^{t} \Gamma(t; s)b(s)\, ds, \quad t \in I,$$

where, for each $t \in I$, $\Gamma(t;\cdot)$ *is the unique* $(n \times n)$*-matrix-valued* *function defined on I which is of bounded variation and also satisfies the equations*

72 $$\Gamma(t;s) + \int_s^t \Gamma(t;\zeta)\Lambda(s,\zeta)\,d\zeta = \textit{the identity matrix}, \quad s \in [t_1,t],$$

73 $$\Gamma(t;s) = 0, \quad s \in (t,t_2].$$

Proof. The existence and uniqueness of the solution of Eq. (67) follow from Theorems (19) and (1), once we rewrite Eq. (67) as an integral equation for x. Let Γ be as described in the theorem statement. (The existence and uniqueness of such a Γ follow from the preceding discussion.)

Let x denote the solution of Eq. (67) satisfying $x(t_1) = \xi$, and let us denote the function $s \to \Gamma(t;s)x(s) \colon I \to R^n$ (for each $t \in I$) by $\tilde{\Gamma}(t;\cdot)$. Since $\Gamma(t;\cdot)$ is of bounded variation and x (as an absolutely continuous function) also is, the same holds for $\tilde{\Gamma}(t;\cdot)$. Evidently, for each $t \in I$,

74 $$\begin{aligned}\int_{t_1}^t d\tilde{\Gamma}(t;s) &= \tilde{\Gamma}(t;t) - \tilde{\Gamma}(t;t_1)\\ &= \Gamma(t;t)x(t) - \Gamma(t;t_1)x(t_1) = x(t) - \Gamma(t;t_1)\xi.\end{aligned}$$

But

$$\int_{t_1}^t d\tilde{\Gamma}(t;s) = \int_{t_1}^t d\Gamma(t;s)x(s) + \int_{t_1}^t \Gamma(t;s)\dot{x}(s)\,ds,$$

so that, by (67) and (74),

75 $$\begin{aligned}x(t) = \Gamma(t;t_1)\xi &+ \int_{t_1}^t d\Gamma(t;s)x(s)\\ &+ \int_{t_1}^t \Gamma(t;s)\int_{t_1}^s d_1\Lambda(\sigma,s)x(\sigma)\,ds\\ &+ \int_{t_1}^t \Gamma(t;s)b(s)\,ds, \quad t \in I.\end{aligned}$$

Since $\Lambda(\sigma,s) = 0$ for $\sigma \geq s$, the double integral in (75) may be rewritten as

$$\int_{t_1}^t \Gamma(t;s)\int_{t_1}^t d_1\Lambda(\sigma,s)x(\sigma)\,ds.$$

Applying the unsymmetric Fubini theorem [see the footnote on page 322] to this integral, we can rewrite it in the form

$$\int_{t_1}^t d\hat{\Gamma}(t;\sigma)x(\sigma),$$

where, for each $t \in I$, $\hat{\Gamma}(t;\cdot)$ denotes the function

$$\sigma \to \int_{t_1}^t \Gamma(t;s)\Lambda(\sigma,s)\,ds \colon I \to R^{n^2}.$$

Since $\Lambda(\sigma,s) = 0$ for $\sigma \geq s$, we have

76 $$\hat{\Gamma}(t;\sigma) = \int_{\sigma}^{t} \Gamma(t;s)\Lambda(\sigma,s)\,ds \quad \text{for} \quad t_1 \leq \sigma \leq t \leq t_2.$$

Thus,

77 $$x(t) = \Gamma(t;t_1)\xi + \int_{t_1}^{t} \Gamma(t;s)b(s)\,ds + \int_{t_1}^{t} d(\Gamma(t;s) + \hat{\Gamma}(t;s))x(s), \quad t \in I.$$

But, by (72) and (76), $\Gamma(t;s) + \hat{\Gamma}(t;s) =$ the identity matrix for $t_1 \leq s \leq t \leq t_2$, so that the last integral in (77) vanishes and (71) holds. |||

1. For Chapter I

Most of the material in Sections 1–6 of Chapter I can be found in the books listed at the end of the chapter, and many specific references are cited in the text itself.

The concept of a set finitely open in itself (I.1.39) was introduced in Neustadt [N.7]. The separation theorem (I.2.25) for convex functionals was proved in a somewhat more general form in Neustadt [N.5], based on an argument found in Halkin and Neustadt [HN.1]. An enormous literature exists on the subject of subgradients for convex functionals [see (1.2.29)], and they have been extensively used in the theory of optimization—for example, see Rockafellar [R.1, R.2, R.3], Dubovitskii and Milyutin [DuM.3], and Pshenichnyi [Ps.1, Ps.2, Ps.3]. The generalization of a convex function as described in (I.2.30 and 32) was apparently first made by Hurwicz [Hu.1].

Tangent cones, as defined in (I.4.25), have been used by many authors, for example, Hestenes [He.3]. Blum [Bl.1] defined a finitely continuous function much as was done in (I.5.26). Simplicial linearizations, as defined in (I.5.39), were introduced in Neustadt [N.5], where a proof of the separation lemma (I.5.43) was also presented.

Gâteaux [Ga.1] was the first to extend the idea of differentiation to functions whose domains lie in vector spaces which are not necessarily finite-dimensional. Indeed, the definition of Gâteaux differentiability given in (I.7.4) is essentially that of Gâteaux. The type of differentiability in normed spaces described in (I.7.37) is due to Fréchet [F.1]. Gâteaux directional differentials, defined much as in (I.7.1), were discussed, for example, by Demyanov and Rubinov [DR.2]. The fact that a convex functional is (Gâteaux) directionally differentiable [Lemma (I.7.7)] goes back to 1893 (see Stoltz [St.1]). The concept of finite differentiability [see (I.7.4)], i.e., differentiability in the "usual" sense when the domain of the function is restricted to finite-dimensional subspaces, was introduced and was shown to be of importance in optimization problems by Halkin and Neustadt [HN.1], Gamkrelidze and Kharatishvili [GK.1, GK.2], Warga [W.6, W.7, W.8], and Neustadt [N.5, N.7]. Finite W-semidifferentials and finite semidifferentials were defined in

Neustadt [N.5] much as in (I.7.13 and 16), and finite semidifferentials were defined in a slightly less general form in Halkin and Neustadt [HN.1].

Dubovitskii and Milyutin, in [DuM.3], introduced a differentiability criterion (which they termed "uniform differentiability") for functionals defined on a Banach space which is slightly stronger than the condition of dually semidifferentiability described in (I.7.26). In Neustadt [N.3, N.5], a differentiability condition for functionals defined on sets in linear topological spaces much like the dual differentiability condition of (I.7.23) was given, while in Neustadt [N.5] dual *W*-semidifferentiability was defined much as in (I.7.25), and results very similar to Lemma (I.7.30) and Theorems (I.7.32 and 44) on properties of dual differentials and dual *W*-semidifferentials were obtained.

The chain rule for Fréchet differentials and the generalized mean-value theorem of the differential calculus [Theorems (I.7.41 and 42)] may be found, for example, in the book by Dieudonné (Reference 3 of Chapter I).

2. For Chapter II

A vast literature exists for the basic optimization problems (II.1.1 and 16), particularly for the special case where $\mathscr{Y}$ is finite-dimensional, i.e., where $\mathscr{Y} = R^n$.

We first discuss the finite-dimensional case of $\mathscr{Y} = R^n$. If $\mathscr{E} = \mathscr{Y} = R^n$ and $\mu = 0$ (i.e., if there are no inequality constraints), and if the constraint functionals φ^i and ϕ^0 are continuously differentiable, then the multiplier rule, Theorem (II.1.4) [see also (II.1.15)], reduces to the classical Lagrange multiplier rule (where it is usually assumed that the $D\varphi^i(e_0;\cdot)$, $i = 1, \ldots, m$, are linearly independent, so that $\beta^0 \neq 0$ and can be set equal to -1)—for example, see Hestenes [He.3]. Strangely enough, optimization problems with nonlinear inequality constraints were not investigated until very recently, and the multiplier rule for inequality constraints only, Theorem (II.1.18) (under the assumption that $\mathscr{E} = \mathscr{Y} = R^n$ and that the ϕ^i are continuously differentiable), was not obtained until 1948, by F. John [J.1]. The case of both equality and inequality constraints had to wait until 1965, when Mangasarian and Fromovitz [MF.1] published the corresponding multiplier rule [Theorem (II.1.4) under the assumptions that $\mathscr{E} = \mathscr{Y} = R^n$ and that the φ^i and ϕ^i are continuously differentiable]. In his book, Mangasarian

[Ma.2] extended these results to the case where $\mathscr{E}$ is either open or convex in R^n and the ϕ^i are differentiable only at e_0. Kuhn and Tucker [KT.1], under an additional assumption (their so-called constraint qualification), strengthened the result of F. John by showing that $\beta^0 \neq 0$. Indeed, the question of finding sufficient conditions under which $\beta^0 \neq 0$ in Theorems (II.1.4 and 18) (which is of interest for sufficiency considerations and computational procedures) has drawn a great deal of attention, and many papers have been written on this subject (see, e.g., Mangasarian [Ma.2]).

The first multiplier rules in more general spaces were obtained in the 1930s by Goldstine [Go.1] and by Lyusternik [Ly.1] (also see the book by Lyusternik and Sobolev [LS.1]) who considered Problem (II.1.1) for the case when $\mathscr{E} = \mathscr{Y}$ is a Banach space, there are no inequality constraints (i.e., $\mu = 0$) and the φ^i and ϕ^0 are Fréchet differentiable. (Goldstine also considered problems with "infinite-dimensional" equality constraints.) Hurwicz [Hu. 1], in 1958, obtained a multiplier rule for the case when $\mathscr{E} = \mathscr{Y}$ is a Banach space and there are no equality constraints [i.e., Problem (II.1.16)]—in fact, he considered more general inequality constraints and "vector-valued" criteria of the type discussed in Chapter III—subject to Fréchet differentiability of the ϕ^i.

The fact that a multiplier rule for Problems (II.1.1 and 16) could be obtained under *semi*differentiability (rather than differentiability)-type hypotheses for the functionals ϕ^i was first observed in 1965 by Dubovitskii and Milyutin [DuM.3]. Indeed, they obtained necessary conditions (in a form which is somewhat different from—though closely related to—our multiplier rule, which they referred to as Euler equations) under the assumption that the ϕ^i are only dually semi-differentiable (in a slightly stronger sense than the one introduced in Chapter I). They confined themselves to the case where $\mathscr{E} = \mathscr{Y}$ is a Banach space, but allowed possibly infinite-dimensional equality constraints. In Neustadt [N.3] a multiplier rule for Problems (II.1.1 and 16) was obtained under the assumption that the ϕ^i are dually semi-differentiable (in the same strong sense), that the φ^i are dually differentiable and continuous, that $\mathscr{Y}$ is a locally convex linear topological space, and that $\mathscr{E}$ is finitely open in itself. Along similar lines is a result of Halkin [H.2]. Finally, in Halkin and Neustadt [HN.1], a general multiplier rule, which essentially includes Theorems (II.1.4 and 18) as special cases, was obtained. A somewhat less general multiplier rule than that of Halkin and Neustadt [HN.1] was obtained earlier by Halkin [H.3]. For related results see [H.4] and the recent paper [H.5].

Hestenes [He.3] has a general multiplier rule which, although it is not stated in terms of differentials of the constraint functionals, subsumes some of the multiplier rules described previously, as well as the multiplier rules Theorems (II.1.4 and 18), if in these we strengthen the requirement on the ϕ^i from semidifferentiability to (finite) differentiability.

Comments on the material of Section 2 will be made later in a more general context. However, maximum principles similar to Theorems (II.2.14 and 36), under a convexity hypothesis much like those in Hypotheses (II.2.9 and 19), were obtained by Warga [W.2, W.7]. For more details of the discussion in (II.2.48) see Warga [W.10] and Young [Y.2].

Convex problems (in the sense of (II.3.1 or 26)), for the case of no equality constraints [i.e., Problem (II.1.16)], $\mathscr{Y} = R^n$, and $\mathscr{E} = \bar{R}^n_+$, were first investigated by Kuhn and Tucker [KT.1] and by Slater [S.1]. They obtained results analogous to the necessary and sufficient conditions of Theorem (II.3.32) and were the first to notice that these conditions could be put in the form of the saddle-point condition (II.3.35 and 36). Slater made use of Condition (II.3.27), i.e., he assumed that the problem is "well-posed." In place of this, Kuhn and Tucker imposed their so-called constraint qualification. Both Slater and Kuhn and Tucker imposed unnecessary requirements on the ϕ^i: the latter required differentiability, while the former only required continuity. Karlin [K.1] derived the more general necessary conditions of Theorem (II.3.28), while Uzawa [U.1] was apparently the first to notice that the necessary conditions with $\beta^0 \neq 0$ are sufficient, whether or not the problem is convex [see Theorem (II.3.30)]. The sufficiency theorem in differential form, Theorem (II.3.41), for the special case of $\mathscr{Y} = R^n$ and $\mathscr{E} = \bar{R}^n_+$, and where the ϕ^i are differentiable (in the usual sense), is due to Kuhn and Tucker [KT.1].

Turning now to convex problems with (affine) equality constraints, the necessary and sufficient conditions in saddle-point form of Theorem (II.3.14) were obtained by Uzawa [U.1], for the special case of $\mathscr{Y} = R^n$ and $\mathscr{E} = \bar{R}^n_+$, under a hypothesis slightly stronger than (II.3.2). The sufficiency conditions in differential form of Theorem (II.3.25), in the special case of $\mathscr{Y} = \mathscr{E} = R^n$ and with the ϕ^i differentiable (in the ordinary sense), were given by Canon, Cullum, and Polak [CCP.1]. The fact that the sufficiency theorem in differential form holds for an arbitrary set $\mathscr{E}$ (even if it is not convex), which is important for applications to optimal control, was first pointed out in Neustadt [N.8].

In Neustadt [N.3], Theorem (II.3.28)—under some mild additional hypotheses—was proved for the case where $\mathscr{Y}$ is a locally convex linear topological space. Essentially all of the necessary conditions and sufficient conditions of Section 3 of Chapter II were obtained in Neustadt [N.8], and the sufficiency theorem in differential form [i.e., Theorem (II.3.25)] was applied much as in Section 4 of Chapter II to obtain sufficient conditions for an optimal control problem very similar to Problem (II.4.14) which are akin to those in Theorem (II.4.32). Earlier, analogous methods had been used by Pshenichnyi [Ps.2] to obtain analogous sufficient conditions for a slightly different optimal control problem. The first sufficiency results of the type of Theorem (II.4.32) were obtained by Rozonoér [Roz.1] and Lee [L.1].

3. General Comments for Chapters II–V

The optimal control problems discussed in Chapter V were originally investigated from a viewpoint quite different from the one taken in this book. The first mathematical result is due to Bushaw [B.1, B.3], who investigated a linear, second-order, time-optimal control problem using geometric arguments in the plane. This was followed in the 1950s by the pioneering work of the Soviet group of mathematicians led by L. S. Pontryagin (see Boltyanskii [Bo.1]; Pontryagin [Po.1]; Boltyanskii, Gamkrelidze, and Pontryagin [BGP.1]; and Pontryagin, Boltyanskii, Gamkrelidze and Mishchenko [PBGM.1]) which led to the celebrated Pontryagin maximum principle [essentially Theorem (V.2.44)]. The methods of proof used by this group were adaptations of methods used for the calculus of variations. In particular, a crucial part of their argument (which is due to Boltyanskii) is based on the construction of a convex set of variations (referred to as the "cone of attainability"), and on the existence of a supporting plane to this cone. This type of proof was first used in 1939 by McShane [Mc.1, Mc.2] to obtain the Weierstrass E-condition (of which the Pontryagin maximum principle is a generalization) for the so-called abnormal problem in the calculus of variations. Thus, it is perhaps fair to say that it was McShane who first noted the importance of convexity for variational problems, and that Pontryagin and his coworkers first exploited convexity for optimal control problems. It should also be pointed out that in 1958 Bellman,

Glicksberg, and Gross [BGG.1] independently obtained necessary conditions for a linear time-optimal control problem based on the fact that there is a supporting plane to a convex set at each of its boundary points.

A different proof of the Pontryagin maximum principle was given by Halkin in 1964 [H.1]. He also made extensive use of convexity, and was the first to appeal to the Brouwer fixed-point theorem, which has been used in almost all subsequent proofs of necessary conditions for optimization problems. Halkin's work was followed by the fundamental paper of Gamkrelidze [G.4], in which the idea of "almost" convexity was first introduced in a precise manner (under the term "quasiconvexity"). More precisely, he noted that, in order to obtain the Pontryagin maximum principle, it was necessary that the set of all admissible "right-hand sides" of the differential equation (V.1.2)—i.e., the set of all functions $(x,t) \rightarrow f(x,u(t),t)$, as u ranges over the set of all admissible controls—be quasiconvex [much as is described in (III.2.9)]. He showed that quasiconvexity was a consequence of convexity under switching [see (IV.5.14)] on the basis of a lemma which, in essence, coincides with Theorem (IV.5.17) and Lemma (IV.5.3). His necessary conditions were in the form of a maximum principle in integral form [see (V.2.29)] [a stronger result than the pointwise maximum principle (V.2.45)], which had previously been obtained (without specific mention) only by Halkin [H.1].

During the 1960s a great deal of research was carried out with the goal of formulating optimal control problems as abstract constrained optimization problems of the type discussed in Section 1 of Chapter III. (Actually, the multiplier rule in Banach spaces developed by Goldstine in 1938 [Go.1] was motivated by, and applied to, the Bolza problem in the calculus of variations.) The principal questions which were posed were the following: *1)* What should be the underlying set of objects over which the optimization is to be carried out, and in what kind of a space should these objects be embedded? *2)* How should the differential equation constraint and the associated restricted class of controls be taken into account? *3)* How should the various constraints on the allowed initial, intermediate, and final values of the phase trajectory, as well as the functional to be minimized, be formulated?

The first results of this type were obtained by Dubovitskii and Milyutin in 1963–65 [DuM.1, DuM.2, DuM.3]. They considered the underlying set of objects to be the set of all pairs (x,u), where u is an admissible control function (in the sense of Section 1 of Chapter V) and

x is a corresponding phase trajectory [i.e., solution of the corresponding differential equation—see Eq. (V.1.2)]. They first confined themselves to the case where the set of all admissible controls, $\mathscr{U}$, coincided with $L_\infty^r(I)$, and thus embedded their underlying set in $\mathscr{C}^n \times L_\infty^r$. The differential equation was considered as an (infinite-dimensional) equality constraint, and the initial and final value constraints in x were considered to be finite-dimensional equality constraints in the evident manner. Phase inequality constraints of the form $\tilde{\chi}(x(t),t) \leq 0$ for all $t \in I$ were replaced by the equivalent *scalar* constraint

1 $$\phi(x) \leq 0$$

where the functional ϕ is defined by

2 $$\phi(x) = \max_{t \in I} \tilde{\chi}(x(t),t).$$

Dubovitskii and Milyutin made the important observation that, although the functional ϕ defined by (2) is not differentiable in the ordinary (e.g., Fréchet) sense, ϕ is dually *semi*differentiable, as a result of which the set of x near x_0 which satisfy (1) can be approximated to first order by a convex cone with a nonempty interior [and defined in terms of $D_1\tilde{\chi}(x_0(t),t)$]. Their abstract necessary condition, which is closely related to a multiplier rule, asserts the existence of linear functionals which (a) separate the origin from certain convex cones which arise from the various problem equality and inequality constraints (which are assumed to be finite in number) as well as from the functional being minimized, and (b) sum to zero. Using this abstract multiplier rule and some properties of subgradients of convex functionals, they obtained necessary conditions for the optimal control problem described above. Minimax problems of the form of Problem (V.4.1) were similarly treated. By means of a complicated device which involved changing the independent variable of the ordinary differential equation, Dubovitskii and Milyutin were also able to treat optimal control problems with constraints on the allowed value of the control variable. A detailed treatment of this approach to optimal control problems can be found in the book of Girsanov [Gi.1].

Hestenes, in [He.2, He.3], developed an abstract (or, in his terminology, "generalized") multiplier rule, not on the basis of differentiability properties of the constraint functionals, but rather on the basis of the range of the constraint functionals (which he assumed to be finite in number) and on the existence of a so-called (convex) derived set for these functionals which approximate this range near the solution point. (As an underlying set, Hestenes took the set of all trajectories x that correspond to admissible controls.) His abstract multiplier rule is applicable to differentiable—but *not semi*differentiable—functionals, as a result of which he was able to obtain necessary conditions for optimal control problems in the absence of phase inequality constraints.

Yet another approach was taken by Neustadt in [N.1, N.2, N.3, N.4]. In these papers, the underlying set was taken to be the set of all admissible trajectories x, i.e., trajectories that correspond to admissible controls, and this set was embedded into $\mathscr{C}^n$. The differential equation constraint was thus automatically taken into account. Phase inequality constraints were treated by means of the device described in (1) and (2), and initial and final value constraints on the trajectory were formulated in the obvious manner. The result was an optimization problem with a set constraint (that the trajectories all had to satisfy to be admissible) and with a finite number of scalar equality and inequality constraints. On the basis of an abstract multiplier rule—which required dual differentiability for the equality constraint functionals and dual *semi*-differentiability for the inequality constraint and the "cost" functionals, and required that the underlying (admissible trajectory) set have a "first-order convex approximation" near the optimum trajectory—the necessary conditions of Sections 2–6 of Chapter V were essentially obtained. In order to ensure that the set of admissible trajectories has the required "first-order convex approximation," it was necessary to assume that the set of all admissible "right-hand sides" of the differential equation (V.1.2) has the quasiconvexity property introduced by Gamkrelidze in [G.4] and discussed previously.

Further results on abstract optimization problems were obtained by Halkin [H.3] and Halkin and Neustadt [HN.1], as described in the notes for Chapter II.

For a comparison of the results of Dubovitskii and Milyutin on the one hand and those of Halkin and of Neustadt on the other, see Halkin [H.7].

An alternate choice for the "underlying set" in the abstract optimization problem was first suggested by Halkin in [H.2]. Namely, he proposed the choice of the collection of all admissible right-hand sides of the differential equation (V.1.2). This idea was extended by

Neustadt in [N.6]. Gamkrelidze and Kharatishvili [GK.1, GK.2] and Gamkrelidze [G.5] also investigated optimal control problems by considering the set of admissible right-hand sides and their properties, but formulated their abstract problem differently using (what they referred to as) quasiconvex filters and critical points of mappings.

Whenever the set of admissible right-hand sides was used as an underlying set, it was necessary to invoke a hypothesis of "almost convexity" made akin to Hypothesis (III.2.14C) on the underlying set in the abstract optimization problem. This required an embedding of the underlying set in some suitable linear topological space and, whereas the choice of the space itself was quite natural and straightforward, no natural choice for the topology seemed to present itself. However, as was pointed out in Neustadt [N.7], when the set of admissible right-hand sides is replaced by the set of associated admissible integral operators [see (V.2.9 and 10)], with which it is algebraically isomorphic, then the proper topology is precisely the very natural topology of pointwise convergence [see (I.6.33)]. Further with this viewpoint, generalizations to equations other than ordinary differential equations (as described in Chapter VII) can very easily be carried out.

In order that it be possible to derive a multiplier rule as a necessary condition for an abstract optimization problem, there is a very delicate balance between the continuity and differentiability properties which one must require for the constraint functions and the way in which one can permit the underlying set to be approximated by a convex set (or a set finitely open in itself). One aspect of this is discussed in (III.2.9 and 10) and (IV.1.20), culminating in Hypotheses (III.2.14) and (IV.1.23). In Neustadt [N.5], a very general joint hypothesis on the set and on the constraint functions, which seems to include most combinations of hypotheses of interest, was discussed. Earlier results of this type can be found in Halkin [H.3] and Halkin and Neustadt [HN.1].

Let us return briefly to the phase inequality constraints. The device described by (1) and (2)—coupled with the observations that semi-differentiability is a sufficient requirement for the inequality constraint functional and that subgradients of convex functionals have certain "nice" properties—was used, as has already been noted, first by Dubovitskii and Milyutin, and then by Neustadt. It was also used by Gamkrelidze and Kharatishvili [GK.2] and by Pshenichnyi [Ps.2, Ps.3]. An alternate method of handling such constraints, namely by means of constraint-defining functions which take on their values in a (not necessarily finite-dimensional) linear topological space and a "generalized" inequality defined in terms of a convex body in this space [as described in Problem (III.1.2) and in (IV.1.12–14)] was first suggested by Neustadt [N.5]—for the special case where the convex body is also a cone—and later by Warga [W.8]—for the case of a general convex body.

A somewhat different approach to optimal control problems was taken by Warga in a series of papers. Namely, Warga observed that, in order to ensure the existence of a solution to optimal control problems, it is generally necessary to enlarge the set of admissible controls into a new class which he referred to as "relaxed" controls [W.1]. The idea of relaxed controls was based on the concept of "generalized curves" first introduced for problems in the calculus of variations by Young [Y.1] in 1937 (see also Young [Y.2]). Warga then obtained necessary conditions for "relaxed" solutions of the optimal control problem [W.2], taking advantage of the convexity of the set of "relaxed admissible right-hand sides" of the differential equation. Warga's results on relaxed controls were somewhat extended by McShane [Mc.3]. A more abstract optimal control problem—still keeping the ideas of "original" and relaxed controls—was introduced by Warga in [W.5]. In that paper, Warga observed that relaxed controls, which are functions whose values are probability measures, could also be identified as continuous linear functionals on some appropriate Banach space, and thereby embedded the relaxed (as well as the ordinary) controls in a linear topological space—namely, the conjugate of the aforementioned Banach space, with the weak* topology. This corresponds almost exactly to the way in which controls are identified with integral operators in (V.2.9 and 10) and the way in which these operators are considered as elements of the linear topological space $\mathscr{T}$ defined in (IV.1.1). Necessary conditions, both for solutions of the original and of the relaxed abstract optimal control problem, were obtained by Warga in [W.6, W.8]. The conditions in the latter paper, which are applicable to problems with "inequality" constraints defined in terms of convex bodies in a general linear topological space, are particularly close to those of Theorem (III.3.1). A more concrete optimal control problem, defined in terms of functional equations, and very much like Problems (IV.1.3) and (IV.2.4), was investigated in Warga [W.7], and necessary conditions (for the relaxed problem) very much like those of

Theorems (IV.1.30) and (IV.2.34) were obtained. A complete development of all of the preceding discussion can be found in the book by Warga [W.10].

4. For Chapter III

The basic optimization problem (III.1.2), the general minimax problem (III.1.8), and the general vector-valued criterion problem (III.1.10) were first stated in more or less these forms in Neustadt [N.5]. A problem closely related to (III.1.2) was stated in Warga [W.8]. In Neustadt [N.5] the concepts of (φ,ϕ,Z)-extremals and (ϕ,Z)-extremals—slightly different from those of Definitions (III.1.14 and 29)—were also first introduced, and results similar to Theorems (III.1.16, 24, and 26) concerning the relation between such extremals and solutions of the aforementioned problems were obtained. The study of minimax problems apparently began with the classical Chebyshev approximation problem. Since then, various particular minimax problems have been extensively investigated. A related historical survey of such problems, with many references, can be found in the book by Demyanov and Malozemov [DM.1]. Problems with vector-valued criteria were first formulated (in an economics setting) in 1896 by Pareto [P.1] and later, in a control theory setting, by Zadeh [Z.1]. Though not as extensively studied as the other two optimization problems, they have nevertheless been the subject of a number of papers.

It is worth noting that if the convex body Z_1 in Problem (III.1.2) is closed, then the generalized "inequality" constraint $\phi_1(e) \in Z_1$ in this problem can be recast in the equivalent form $\tilde{\phi}_1(e) \leq 0$, where $\tilde{\phi}_1$ is some suitable function*al* defined on $\mathscr{E}'$, so that, effectively, Problem (III.1.2) can be put in the form of the seemingly more specialized problem (III.1.3) (even with $\mu = 1$). Indeed, if ζ_1 is an arbitrary interior point of Z_1, let $\tilde{\phi}^0$ be the *gauge* function of $(Z_1 - \zeta_1)$, i.e.

$$\tilde{\phi}^0(z_1) = \inf\{\lambda\colon \lambda > 0, \lambda^{-1}z_1 \in (Z_1 - \zeta_1)\} \quad \text{for all} \quad z_1 \in \mathscr{Z}_1.$$

Then it is not hard to show [see Reference 4 of Chapter I, Lemma V.1.8 on p. 411] that $0 \leq \tilde{\phi}^0(z_1) < \infty$ for all $z_1 \in \mathscr{Z}_1$; $\tilde{\phi}^0(\lambda z_1) = \lambda\tilde{\phi}^0(z_1)$ for all $\lambda \geq 0$ and $z_1 \in \mathscr{Z}_1$; $\tilde{\phi}^0(z_1' + z_1'') \leq \tilde{\phi}^0(z_1') + \tilde{\phi}^0(z_1'')$ for all $z_1', z_1'' \in \mathscr{Z}_1$; $z_1 \in \operatorname{int}(Z_1 - \zeta_1)$ if and only if $\tilde{\phi}^0(z_1) < 1$; $z_1 \in (Z_1 - \zeta_1)$ if and only if $\tilde{\phi}^0(z_1) \leq 1$. It follows from the preceding that $\tilde{\phi}^0$ is convex and continuous and that, if we define the functional $\tilde{\phi}_1$ on $\mathscr{E}'$ by the formula

$$\tilde{\phi}_1(e) = \tilde{\phi}^0(\phi_1(e) - \zeta_1) - 1 \quad \text{for all} \quad e \in \mathscr{E}',$$

then $\phi_1(e) \in Z_1$ if and only if $\tilde{\phi}_1(e) \leq 0$.

By the same token, the minimax problem (III.1.8) can be cast in the form of Problem (III.1.2) if we define the functional $\hat{\phi}^0$ on $\mathscr{Z}_2$ by the formula $\hat{\phi}^0(z_2) = \sup_{\ell \in \mathscr{L}} \ell(z_2)$, and set $\phi^0 = \hat{\phi}^0 \circ \phi_2$. It is easily seen that $\hat{\phi}^0$ is convex and that it is also continuous if the sets V_2^κ defined by (III.1.18) are open for every $\kappa \in R$.

In yet the same vein, if e_0 is a (φ,ϕ,Z)-extremal, then e_0 is at the same time a $(\varphi,\tilde{\phi},\tilde{Z})$-extremal, where $\tilde{\phi}$ is a suitable function*al* defined on $\mathscr{E}'$ and $\tilde{Z}$ is $\bar{R}_-$ (in the space $\tilde{\mathscr{Z}} = R$). Indeed, it is only necessary to set

$$\tilde{\phi}(e) = \tilde{\phi}^0(\phi(e) - \zeta) - 1 \quad \text{for all} \quad e \in \mathscr{E}',$$

where ζ is some arbitrary fixed interior point of Z and $\tilde{\phi}^0$ is the gauge function of $\bar{Z} - \zeta$. As before, $\tilde{\phi}^0$ is convex and continuous. Hence, if $\mathscr{E}' \subset \mathscr{Y}$, where $\mathscr{Y}$ is a linear vector space, if $\mathscr{E}'$ is finitely open in itself, and if ϕ is finitely differentiable at e_0, then it follows from Corollary (I.7.33) and Theorem (I.7.34) that $\tilde{\phi}$ is finitely semidifferentiable at e_0. In fact, $\tilde{\phi}$ is finitely semidifferentiable at e_0 even if ϕ only satisfies the weaker differentiability requirement of Hypothesis (III.2.3), as is easily verified. Thus, if e_0 satisfies the hypotheses of Theorem (III.2.18), then the same hypotheses hold with ϕ and Z replaced, respectively, by $\tilde{\phi}$ and $\bar{R}_-$. As one would suspect, and as it is not too hard to show, the necessary conditions that Theorem (III.2.18) yields for e_0 being a $(\varphi,\tilde{\phi},\bar{R}_-)$-extremal are completely equivalent to those for e_0 being a (φ,ϕ,Z)-extremal. In case ϕ is finitely differentiable, this follows directly from Theorem (III.4.2).

The equivalence of the necessary conditions in the two possible problem formulations carries over to Problems (III.1.2, 8, and 10) with regard to Theorems (III.3.1 and 12), and to (ϕ,Z)-extremals and the corresponding problems without equality constraints as well.

Abstract multiplier rules which essentially coincide with those of Theorems (III.2.18 and 60)—for the special case where Z is a (convex) cone (with nonempty interior)—were proved in Neustadt [N.5] under very general hypotheses which subsume those made in Theorems (III.2.18 and 60). The proofs given here are essentially the same as those in Neustadt [N.5]. The proof of Theorem (III.2.18) [the proof of the other multiplier rule is really a simplified version of the proof of Theorem (III.2.18)] rests on the following: *1*) the construction of a convex set of "variations" in a finite-dimensional space whose dimension coincides with the dimension of the equality constraint, *2*) a separation theorem for convex sets in this finite-dimensional space, *3*) a separation theorem in the linear topological space $\mathscr{X}$, and *4*) a fixed-point theorem, which is used essentially as an implicit functions–type theorem [see Theorem (I.5.38) and the proof of Lemma (I.5.43)]. As was noted earlier, the idea of a convex set of variations was originally due to McShane [Mc.1, Mc.2]. The fact that the inequality and cost functionals did not have to be included in this set of variations—which is very crucial in order for one to be able to handle the general type of inequality constraints previously described as well as minimax problems—was first noted in Neustadt [N.3]. Halkin [H.1] originated the use of a fixed-point theorem in this context.

Essentially all of the derivations of necessary conditions for nonlinear optimization problems which have been obtained by the various authors, as previously described, are similarly based on the construction of a convex set of variations, on a separation theorem, and on a fixed-point theorem. In some simple cases, the fixed-point theorem was replaced by the standard (finite-dimensional) implicit functions theorem.

A result similar to Lemma (III.2.51) was obtained in Neustadt [N.8].

The abstract multiplier rule for the basic optimization problem, Theorem (III.3.1), was presented in more or less this form in Neustadt [N.7], where a hypothesis essentially the same as Hypothesis (III.2.14) also appeared. A closely related result can be found in Warga [W.8]. A result of the type of Corollary (III.3.6) was obtained in Neustadt [N.3]. For the basic optimization problem without equality constraints, with all spaces Banach spaces, results similar to the multiplier rule in Theorem (III.3.34) were obtained by Hurwicz [Hu.1] and Demyanov and Rubinov [DR.1, DR.2]. Demyanov [D.1], Malozemov [M.1] and Demyanov and Malozemov [DM.1] investigated the following finite-dimensional minimax problem without equality constraints: Find $\min_{x \in \Omega} \max_{y \in G} F(x, y)$, where Ω is a closed convex set in R^n, G is a compact set in R^m, and F is a continuous real-valued function which is continuously differentiable in its first argument. In particular, they considered the case where $\Omega = \{x\colon h(x,\beta) \leq 0 \text{ for all } \beta \in \omega\}$, where ω is some compact set in R^s and h has the same properties as F. As necessary conditions, they obtained multiplier rules which are special cases of Theorem (III.3.37). Similar results were obtained by Pshenichnyi [Ps.3]. A result similar to the multiplier rule for vector-valued criterion function problems without equality constraints in Theorem (III.3.37) was obtained by Hurwicz for the special case where all spaces are Banach and all differentials are Fréchet.

There are two basic notions which underlie the results of Section 4 of Chapter III: *1*) If a convex functional ϕ (defined on some linear vector space) attains its minimum over some convex set $\mathscr{E}$ at the origin, then there is a subgradient of ϕ at 0 which also attains its minimum over $\mathscr{E}$ at 0. *2*) If $\phi^1, \ldots, \phi^\mu$ are convex functionals defined on some linear vector space, $\phi = \sum_{i=1}^{\mu} \alpha^i \phi^i$ with the $\alpha^i \geq 0$, and Λ is a subgradient to ϕ at 0, then $\Lambda = \sum_{i=1}^{\mu} \alpha^i \Lambda^i$ for some suitable subgradients Λ^i of the ϕ^i at 0. Assertion *2*), in a Banach space setting, seems to have been first proved by Dubovitskii and Milyutin [DuM.3]. It was proved under somewhat less restrictive hypotheses (but still in a Banach space), together with Assertion *1*), by Pshenichnyi [Ps.1] (see also [Ps.2, Ps.3]), who exploited these results to obtain multiplier rules for convex optimization problems in Banach spaces. A result of the type of *2*), in locally convex linear topological spaces, is found in Rockafellar [R.1]. Assertions *1*) and *2*)—in an arbitrary linear vector space—follow at once, respectively, from Theorem (I.2.25) and Lemma (III.4.25). The induction argument employed in the proof of Lemma (III.4.25) was employed by Halkin and Neustadt [HN.1]. The first part of Theorems (III.4.2 and 21) were proved, in a slightly modified form, in Neustadt [N.5]. Theorem (III.4.10) is a special case of the general maximum principle in Halkin and Neustadt [HN.1].

Theorems (III.5.4, 7, 9, 12, and 14) on necessary conditions and sufficient conditions for optimality were proved in a slightly less general form in Neustadt [N.8]. Pshenichnyi, in [Ps.1, Ps.2, Ps.3], developed a multiplier rule in terms of subgradients essentially the same as that of Theorem (III.5.16) for the case where $\mathscr{E}$ is a closed (convex) set, $\mathscr{Y}$ is a Banach space, and the constraint functionals φ^i and ϕ^i are continuous. Rockafellar [R.1] obtained a similar result—on the assumption that the φ^i and the ϕ^i for $i > 0$ are continuous and Gâteaux differentiable—in a locally convex linear topological space. Demyanov

[D.1], Malozemov [M.1], and Demyanov and Malozemov [DM.1] obtained sufficient conditions and necessary and sufficient conditions for special finite-dimensional minimax problems (as described earlier) which are essentially subsumed in Theorems (III.5.29, 31, and 32). For the special vector-valued criterion problem (III.1.12) with $\mathscr{Y}$ finite-dimensional, Karlin [K.1] obtained the necessary conditions of Theorem (III.5.24) as well as the sufficient conditions of Theorem (III.5.29). Earlier results along this line were obtained by Slater [S.1] and by Kuhn and Tucker [KT.1].

The convex vector-valued criterion problem [see (III.1.10) and (III.5.21)] without equality constraints was extensively investigated (in quite general spaces) as early as 1958 by Hurwicz [Hu.1] (although he defined this problem slightly differently from the way it is done in this book). For example, in that paper, theorems much like the necessity theorem (III.5.24) and the sufficiency theorem (III.5.29) were proved.

5. For Chapter IV

Optimal control problems formulated in terms of sets of differentiable operators acting on a Banach space and their fixed points, much like Problems (IV.1.3) and (IV.2.4), were first investigated by Warga in [W.7]. In that paper, Warga confined himself to relaxed controls—which meant that the underlying set of operators is convex—and to the case where the space $\mathscr{Z}_1$ (which defines the "inequality" constraint) is finite-dimensional. These restrictions were removed in Warga's book [W.10], and necessary conditions similar to those in our Theorems (IV.1.30) and (V.2.34) were obtained. Warga's hypotheses were very much like those made in Sections 1 and 2 of Chapter IV, except that, in the problem with parameters of Section 2, he required only finite differentiability of the operators $\hat{T}$ and of the constraint functionals $\hat{\varphi}$, $\hat{\phi}^0$, and $\hat{\phi}_1$ with respect to the parameters. The results of Section 2, Chapter IV, can also be shown to hold under such weakened hypotheses.

In Neustadt [N.7], a problem which is a slight specialization of Problem (IV.1.3) was discussed together with its special form (IV.1.15), and the corresponding necessary conditions [as in Theorem (IV.1.30)] were presented. An abstract problem with parameters essentially the same as Problem (IV.2.4) was described in Neustadt [N.9], and necessary conditions much as those in Corollary (IV.2.47) were presented.

Necessary conditions for an optimal control problem with a vector-valued criterion function which is a special case of Problem (IV.1.94) were obtained in Da Cunha and Polak [DP.1].

The following variant of Problem (IV.1.15) has been investigated by Warga [W.4, W.8, W.9, W.10] (also see Baum and Cesari [BC.1]). In the notation of Section 1 of Chapter IV, let $\mathbf{K}$ be a given compact subset of R^{ℓ}, let $\mathscr{X}$ denote the Banach space of all continuous functions $x:I \times \mathbf{K} \to R^n$ with the sup norm (i.e., $\|x\| = \max_{t\in I, s\in \mathbf{K}} |x(t,s)|$), and let $\mathscr{W}$ be a given subset of $\mathscr{T}_0$ (defined as in (IV.1.1 and 2)). Further, let there be given continuous functions $\chi = (\chi^0, \chi^1, \ldots, \chi^{\mu+m}): (R^n)^k \times \mathbf{K} \to R^{\mu+m+1}$ and $\tilde{\chi} = (\tilde{\chi}^1, \ldots, \tilde{\chi}^{\nu})$: $(R^n)^k \times \mathbf{K} \to R^{\nu}$, a finite subset $\{\tau^1, \ldots, \tau^k\}$ of I, closed subsets $\mathbf{K}_1, \ldots, \mathbf{K}_\nu$ of $\mathbf{K}$, and countably additive, real-valued measures ν^i $(i = 0, 1, \ldots, \mu + m)$ defined on the Borel subsets of $\mathbf{K}$. Then find a function $x \in A$ (where A is a given open set in $\mathscr{X}$) satisfying the constraints

$$x = Tx \quad \text{for some} \quad T \in \mathscr{W}$$

$$\int_{\mathbf{K}} \chi^i(x(\tau^1,s), \ldots, x(\tau^k,s),s)\, d\nu^i(s) \leq 0 \quad \text{for} \quad i = 1, \ldots, \mu,$$

$$\int_{\mathbf{K}} \psi^i(x(\tau^1,s), \ldots, x(\tau^k,s),s)\, d\nu^i(s) = 0 \quad \text{for} \quad i = \mu + 1, \ldots, \mu + m$$

$$\tilde{\chi}^i(x(\tau^1,s), \ldots, x(\tau^k,s),s) \leq 0 \quad \text{for all} \quad s \in \mathbf{K}_i \quad \text{and each} \quad i = 1, \ldots, \nu,$$

and which, in so doing, achieves a minimum for

$$\int_{\mathbf{K}} \chi^0(x(\tau^1,s), \ldots, x(\tau^k,s),s)\, d\nu^0(s).$$

If we set $\mathscr{Z}_1 = R^{\mu} \times \mathscr{C}^{\nu}(\mathbf{K})$—where $\mathscr{C}^{\nu}(\mathbf{K})$ denotes the Banach space of continuous functions from $\mathbf{K}$ into R^{ν} with the sup norm—and set $Z_1 = \bar{R}^{\mu}_{-} \times Z'$, where $Z' = \{z: z = (z^1, \ldots, z^{\nu}) \in \mathscr{C}^{\nu}(\mathbf{K}),\ z^i(s) \leq 0$ for all $s \in \mathbf{K}_i$ and each $i\}$, and define the functions $\hat{\varphi}:A \to R^m$, $\hat{\phi}^0:A \to R$, and $\hat{\phi}_1:A \to \mathscr{Z}_1$ in the obvious manner, then this problem takes the form of Problem (IV.1.3), with $A = A'$. One may then make use of Theorem (IV.1.30) in order to obtain necessary conditions that solutions of this problem must satisfy. These necessary conditions may of course then be specialized, depending on the character of $\mathscr{W}$, as was done in Chapters V–VII.

The preceding problem may be modified in various ways. For example, the "cost" functional may be replaced by

$$\max_{s\in\mathbf{K}_0} \tilde{\chi}^0(x(\tau^1,s), \ldots, x(\tau^k,s),s),$$

where $\tilde{\chi}^0$ and $\mathbf{K}_0$ have the same properties as the $\tilde{\chi}^i$ and $\mathbf{K}_i$ for $i \geq 1$. In this case, we obtain a problem of the form of the minimax problem (IV.1.65). Alternatively, we may generalize the problems by allowing "free times" as in Problem (IV.2.20).

"Volterra type" operators, defined much as in (IV.3.3), were introduced in Neustadt [N.10]. Most of the concepts and results described in (IV.3.1–47), almost all of which pertain to Volterra operators, were presented in Neustadt [N.10, N.7].

The importance of the property of "convexity under switching" for the class of admissible controls in optimal control problems was first pointed out by Boltyanskii, Gamkrelidze, and Pontryagin [BGP.2]. Gamkrelidze [G.4] noted that it is really the set of "admissible right-hand sides" in the optimal control problem which must be convex under switching [in the sense of (IV.5.14)] if one is to be able to obtain necessary conditions in the form of a maximum principle. The term "convexity under switching" itself is due to Halkin [H.6]. The relation between convexity under switching and "almost convexity" (or quasiconvexity, in the terminology of Gamkrelidze) was proved by Gamkrelidze in a fundamental lemma in [G.4], although the idea of the lemma and its proof can be found in an earlier paper of Gamkrelidze [G.3]. A similar result, in a slightly different context, was obtained by Warga in [W.2]. Lemma (IV.5.3) and Theorem (IV.5.17) are closely related to these aforementioned results, as are Theorems (IV.5.26, 28, and 33). Theorems (IV.5.34 and 35) are generalizations of results on quasiconvexity in Gamkrelidze [G.4]. In Neustadt [N.7], a result which essentially subsumes Theorems (IV.5.17 and 28) was presented.

The sufficient conditions for optimal control problems discussed in Section 6 of Chapter IV have previously only been obtained for special cases, and these will be discussed in our comments for Chapter V.

6. For Chapter V

An optimal control problem was apparently first formulated mathematically in 1950 by Hestenes [He.1]. Indeed, in this report Hestenes investigated a free-time optimal control problem which bore a close resemblance to Problem (V.2.1) and, using results from the calculus of variations for the problem of Bolza (particularly the Weierstrass necessary condition), obtained essentially the necessary conditions of the pointwise maximum principle, Theorem (V.2.44). For his class of admissible controls, Hestenes took the set of all piecewise-continuous functions from I into V, where V is either an open set in R^r or a closed subset of R^r defined by a finite number of equalities or inequalities. Finally, Hestenes imposed two very strong restrictions on his optimal control, neither of which can be verified a priori in most cases, and the second of which was subsequently shown to be violated in many problems of interest: *1*) that the optimal trajectory and control are "normal" in a certain technical sense; *2*) that the optimal control is a continuously differentiable function of time.

The first mathematical investigation of an optimal control problem with a discontinuous optimal control function was made by Bushaw [B.1, B.2] who solved a rather simple, linear, second-order time-optimal control problem. Bushaw's proofs were based on geometric arguments and were not generalizable to more complex problems.

The first general results for free-time (as well as fixed-time) optimal control problems with ordinary differential equation constraints (and in the absence of restrictions on the phase coordinates) were obtained in the late 1950s by a group of Soviet mathematicians under the leadership of L. S. Pontryagin—see Boltyanskii [Bo.1], Pontryagin [Po.1], Boltyanskii, Gamkrelidze, and Pontryagin [BGP.2], and Pontryagin, Boltyanskii, Gamkrelidze, and Mishchenko [PBGM.1]. They were the first to obtain the pointwise maximum principle essentially as given in Theorem (V.2.44). More or less the same result, using somewhat different methods of proof, was obtained subsequently by Halkin [H.1], Hestenes [He.2, He.3], Dubovitskii and Milyutin [DuM.3], and Warga [W.5]. A maximum principle in integral form (and its relation to the pointwise maximum principle), much like Theorem (V.2.34), was first obtained by Gamkrelidze [G.4] and subsequently, by slightly different arguments, by Neustadt [N.4]. Additional considerations on the pointwise maximum principle much like those in (V.2.48) were made by Baum and Cesari [BC.1].

Problems with restricted phase coordinates, much like Problem (V.3.1), were first studied by Gamkrelidze [G.1, G.2] (see also Pontryagin, Boltyanskii, Gamkrelidze and Mishchenko [PBGM.1, Chapter VI]), who obtained necessary conditions which are somewhat weaker than those in Theorem (V.3.70), and did so under relatively strong "regularity" hypotheses. Stronger necessary conditions, much like those in Theorem (V.3.70), were obtained a few years later by Chang [C.1] with the use of arguments that were, in part, formal. Dubovitskii and

Milyutin [DuM.2, DuM.3] were the first to employ functional analysis in order to investigate the phase inequality constraints, and obtained the necessary conditions of Theorem (V.3.70) in a slightly different but equivalent form. Results much like those in Theorems (V.3.44 and 70) were obtained by Neustadt in [N.1, N.4]. Other significant early work for the restricted phase coordinate problem was done by Warga [W.3, W.4] who obtained necessary conditions similar to those in Theorem (V.3.70) for so-called relaxed controls. More recently Schwarzkopf [Sc.1] has also treated restricted phase coordinate problems in a relaxed setting similar to that of McShane [Mc.3]. Later work on this problem, resulting in essentially the same conditions as those in (V.3.70), was carried out by Neustadt [N.5], Gamkrelidze and Kharatishvili [Gk.2], and Warga [W.9], using techniques very similar to those of this book. Finally, Rockafellar [R.3] has shown that methods of convex analysis [R.2] can be applied to certain problems with restricted phase coordinates.

A minimax problem of the form of Problem (V.4.1) was apparently first investigated in 1963 by Dubovitskii and Milyutin [DuM.1] and later, in more detail, in [DuM.3], where necessary conditions essentially the same as those in Theorem (V.4.11) were obtained. They also pointed out the close connection between the minimax problem and the problem with restricted phase coordinates, and the resultant close similarity of the necessary conditions. Similar observations were made by Neustadt [N.4], Warga [W.3, W.10], and Gamkrelidze and Kharatishvili [GK.2].

The considerations, as well as the example, in Section 5 dealing with convexity as an alternate to convexity under switching were essentially due originally to Gamkrelidze [G.4]. The same is true of the material in (V.6.1) on alternate forms of the set of admissible right-hand sides in the differential equation. Rockafellar [R.2, R.3] extensively studied optimal control problems in which the constraint functionals χ^i are convex and not necessarily differentiable. Generalized differentiability requirements similar to those described in (V.6.8) were discussed by Hestenes [He.3]. Necessary conditions for an optimal control problem with a vector-valued criterion function were obtained by Da Cunha and Polak [DP.1].

The necessary conditions of Theorem (V.7.20) for the "unorthodox" optimal control problem (V.7.1)—in a slightly less general form—were first obtained, by methods essentially the same as those used here, by Neustadt [N.5].

The sufficiency result of Theorem (V.8.9) was obtained by Neustadt [N.8], though for a somewhat more restricted problem than Problem (V.8.1). Earlier, less general results of this type were obtained by Mangasarian [Ma.1], Pshenichnyi [Ps.2], and Funk and Gilbert [FG.1].

7. For Chapter VI

Hestenes in [He.2] considered a very general problem of Bolza, with the payoff, system equations, mixed control-phase equality and inequality constraints [much like those of Problem (VI.3.10)], boundary conditions, and isoperimetric constraints all depending explicitly on control parameters. Under very stringent regularity assumptions on the constraints (which allow use of implicit function theorem arguments), he derived necessary conditions in the form of a pointwise maximum principle similar to the pointwise version of Theorem (VI.1.28). In [He.3] Hestenes considered a Mayer problem (with control parameters explicit only in the boundary conditions and constraints of the form $g_i(\pi) \leq 0$, $i = 1, \ldots, m'$, $g_i(\pi) = 0$, $i = m' + 1, \ldots, m$) which can be considered a special case of Problem (VI.1.10).

Warga [W.7, W.10] also considered (in both relaxed and original control settings) very general optimal control problems with the differential and integral equation constraints depending explicitly on control parameters as well as control functions.

Gamkrelidze [G.2, PBGM.1] considered mainly problems with mixed control-phase equality constraints of special types and derived necessary conditions in the form of a global maximum principle. He did, however, indicate how mixed control-phase inequality constraints can be treated using his results.

Lagrange control problems with mixed control-phase constraints of both inequality and equality type were treated by Hestenes in [He.2, He.3]. Under quite strong assumptions he presented necessary conditions which are essentially equivalent to those of Theorem (VI.3.115) when specialized to problems containing only inequality constraints of mixed type. Hestenes restricted his investigations to problems with piecewise-continuous controls, whereas a much more general class of controls is allowed in Problem (VI.3.10). Dubovitskii and Milyutin [DuM.4] also considered a Lagrange control problem with mixed control-phase inequality constraints and obtained a "local" maximum principle. Makowski and Neustadt [MN.1] treated problems with mixed control-phase equality and inequality constraints similar to those of Hestenes in [He.2, He.3]. They presented necessary conditions in the form of a global maximum principle (integral and pointwise forms) of which Theorems (VI.3.108 and 115) can be essentially considered as special cases.

Dubovitskii and Milyutin [DuM.4] in their treatment of general problems involving mixed control-phase inequality constraints also considered a number of special examples. One of them was a minimax problem involving a payoff of mixed control-phase type such as that treated in Problem (VI.4.1) where $\nu' = 1$ and $\xi = 1$.

8. For Chapter VII

Necessary conditions [which are subsumed by Theorem (VII.1.62)] for optimality in problems involving general functional differential equations of retarded type [as described in (VII.1.21–23)] with initial and terminal constraints given in terms of a C^1 target manifold in Euclidean space were first obtained by Banks [Ba.2] using generalizations of the quasiconvexity ideas of Gamkrelidze [G.4]. The nontriviality conditions and arguments of Lemmas (VII.1.71 and 75) contain, as special cases, the conditions and arguments in [Ba.2] (see also [Ba.1]) for nontriviality of the multipliers in the maximum conditions obtained there for problems described by systems as in (VII.1.21–23).

Buehler [Bue.1], reformulating the problem in terms of operator equations as in (VII.1.24) and using the general extremal approach of Neustadt [N.5, N.6], considered systems similar to those of Banks [Ba.2] but allowed inequality phase constraints, boundary conditions similar to those in (VII.1.13 and 14), and classes of initial functions which are open and/or convex in $\mathscr{C}^n([t_1,t_3])$. He assumed the continuity in Hypothesis (VII.1.76.iiia) and obtained a maximum principle in pointwise form.

Banks and Kent [BK.1] considered problems with target set (terminal constraints) in a function space and general functional differential equation systems of neutral type which include as a special case those described in (VII.1.21–23). They used the abstract extremal approach of Neustadt [N.4, N.5] to give necessary conditions [similar to those in Theorem (VII.1.62) with additional multipliers corresponding to the function space terminal restrictions] which they showed are sufficient for optimality under the usual strengthening (normality) in problems with linear-in-the-state systems and cost functionals possessing certain convexity properties.

Friedman [Fr.1] used arguments similar to those of Pontryagin et al. [PBGM.1] to derive a pointwise maximum principle for problems with processes governed by Volterra equations on $[t_1,t_2]$,

3 $$x^i(t) = x_0^i + \int_{t_1}^{t} h^i(t-\tau)f^i(\tau,x(\tau),u(\tau))\,d\tau, \quad i = 1,\ldots,n,$$

which are hereditary in the state and the control and which are equivalent to a certain class of integro-differential equations. In [Ha.1] Halanay considered optimal control of very general systems that are hereditary in the state and control and include as special cases the systems of Friedman and a large class of functional differential equations. Using the multiplier rule of Hestenes [He.2, He.3], he obtained a pointwise maximum principle. Halanay gave necessary conditions for similar systems in a relaxed setting in [Ha.2].

Warga in [W.10] developed necessary conditions in both relaxed and original (control) settings for problems involving very general functional-integral equations that include a large class of integro-differential equations and functional differential equations as special cases. In [W.11] he treated systems of an even more general type which also allows delays in the controls.

Kharatishvili [Kh.1, PBGM.1] was one of the first to treat control problems governed by differential-difference equations and give conditions included in those of Theorem (VII.2.31). Oshiganova [O.1] considered control of systems with a single time-varying delay which, because of special assumptions on the delay functions, could be replaced by equivalent systems of ordinary differential equations by use of what is essentially the so-called step method. Necessary conditions were then obtained directly by consideration of the equivalent ordinary differential equation problem. Banks [Ba.1] used the quasi-convexity ideas of Gamkrelidze to give necessary conditions for optimality of systems with time-varying delays which could not be reduced to equivalent systems of ordinary differential equations. Kharatishvili [Kh.2], using the same approach, first presented [with incorrect proofs—see our remarks in (VII.2.36) and those in *Math. Rev.* **41** (1971), No. 2499, 461] necessary conditions for systems with delays in the control as well as state. Hughes [Hug.1], using the abstract multiplier rule of Hestenes, gave a correct derivation of the maximum principle for problems involving systems with delays in both the state and the control.

Huang [Hua.1] used the abstract variational approach of [N.4] to treat essentially Problem (VII.2.10) with, in addition, constraints on the phase coordinates.

The literature on control of hereditary differential systems (and differential-difference equations in particular) is by now rather vast, and consideration of very special types of hereditary processes may be found. For example, necessary conditions for optimality in problems with systems where the delays in the state depend on both the state and the control were presented in [GC.1] (see also [GaK.2]). We shall not attempt a thorough review here, but instead refer the reader to [BM.1] and the survey papers referenced there for a more complete bibliography.

In contrast, necessary conditions for optimality in problems involving systems of Volterra integral equations have been discussed by a relatively small number of authors. Among the first of these was Friedman [Fr.1], who considered problems with Volterra integral equations (3) and obtained a pointwise maximum principle as mentioned above. Friedman's problem can be considered as a special case of our Problem (VII.3.9), although, because of additional smoothness assumptions on the problem data, his necessary conditions can be given in terms of absolutely continuous adjoint variables ψ^i satisfying the integro-differential equations

$$\dot{\psi}^i(t) = -\sum_{j=0}^{n} f^j_{x^i}(t)\left[h^j(0)\psi^j(t) + \int_t^{t_2} \dot{h}^j(\sigma - t)\psi^j(\sigma)\, d\sigma\right].$$

One of the problems considered by Huang [Hua.2] (he also considered a bounded phase coordinate problem for these Volterra systems) is essentially the same as Problem (VII.3.9). Under more smoothness assumptions than those in (VII.3.9), Huang derived a maximum principle in integral form which was also stated in terms of adjoint variables. The general problems investigated by Halanay [Ha.1] include as a special case this problem of Huang, and the necessary conditions of Halanay reduce to those given by Huang in this situation. However, the results of both Huang and Halanay are, even under the added smoothness assumptions, in part formal, although it appears that in certain cases one can make rigorous arguments for these results. To our knowledge the necessary conditions of Theorems (VII.3.31 and 41) in terms of resolvent kernels for Problem (VII.3.9) under the weak smoothness requirements of (VII.3.1–4) are new.

Butkovskii [Bu.2] gave a formal presentation of a maximum principle (again involving adjoint variables) for a control problem described in terms of a very general class of nonlinear integral equations which include as special cases certain types of Volterra and Fredholm equations. Vinokurov [V.1] studied control problems with systems such as those in (VII.3.9), but in a somewhat different context (see [NW.1, V.2]). Finally, Warga [W.7, W.9, W.10] treated in both original and relaxed settings control problems for a general class of integral equations which includes those of Volterra type as in (VII.3.9).

The development of necessary conditions for discrete optimal control problems such as those discussed in Section 4 of Chapter VII has a rather interesting history, which is due in part to the fact that, as we saw in Section 4, the extension of the usual global maximum principle to cover discrete systems is possible only under very strong assumptions. Early contributors in this area did not always recognize this, and there was some confusion as to what form the necessary conditions might take under various hypotheses. Among the first to consider a maximum principle for discrete control problems was Chang [C.2], who gave a formal presentation of a "digitized" maximum principle which was a local condition somewhat similar to the condition (VII.4.96). While Chang's main interest was in problems arising from discretizing continuous systems, other investigators showed a great interest in discrete systems as the underlying model for a control process. (See, for example, the monograph by Fan and Wang [FW.1], where examples of multistage processes arising in such diverse areas as biochemical reactions, transportation problems, refrigeration systems, and stirred tank reactors are discussed.) Wide use (often without validity) was made of a global maximum principle for these problems analogous to that first hypothesized by Pontryagin et al. for continuous systems control problems. Rozonoér [Roz.2] was the first to question the validity of the extension of the global Pontryagin-type condition for problems involving nonlinear discrete systems. He was the first to recognize the need for some stringent convexity hypotheses in order to obtain global necessary conditions such as those presented in Theorem (VII.4.44). While certain parts of Rozonoér's presentation were formal, the form of his hypotheses and necessary conditions for linear-in-the-state, nonlinear-in-the-control systems is similar to that in Theorems (VII.4.56 and 51). While Rozonoér only speculated that one might have difficulty extending the global maximum principle to treat general nonlinear discrete problems, Butkovskii [Bu.1] gave a counterexample to establish the invalidity of the global maximum principle for certain nonlinear problems. He also gave a formal derivation of a local maximum principle. In a later account [Bu.2] Butkovskii gave a somewhat more rigorous derivation of these conditions, though even here not all of the assumptions needed to carry out the arguments are stated explicitly. Butkovskii's results, when stated with complete and verifiable hypotheses on the problem, are essentially subsumed by Theorem (VII.4.89) [with the corresponding pointwise condition (VII.4.96)].

Jordan and Polak [JP.1] were among the first to give a rigorous development of a local maximum principle for nonlinear discrete systems problems. Under assumptions similar in spirit to Hypothesis (VII.4.61), they derived a maximum principle for Lagrange problems with nonlinear discrete systems that can be considered a special case of the conditions of Theorem (VII.4.89). In fact, the derivation of Jordan and Polak was made under slightly weaker assumptions than those in Hypothesis (VII.4.61). Treating the problem for control sets $\mathscr{U}$ satisfying the Hypothesis (VII.4.50) with $U_j = U$, U a fixed subset of R^r, they assumed essentially that (i) any $v \in U$ is not isolated (i.e., there exist allowable perturbations along some ray emanating from v), and (ii) the set of allowable perturbations at such a point v forms a convex set. It is easy to see that $\mathscr{U}$ finitely open in itself for such classes $\mathscr{U}$ implies that U is finitely open in itself; and, further, it is easy to give simple examples of a set U which is not finitely open in itself but does satisfy the assumptions (i) and (ii) of Jordan and Polak.

Almost simultaneously with the appearance of the Jordan and Polak paper, Propoi [Pr.1] published an article in which he also pointed out why one should not in general expect in the case of nonlinear discrete problems to obtain necessary conditions as strong as the global Pontryagin-type condition for continuous systems. Propoi considered a nonlinear problem with Lagrange-type cost functions and no phase constraints, and while his arguments were in part formal, he derived conditions similar to those given in Theorems (VII.4.56 and 51), His convexity assumptions, viewed in light of the comments in (VII.4.57), are similar to those assumed in obtaining (VII.4.56 and 51).

Halkin [H.8, H.9] was one of the first to give a complete and rigorous presentation of a global maximum principle for problems with nonlinear discrete systems. In [H.8] Halkin used a geometric approach (attainable sets) combined with elementary topological arguments (Jordan and Polak [JP.1] also employed similar separation arguments) to give a derivation of a global maximum principle for systems linear in the state, nonlinear in the controls under convexity assumptions that are a special case of Hypothesis (VII.4.54). Holtzman [Ho.1] considered the same problem as Halkin and showed (assuming the existence of certain tangent hyperplanes but claiming that this assumption was really not needed) that the convexity assumptions could be relaxed to an assumption of "e^n-directional convexity." Holtzman's arguments were essentially those presented in (VII.4.57–60), and his assumption of "directional convexity" was just Hypothesis (VII.4.58). The problem formulation (VII.4.3) includes the problem of Halkin and Holtzman as a special case, and their results are contained in Theorems (VII.4.56 and 51). Holtzman and Halkin [HH.1] later showed that the existence assumption for the tangent hyperplane made in Holtzman's development [Ho.1] was indeed unnecessary.

Employing arguments that involve generalizations of his earlier attainable sets approach, Halkin [H.9] showed that the results of [H.8] could be extended (again under strong convexity assumptions) to treat problems involving general nonlinearities in the right side of the system equations. While he made effective use of attainable sets for an associated linearized system, he pointed out that some naive linearization ideas used prior to that time to treat nonlinear problems were in fact incorrect. Holtzman [Ho.2] once again argued that Halkin's strong convexity assumptions could be replaced by directional convexity, and, as before, the results obtained are a special case of Theorems (VII.4.56 and 51) where the problem is formulated as in (VII.4.57–60) and Hypothesis (VII.4.58) is invoked. Holtzman also used the idea [described in (VII.4.55)] of "relaxing" the original problem and observing that this relaxed problem is equivalent to the original problem [in the sense explained in (VII.4.55)].

Rosen [Ro.1] and Pearson and Sridhar [PS.1] considered nonlinear discrete optimal control problems [special cases of Problem (VII.4.62)] with inequality phase constraints and used a programming approach (the Kuhn-Tucker multiplier rule) to obtain necessary conditions that are included in Theorem (VII.4.89) with Condition (VII.4.97).

Canon, Cullum, and Polak [CCP.1, CCP.2] formulated a general constrained optimization problem in finite-dimensional spaces and used geometric arguments (ideas similar to the calculus of variations techniques of McShane and those used by Pontryagin et al. in their early derivation of the global maximum principle for continuous systems—see the remarks and references under General Comments for Chapters II–V above) to derive basic necessary conditions for this problem. They argued that essentially all of the results of Jordan and Polak [JP.1], Halkin [H.9], Holtzman [Ho.2], and Rosen [Ro.1] for discrete control problems can be obtained from their basic necessary conditions. In addition, Canon, Cullum, and Polak presented new results for nonlinear bounded phase-space discrete problems. One problem considered by Canon, Cullum, and Polak is a special case of Problem (VII.4.62), and while their assumptions (involving radial cones and linearizations of the first kind) were weaker than the assumption of $\mathscr{U}$ finitely open in itself made in Problem (VII.4.62), their results [a local maximum principle subsumed by Theorem (VII.4.89) with Condition (VII.4.96)] were weaker in the sense that they did not allow phase and boundary constraints which also depended on the control variable as we do in Problem (VII.4.62). Using the concept of linearization of the second kind (the "first-order convex approximation" concept mentioned above in the general comments for Chapters II–V), Canon, Cullum, and Polak pointed out that the global maximum principles of Halkin [H.9] and Holtzman [Ho.2] could be easily obtained from their basic necessary conditions.

In a related effort but with a completely different approach from that of Canon, Cullum, and Polak, Mangasarian and Fromovitz [MF.2] took a mathematical programming approach (a rigorous use of the generalized Fritz John multiplier rule applied to nonconvex programming problems) to discrete problems and presented both local [a special case of Theorem (VII.4.89)] and global [a special case of Theorems (VII.4.56 and 51)] maximum principles under somewhat different assumptions from any of those given in Section 4 of Chapter VII. They argued that their necessary conditions, derived under assumptions different from the convexity hypotheses of Halkin [H.9] or Canon, Cullum, and Polak [CCP.2], contain the necessary conditions of those latter two references as special cases.

Hautus [Hau.1] used the notion of derived cones to obtain necessary conditions for a constrained optimization problem that include the multiplier rules of John [J.1] and Mangasarian and Fromovitz [MF.1]. He applied these results to discrete optimal control problems and obtained necessary conditions that contain those of Halkin [H.9], Holtzman [Ho.1], Jordan and Polak [JP.1], and Canon, Cullum, and Polak [CCP.1, CCP.2] as special cases.

Pshenichnyi [Ps.3] also applied a generalized multiplier rule to a state-constrained discrete optimal control problem that, along with his assumptions, is a particular case of Problem (VII.4.3) with Hypotheses (VII.4.47 and 54). His global (pointwise) maximum principle is included in Theorems (VII.4.51 and 56) and Lemma (VII.4.49). A global maximum principle included in Theorem (VII.4.56) for a special case of Problem (VII.4.3) under Hypothesis (VII.4.54) was also given in [N.4].

A number of the contributors mentioned in the historical sketch above also discussed in their papers sufficiency of their respective necessary conditions for discrete optimal control problems.

We mention several other papers on discrete control problems which, while not directly related to the results of Chapter VII, may be of interest to readers. Chyung [Ch.1] has shown that for the linear regulator problem with discrete systems (linear system, quadratic cost functional), global necessary conditions similar to those in Theorem (VII.4.44) [reformulated, of course, with the directional convexity assumption (VII.4.58)] lead to an expected two-point boundary value problem for the state-multiplier equations. Recently, Boltyanskii [Bo.2] has shown that techniques involving the concept of "local sections" can also be used to treat nonlinear discrete system Lagrange cost control problems with variable (state-dependent) control constraints. His necessary conditions include a maximum principle of the global type. Finally, Gabasov and Kirillova [GaK.1] stated without proof a set of necessary conditions (their quasi-maximum principle) which they claimed were stronger than any known (at that time) local maximum principle for nonlinear discrete system problems. A further discussion of these conditions may be found in [GaK.2].

9. For the Appendix

The results for Volterra-type operators can be found in an earlier paper by Neustadt [N.10]. The local existence, continuation, and maximal continuation results of Theorems (A.13, 15, and 18) are essentially the same as those in Theorems 3.1, 3.2, and 3.3 of [N.10].

Theorem (A.19) combines Theorems 3.4, 4.1, and 4.2 of that reference. Theorems (A.29 and 31) are special cases of the results of Theorems 4.4, 4.5, and 4.6 of [N.10]. Finally, Theorems 6.1 and 6.2 of [N.10] give continuous dependence results like those in Theorem (A.32).

The statements in Theorems (A.47 and 51) for linear ordinary differential equations are quite standard and may be found in any number of texts (e.g., see [CL.1]). The results and arguments involved in Theorem (A.65) are very similar to those found in well-known and classical presentations—see, for example, [T.1] where these arguments are carried out in L_2, and [Pog.1] where a pointwise bound $|K(t,s)| \leq M$ is assumed in place of the weaker assumption (A.54).

The representation results of Theorem (A.70) were first given correctly by Banks [Ba.2, Ba.3].

REFERENCES

Banks, H. T.

[Ba.1] Necessary conditions for control problems with variable time lags, *SIAM J. Control* **6** (1968), 9–47.

[Ba.2] Variational problems involving functional differential equations, *SIAM J. Control* **7** (1969), 1–17.

[Ba.3] Representations for solutions of linear functional differential equations, *J. Differential Equations* **5** (1969), 399–409.

Banks, H. T., and Kent, G. A.

[BK.1] Control of functional differential equations of retarded and neutral type to target sets in function space, *SIAM J. Control* **10** (1972), 567–593.

Banks, H. T., and Manitius, A.

[BM.1] Application of abstract variational theory to hereditary systems—a survey, *IEEE Trans. Automatic Control* **AC-19** (1974), 524–533.

Baum, R. F., and Cesari, L.

[BC.1] On a recent proof of Pontryagin's necessary conditions, *SIAM J. Control* **10** (1972), 56–75.

Bellman, R., Glicksberg, I., and Gross, O.

[BGG.1] *Some Aspects of the Mathematical Theory of Control Processes*, The RAND Corporation, R-313, 1958.

Blum, E. K.

[Bl.1] Minimization of functionals with equality constraints, *SIAM J. Control* **3** (1965), 299–316.

Boltyanskii, V. G.

[Bo.1] The maximum principle in the theory of optimal processes, *Doklady Akad. Nauk SSSR* **119** (1958), 1070–1073.

[Bo.2] Discrete maximum principle (method of local sections), *Differencial' nye Uravneniya* **8** (1972), 1927–1935 [English translation in *Differential Equations* **8** (1972), 1490–1496].

Boltyanskii, V. G., Gamkrelidze, R. V., and Pontryagin, L. S.

[BGP.1] On the theory of optimal processes, *Doklady Akad. Nauk SSSR* **110** (1956), 7–10.

[BGP.2] Theory of optimal processes. I. Maximum principle, *Izv. Akad. Nauk SSSR, Ser. Mat.* **24** (1960), 3–42.

Buehler, H. H.

[Bue.1] Application of a general theory of extremals to optimal control problems with functional differential equations, Ph.D. dissertation, Univ. Southern California, June 1971.

Bushaw, D. W.

[B.1] Differential equations with a discontinuous forcing term, Ph.D. dissertation, Dept. of Mathematics, Princeton University, 1952.

[B.2] Optimal discontinuous forcing terms, *Contr. Theory Nonlinear Oscillations* **4**, Princeton Univ. Press, 1958.

[B.3] Dynamical polysystems and optimization, *Contributions Diff. Equations* **2** (1963), 351–365.

Butkovskii, A. G.

[Bu.1] The necessary and sufficient conditions for optimality of discrete control systems, *Avtomat. i Telemekh.* **24** (1963), 1056–1064 [English translation in *Automat. Remote Control* **24** (1963), 963–970].

[Bu.2] *Distributed Control Systems*, American Elsevier, New York, 1969.

Canon, M., Cullum, C., and Polak E.

[CCP.1] *Theory of Optimal Control and Mathematical Programming*, McGraw-Hill, New York, 1970.

[CCP.2] Constrained minization problems in finite dimensional spaces, *SIAM J. Control* **4** (1966), 528–547.

Chang, S. S. L.

[C.1] Optimal control in bounded phase space, *Automatica* **1** (1963), 55–67.

[C.2] Digitized maximum principle, *Proc. Inst. Radio Engrs.* **48** (1960), 2030–2031.

Chyung, D. H.

[Ch.1] Discrete linear optimal control systems with essentially quadratic cost functionals, *IEEE Trans. Automatic Control* **AC-11** (1966), 404–413.

Coddington, E., and Levinson, N.

[CL.1] *Theory of Ordinary Differential Equations*, McGraw-Hill, New York, 1955.

Da Cunha, N. O., and Polak, E.

[DP.1] Constrained minimization under vector-valued criteria in linear topological spaces, in *Mathematical Theory of Control* (A. V. Balakrishnan and L. W. Neustadt, editors), Academic Press, New York, 1967, 96–108.

Demyanov, V. F.

[D.1] On the solution of some minimax problems I, II, *Kibernetika* **2** (1966), No. 6, 58–66; **3** (1967), No. 3, 62–66 [English translation in *Cybernetics* **2** (1966), No. 6, 47–53; **3** (1967), No. 3, 51–54].

Demyanov, V. F., and Malozemov, V. N.

[DM.1] *Introduction to Minimax*, Nauka, Moscow, 1972.

Demyanov, V. F., and Rubinov, A. M.

[DR.1] The minimization of a smooth convex functional on a convex set, *SIAM J. Control* **5** (1967), 280–294.

[DR.2] Minimization of functionals in normed spaces, *SIAM J. Control* **6** (1968), 73–88.

Dubovitskii, A. Y., and Milyutin, A. A.

[DuM.1] Extremum problems with constraints, *Doklady Akad. Nauk SSSR* **149** (1963), 759–762 [English translation in *Soviet Math. Dokl.* **4** (1963), 452–455].

[DuM.2] Certain optimality problems for linear systems, *Avtomat. i*

Telemekh. **24** (1963), 1616–1626 [English translation in *Automat. Remote Control* **24** (1963), 1471–1481].

[DuM.3] Extremum problems in the presence of restrictions, *Zh. Vychisl. Mat. i Mat. Fiz.* **5** (1965), 395–453 [English translation in *USSR Comput. Math. and Math. Phys.* **5** (1965), No. 3, 1–80].

[DuM.4] Necessary conditions for a weak extremum in optimal control problems with mixed inequality-type constraints, *Zh. Vychisl. Mat. i Mat. Fiz.* **8** (1968), 725–779 [English translation in *USSR Comput. Math. and Math. Phys.* **8** (1968), No. 4, 24–98].

Fan, L. T., and Wang, C. S.

[FW.1] *The Discrete Maximum Principle*, John Wiley & Sons, New York, 1964.

Fréchet, M.

[F.1] La notion de differentielle dans l'analyse générale, *Ann. Ecole Norm. Sup., Paris* **42** (1925), 293–323.

Friedman, A.

[Fr.1] Optimal control for hereditary processes, *Arch. Rational Mech. Anal.* **15** (1964), 396–416.

Funk, J. E., and Gilbert, E. G.

[FG.1] Some sufficient conditions for optimality in control problems with state space constraints, *SIAM J. Control* **8** (1970), 498–504.

Gabasov, R., and Churakova, S. V.

[GC.1] Necessary optimality conditions in time-lag systems, *Avtomat. i* Telemekh. **29** (1968), 45–64 [English translation in *Automat. Remote Control* **29** (1968), 37–54].

Gabasov, R., and Kirillova, F. M.

[GaK.1] Extending L. S. Pontryagin's maximum principle to discrete systems, *Avtomat. i Telemekh.* **27** (1966), 46–51 [English translation in *Automat. Remote Control* **27** (1966), 1878–1882].

[GaK.2] *Qualitative Theory of Optimal Processes*, Nauka, Moscow, 1971.

Gamkrelidze, R. V.

[G.1] Optimum-rate processes with bounded phase coordinates, *Doklady Akad. Nauk* **125** (1959), 475–478.

[G.2] Optimum control processes with bounded phase coordinates, *Izvestia Akad. Nauk SSSR, Ser. Mat.*, **24** (1960), 315–356.

[G.3] On sliding optimal regimes, *Doklady Akad. Nauk SSSR* **143** (1962), 1243–1245 [English translation in *Soviet Math. Dokl.* **3** (1962), 559–562].

[G.4] On some extremal problems in the theory of differential equations with applications to the theory of optimal control, *SIAM J. Control* **3** (1965), 106–128.

[G.5] First-order necessary conditions for axiomatic extremal problems, *Akad. Nauk SSSR, Math. Institute, Trudy* **112** (1971), 152–180.

Gamkrelidze, R. V., and Kharatishvili, G. L.

[GK.1] Extremal problems in linear topological spaces. I, *Math. Systems Theory* **1** (1967), 229–256.

[GK.2] Extremal problems in linear topological spaces, *Izv. Akad. Nauk SSSR, Ser. Mat.* **33** (1969), 781–839 [English translation in Math. USSR-Izv., **3** (1964), 737–794].

Gâteaux, R.

[Ga.1] Sur les fonctionelles continues et les fonctionelles analytiques, *Bull. Soc. Math. France* **50** (1922), 1–21.

Girsanov, I. V.

[Gi.1] *Lectures on Mathematical Theory of Extremum Problems*, Lecture Notes in Economics and Math. Systems, Vol. 67, Springer-Verlag, New York, 1972.

Goldstine, H. H.

[Go.1] A multiplier rule in abstract spaces, *Bull. Amer. Math. Soc.* **44** (1938), 388–394.

Halanay, A.

[Ha.1] Optimal controls for systems with time lag, *SIAM J. Control* **6** (1968), 215–234.

[Ha.2] Relaxed optimal controls for time lag systems, *Rev. Roum. Math. Pures et Appl.* **16** (1971), 1059–1072.

Halkin, H.

[H.1] On the necessary condition for optimal control of nonlinear systems, *J. Analyse Math.* **12** (1964), 1–82.

[H.2] Finite convexity in infinite dimensional spaces, in *Proceedings of the Colloquim on Convexity* (W. Fenchel, editor), Copenhagen, 1965, 126–131.

[H.3] An abstract framework for the theory of process optimization, *Bull. Amer. Math. Soc.* **72** (1966), 677–678.

[H.4] Nonlinear nonconvex programming in an infinite dimensional space, in *Mathematical Theory of Control* (A. V. Balakrishnan and L. W. Neustadt, editors), Academic Press, New York, 1967, 10–25.

[H.5] Implicit functions and optimization problems without continuous differentiability of the data, *SIAM J. Control* **12** (1974), 229–236.

[H.6] Finitely convex sets of nonlinear differential equations, *Math. Systems Theory* **1** (1967), 51–53.

[H.7] A satisfactory treatment of equality and operator constraints in the Dubovitskii-Milyutin optimization formalism, *J. Optimization Theory Appl.* **6** (1970), 138–149.

[H.8] Optimal control for systems described by difference equations, *Advances in Control Systems* **1** (1964), 173–196.

[H.9] A maximum principle of the Pontryagin type for systems described by nonlinear difference equations, *SIAM J. Control* **4** (1966), 90–111.

Halkin, H., and Neustadt, L. W.

[HN.1] General necessary conditions for optimization problems, *Proc. Nat. Acad. Sci. USA* **56** (1966), 1066–1071.

Hautus, M. J. L.

[Hau.1] Necessary conditions for multiple constraint optimization problems, *SIAM J. Control* **11** (1973), 653–669.

Hestenes, M. R.

[He.1] *A General Problem in the Calculus of Variations with Applications to Paths of Least Time*, Rand Corporation RM-100 (1950), ASTIA Document No. AD-112382.

[He.2] On variational theory and optimal control theory, *SIAM J. Control* **3** (1965), 23–48.

[He.3] *Calculus of Variations and Optimal Control Theory*, John Wiley & Sons, New York, 1966.

Holtzman, J. M.

[Ho.1] Convexity and the maximum principle for discrete systems, *IEEE Trans. Automatic Control* **AC-11** (1966), 30–35.

[Ho.2] On the maximum principle for nonlinear discrete-time systems, *IEEE Trans. Automatic Control* **AC-11** (1966), 273–274.

Holtzman, J. M., and Halkin, H.

[HH.1] Directional convexity and the maximum principle for discrete systems, *SIAM J. Control* **4** (1966), 263–275.

Huang, S.-C.

[Hua.1] Optimal control problems with retardations and restricted phase coordinates, *J. Optimization Theory Appl.* **3** (1969), 316–360.

[Hua.2] Optimal control problems with a system of integral equations and restricted phase coordinates, *SIAM J. Control* **10** (1972), 14–36.

Hughes, D. K.

[Hug.1] Variational and optimal control problems with delayed argument, *J. Optimization Theory Appl.* **2** (1968), 1–14.

Hurwicz, L.

[Hu.1] Programming in linear spaces, in *Studies in Linear and Nonlinear Programming*, by K. J. Arrow, L. Hurwicz, and H. Uzawa, Stanford Univ. Press, Stanford, 1958, 38–102.

John, F.

[J.1] Extremum problems with inequalities as side conditions, in *Studies and Essays, Courant Anniversary Volume* (K. O. Friedrichs, O. E. Neugebauer, and J. J. Stokes, editors), John Wiley & Sons, New York, 1948, 187–204.

Jordan, B. W., and Polak, E.

[JP.1] Theory of a class of discrete optimal control systems, *J. Electronics and Control* **17** (1964), 697–711.

Karlin, S.

[K.1] *Mathematical Methods and Theory in Games, Programming, and Economics, I, II*, Addison-Wesley, Reading, Mass., 1959.

Kharatishvili, G. L.

[Kh.1] Maximum principle in the theory of optimum time-delay processes, *Doklady Akad. Nauk SSSR* **136** (1961), 39–42.

[Kh.2] A maximum principle in extremal problems with delays, in *Mathematical Theory of Control* (A. V. Balakrishnan and L. W. Neustadt, editors), Academic Press, New York, 1967, 26–34.

[Kh.3] Extremal problems in linear topological spaces, Ph.D. dissertation, Tbilisi State Univ., USSR, 1968.

Kuhn, H. W. and Tucker, A. W.

[KT.1] Nonlinear programming, in *Proceedings of the Second Berkeley Symposium on Mathematical Statistics and Probability* (J. Neyman, editor), Univ. California Press, Berkeley, California, 1951, 481–492.

Lee, E. B.

[L.1] A sufficient condition in the theory of optimal control, *SIAM J. Control* **1** (1963), 241–245.

Lyusternik, L. A.

[Ly.1] On conditional extrema of functionals, *Mat. Sbornik* **41** (1934), 390–401.

Lyusternik, L. A., and Sobolev, V. I.

[LS.1] *Elements of Functional Analysis*, Frederick Ungar, New York, 1961.

Makowski, K., and Neustadt, L.

[MN.1] Optimal control problems with mixed control-phase variable equality and inequality constraints, *SIAM J. Control* **12** (1974), 184–228.

Malozemov, V. N.

[M.1] Concerning the theory of nonlinear minimax problems, *Kibernetika* **6** (1970), No. 3, 121–125 [English translation in *Cybernetics* **6** (1970), 334–339].

Mangasarian, O. L.

[Ma.1] Sufficient conditions for the optimal control of nonlinear systems, *SIAM J. Control* **4** (1966), 139–152.

[Ma.2] *Nonlinear Programming*, McGraw-Hill, New York, 1969.

Mangasarian, O. L., and Fromovitz, S.

[MF.1] The Fritz John necessary optimality conditions in the presence of equality and inequality constraints, *J. Math. Anal. Appl.* **17** (1967), 37–47 (see also Shell Development Company Paper P1433, 1965).

[MF.2] A maximum principle in mathematical programming, in *Mathematical Theory of Control* (A. V. Balakrishnan and L. W. Neustadt,editors), Academic Press, New York, 1967, 85–95.

McShane, E. J.

[Mc.1] On multipliers for Lagrange problems, *Amer. J. Math.* **61** (1939), 809–819.

[Mc.2] Necessary conditions in generalized-curve problems of the calculus of variations, *Duke Math. J.* **7** (1940), 1–27.

[Mc.3] Relaxed controls and variational problems, *SIAM J. Control* **5** (1967), 438–485.

Neustadt, L. W.

[N.1] Optimal control problems as extremal problems in a Banach space, *Proceedings, Symposium on System Theory*, Polytechnic Institute of Brooklyn, New York, 1965, 215–224.

[N.2] A general theory of variational problems with applications to optimal control problems, *Doklady Akad. Nauk SSSR* **171** (1966), 48–50.

[N.3] An abstract variational theory with applications to a broad class of optimization problems I: General theory, *SIAM J. Control* **4** (1966), 505–527.

[N.4] An abstract variational theory with applications to a broad class of optimization problems II: Applications, *SIAM J. Control* **5** (1967), 90–137.

[N.5] A general theory of extremals, *J. Comput. System Sci.* **3** (1969), 57–92.

[N.6] Optimal control problems as mathematical programming in an unorthodox function space, in *Control Theory and the Calculus of Variations* (A. V. Balakrishnan, editor), Academic Press, New York, 1969, 175–207.

[N.7] Optimal control problems with operator equation restrictions, in *Symposium on Optimization*, Lecture Notes in Mathematics, Vol. 132, Springer-Verlag, New York, 1970, 292–306.

[N.8] Sufficiency conditions and a duality theory for mathematical programming problems in arbitrary linear spaces, in *Nonlinear Programming* (J. B. Rosen, O. L. Mangasarian, and K. Ritter, editors), Academic Press, New York, 1970, 323–348.

[N.9] Optimal control: a theory of necessary conditions, *Proceedings of the 1970 International Congress of Mathematicians*, **3** (1970), 183–185.

[N.10] On the solutions of certain integral-like operator equations. Existence, uniqueness, and dependence theorems, *Arch. Rational Mech. Anal.* **38** (1970), 131–160.

Neustadt, L. W., and Warga, J.

[NW.1] Comments on the paper "Optimal control of processes described by integral equations. I," *SIAM J. Control* **8** (1970), 572.

Oshiganova, I. A.

[O.1] On the theory of optimal regulation of systems with time-lag, *Proc. Seminar on the Theory of Differential Equations with Deviating Arguments* **2** (1963), 136–145.

Pareto, V.

[P.1] *Cours d'Economie Politique*, Lausanne Rouge, 1896.

Pearson, J. B., Jr. and Sridhar, R.

[PS.1] A discrete optimal control problem, *IEEE Trans. Automatic Control* **AC-11** (1966), 171–174.

Pogorzelski, W.

[Pog.1] *Integral Equations and Their Applications*, Vol. I, Pergamon Press, New York, 1966.

Pontryagin, L. S.

[Po.1] Optimum control processes, *Uspekhi Mat. Nauk* **14** (1959), No. 1, 3–20.

Pontryagin, L. S., Boltyanskii, V. G., Gamkrelidze, R. V., and Mischenko, E. F.

[PBGM.1] *The Mathematical Theory of Optimal Processes*, Fizmatgiz, Moscow, 1961 [English translation, John Wiley & Sons, New York, 1962].

Propoi, A. I.

[Pr.1] The maximum principle for discrete control systems, *Avtomat. i Telemekh.*, **26** (1965), 1177–1187 [English translation in *Automat. Remote Control* **26** (1965), 1167–1177].

Pshenichnyi, B. N.

[Ps.1] Convex programming in a normed space, *Kibernetika* **1** (1965), 46–54 [English translation in *Cybernetics* **1** (1965), 46–57].

[Ps.2] Linear optimal control problems, *SIAM J. Control* **4** (1966), 577–594.

[Ps.3] *Necessary Conditions for an Extremum*, Marcel Dekker, Inc., New York, 1971.

Rockafellar, R. T.

[R.1] An extension of Fenchel's duality theorem for convex functions, *Duke Math. J.* **33** (1966), 81–90.

[R.2] Conjugate convex functions in optimal control and the calculus of variations, *J. Math. Anal. Appl.* **32** (1970), 174–222.

[R.3] State constraints in convex control problems of Bolza, *SIAM J. Control* **10** (1972), 691–715.

Rosen, J. B.

[Ro.1] *Optimal Control and Convex Programming*, MRC Tech. Report 547, Madison, Wisconsin, 1965.

Rozonoér, L. I.

[Roz.1] The maximum principle of L. S. Pontryagin in optimal system theory. I; II, *Avtomat. i Telemekh.* **20** (1959), 1320–1334; 1441–1458 [English translation in *Automat. Remote Control* **20** (1959), 1288–1302; 1405–1421].

[Roz.2] The maximum principle of L. S. Pontryagin in optimal system theory III, *Avtomat. i Telemekh.* **20** (1959), 1561–1578 [English translation in *Automat. Remote Control* **20** (1959), 1517–1532].

Schwarzkopf, A. B.

[Sc.1] Optimal controls for problems with a restricted state space, *SIAM J. Control* **10** (1972), 487–511.

Slater, M.

[S.1] Lagrange multipliers revisited: a contribution to nonlinear programming, Cowles Commission Discussion Paper, Mathematics 403, November, 1950.

Stoltz, O.

[St.1] *Grundzüge der Differential- und Integralrechnung*, Bd. 1, B. G. Teubner, Leipzig, 1893.

Tricomi, F.

[T.1] *Integral Equations*, John Wiley & Sons, New York, 1957.

Uzawa, H.

[U.1] The Kuhn-Tucker theorem in concave programming, in *Studies in Linear and Nonlinear Programming*, by K. J. Arrow, L. Hurwicz, and H. Uzawa, Stanford Univ. Press, Stanford, 1958, 32–37.

Vinokurov, V. R.

[V.1] Optimal control of processes described by integral equations. I, *SIAM J. Control* **7** (1969), 324–336.

[V.2] Further comments on the paper "Optimal control of processes described by integral equations. I," *SIAM J. Control* **11** (1973), 185.

Warga, J.

[W.1] Relaxed variational problems, *J. Math. Anal. Appl.* **4** (1962), 111–128.

[W.2] Necessary conditions for minimum in relaxed variational problems. *J. Math. Anal. Appl.* **4** (1962), 129–145.

[W.3] Minimizing variational curves restricted to a preassigned set, *Trans. Amer. Math. Soc.* **112** (1964), 432–455.

[W.4] Unilateral variational problems with several inequalities, *Michigan Math. J.* **12** (1965), 449–480.

[W.5] Functions of relaxed controls, *SIAM J. Control* **5** (1967), 628–641.

[W.6] Restricted minima of functions of controls, *SIAM J. Control* **5** (1967), 642–656.

[W.7] Relaxed controls for functional equations, *J. Functional Anal.* **5** (1970), 71–93.

[W.8] Control problems with functional restrictions, *SIAM J. Control* **8** (1970), 360–371.

[W.9] Unilateral and minimax control problems defined by integral equations, *SIAM J. Control* **8** (1970), 372–382.

[W.10] *Optimal Control of Differential and Functional Equations*, Academic Press, New York, 1972.

[W.11] Optimal controls with pseudodelays, *SIAM J. Control* **12** (1974), 286–299.

Young, L. C.

[Y.1] Generalized curves and the existence of an attained absolute minimum in the calculus of variations, *Compt. Rend. Soc. Sci. et Lettres, Varsovie, Cl. III* **30** (1937), 212–234.

[Y.2] *Lectures on the Calculus of Variations and Optimal Control Theory*, W. B. Saunders Co., Philadelphia, 1969.

Zadeh, L. A.

[Z.1] Optimality and non-scalar-valued performance criteria, *IEEE Trans. Automatic Control* **AC-8** (1963), 59–60.

SUBJECT INDEX

Adjoint variable, (II.2.37), (V.2.36)
Admissible control, (II.2.5)

Banach space, (I.4.35)
Barycentric coordinates, (I.1.33)
Base for a topology, (I.3.10)
Basis of a vector space, (I.1.25)
Boundary point, (I.3.6)
Bounded set, (I.3.45), (I.4.37)

Cartesian product, (I.1.5)
Cauchy sequence, (I.4.35)
Causal operator, (IV.3.2)
Chattering combination, (IV.5.1)
Chattering lemma, (IV.5.3)
Closed set, (I.3.4)
Closure of a set, (I.3.8)
Compact set, (I.3.14)
—, conditionally, (I.3.14)
—, sequentially, (I.3.14)
Complete space, (I.4.35)
Cone, convex, (I.1.13)
—, dual, (I.5.23)
—, pointed, (I.1.14)
—, tangent, (I.4.25)
Conical hull of a set, (I.1.28)
Conjugate space, (I.5.4)
Constraints, equality, (II.2.17)
—, inequality, (II.2.17)
—, mixed control-phase variable, (VI.3.1)
—, operator equation, (II.2.1), (IV.1.3)
—, restricted phase coordinates, (V.1.7)
Continuous mapping, (I.3.24)
Control problem, discrete, (VII.4.1)
—, fixed time, (V.1.7), (V.3.1), (V.8.1)
—, free time, (V.1.7), (V.2.1)
—, minimax, (V.4.1), (VI.4.1)
—, time optimal, (V.2.53)
—, with functional differential equations, (VII.1.10)
—, with parameters, (VI.1.1)
—, with retarded differential equations, (VII.2.10)
—, with Volterra integral equations, (VII.3.9)
Control variable, (V.1.6)
Convex body, (I.4.18)
Convex combination of vectors, (I.1.24)
Convex conical hull, (I.1.29)
Convex function, (I.2.22)
Convex hull of a set, (I.1.27)
Convex polyhedron, (I.1.35)
Convex problem, (II.3.1), (II.3.26), (III.5.1), (III.5.21), (V.8.1)
Convex set, (I.1.11)
Convex under switching, (IV.5.14)
Core point, (I.1.42)
Covering of a set, (I.3.14)

Dense set, (I.3.8)
Differential of a function, (I.7.2)
—, dual, (I.7.24)
—, finite, (I.7.4)
—, finite directional, (I.7.2)
—, formal dual partial, (I.7.56)
—, Fréchet, (I.7.37)
—, Fréchet partial, (I.7.52)
—, Gâteaux, (I.7.4)
—, Gâteaux directional, (I.7.2)
Dimension of a convex set, (I.1.34)
Dimension of a simplex, (I.1.31)
Dimension of a vector space, (I.1.25)
Direct product, (I.1.5)

Equicontinuous set of functions, (I.5.28)
—, uniformly, (I.5.30)
Equivalent norms, (I.4.33)
Extremal, (φ,ϕ,Z)-, (III.1.14)
—, (ϕ,Z)-, (III.1.29)
Extremality, (III.1.7)

Finitely continuous mapping, (I.5.26)
Finitely open in itself, (I.1.39)
Finitely open set, (I.1.37)
Fixed point, (I.5.37)
Flat, (I.1.10)
Functions, (I.2.1)
—, affine, (I.2.16)
—, Chattering combination of, (IV.5.13)
—, composition of, (I.2.4)

—, continuous Fréchet differentiable, (I.7.48)
—, convex, (I.2.22)
—, differentiable, (I.7.4)
—, directionally differentiable, (I.7.3)
—, dual differentiable, (I.7.24)
—, finitely differentiable, (I.7.4)
—, Fréchet differentiable, (I.7.37)
—, Gâteaux differentiable, (I.7.4)
—, inverse, (I.2.3)
—, measurable, (V.1.20)
—, semidifferentiable, (I.7.19)
—, *W*-semidifferentiable, (I.7.15)
Functional, linear, (I.2.8)
—, support, (I.2.29)

General position, (I.1.22)

Hamiltonian, (II.2.44)
Hausdorff space, (I.3.29)
Homeomorphism, (I.3.28)

Implicit function theorem, (I.7.55)
Interior of a set, (I.3.5)
Interior point, (I.3.5)
Internal point, (I.1.42)
Isometrically isomorphic normed spaces, (I.4.34)
Isomorphic vector spaces, (I.1.4)

Jacobian matrix, (I.7.58)
Jump conditions, (V.2.36), (V.3.49), (VI.3.103)

Lagrange multipliers, (VI.3.123)
Lagrangian, (II.3.12)
Limit, (I.3.17)
Limit point, (I.3.6)
Linear combination of vectors, (I.1.23)
Linear manifold, (I.1.9)
Linear topological space, (I.4.1)
Linear vector space, (I.1.1)
Linearly dependent vectors, (I.1.22)
Locally convex topological vector space, (I.4.30)

Maximum principle, (II.2.18), (III.4.16), (IV.1.57), (IV.1.89)
—, integral form of, (II.2.37), (V.2.36), (V.3.49), (VI.3.106), (VII.1.46), (VII.2.29)
—, pointwise, (II.2.41), (V.3.85), (VII.1.77)
—, local, (VI.1.51), (VII.4.90)
Mean-value theorem, (I.7.43)
Minimax problems, (III.1.8)
Multiplier rules, (II.1.4), (III.2.18), (III.3.1), (III.4.2), (III.5.4)

Neighborhood, (I.3.3)
Norm, (I.4.31)
Normed vector space, (I.4.33)

Open set, (I.3.3)
Operators, (I.2.5)
—, causal, (IV.3.2)
—, identity, (I.2.9)
—, linear, (I.2.6)
—, locally compact, (IV.3.2), (IV.4.1)
—, Volterra, (IV.3.3), Al
Optimal control, (V.2.7)
Optimal parameter, (VI.1.18)

Phase trajectory, (II.2.5)
Phase vector, (V.1.6)
Point of density, (VI.3.118)

Regular set of operators, (IV.3.43), (IV.4.39)

Semidifferential of a function, (I.7.17)
—, dual, (I.7.26)
—, finite, (I.7.19)
—, Gâteaux, (I.7.19)
Separation theorem, (I.2.19), (I.2.25)
Simplex, (I.1.31)
Simplicial linearization, (I.5.39)
Span, (I.1.23)
Strongly convex under switching, (IV.5.15)
Subgradient, (I.2.29)

Topological space, (I.3.1)
Topologically isomorphic spaces, (I.4.12)
Topology, (I.3.1)
—, Euclidean, (I.3.40)
—, pointwise convergence, (I.6.33)
—, product, (I.3.33)
—, relative, (I.3.9)
—, stronger, (I.3.2)
—, weak, (I.5.7)
—, weak*, (I.5.5)

Transversality conditions, (V.2.36), (V.3.49), (VI.3.104), (VII.2.22)

Uniformly continuous mapping, (I.5.27)
Uniformly equicontinuous family, (I.5.30)
Uniformly locally compact operators, (V.3.42)

Vector-valued problems, (III.1.10)

W-convex, (I.2.32)
W-semidifferential of a function, (I.7.14)
—, dual, (I.7.25)
—, finite, (I.7.15)
—, Gâteaux, (I.7.15)
Well-posed problems, (II.3.2), (II.3.26), (III.5.3), (III.5.22)

Library of Congress Cataloging in Publication Data

Neustadt, Lucien W
Optimization.

Bibliography: p.
Includes index.
1. Mathematical optimization. I. Title.
QA402.5.N464 515′.64 76–3010
ISBN 0-691-08141-7